T0144683

Herbert Spencer
and the Invention of Modern Life

Herbert Spencer
and the Invention of Modern Life

Mark Francis

ACUMEN

First published in 2007 by Acumen

Acumen Publishing Limited
Stocksfield Hall
Stocksfield
NE43 7TN
www.acumenpublishing.co.uk

ISBN: 978-1-84465-086-6 (hardcover)

British Library Cataloguing-in-Publication Data
A catalogue record for this book is available from the British Library.

Designed and typeset in Classical Garamond
by Kate Williams, Swansea.
Printed and bound by Cromwell Press, Trowbridge.

Contents

Preface

Writing about Herbert Spencer has made me aware of the narrowness of academic disciplines. It has also increased the number of my scholarly debts. Since Spencer cut across academic boundaries in framing his system of knowledge, this has forced me to do the same. I have had to borrow freely from recent work in Victorian history and literature as well as from more specialized scholarship in history of science and philosophy, political theory, anthropology and sociology. My obligations are too many to enumerate fully here, but I do feel that I must mention how much I have gained from the lucid and stimulating ideas of Robert J. Richards, George W. Stocking Jr and Frank M. Turner. Most of my other scholarly debts are simple ones of gratitude to those whose work has provided mine with insight. Sometimes, however, I have received great benefit from disagreeing with an author's opinion, and using it as a stimulus for research and reflection. While I am not grateful on these occasions there is a sort of perverse obligation for which I give credit in the references to various chapters.

Original research on Spencer has been a painstaking business, forcing me to examine, and re-examine, all records with extraordinary scrupulousness and suspicion. It was not that the material was forged or tampered with, but, nonetheless, it was clear that much of it had been, in a sense, manufactured by Spencer in such a way that it is now difficult to test for the truth. The root of the problem is that, when old, he recalled the vast bulk of his correspondence and had it destroyed, preserving only that material by which he wanted to be remembered. While he exerted control over his posthumous reputation in order to emphasize his originality and importance, his plan went awry and the opposite has happened. Instead of his efforts leading to admiration, the result was that his boasts excited ridicule while his occasional ironic disclaimers were received as confessions of naivety. In brief, rather than enhancing his fame Spencer's management of his manuscripts trapped his work and his reputation in a mausoleum whose significance could not be grasped by succeeding generations. To remedy this problem was difficult because one could not overthrow the edifice; it was often the sole authentic repository of Spencerian material. However, the vetting of this source was onerous, not impossible, and in doing it I have restored to the man and his ideas some of the brilliance they possessed in mid-Victorian England.

The research for this book has entailed my critically reading and sifting thousands of Spencer letters that were entombed in the publications he arranged at the end of his life. I have also compared his versions with the few hundred originals that escaped the conflagration that consumed the others. In addition, I have sought out any letters that he was unable to find, together with any manuscripts by acquaintances that mentioned him. Most of this material was deposited in a couple of dozen archives – mostly in the United Kingdom – and references to it are given in my footnotes. With the exception of the long-embargoed George Eliot letters referred to in Chapter 3, surviving individual manuscripts hold few surprises, but, collectively, they alter the picture of Spencer. My analysis of familiar and unfamiliar manuscript material has been integrated with a thorough examination of any nineteenth-century philosophical and scientific literature that could throw light on either Spencer's intentions or his arguments. At times, as I examined old accession lists of books as they accumulated year by year in a library, or as I checked every edition of a popular mid-nineteenth-century scientific textbook, my efforts seemed too exhaustive, even though they have not left me confident that I have understood everything about Spencer's contexts and texts. However, I can at least make the weaker claim to have read everything that could be relevant. Such modesty aside, I feel that my manuscript work, together with my labour on nineteenth-century books, pamphlets and periodicals will permanently change existing discussions of Spencer's philosophy, and create new interpretive patterns where none existed. Although I have discarded Spencer's key claim – that he was the autodidact who explored evolution – and showed that he was a syncretic thinker rather than an original one, this will do his reputation a good service. Some of his ideas are valuable precisely because they recycled and elevated the more valuable insights of his contemporaries. While his theories were not always the products of his imagination and logic, they possessed a consistent and clear formulation, which is a permanent legacy from the nineteenth century.

On a personal note I must express my thanks to a number of scholars who helped me understand Victorian culture and intellectual history. Beginning with efforts by Owen Chadwick, G. Kitson Clark and Duncan Forbes, various scholars have attempted to educate me or, at least, remind me of literature I would have otherwise missed. At different times this task has been taken up by Sam Adshead, Ruth Barton, Dinah Birch, Pietro Corsi, Paul Crook, Kate Flint, Knud Haakonssen, Jose Harris, Ken Minogue, Maureen Montgomery, John Morrow and Paul Weindling. If they have been unsuccessful in keeping me informed, the fault is all mine.

I should also thank librarians and archivists at a number of depositaries, especially the Bodleian Library, the University of Cambridge Library, the British Library, the National Library of Scotland, the Birmingham Central Library, the Manchester Central Library and the libraries of the London School of Economics, Imperial College and the Royal Institution of Great Britain for their patience and kindness. The interloan librarians at my own institution have been assiduous and energetic, seldom failing to find a rare title. Thanks also to Shari Wigley, Craig Shatford and Felicity Hattrell for their assistance in proofreading and editorial suggestions, Carol Davison for providing an index and Jill Dolby for patiently typing many manuscript drafts. I should also add that I am grateful for suggestions from my publisher's readers and from my editors, especially Steven Gerrard and Kate Williams. Last, but not least, I should express my thanks to Angela for encouraging me to finish this book.

Part of the research for this book was conducted while I was a visiting fellow at Magdalen College and, subsequently, Hamilton Fowler Senior Research Fellow at Christ Church, Oxford. My own university, the University of Canterbury has also been helpful by granting me two sabbatical leaves during the period I was working on this subject. Without the support given to me by these institutions the ten years I have spent writing this book would have stretched to the length of time Spencer took to write his philosophy.

MF

Chronology

27 April 1820   Herbert Spencer's birth in Derby.

1830   Herbert's father, George, felt unable to educate Herbert so placed him under the tutelage of one of his brothers, William, who ran a school in Derby.

June 1833   Herbert was placed with another uncle, Thomas, an Anglican clergyman who taught boarders at Hinton Charterhouse.

November 1837   William Spencer obtained Herbert a post with a railway engineer, Charles Fox.

April 1841   Spencer gave up his job with the Birmingham and Gloucester Railway, and returned to Derby to help his father design an electromagnetic engine. Within a month it became obvious that this scheme would be fruitless, and Herbert spent the next year in desultory study of engineering and science.

1842   Spencer's first political publication, "The Proper Sphere of Government", appeared in twelve issues of the *Nonconformist* between June and December. Spencer also became the Derby delegate for the Complete Suffrage Union, a constitutional and moderate party allied to the Chartists.

August 1844   After three years without an income, Spencer's support for the Complete Suffrage Union led to a moment of employment as editor of a short-lived radical periodical in Birmingham, the *Pilot*. Following this Spencer returned to working as a railway surveyor.

November 1848   Spencer obtained a position as sub-editor on *The Economist*, the London weekly newspaper founded by James Wilson.

1849   Spencer began to attend John Chapman's weekly soirées where, for the first time, he met philosophers and writers. It was in this milieu that Spencer met George Eliot, who worked for Chapman as an editor.

1851   His first major book, *Social Statics*, was published by Chapman and received favourable reviews.

1852   Spencer began to write for the *Leader*, a weekly newspaper edited by G. H. Lewes and Thornton Hunt. Spencer's article "A Theory of Population" appeared in the *Westminster Review*. The latter was his first serious attempt at a scientific topic.

1852–53   Spencer had a love affair with George Eliot. It was his only passionate relationship, but was unconsummated because he was unable to cope with the demands of physical affection.

1853   His first philosophical article, "The Universal Postulate", appeared in the *Westminster Review*.

1855   The year saw the appearance of Spencer's first scholarly work, *The Principles of Psychology*, which founded his reputation as an evolutionary thinker.

1856   Spencer's depression deepened and became a "breakdown". He blamed this more serious disability on his attempt to write a testimonial on behalf of A. C. Fraser, which meant that he had to read the latter's philosophy works in depth. This was taxing as serious reading always affected him adversely.

December 1856   Deciding to stop living alone, Spencer took lodgings with a family in St John's Wood in North London. The family had children and he was pleased to exercise his "philoprogenitive instincts" for the first time.

January 1859   Spencer produced a programme for a "A System of Synthetic Philosophy", which proposed to analyse all aspects of organic knowledge.

1862   Spencer published *First Principles*, the metaphysical foundation of his philosophical system. He also published *Education*, a progressive educational work that emphasized the need to avoid coercion and unkindness when teaching children.

1864   The publication of volume I of *The Principles of Biology*, in which Spencer distanced himself from the vitalism that permeated his scientific ideas during the 1850s.

1867   The publication of volume II of *The Principles of Biology*.

1868   Spencer was elected to the Athenaeum, a club at which he became a habitué. The election to the Athenaeum was one of the few signs of recognition that he ever accepted.

1870   Spencer edited and reissued his 1855 *The Principles of Psychology* as volume I of the psychology volumes in "A System of Synthetic Philosophy".

1872   Spencer completed volume II of *The Principles of Psychology*.

1873   Spencer published his first social science book, *The Study of Sociology*, a popular text written for the International Scientific Series.

1875   Spencer began writing his autobiography, a task that occupied much of his leisure time until 1889.

1876   The first volume of *The Principles of Sociology* appeared.

June 1879   Spencer issued *The Data of Ethics*, part 1 of *The Principles of Ethics*. The remainder of this work did not appear until the early 1890s.

November 1879   Spencer issued *Ceremonial Institutions*, part 1 of volume II of *The Principles of Sociology*.

April 1882   The publication of *Political Institutions* (part 2 of volume II of *The Principles of Sociology*). This book, which he viewed as his most important, established Spencer as a doctrinaire anti-militarist.

Feb–Jul 1884   The publication of the articles for the *Contemporary Review* that were republished as *The Man "versus" the State*. This powerful attack on socialism often misled contemporaries into thinking that Spencer was an individualist.

1885   The publication of *Ecclesiastical Institutions*. This is part 1 of volume III of *The Principles of Sociology* (also listed as part VI of *The Principles of Sociology*), and his last significant writing.

1886   Spencer suffered a severe attack of depression that caused him to abandon writing. Even the constant re-editing of *An Autobiography* was temporarily suspended. He claimed to have been periodically ill over the next three years.

1889   Tired of living at boarding houses Spencer finally set up his own house. Unfortunately for his posthumous reputation, he did this in company with the Sickle sisters, with whom he lived for nine years. They later published a book called *Home Life with Herbert Spencer* "by Two", which dwelt on his eccentricities.

1893   Spencer arranged his will so that one of his former secretaries, David Duncan, would have the use of any private papers that he had not already used in *An*

Autobiography. This was to allow Duncan to write an official biography after Spencer's death.

1894 Spencer gave *An Autobiography* to his trustees and arranged for it to be published after his death.

8 December 1903 Spencer died, aged 83. His body was cremated and the ashes were deposited at Highgate Cemetery.

List of illustrations

Introduction

"When I say 'we' my dear," returned her father, "I mean mankind in general; the human race, considered as a body, and not as individuals. There is nothing personal in morality, my love. Even such a thing as this," said Mr. Pecksniff, laying the forefinger of his left hand upon the brown-paper patch on the top of his head, "slight casual baldness though it be, reminds us that we are but" – he was going to say "worms", but recollecting that worms were not remarkable for heads of hair, he substituted "flesh and blood".
Charles Dickens, *The Life and Adventures
of Martin Chuzzlewit*, 1844

To me, science, in its most general and comprehensive acceptance, means the knowledge of what I know, the consciousness of human knowledge. ... The man of science values an object because of the place he knows the object holds in the general universe by the relations it bears to other parts of knowledge. To arrange and classify that universe of knowledge becomes therefore the first, and perhaps the most important object and duty of science.   Prince Albert's speech to the meeting of the British Association for the Advancement of Science, 14 September 1859

It has been more than thirty years since a general book on Herbert Spencer has appeared.[1] Although there have been monographs and articles by sociologists, anthropologists, historians of science, philosophers and political theorists, there is now a need for a fuller study that takes account of recent advances in scholarship.[2] The need is for a coherent portrait of a man whose ideas have been selectively splintered into fragments and then press-ganged into narratives on the development of scholarly disciplines. Spencer scholarship has concentrated on producing a series of vignettes of why emerging academic subjects have responded to, or ignored, a great Victorian. As a result his ideas have been forced into explanatory models lacking historical sophistication. This, in turn, has led to the creation of contradictory patterns where none exist.[3] Instead of being portrayed

1

as an intrinsically interesting and coherent figure, Spencer has been lauded, or belaboured, as an intellectual who opposed Darwinism, utilitarianism, socialism or rights theory. For many academic writers, Spencer has served only as a plausible straw man who unsuccessfully resisted scientific and philosophical progress.

Spencer's popular image is even more inchoate than his academic one. Journalistic writers depict him as an improbable figure, whose arch-reactionary philosophy aligned liberalism with biological science so that together they would support imperialism and a competitive business ethos. Interpretative confusion reigns here because Spencer, who continually protested against war, cruelty, domination and inequality, has been unfairly conscripted as a voice favouring racial supremacy, libertarianism and the defence of wealth. In these guises he appears to support unnecessarily harsh corrective sanctions in the domestic and foreign affairs of states. Sometimes Spencer is even labelled an apologist for the capitalists whom, in fact, he detested as workaholics lacking the sense to pursue a healthy life.

Some of the plethora of misinformation about Spencer is a consequence of the English habit of ignoring their own intellectuals in favour of other European ones.[4] An extreme example of this can be found in Alan Macfarlane's attempt to analyse an indigenous phenomenon, English individualism. He takes his ideological stance from outsiders such as Max Weber and Alexis de Tocqueville, neither of whom had thought deeply about English civilization, while ignoring English writers such as Spencer, J. S. Mill, and H. T. Buckle, who might have informed his thesis.[5]

However, the greatest source of popular confusion about Spencer does not arise from national prejudice, but from writers who have explained his theories by reference to those of Charles Darwin as if the former were a simple version of the latter. This misidentification has been so common that its correction would be an obligatory as well as unpleasant task for any Spencerian scholar. There are two reasons why it is painful. First, it forces me to write about Darwinism even though this should have been unnecessary. Also, it is slightly obtuse to explain an intellectual phenomenon such as Spencer's evolutionary theory by reference to something it is not. Secondly, my attempt to clarify the difference between Spencer and Darwin might bring me into conflict with Darwinian scholars who indulge in intemperate polemics against the suggestion that it was Darwin not Spencer who was the more responsible for the harsh reductionist biological theory that has coarsened political and ethical discourse. However, since I cannot avoid being a target I might as well boldly canvass my views in the introduction. In addition, since I do not provide an extensive treatment of the relationship between Spencer's ideas and Darwinism until Parts III and IV of this book, it would be useful to foreshadow, in a simple form, the complex arguments I make later in case any reader lacks patience. In essence, I claim that Spencer was not Darwinian, either in his biological writing or in his account of human evolution. My reasons for distinguishing between Spencer's theories and Darwinism are: (i) Spencer's evolutionary theory did not focus on species change; (ii) Spencer's faith in progressive evolution did not draw on natural selection or competition; and (iii) Spencer did not accept that modern individuals and societies would continue to make progress through struggle for survival. My insistence on the non-Darwinian quality of Spencer's ideas does not refer to anything intrinsically interesting about Spencer's evolutionary stance. The propositions I have asserted are merely blunt instruments designed to dispose of facile interpretations by Darwinian enthusiasts who use Spencer as a whipping boy who can be credited with unattractive or simplistic comments on natural selection. I have had to be plain-spoken on this point because disinformation of this kind is so entrenched that it could almost be called tradition. Although fake, it has dominated popular discourse on Spencer during the twentieth century. A thorough analysis of this is beyond the scope of an intellectual biography, but it would be a useful study in twentieth-century intellectual history. That is, I hope that someone will analyse the need by social scientists and scientists to claim that Darwin, whom they admire, is free of historical culpability, when in general they do not care about the past.

In lieu of this study I will offer two suggestive examples. First, there was Graham Wallas, a professor of political science at the University of London, who was considered a penetrating observer of the intricacies of Victorian Britain. To him Spencer was merely an early and hasty generalizer on the subject of evolution,

> who seemed to suggest that if shopkeepers were encouraged to compete for business and clergymen for congregations, a process of the "survival of the fittest" would automatically set in, which would rapidly improve the race without the necessity of further thought or the starting of new moral difficulties.[6]

Secondly, there is Richard E. Leakey, the celebrated speculator on the ancestry of *Homo sapiens*. He possessed the same information as Wallas except that in using it he was praising, not condemning, Spencer. Leakey explained Darwin's term "natural selection" in this way: "But the expression often used by Mr Herbert Spencer of Survival of the Fittest is more accurate, and is sometimes equally convenient".[7] Leakey's memory had failed here; instead of referring to Spencer's usage of an expression, he should have cited the well-known anecdote that originated from a brief discussion between Spencer and Darwin prior to the publication of the latter's *The Origin of Species*. After Darwin had explained his theory of natural selection, Spencer quipped that it might as well be called "survival of the fittest". Subsequently, Darwin adopted this phrase as describing evolutionary theory while its originator did not. If either Wallas or Leakey had read Spencer's *The Principles of Biology* they would have realized that its author was unsympathetic to Darwin's natural selection theory. Alternatively, if either had been well acquainted with Spencer's philosophy or sociology, they could not have supposed that Spencer argued that future human evolution would be governed by the survival of the fittest. When I correct such crude mistakes about Spencer's views,[8] I do not intend to question the scholarship of those sociobiologists and economists who continue to place their faith in natural selection or competition as a progressive force in social change. Instead, my goal is only to warn them that they would be unlikely to find support from Spencer, who resisted these ideas, and whose originality was greatest on non-biological subjects such as the classification of knowledge and institutional change. Others, however, may benefit from my reconstruction of Spencer's personality and ideas and my belief that these have to be taken together to be properly understood.

A coherent picture of Spencer can be achieved by an intellectual biography, but such an endeavour is not a straightforward one. There are two challenges to overcome. First, the reader must be convinced of the possibility that any Briton might have had a life of the mind when he lived in a period that was supposedly devoid of philosophical interest. More particularly, it has to be demonstrated that such a character is in need of a substantial record of his existence beyond the fragmentary references in scholarly literature. Secondly, this general book on Spencer is a contribution to a genre entitled intellectual biography, which is, in itself, problematic. It is a hybrid between biography and textual analysis, which must achieve a precarious balance between chronicling an often-uneventful personal life and offering a substantial account of ideas that frequently seem to have no current interest. Any intellectual biography would face similar problems, but there is an extra difficulty here because Spencer's life appears especially empty. It did not contain the dramatic tensions that have often accompanied the lives of memorable people. Spencer's days were not coloured by a tincture of villainy, a scandalous affair or a tragic early death. The kind of incidents that enliven tales of Byron, Shelley and Galois are absent from Spencer's existence. At first sight this dramatic void suggests the lives of poets and mathematicians would be more attractive subjects to recreate. Yet, this is not the case because such biographies are less necessary as their subjects had lives that were more distant from their writings than Spencer's. Many texts can be intelligently read without knowledge of authorial intentions or behaviour, while Spencer's cannot. He lived his philosophy, and without a grasp of his aspirations, dislikes and personal responses his ideas appear uninspired and disconnected. The only window into Spencer's world is through intellectual biography: this approach can accurately interpret books that are embedded in lost technical debates, or that are tangled in self-reflections on failed desires and lingering hopes. Of course, a biography constricts interpretation as well as enhancing it. It privileges a particular contextual explanation of Spencer's ideas, while excluding others as historically naive, inconsistent with contemporary technical literature or incompatible with the central preoccupations of the author. Generally, an intellectual biography favours a historically based explanation of Spencer at the expense of recent philosophical and scientific preoccupations. While this might seem arbitrary or presumptuous, it is a bias that should be accepted because there is no other fair and credible way to discuss the empirically minded Spencer. Alternative explanations of his ideas would have to analyse them awkwardly from the perspective of a system of philosophy, such as rule utility, that he would have rejected, or from the standpoint of absolute principles he believed were always baseless. In any case, the adoption of such strategies to explain Spencer would be perverse: they ignore the scrupulous autobiographical and philosophical rationales with which he accompanied his own actions and utterances. This is not to say that my book faithfully re-inscribes Spencer's views; that would be impossible and unwise. He was not infallible, and was capable of deliberately misleading his readers, or even of losing sight of what he saw as the truth. Nevertheless, his autobiographical insights must be the primary source for a plausible account of his life and ideas. Spencer was first and foremost a psychologist who forensically dissected his own subjective state of mind. Rather than being arrogant, he was almost excessively conscious of his metaphysical limitations and his momentary lapses in scientific objectivity. However, he perceived these lapses as strengths, rather than weaknesses, and, as a result, focused on the "Unknown" as an object of religious sentiment, and on moral indignation as essential to any enquiry in social science.

When Spencer was thirty-four years old his friend, George Eliot, wrote an entry for an imaginary biographical dictionary to be published a century later.

> Spencer, Herbert, an original and profound philosophical writer, especially known by his great work xxx which gave a new impulse to psychology and has mainly contributed to the present advanced position of that science, compared with that which it had attained in the middle of the last century. The life of the philosopher, like that of the great Kant, offers little material for the narrator. Born in the year 1820 etc.[9]

The great work proved to be *The Principles of Psychology*, which appeared in 1855. It did not transform the discipline, although it did give Spencer an equal status with Alexander Bain and Alexander Campbell Fraser, the other British experts in mid-Victorian philosophy of mind. Eliot's sketch of Spencer was not only narrow in scope; it failed to predict accurately any aspect of his life. She did not forecast that her subject would be hailed as a religious prophet for the uplifting quality of his metaphysical speculation, that he would write systematic works on biology and sociology, or that the general public would take his political pronouncements seriously. Eliot was surprisingly blind to the specific qualities of a man who had been her lover only a few years before. It remains a curiosity that an observer on the verge of becoming a magisterial novelist could fail to perceive nuances in the character in a man with whom she had been intimately acquainted and who remained her friend. Instead of seeing into his soul, she saw only his outward traits: Eliot did not sense that Spencer was more tormented with anxiety than herself. The crudeness of her sketch suggests that her love for him had been indiscriminate; she had not seen the man hidden behind his defences of emotional reserve. Initially her hopes had blinded her to the fact that he was incapable of accepting affection even though this was his greatest flaw as a human being. In reality, Spencer was unable to experience love: worse, he believed that if he surrendered to this passion it would cause the re-emergence of the inherited aggression towards women that his father had displayed. Far from guessing the nature of his torment, Eliot speculated that the young Spencer would benefit from an emotional catharsis when he let down his defences and surrendered to his feelings. This was a fantasy because of Herbert's belief that if he were to give free rein to his feelings he would be cruel. He also came to realize that his affection for George Eliot had no true basis. He found her intellectually fascinating but sexually unattractive, which, from the perspective of his new hedonistic psychology, was an unpalatable combination.

The tensions in Spencer's personality were usually visible to close contemporaries. His earliest official biographer, Hector Macpherson, remarked that only those privileged with the friendship of the philosopher knew the personal difficulties he had to overcome in order to conceptualize the natural world and human destiny.[10] The philosopher's gift to humanity had only been possible because of his pain, suffering and isolation. There were other dreamers besides Spencer, but to Macpherson these had lost their way in the pleasures, anxieties and ambitions of society. Only "Herbert Spencer had refused to soil his robe in Vanity Fair".[11] This was hyperbole; Spencer's isolation and pain were not heroic and proud, but distant and quirky. There was something about him that disturbed even his friends. Leonard Courtney's funeral oration, which dwelt on the philosopher's "coherent, luminous, and vitalizing conception of the evolution of the world", also chided Spencer for being overly concerned about the delicacy of his constitution.[12] His dreams about the future and his humanity were unaccompanied by the gravitas or sensible demeanour that great Victorians were supposed to possess. When the journal *Vanity Fair* lampooned him, it did not focus on his unsoiled robes, but on his odd behaviour and difficult opinions: "Mr. Herbert Spencer is believed by many to be a companionable, cheerful man. He has been more than once to a shareholders' meeting to war with railway directors; he delights in children; and he holds that suicide should rather be encouraged. Yet he goes on living".[13] The alarm Spencer caused on topics such as the sacredness of life was a sign that he was not quite respectable.

Although philosophers are supposedly notable for having calm and methodical personalities, Spencer's was overheated and hypersensitive. However, this emotional core was sealed within a cool exterior. He could only feel sympathy when he held the object of his pity at a distance. A distant trauma could be lamented; this is how he could feel sympathy for indigenous people and their mistreatment at the hands of colonists and the military. Spencer also concerned himself about cruelty to animals, agitated on the effect on children of coercion, and even mourned the fate of the Liberal party, as it lay helpless in the imperious grasp of W. E. Gladstone.[14] However, in Spencer's dealings with personal acquaintances he was far less susceptible to feelings. It was not that he excluded people; on the contrary, he went to his club to play billiards, he entertained and amused his married friends at picnics, he gave their children holidays and, on occasion, he even went in for matchmaking. However, much of this conviviality was contrived. Spencer forced his involvement in social situations because he believed that a measure of amicability was necessary for his internal harmony. Similarly, when he considered that he should exercise parental instincts he borrowed children from acquaintances. Most of Spencer's personal interactions were, in fact, the severe self-imposed tasks of a man whose shyness and dogmatism left him unequipped to deal with the world. In addition to this disability, he imported sociological and scientific theories into his personal life, which made him either harshly didactic or amusingly eccentric. Sometimes, rather than avoiding predictable comment on his idiosyncratic behaviour, he courted negative reactions and erected an artificial individuality with which to withstand it. Spencer was a carefully constructed individual, whose constant sense of emotional peril meant that he was unsettled by every event in his intellectual and his social life. Each day produced its quota of embarrassment and stress that required all his reserves of courage to combat.

Spencer himself was well aware of his shortcomings. Rather than concealing his personal failures, he examined them openly; after all, he was an empirically minded psychologist. He felt it was important to acknowledge his defects because they would only be justified if he were compensated by equivalent gains. The reward had to be a personal one: the social progress he had forecast for the human species as a whole did not bring him any pleasure. In any case, for Spencer no one's personal pain could be traded off against a societal or species benefit. This was why he believed that utilitarian calculus was always a vulgar mistake. Spencer also could not be humbugged into believing that long-term happiness for humanity was beneficial to people now. His social views echoed his personal ones. His unhappiness about the way in which he had conducted his life was a consequence, which had to be borne; all he could do was altruistically warn others not to develop like himself.

Spencer believed that his unhappiness was caused by his Christian upbringing, which had emphasized self-denial and virtuous industry. This, he believed, had forced him to ignore biological urges and emotional sentiments. Spencer's *An Autobiography* is a testament to a harsh upbringing; he had been reared without family affections and imprinted to avoid sex, play and gossip. In their place were work, scientific learning and political reform. While he lived true to his programming, as he grew older he increasingly regarded such constancy as a personal tragedy. This plunged him into an alienation that was so great that he was never fulfilled by his philosophical and publishing successes. His adult experiences reinforced his sense of inadequacy. His sole romantic venture, an attempt to love George Eliot, produced the bitter lesson that happiness was an instinctive satisfaction: a condition from which he was excluded because of his inability to follow his feelings.

Spencer was unable to stop himself from observing the inner workings of individuals. He was an armchair psychologist who vicariously experienced the pain and suffering of others. For him clinical observations served as real sensations; his friend Eliot had been mistaken to believe that Spencer's emotional life offered little of interest to the narrator. Her perception of him focused on his neutered, crotchety and overly rational qualities, and this was rather a reflection of her own severe way of depicting humanity. Eliot's special gift was not insight into a particular person, but the ability to provide a pseudo-scientific and objective commentary on the social scene, which is why her fictional characters are so often representations of ideal types. It was not Eliot, but Spencer, with his exaggerated empathy, his hidden sense of irony and his finely honed psychological tools, who saw individuals distinctly, and felt compassion for them. He was not protected by an aura of philosophical calm because he was never inured to emotion and pain. The passionless neutrality, which he wrote about in his texts, cost him dearly.

Spencer's life was unusual: he had no children, nor did he ever possess a wife or mistress. It is probable that he also died a virgin. He was a hypochondriac, and while his condition was not viewed as extreme in the Victorian period, it did excite some adverse comment. His inherited fortune was modest, but not so insignificant as to force him into paid employment during the last half century of his life. Spencer had no dependents, detested travelling and was content to live in a boarding house until he was quite elderly: all of this meant his expenses were slight. Finally, since he possessed no conventional vices such as gambling, and believed sexual activity to be debilitating, he was at little risk from emotional turmoil. At first glance these comments seem to support a suggestion that Spencer's life was too dull to need a narrator. Yet, somehow, without adventure, emotional drama or domestic bliss, Spencer's life overflowed with meaning both for himself and his contemporaries. Inwardly he sailed on the edge of the known and conscious worlds; it was his imaginative accounts of these voyages, rather than his philosophical rigour, that attracted admiration. Spencer sensed the coming of the modern world and through his writings transmitted this excitement to others.

Herbert Spencer was the quintessentially English thinker of the Victorian period. Other famous nineteenth-century figures were not so centred on England. Jeremy Bentham and J. S. Mill were as comfortable with French literature as they were with English traditions. Samuel Taylor Coleridge's ideas were so dependent on those of German writers, such as Schelling and Schiller, that no one could tell them apart. Even the greatest London literary figure of the nineteenth century, Thomas Carlyle, was not English. In his soul Carlyle remained a Scot, and if he empathized with any other nationality it was the Teutonic one. In contrast with these cosmopolitan Britons, Spencer seems irredeemably provincial. He was so bounded by his insularity that he did not view the ocean until he was twenty-one years old.[15] On seeing Paris his astonished delight at its beauty was insufficient to prevent him from feeling "cut off from my roots".[16]

Yet, for all his provincialism, Spencer became the world philosopher of the late-nineteenth century. His works were translated into languages as distant from each other as Chinese and Mohawk and his philosophy of altruism found a home in ancient lands, such as India and China, and new worlds, such as the Americas. He was especially idolized in the United States, where he had many more followers than Darwin or Marx. Finally, Spencer invented a kind of liberalism that was not swallowed up by either individualism or collectivism. While other nineteenth-century liberals were captured by one of the polarized languages of negative and positive liberty, he forged a political theory that bravely ignored this seemingly inescapable antinomy. It seems a curiosity that a parochial Victorian Briton became a colossus. It cannot be explained by attendance at a great public school or at Cambridge and Oxford: Spencer was not a product of such institutions and, had he been, he would not have been driven to the unrelenting pursuit of science, nor would he have been so critical of the classical learning of the establishment. Spencer was, first and foremost, a proponent of things modern. It was the possession of an exclusively scientific education that gave him the advantage over his more formally educated countrymen. He should have been less unique because, as Pietro Corsi has observed, Britain still possessed a tradition of natural history. Spencer was thoroughly imbued with this from his youth while avoiding the Latin and Greek that distracted most of his contemporaries. Unlike the professionalized sciences that replaced it, natural history was both all-encompassing and immediately appealing to the general public. Rather than being a series of reservoirs of specialized knowledge, it reinforced the community's links with practical pursuits such as horticulture, animal husbandry, pharmacology and medicine.

Spencer had grown up within the confines of one of the nascent centres of scientific learning that were dotted about provincial England. While science had not yet gained a foothold in the universities, it was alive and well in urban centres such as Bristol, Leeds and Derby. As the son of the secretary of the Derby Philosophical Society, Spencer's early learning was moulded by attendance at lectures in phrenology, chemistry, electricity and pneumatics. Lying around his father's house were copies of the *Lancet*, the *British and Foreign Medical Review*, the *Medico-Chirurgical Review*, the *Athenaeum* and other journals.[17] Spencer's father and uncles, all skilled teachers of mathematics, taught him geometry, algebra, trigonometry and mechanics. His father also encouraged him to explore biology. As a child Herbert was an enthusiastic collector of bird nests, plants, larvae and other "hedge-side treasures".[18] He regretted that his early knowledge of formal botany was weak,[19] but, like all his scientific explanations, it never relied on the miraculous. From his earliest days he saw natural phenomena as having a physical cause. Spencer's universal philosophy originated in his unique education: his was the first generation possessing an exclusively scientific explanation of the universe. While the abandonment of a theocentric world was painful for many of his contemporaries, for Spencer it slipped away without a struggle. He effortlessly shrugged off his orthodox cosmology, and slipped into a secular worldview without a sense of loss. There was no scarring because Spencer believed that Life,[20] which he spied through his scientific spectacles, was filled with religious significance. The sight of it never failed to inspire him with hope for the future; as a result he was not forced into a grim acceptance of materialism like the secularist pioneer Ernest Renan. Spencer's universe was not bleak and vacuous, but the living promise of a benign future. For him progress ensured that humanity would not remain a collection of isolated individuals, but would evolve into the sentient completion of organic unity in harmony with itself and with the universe.

Herbert Spencer's metaphysics was generated in a different intellectual milieu than his science. It developed from a formal philosophical discourse that was popular during the 1850s, but that vanished shortly after the publication of *First Principles*, the foundation volume of his philosophical system.[21] After the 1860s scarcely anyone remembered the thinkers who had been Spencer's inspiration; their ideas were familiar only as rephrased by him. This effect of novelty has been long-lasting. Even today, in spite of an abundance of philosophical research, a search of *The Philosopher's Index* will uncover only sparse citations of a couple of Spencer's metaphysical sources, leaving the remainder in total oblivion. As a result commentary on Spencer often lacks scholarly depth. For example, crude and implausible connections are drawn between Spencer and "classical" writers, such as Plato, Locke and Kant. In this way John Cartwright argues that Spencer's view of the human mind was "a sort of evolutionary Kantianism" that answered

Locke. Cartwright also identifies Spencer's ideas as part of a controversial revival of the nineteenth-century debate between J. S. Mill's Lockeanism and William Whewell's Kantianism.[22] Such speculative theory of ideas obscures more than it illuminates: in reality, Spencer's evolutionary views of the mind responded neither to Locke nor Kant, but to philosophical problems that were specific to the mid-nineteenth century. Further, in his philosophy of science Spencer did not recognize a division between Mill and Whewell, but incorporated values from both into his own philosophical system.

The school of philosophy that generated Spencer is now a lost world whose structures of metaphysics and sense of religious enquiry are enveloped by mist. Only Spencer's writings remain visible; they loom prominently as curious features in an otherwise apparently flat and unoccupied Victorian landscape. This appearance of novelty is a misapprehension caused by the erosion of his intellectual context. This is not merely a historical problem; it has caused continuing antithetical mistakes in the interpretation of Spencer's philosophy. First, his principles often appear to be idiosyncratic even on subjects where they were conventional. Secondly, when falsely taking on this guise of novelty they become hard to understand because his borrowings appear as oblique as his original contributions. Spencer's philosophical context was not synchronized with the waves of idealistic and analytical thought that shaped later British philosophy. Instead, his arguments are intertwined with forgotten nineteenth-century debates. The result is to make Spencer's philosophy appear homespun and mysterious, even in places where it was derivative or sophisticated and clearly articulated.

Spencer's reputation as an isolated thinker began while he was still alive.[23] Initially he had been stylish, but his approach was such that he could not remain at the forefront of fashion. His work habits were so empirically painstaking and exhaustive that he was unable to continue harnessing new currents of discourse. He was also slow to complete his plan; by the end of the 1860s only three volumes of "A System of Synthetic Philosophy" had been published. The remaining seven appeared sporadically until the 1890s, after the deaths of most of the original subscribers who had initially bankrolled his endeavours. After 1886 Spencer's writings were mechanical and flat because age and illness had robbed him of ability. Delays in publication also had an impact on the coherency of the system. While avoiding new philosophical discourse, Spencer was continually adding extra scientific data, some of which were incompatible with his initial ideas.

Beginning with the volumes of *The Principles of Biology* in 1864 and 1867, contradictions appeared. This was a continuing problem: successive parts of Spencer's "A System of Synthetic Philosophy" always interpolated extra and variant arguments into his system. The results troubled his metaphysical foundations. For instance, his philosophy of science in *The Principles of Biology* relied heavily on the work of Whewell, even though this caused Spencer to appear more sympathetic towards an idealistic Platonism than was easily reconcilable with his *First Principles*. In addition, after 1865, in an attempt to defend his synthesis against an attack by J. S. Mill, he emphasized a more extreme empiricism. These were tactical moves designed to placate philosophers to whom he owed much. However, his efforts to avoid unpleasant disputes would confuse later philosophers, whose increasing professionalism made them hypercritical of inconsistency. In addition, Spencer was unpalatable to late-nineteenth-century intellectuals because he preferred data over logic while their taste began to run increasingly on non-empirical lines towards Kant and G. W. F. Hegel. Finally, the continual emendations he made to his texts failed to clarify his doctrines because it was his habit never to withdraw his original statement while responding to a challenge. Often he simply appended modifications to a position to the passage that had attracted critical attention. Spencer himself seemed unfazed by the resultant confusion. He practised the kind of defence the Russians had used against Napoleon by withdrawing into the vastness of his increasingly isolated philosophical system, confident that opponents would never muster enough empirical resources to threaten him. Spencer was also certain that, unlike Whewell and Mill, he had forged a philosophy of science whose principles accurately reflected the realities of scientific knowledge.

Spencer's politics and sociology have suffered a different fate than his metaphysics, psychology and biology. The latter have been neglected, while the former have been remembered in the form of a caricature. The most grotesque comments on Spencer are issued by supporters and critics of liberalism, who, with a disregard for his general philosophical system, treat him as if he had only written political works advancing an ideology of individualism. This practice has obscured Spencer as a person, while at the same time it permits the misreading of his political and social theory. He was not a strong individual himself, nor did he envisage an anarchic future peopled by giant egos. Such a vision would have been foreign to him; Spencer lacked the hyper-masculine traits that are often associated with powerful men. Instead he was passive, non-aggressive and conflict-avoidant. When faced with any altercation or cruelty he immediately withdrew from the threatening situation. His inability to tolerate offensive behaviour meant that he was equally repelled by aristocratic ceremony and working-class vulgarity. His over-sensitivity also led him to react negatively to both criticism and praise. Much of his well-known eccentricity was due to the extreme measures he took when protecting himself from feeling distress. Similarly, his dreams of the future grew from his desire to avoid contact with social attitudes he found unpleasant. He imagined that he would have been less emotionally disturbed if he had lived in the future when regressive behaviour would have diminished. Then everyone would be socialized to avoid violence, rudeness and intrusive enquiry into private matters. His vision was not utopian – that would have smacked of the rational planning he disliked – but of a gradual evolution of customs so that negative social attributes would fall away exposing a harmonious corporate society where pleasure would be unalloyed with pain. Emphatically neither he nor his philosophy was a product of individualism.

Spencer should bear some of the responsibility for being incorrectly labelled as an individualist. In his autobiography he frequently portrayed himself as a self-sufficient, courageous and original thinker owing few intellectual debts. He also claimed to have inherited genetic traits from independently minded and nonconformist ancestors. His statements have often been accepted at face value as naive admissions of a lack of education and technical knowledge. Spencer is taken to be a self-taught figure, like John Clare or James Hogg, when he was perpetrating a jest. His artful disguise has not been easy for social scientists and philosophers to penetrate. They have failed to observe that Spencer's self-reflections were his ironical entertainment, the writing of which periodically amused him for almost two decades: the reflections were not the work of a disingenuous self-made man. Like his practical jokes they were meant to mystify and astonish but, unlike them, the autobiographical writings are infused with irony and confessional self-deprecation mocking his existence and works. Not only was he sceptical about his own success, he was dismayed by the cult of great men that set the tone of many nineteenth-century biographies. These were palladiums that supposedly contained the spirit of culture and science, while Spencer's account of himself was deliberately ambiguous. This leaves his biographer with a dilemma; Spencer's claims to originality and eminence were sometimes deliberately obtuse. It is difficult, therefore, to either accept or dismiss them. Still, the ambiguities cannot be ignored; he intended his autobiographical reminiscences as his bequest to humanity. It is not as a great man that Spencer should be remembered, but as a precursor of the modern taste for self-doubt and alienation.

To Spencer, an individual human being was primarily a biological unit whose behaviour and feelings were modified by evolution. At first sight such a definition of an individual seems to be innocuous and scientific: too bland to have much impact on political or moral judgements. However, Spencer possessed two additional ideas that gave his organicism a cutting edge. First, he believed that the ultimate goal of human evolution was happiness; to a large extent, he considered this as a natural phenomenon that people would achieve through exercising their faculties, avoiding stressful work and engaging in play. Spencer believed that only on rare occasions could pathways to pleasure be provided by political institutions or other organizations; usually these routes fell in the exclusive domain of the individuals. This was problematic because Spencer also believed that people could not be expected to have a rationally informed will, nor aim towards a happiness that had yet to be conceptualized. His solution to this quandary was his expectation that socialization would provide a guide for behaviour. Secondly, Spencer perceived individual development as a process in conflict with social evolution: the progress of a social organism would often be at the expense of its individual components. In his view, there was no sense in fooling oneself by imagining that individuals' well-being would be furthered when they were sacrificed for collective gains. Spencer's idealized future did not imply that the spreading of knowledge to the masses would enable them to share power. That particular modern belief should not be assigned to Spencer, but to John Ruskin[24] and to other socialists as well as to liberal democrats. This, of course, begs the question of what modernism is in its political guise. Usually the answer has referred to the end of aristocracy and the rise of popular power, but Spencer insisted on the novel idea that it should primarily signal an improvement in the ethical quality of life. He emphasized this ethical feature of evolutionary progress in order to avoid the grossness of those utilitarian calculations that focused solely on increasing general happiness. For him that sort of analysis ignored the natural brutality of life, which he found evidenced in biological and sociological studies. Ethical change was a social phenomenon and seemed more fundamental to Spencer than moral enquiry, which interrogated people's internal senses of satisfaction and later aggregated them into a democratic general good.

Spencer's biological studies had led him to the repellent discovery that propagation in some species was accompanied by pain and death for the parents: parenthood, it seemed, was not necessarily productive of happiness. It was often the case that the individual parent would suffer in order for the species to carry on. When Spencer applied this idea to the human condition it left him ambivalent about the results of social growth. He speculated that progress would probably be accompanied by individual pain and death: a tragedy that could not be papered over by talk of artificial political contrivances in which duties to the state were balanced by correlative rights, or in which Benthamite legislation traded off conflicting pleasures in order to maximize the happiness of the greatest number. Such strategies did not effectively protect individuals, or allow for their pleasurable survival. Further, Spencer considered that political philosophers, such as Plato, Hobbes and Rousseau, had erred in placing the individual person at the centre of their conceptions of justice. Their tendencies had been to interrogate the individual's reason as if each person participated effectively in the polity; this was, in Spencer's eyes, archaic. Individual political personae had ceased to matter; like heroes and sovereigns they belonged to humanity's barbaric past. Tomorrow, he believed, would herald a new pattern of human behaviour; well-adjusted people would interact consensually without bowing to the crowned heads who had customarily forced agreements on their subjects. In the future people would also experience greater happiness as struggle, pain and death diminished. Negative impacts on life would increasingly diminish because they would be a development of corporate organizational structures, which would function without use of force. As a result of these changes people would have longer and more contented lives. This benefit would be a consequence of their successful adjustment to improved social mores and organizations. Spencer's sentiments here were not a plea for collectivism – in his time that would have meant advocacy of socialism – but neither did they support individualism.[25] Instead he focused on how *individuation* had structured politics, sociology and ethics in the past, and on why the perpetuation of such primitive values in the present led to aggression and domination. These ideas put him at loggerheads with his contemporaries, whose political discourses were largely a matter of erecting popular sovereigns, sustaining empires and justifying aggressive national policies.

The understanding of Spencer's politics has suffered because he has been wrongly classified as individualist, and then been dragged into this great nineteenth-century dispute between individualism and the state. Political theorists have characterized this debate in linguistic terms so that individualism is treated as a concept paired with state theory. That is, the word "individual" is treated as a natural antonym of "state"; it then becomes a historical task to discover at what point in the eighteenth or nineteenth centuries these polar opposites appeared, and whether they captured already existing connotations of social well-being and the public good.[26] In this fashion one can examine the writings of eighteenth-century figures, such as Henry St. John Bolingbroke and Bishop Hoadly, to search for the linguistic origins and primary meanings of concepts of the state with some assurance that this procedure would also elicit contrasting definitions of the individual.[27] Such an explanation is essentially an appeal to linguistic and conceptual traditions; but it is seriously flawed when it is applied to a modernizing nineteenth-century writer, such as Spencer, who stood outside the dual languages of state and individual and whose political values were based on empirical data. It is not that he lacked the notion of a conceptual tradition, or that he had freed himself from the grip of language: on the contrary, Spencer, like his contemporaries, incorporated tradition into his political theory, and frequently engaged in disputes with dead authors. Rather, the Victorian notion of tradition had shifted from that of the eighteenth-century. Like his contemporaries Spencer had adopted the new strategy of conceptualizing political theory as a collection of great classical texts, divorced from the legal and religious contexts in which they were written. Spencer is modern because, like T. H. Huxley and T. H. Green, he was arguing with Plato, Hobbes and Rousseau as if they offered coherent, although misleading, accounts of political principles. Although Victorian critics possessed less piety, as well as less scholarship, than their twentieth-century successors they similarly selected classical discourses as representing the significant tradition of politics. The result was that they echoed or contested the ideas of distant philosophical figures, not of eighteenth-century British authors with their densely packed discourses of a political theology and natural law.[28] In brief, Spencer's corporate theories of sovereignty and citizenship did not draw on the finely worked conceptual traditions of the preceding century. Even if he had been familiar with these, he would not have mentioned them when he had a chance to pillory classic texts: to him it was vital that he assault what he saw as the purveyors of absolute values in order to show that the value of science was greater than that of classical philosophy.

I
An individual and his personal culture

ONE

A portrait of a private man

There are two Herbert Spencers. One is the public man, the didactic philosopher who laboured hard to make sense of the fields of metaphysics, biology, psychology and sociology. This Spencer wished to be seen as neutral, passionless and objective. His formal prose reflected this wish. His language – studded with technical terms – crackled with a stiff phraseology that alternately inspired and horrified his contemporaries. When he was in his didactic mode, he was careless of how he was perceived as well as hostile to the manners and habits of polite society.

The other Spencer was an intensely private and overly sensitive man who was afraid of conflict and incapable of accepting criticism. He desired passion and affection, but was afraid of the consequences of these or of any other emotions. As this private person, he was appalled at the way in which modern society harshly employed the work ethic and political force to control individual lives so as to harden them against grief, pain and desire. His despair was deepened because while he remained hopeful that there would be social improvements, he was helpless to reform his own personal failings. Progress might correct flaws in future generations, but this would come too late for him. He could not rebel and he would always be isolated from companionship and physical love. Like the other philosophers who peopled Havelock Ellis's utopia, he was outside the charmed grove. These sages spoke of life, but did not hear it: "And as they passed up the slope still discoursing of life and death, they heard the soft laughter of young men and maidens among the trees, as it always has been, as it always will be through the brief day of Man's life on earth".[1] Spencer regretted that his day had been so brief that he had missed hearing the laughter. The result of his retarded discovery of emotion was that he felt no love and was left only with discourse on life and death. This he regarded as a crime against his own nature. Spencer's *An Autobiography* was an act of expiation, and a warning to others to avoid his fate.

In a sense, *An Autobiography* is Spencer's ultimate philosophical achievement as it is the only place where he completely abandoned science for a subjective investigation of his own intentions and meaning. This work was not only intended as a philosophical one; it was also a database for the lives of the two Spencers. The first of these, the public man, had a character that has always been easily delineated. That is, the reader has always seen the first of the images Spencer put before them. This was a portrait of an independently minded and original thinker who, from his youth to his old age, had successfully struggled against the dead hand of conservatism. Here was a self-made and self-educated colossus whose independence from social conventions was as great as his lifelong opposition to those primitive legacies from the past: war, cruelty and ceremony. Whether it was Spencer's refusal to learn Latin, his distaste for formal attire, his absenting himself from formal occasions such as levees or his rejection of honorary degrees, he was thoroughly modern in his attitudes. Even in death he refused ceremony: his remains were cremated at Golder's Green Crematorium and, according to his wishes, there were no flowers and those attending the ceremony did not wear mourning.[2]

The second Spencer – the private man whose life had been a tragedy – has usually been ignored. However, this omission will be remedied here. This is a kindness to Spencer's memory; the private persona, like the public one, was a lesson for the modern world, but has been overlooked because, uncharacteristically, Spencer spoke it in an ironic voice. The result would have seemed muffled to Spencer's audience because it was not habituated to artfulness and doubts. Anyone who empathized with the monotonous, and seemingly honest, scientific prose of works such as *First Principles* and *The Principles of Psychology*[3] would not have guessed that he was capable of deliberate ambiguity. Nonetheless, Spencer's self-mocking *alter ego* must be restored because his notions of individuality and modernity were always meant to be two-sided. That is, Spencer always intended that his boasts about the virtues of independence and originality should be counterpointed by a warning that these qualities were accompanied by a passionless rationality that had trapped him into a cruel self-denial of his emotions. It was this warning, not his often repeated admiration for independence, that was to have been his chief gift to the future: the important revelation in his personal testament was his late realization that it was not an isolated spiritualism that would redeem humanity, but a recognition that everyone shared the same biologically driven passions and urges. This message was ignored because Spencer had arranged for his memoirs to be published posthumously when his reputation had waned. They appeared in 1904 when all of Spencer's hopes and doubts were being discarded by a generation who detested their past so strongly that they could not bother distinguishing between one Victorian prophet and another. Suddenly everything from the nineteenth century was yoked together by neglect. Few possessed enough imagination to see that the first modern man had had the altruism to bare his soul for the good of succeeding generations. Spencer had written an elegy, not a confession, but it still demands respect.

Herbert Spencer's *An Autobiography*, a massive compilation of over one thousand pages, is strongly marked by his desire to prove that he himself was personally independent. His late political book, *The Man "versus" The State*, had proclaimed the value of individuals struggling to obtain their freedom from the institutions of church and state and from the dead hand of tyrannical custom. In the same way, his account of his own life was designed to illustrate the novel achievement of a natural man freeing himself from parental authority without succumbing to the artificial parents of established authority. This was a futuristic vision. His youthful self-creation owed few debts to history. In his view, he was unburdened by legacies from the past. This was a blessing because history was only a matter of war, ritual and aristocratic privilege, the records of which bored him. They were repetitive epics about great men. Anything from the past – symbols, titles, institutions or narratives – was to be rejected. To admire a coronation procession, a general, a titled lord, a medieval university or an ancient legend about Greek heroes was to subject oneself to the archaic values of the past. They were all shackles. As a youth Spencer believed that he was free from myth and owed allegiance only to science and progress. The narrative of his life was to be the erosion of his youthful vision.

In his youth he had seen the future of humanity as functional, peaceful and rational. The last quality was particularly potent to Spencer because rationality meant the avoidance of passion. He believed that if we were controlled by surges of emotion, Dionysian behaviour, love, fear or rage, then we would slip away from the modern world, back into the concealed caverns of memory that still featured in our psychological and sociological landscapes. Hidden in the recesses of both our personas and our society were passions that might throw us back into the past. Spencer constructed his personality to avoid these lingering, and sometimes enticing, inheritances. His public achievements and his personal tragedies all flowed from his hope for a modern and rational world and his failure to reach it.

A metaphor for Spencer's life is captured by his attempt as a young man to invent some steel-soled shoes.[4] These were to have a permanence that he would deny to the familiar and social structures around him. If his environment seemed a collection of decaying and dated customs for which no reasonable explanation could be given, then it was only his personal existence that carried meaning. However, this made the problem of permanence more pressing. If his own personality, and the society with which it interacted, was an evolving set of characteristics, how could he be assured that the often transitory qualities of life had a timeless objectivity? How could he establish a metal-shod sense of self? Spencer's answer to this question is to be found in *An Autobiography*. There, and in the accompanying *The Life and Letters of Herbert Spencer* (the material for which Spencer arranged before his death),[5] he built and reinforced the metallic shell of his personality. For almost two decades he shuffled and polished the story of his life so it would display and emphasize the independence that he believed was the indispensable feature of modernity.

For Spencer the creation of his autobiography was a Herculean labour; it was not that there was much dirt to be cleaned away, but the process of self-creation required much strength in an area in which he was naturally weak. As he quite truthfully remarked, his memory was poor.[6] When he explored his family's letters and his own correspondence he constantly surprised himself by finding incidents and attitudes of which he had no remembrance. To echo Spencer, he was constructing his own "natural history", a sort of personal sociobiology, but he had to do so relying heavily on written records. Instead of the process being one of retrieval with the archival material prompting hidden memories, it was a bare editorial and philosophical task. Spencer's autobiography lacked subjectivity. In his accounts of himself and of his closest relations there was nothing below the written page. Instead of self-revelations, there were stories that he could only assess in terms of their probability or consistency, not in terms of their inner truth.[7] At the heart of Spencer's reconstruction of his own individuality is an enormous sense of impersonality. He wanted his character to demonstrate a naturally evolved independence, but he often could not recall his past. As a substitute, he busied himself with providing objective commentaries on the evidence. The language he used in this process was sometimes opaque and harsh. It was also more often drawn from his political views than from his psychological or sociological works. That is, instead of structuring his own natural history around explanations from association psychology and ethnography, the languages with which he explained the logic of individual and social reasoning, it was docketed in categories taken from his early radical politics. These political views originated in the period before 1850 when Spencer was a committed activist in movements striving to provide increased democracy, and to destroy the control of aristocratic elites in the areas of education and religion.

The foremost political term in Spencer's autobiographical lexicon was "non-conformity". In using this, Spencer was mischievously toying with puritan religious history and the beliefs associated with it. In providing himself with a distinguished rebellious pedigree he was offering a parody of aristocratic autobiographies in which family characteristics were traced to an ancient Norman or Saxon notable. His own lineage through his mother, Harriet Holmes, was fashioned in such a way as to mock contemporary biographical tales of noble ancestors. After supplying a curious and speculative genealogy of Huguenot and possibly Bohemian, and therefore Hussite, pre-Reformation, rebellious bloodlines, Spencer remarked in a passive voice, "If any ingrained nonconformity of nature is to be hence inferred, it may have gone some way to account for that nonconformity which, however derived, was displayed by the children of my great-grandfather Brettell".[8] In case anyone missed the fictive point here, Spencer reformulated it as a question: "Has there not been inheritance of these ancestral traits, or some of them? That the spirit of nonconformity is shown by me in various directions, no one can deny: the disregard of authority, political, religious, or social, is very conspicuous".[9] Later he added that Harriet showed no sign of the Huguenot and Hussite blood he had "inferred", and that, "So far from showing any ingrained nonconformity, she rather displayed an ingrained conformity".[10] This is the territory of self-conscious imaginative writing and of Tristram Shandy, not historical truth. It has been Spencer's posthumous misfortune to have been studied chiefly by social scientists who have brought little humour or critical appreciation to their work.[11] If, instead, Spencer had been frequently chronicled by literary scholars, his memoirs would have been recognized for what they were: an exercise in ironic self-mockery in which historic detail was manipulated the better to illustrate artistic and political metaphor. Spencer knew only too well that "independence" was not a word that existed in its modern sense before the middle of the eighteenth century.[12] He was inviting his reader to know him – not his imaginary ancestry. As has been remarked by John Sturrock, autobiographical language is a self-conscious kind of writing that invites intimacy. It also marks off the individual from his or her tradition.[13] However, Spencer's literary model was not one of the serious prototypes offered by Descartes, Alfieri or Rousseau; it was Laurence Sterne's classic novel, *The Life and Opinions of Tristram Shandy, Gentleman*, in which memory was overcome by genetic parody. The invitation that Spencer's *An Autobiography* extended was not to intimacy, but to a self-mocking ambiguity. This was a kind of modernity because the deliberate reference to the fictional aspect of one's own identity placed Spencer outside the comfortable framework of contemporary self-affirmation. While Victorian novelists delighted in extolling moral certainties, and in the inheritance of ancestral characteristics, Spencer's lack of certainty was borrowed from an earlier period. His library was deficient and mostly consisted of works of philosophy, but significantly, in the area of pure literature the only work was a much prized copy of *Tristram Shandy*.[14] Without his attempts at humour and ambiguity Spencer's self-proclaimed nonconformity does not make sense. Spencer's dour Scottish biographer Duncan sensed that there was something wrong as he seized the point while missing the joke. There were, he soberly intoned, several respects in which the mother's "character belied the Hussite and Huguenot extraction her son was at some pains to make out".[15] In other words, as Spencer himself had observed, Harriet, whom he credited with improving the family by the addition of nonconforming ancestors, was passive and conformist in the face of authority, and even in the face of cruelty. (It was only with the help of humour that her son could empathize with her.) An inexplicable quietism was such a marked feature of her character that it tormented Spencer throughout his life, and distorted his relations with women. He felt responsible for his mother's destruction; although he had not caused this he took the guilt upon himself.

Spencer's claim to nonconformity has to be handled with care. By making it he was signalling his lifelong advocacy of the political principles that underlay his own contributions to the mid-century journal, *The Nonconformist*. This newspaper had agitated for such radical causes as the reduction of aristocratic power, the introduction of democracy and the attack on the Church of England's "state monopoly" in education and religion. However, Spencer's political constancy towards "nonconformist causes" should not be followed through to those other connotations of nonconformity that he sometimes invoked in a jest. In particular, Spencer's background and his personality should not be confused with those belonging to Unitarians, English Presbyterians, Quakers, Plymouth Brethren and other nineteenth-century remnants of "Old Dissent". The Independents and their like were not truly his ancestors. When Herbert Spencer had something precise to say about his family's religious background it was not a reference to seventeenth-century Protestants. His grandparents were Wesleyan Methodists and, as he carefully noted, the Wesleyanism of his relations did not imply dissent from church doctrine, but only from church discipline.[16] This distinction between kinds of dissenters was important for it meant that Wesleyans were not excluded from established institutions. For example, a member of this congregation – without much harm to his conscience – could attend one of the old universities. Spencer's immediate relations consisted of his mother, who was a devout Wesleyan, his father, a nominal one, and his uncle Thomas, an Anglican clergyman whose Cambridge education had been paid for by Herbert's father, George.[17] The uncle's religiosity took a political form during the period he was teaching the young Herbert. Taken as a whole the family would be more accurately described by the contemporary term "evangelical" rather than nonconformist.

George Spencer's own social and religious affiliations were scarcely typical Wesleyan ones. While he was on a committee for the Methodist library in his town,[18] his greatest involvement with an organization was as honorary secretary to the Derby Philosophical Society.[19] This institution had been founded late in the eighteenth century by Erasmus Darwin and some associates. A portion of the original membership survived to the time in which George became active in the society, when the whole group were regarded as irreligious by its only clerical member. The Reverend Thomas Mozley called his fellow members of the Derby Philosophical Society "Darwinians". In this sense, the word "Darwinian" was not a scientific label referring to the theories of Charles Darwin, but one that designated followers of Erasmus Darwin, the author of *Zoonomia*, in which, according to Mozley, it was maintained that creatures were created by themselves.[20] The scientific society that Erasmus Darwin left in Derby to continue his work propagated such ideas as a matter of natural theology as well as of natural history. Self-creation removed generative power from God, and placed it in the physical universe as a natural process. This bold statement relied on the power of reason rather than the authority of scripture. Such a posture was not a conventional theological one. The society and its secretary were a scientifically informed way station between Anglicanism and remnants of eighteenth-century heterodoxy. Some members had read writers such as Rousseau rather too carefully for the society to be regarded as a responsible bastion of orthodoxy.[21] To be the secretary of such a society was to be mildly dissident. While George Spencer did not regard his own views as redolent of godless materialism, they were part of an ambiance that was difficult to reconcile with devout Christianity.

In educating his son, George Spencer simply combined his emphasis on the need for a scientific explanation of all phenomena with a reliance on standard evangelical works for children. These latter seem to have made a permanent impression on Herbert Spencer out of all proportion to his father's intentions and to his family's quite conventional, and even lax, sense of religious devotion. Herbert was taught to read by memorizing stanzas from Isaac Watts's book *Divine and Moral Songs, Attempted in Easy Language*, and from its imitator Anne Taylor's *Hymns For Infant Minds*.[22] While Spencer later claimed to have been uninfluenced by the sentiments expressed in these works,[23] this was merely an attempt to make his later materialistic views fit with his youth. The echoes of Watts and Taylor in *An Autobiography* suggest that their moral sentiments retained a place in his subconscious mind. In other words, despite his systematic rejection of Christianity in his adult life, many of his deepest attitudes and beliefs are hauntingly similar to the childish refrains of the songs and hymns. They posed a model of benign speech and morals that contrasted sharply with the turbulent, critical and passionate utterances of the older men in his family. This model of goodness was a general Christian one, not a projection of nonconformity. For Herbert Spencer, the ideal of calm goodness could only be found in books because the reality of his family life lacked this quality. When, as a child, he adopted the language of Christian morals, it was far from being an acceptance of familial belief and behaviour. Rather, it arose from a reaction against his family's fervid emotional behaviour.

Watts's *Divine and Moral Songs* had been a perpetual favourite since its first publication in 1715. In the early-nineteenth century several towns, including Derby, had their own editions, each differently illustrated with simple woodcuts. Watts had originally produced his work so it would possess nothing "savouring" of party. He hoped that in reading it children of high or low degree, whether of Church of England or dissenters, whether baptised in infancy or not, would all be enabled to join together.[24] Anne Taylor, in some sense a continuator of Watts, had no complaint about his expression of ecumenical piety; she just wished to add some new evangelical truths.[25]

Watts's "Against Quarrelling and Fighting" contrasted animal brutes such as dogs, lions and bears with calm human beings. Animals had "angry passions" in their natures, whereas human beings were, by nature, loving and mild:

> Let love through all your actions run.
> And all your words be mild;
> Live like the blessed virgin's son,
> That sweet and lovely child.[26]

Animals are opposed to human beings in another song as well. In "Obedience to Parents" Watts reminded a child of what would befall him if he broke his father's law or mocked his mother's word:

> What heavy guilt upon him lies!
> How cursed is his name!
> The ravens shall pick out his eyes,
> And eagles eat the same.[27]

In case the child had not understood the message that God used brute animals to punish disobedience, Watts regaled his little readers with the story of Elisha, who was mocked by children. Those juvenile offenders were punished by God, who stopped their wicked breath and then sent two raging bears that "tore them limb from limb to death, with blood, and groans and tears".[28] Again, the lesson was that rage and violence do not properly belong to human nature; they were animal forces visited upon human beings from without as a punishment for the possession of other faults.

Anne Taylor's later and more insipid moralizing could hardly hope to match the inspired strenuousness of Isaac Watts. However, she too could remind her childish readers that passion should not be part of their nature. With Christ as an example before her she instructed her infants to repeat,

> All my nature is unholy;
> Pride and passion dwell within:

> But the Lord was meek and lowly,
> And was never known to sin.[29]

In case the children were not proud, and therefore failed to link that sin with passion, they were also reminded that if they forgot God's precepts they would become "Idle, passionate and fretful". The homily was clear; morality was a matter of avoiding emotion. Meekness and controlled behaviour were preferable to emotive and proud grasping after superiority. This was an ancient Christian message, but in Herbert Spencer's childhood it was reinforced by an unusual secular source: the 1780s compendium for children by Thomas Day, which was also popular throughout the early nineteenth century.[30] This work, *The History of Sandford and Merton*, was the first book Spencer read without help.[31] It offered a naturalistic pattern for behaviour copied from the ideas of Rousseau. In this work, the narrator, Mr Merton, appeals to natural behaviour *outside* civil society when he wishes to base manners on something solid. *Inside* society one finds only superficial manners and hollow forms of address. If, however, one looks within the mind for the true sources of virtue then one finds dignified sentiments and superior courage, which were accompanied by genuine and universal courtesy.[32]

On a simple level Spencer's early guides to morals emphasized the need to adopt sincerity and meekness. Whether Christian or secular, the model of behaviour enjoined the avoidance of pride, quarrelling and passion. There was nothing unique in the content of Spencer's reading; the works of Watts, Taylor and Day went through many editions, and were immensely popular during his youth. What was significant was how sharply their readings contrasted with the day-to-day turmoil in his family life. Neither his childhood literature nor his family were nonconformist, but the clash between the two eventually forced the adult Spencer to scrutinize both his family and his own personal life. He conducted this examination with the moral tools he had been given with his early reading. His ultimate rejection of his family's mores was made in the name of a passionless neutrality that outlawed rage and anger. This was a permanent feature of his character. He never completely freed his personal life from his sense that passion was immoral, although late in life he began to erect a more humane psychology that did find a place for emotion. Yet even then he was still partly captivated by his early ideas. His striving for a humane psychology was a recapitulation of the yearning for love reinforced by a traditional teaching of Christian dogma.

The sources of Spencer's morals, and his belief in the importance of being passionless, are concealed by his claim to nonconformity in *An Autobiography*. His artfulness blurred the edges of one of the great intellectual divides in the English past. Children of "Old Dissenters" among Spencer's contemporaries demonstrated intellectual and emotional profiles that were different from his. They belonged to a Protestant minority marked by an exclusiveness that had originally been forced upon them. Their exclusive identity had not been a matter of choices and options as it had been for Spencer's relations. For example, when Herbert's uncle, Thomas, became an Anglican parson and married the evangelical granddaughter of a governor of St Helena, it caused no disruption in the family. Nor did it matter that

Thomas's relations were middle-class folk from Derby, and hers were Anglo-Indians in Bath.[33] Religious and class demarcations existed, but they could be put aside.[34] If Thomas had married a dissenter the case might have been very different. When one of Herbert Spencer's contemporaries, the educational reformer W. E. Forster, married an Anglican he was expelled from the Society of Friends.[35] Sophisticated dissenters among Herbert Spencer's contemporaries, such as W. J. Fox, Caroline Fox, P. H. Gosse, W. R. Greg and Mark Rutherford (W. Hale White), could not easily choose conformism; they were marked for life. Herbert Spencer's background was less stringent. Even though his origins were on the radical edge of Protestantism, it was an edge from which one could move smoothly towards the mainstream of English culture. Like his uncle he could have chosen to go to Cambridge, or, like his father, he could take part in an established scientific society that possessed a wealthy establishment patron.[36] Herbert's father, George, may have been "independent" in his manners, but this did not exclude him from the community in which he lived. His private pupils included children from all the wealthy and aristocratic homes around Derby. Spencer's grandfather had been a master in the local grammar school. None of the several teachers who were closely related to Herbert Spencer either studied or taught in a dissenting school or academy.

Herbert's own attitudes bore little resemblance to the qualities that he supposed nonconformists to have. Rebellion was not admired by him, nor was an excessively Protestant reliance on reason. His hero or anti-hero was not a Christian one. The ideal was not Milton's Abdiel or Satan, but the pre-Christian figure of Prometheus. In Shelley's poem *Prometheus Unbound* Spencer found an object of worship who was without the faults of Milton's Satan. As Shelley himself had observed, his Prometheus did not suffer from ambition or envy, nor from the desire for revenge that had so disfigured the fallen archangel in *Paradise Lost*.[37] That is, unlike Satanic virtues, Promethean courage and majesty were not disfigured by the accompaniment of passions. The half-Titan admired by Shelley did not revolt, but bore his tribulations patiently. Like Shelley, Spencer felt removed from the darker forces that lay under the puritan sense of revolt. This was closely observed by Beatrice Webb. In reading one of the unpublished drafts of *An Autobiography* she was struck by one central strand of her old friend's personality. Spencer may have struggled with outside circumstances "yet there has been, and is, no sign of a struggle with his own nature".[38] The nineteenth-century Prometheus did not question his own value, or struggle with his own nature. He was not a puritan rebel, but the benefactor of humanity, who had suffered because he possessed insight into the future. Passively he waited for his gifts to humanity to change the species for the better. Struggle was futile; it would not bring personal redemption.

With such a hero in mind, Spencer did not feel drawn to any features of dissent. He could not be bothered reading Carlyle's *Oliver Cromwell's Letters and Speeches*,[39] and dismissed Emerson's musings about the "universal soul" because they were analogous to the Quaker "promptings of the spirit".[40] Behind Spencer's aversion to dissent was a rejection of the great Protestant belief in the supremacy of reason that so distinguished writers such as William Godwin, who did possess a dissenting background. While the great high tide of rationality had receded by Spencer's time, an unruffled faith in reason could be still found among some of his contemporaries such as the Unitarian writer, W. R. Greg.[41] Spencer was a social and cultural critic in a way that would have been alien to a dissenter; his critical stance was accompanied by a belief that rationality itself had no special standing. As he reminded his readers, his *The Principles of Psychology* set forth his dictum "that belief in the unqualified supremacy of reason is the superstition of philosophers".[42] Human beings had to proceed on the basis of beliefs that in turn were based on intuitions; reasoning about such intuitions made them even weaker than they were to begin with. Rational thought was no more trustworthy than any of the passions. Those who believed in the sovereignty of reason believed that this supremacy depended on the will, yet there was no basis for this in psychology. Reason was on the same level as the warring passions, and had no more validity than they had.[43]

Spencer's half-humorous depiction of himself as a nonconformist has proved a trap for unwary historians, but its correction is not a simple matter of adjusting antiquarian detail. In pressing the term "nonconformity" too hard, and in ignoring the mixture of evangelical and secular aspects of Spencer's background, his chroniclers have misunderstood the texture of his individuality. His links with his family, his failed attempt to rebel from them and his invention of a *persona* have to be seen in the context of a man's endeavour to get to grips with a passionate family that had overwhelmed him. The process was not a religious one. The method by which he formed his own character did not follow the normal nineteenth-century pattern of acceptance of an orthodox creed, followed by doubts and rejection. As his first and semi-official biographer put it, there is nothing to show that religion ever had a vital hold on him as it had on many of his contemporaries.[44] The analytical comparison in *An Autobiography* between his own personal qualities and those of dissenters is a detached exercise, not an expression of buried religious yearnings.

In *An Autobiography* Spencer often summoned up qualities that he associated with imaginary puritans. He carefully examined himself and his family for thrift, prudence, fortitude under provocation and, most centrally of all, rebellion. Yet only the first of these, thrift, was consistently maintained by Spencer. His careful and measured expenditure of money was re-examined at each period of his life. He regretted each lost shilling and pound; as both a youth and an old man he was consistently frugal.

Spencer always kept meticulous financial records. The lament for 1848 was that his life up to that point had been "unprofitable".[45] This did not mean that he had wasted a portion of his first twenty-eight years in taverns, in sport or with idle acquaintances. Quite literally it was a complaint that he often earned too little in return for his efforts. While his memory was generally poor, he managed to recall his feelings about every sum spent to aid his inventions and publications from the 1840s to the 1890s. Each of these, even if they were small, recreational or motivated by public spirit, were kept track of as a financial loss or gain. The "economic" events of his life – the invention of a paper clip, the publication of a political pamphlet or the production of personally subsidized compilations of data in social science[46] – were at some level just a matter of expenditure, which should have been modest.

Spencer's concern about finances had a perennial poignancy. When he was living in youthful hardship in London he lacked money to pay his tailor's bill. He was without a pair of trousers or a coat that could be seen in public. His father, George, lent him the money to relieve this embarrassment. Herbert's reactions to the loan are revealing; they are worth examining because they show a prosaic and cold quality in his character. He showed no gratitude, excitement or pleasure. It was merely a matter of dutifully recording that his father could not easily afford the extra expense at that time.[47] Later, when his father, then retired, bought several copies of an educational book to lend to his friends, Herbert soberly noted that this action displayed a lack of sense rather than generosity.[48] There is no reason to believe that Spencer's concerns about thrift and care in financial matters is traceable to a puritan consciousness about the need to avoid luxury. His attitude is more accurately explained by the fact that his father, George, and his uncle Henry had experienced losses as manufacturers that placed their families in straightened circumstances. Herbert himself had occasional vicarious experiences of improvidence. He had been the chief assistant to a railway promoter whose projects were almost criminally facile, especially when it was a matter of acting as a steward for the interests of his investors. These experiences, and his own difficulties in marketing inventions, had rendered Spencer intolerant of any Micawberish tendencies.

Spencer's claims to other "nonconformist" feelings – prudence, fortitude and rebellion – were even more ambivalent and artful than his claim to thrift. He suggested that his ancestors showed "prudential" feelings by marrying late, yet his own early hopes about marriage show that his lack of a wife was as much a matter of regret or lost opportunities as of self-denial. Through his twenties he frequently regretted the growing length of time he spent as a bachelor. He was often subject to a lack of prudence, such as the times he quit engineering positions. Once he turned down a permanent position without even asking what it was, and at another time he left his position when he was asked to assume extra duties that had not been previously specified.[49] Yet he took up a place as an editor with a short-lived Birmingham newspaper, the *Pilot*, with no information of any kind about remuneration; it was his uncle Thomas who worried about the imprudence of his actions.[50] The reality was that the young Herbert Spencer cared little for prudence, and certainly did not possess it as a central tenet of his character. Instead of being enamoured with this Christian virtue, he resented it. When his uncle Thomas was appointed chairman of the Board of Guardians at Bath, Herbert was quick to notice that Thomas was too hard-fisted with the poor. Thomas Spencer, in sympathy with Thomas Chalmers and other Poor Law reformers of the period, devoutly believed that each person's misfortune came from their own improvidence and lack of virtue. Consequently, he reduced the local tax that supported the poor.[51] While Thomas Spencer was efficient in his administration, his nephew thought him lacking in compassion. Herbert believed that when acting as a philanthropist his uncle should have used his heart, and observed that fate, or circumstances beyond their control, played a role in the distress of some in the community.[52] Later, when Thomas Spencer made an unwise investment that greatly diminished his own income, Herbert took a grim satisfaction in noting how misfortune had mellowed his uncle's views of his fellow human beings, and given him enough humanity to become a popular preacher in his last years. Herbert Spencer first attempted to tell this story in an obituary he wrote in 1853 after his uncle's early death. The tale was a parable against excessive reliance on one's own sense of fortitude and prudence. Its message was *be charitable, lest you yourself suffer unpredictable misfortune.*[53] He was consistent in this position in his later life; when he argued against public assistance he always supported private or voluntary aid even though he suspected that this too would reduce the quality of the individual's existence. His belief was that even if charity produced ill consequences, it should be tolerated because the exercise of sympathy towards the worst off would, by and large, be beneficial.[54]

Herbert Spencer's attitude towards fortitude was complex and ambiguous. He would have liked to admire calmness and suavity. His father had always displayed these qualities in public but, unfortunately, they disappeared with his nervous breakdown shortly after Herbert's birth in 1820.[55] Herbert was able to admire the amiable and equable characters of his lifelong friends Richard Potter and Edward Lott, but calm fortitude was not in his family tradition. His father, and several of his father's brothers, had the tendency to quarrel easily, and regarded themselves to be in need of dispute procedures.[56] The possession of a strong temper was so common in the family that Herbert regarded its absence in one uncle, William, as signifying that he was generally inadequate.[57] Herbert's own behaviour was constructed in opposition to that of his father and uncles. In other words, rather than relying on the possession of fortitude to protect him from emotional turmoil, he would go to great lengths to avoid arguments. His father and his uncle Thomas both complained of his reticence. For example, his father's letters written about 1840 called attention to Herbert's lack of response on religious questions at a time when the son was adopting a form of unbelief that was more militant than his own. Herbert smugly ignored such appeals for a response: "So far as I can remember they met with none, simply from inability to say anything which would be satisfactory to him, without being insincere".[58] At other times, when he was speaking to other young men, he was "constitutionally wanting in reticence".[59] There was no moral consistency to his silence or loquacity. Herbert lacked the bravery to accept the criticism that would have resulted had the father discovered the full extent of his heterodoxy. George Spencer criticized him for referring to Voltaire,[60] and would have reacted badly if Herbert had become a free-thinker.

What consistency Herbert Spencer displayed in controlling his temper was a psychological rather than a moral feature of his personality. That is, he feared conflict with those to whom he was close. This lasted throughout his adult life, and had a deleterious effect on his love affair with George Eliot and, as a result, condemned him to a life of permanent solitude that he found very burdensome. As he grew older he became increasingly intemperate in quarrels on the printed page, but he never developed the ability to survive an emotional encounter with someone with whom he was face-to-face. His fragility began with a disinclination to discuss contentious issues with friends, and eventually progressed to a point where he would even refuse to grant audiences to admirers of his writings. Although he enjoyed companionship, his need to avoid emotive interactions meant that his behaviour occasionally became extreme, and was the subject of those gossips who sought out examples of eccentric behaviour with which to decorate their memoirs. One story that was current in the closing years of the nineteenth century concerned a breakfast party given by his friends, G. H. Lewes and George Eliot. At this party the young Frederic Harrison was seated between Tennyson and Spencer. After annoying the poet, Harrison began to quarrel with Spencer about his philosophy. Spencer reputedly covered his ears and had to be carried from the room while he behaved as if he were having an epileptic fit.[61] Harrison, although rebuked by his hosts, was delighted with the opportunity of exposing such pride and perverted arrogance. However, the reality behind the story was that Spencer's delicacy had a less moral cause: he had been rendered incapable of quarrelling in his youth by his boisterous and interfering father and uncles. In later life he remained incapable of verbal violence or of any other interaction that was accompanied by warmth. It did not matter to him whether the enthusiastic reactions of others were hostile or sympathetic; they always provoked distressing feelings. It was for this reason that Spencer avoided admirers of his work. He explained his hostility to fans by attributing it to the pressure of work. However, this was inadequate since he behaved in the same way when he was on holiday. For example, when he travelled in Egypt in 1879 he rebuffed all attempts to lionize him. He would be equally rude to a nonentity or to a distinguished professor who was carrying a letter of introduction.[62]

He also took other opportunities to restrict the amount of contact with those around him. If Spencer decided he possessed a working relationship with someone rather than a personal one, he could be extremely reserved. In this way he offended one of his secretaries, W. H. Hudson, who repaid Spencer after his death by giving the world an accurate, but unkind, sketch of this feature of his character. Hudson's pen was sharpened by his trade as a professional writer on natural history; he was extraordinarily sensitive and had no difficulty in analysing how Spencer's feelings were so restrained that warmth and spontaneity were even leached out of his prose style. Reserve and austerity, Hudson continued, unfavourably impressed even those who came casually into contact with Spencer. Using Spencer's own style of analysis Hudson applied the "development hypothesis" to his subject, and "deplored" that the whole course of Spencer's life from childhood onwards should have tended towards emotional repression; he had cut himself off from common human relationships and responsibilities.[63] There was an element of truth mixed with the venom in Hudson's assessment. Spencer's childhood had affected him in such a way as to severely restrict any display of warmth and spontaneity. Instead of these qualities flowing naturally from his character, they had to be produced by conscious effort on his part. People for whom Spencer made no effort saw him as reserved and austere, and Hudson was correct in assigning the cause of this to Spencer's family background.

George Spencer and his brothers were, as Herbert put it, in the habit of very much saying what they thought whether on personal questions or on impersonal ones. "Necessarily there continually arose differences of opinions among them, which, expressed without much restraint, caused disputes."[64] Herbert dwelt meaningfully on the pathology of family arguments, which began with heated words and controversy and usually ended in a sentimental expression of mutual regard.[65] The young Herbert, as an only child,[66] grew up surrounded by men who contended passionately for recognition. Women were excluded from his early environment; his uncle Thomas would write to George Spencer that he had Herbert under control now, but if Herbert's mother should again have influence the boy would relapse into his old ways of rebellion.[67] The harsh reality of this comment can only be fully appreciated if it is understood that the mother was a decent, humble and devout Wesleyan who had been bullied into submission by her husband. Rather than being able to aid a youthful rebel she could not even protect herself. Harriet had failed to make an impression in the masculine world dominated by a husband whose mental collapse still left him able to reprove her daily for her stupidity.[68] This family tragedy echoed through the pages of Spencer's political works as well as his autobiography. One of the passages of the 1851 edition of *Social Statics*, which stayed in the abridged version of 1892, was: "Command is a blight to the affections. Whatsoever of refinement – whatsoever of beauty – whatsoever of poetry, there is in the passion that unites the sexes, withers up and dies in the cold atmosphere of authority".[69] When George's illness became so severe as to prevent him from exercising his parental duty, he did not rely on his wife as a substitute. Instead, one of his brothers, Thomas or William, would stand in his place, exercising sufficient authority to frustrate rebellion. At times, they intruded on the young Herbert's life in such a way as to supplant his father.

Herbert later glossed over this superabundance of paternal authority by calling it "nonconformity". The severity with which he had been treated could be excused if the father and uncles had had their eyes turned upwards towards the heavens. That is, the explanation might have been plausible if they had been thoughtless religious conservatives who were unaware of the consequences of their actions. However, they were informed liberal educators who thought deeply about what they were doing. Herbert could find no plausible reason that condoned their treatment of him, and retained an abiding sense of resentment about the neglect he had suffered while a child. He was left with a sense of betrayal that made him permanently ungracious about his father's mental illness. He blamed this for causing the failings of his early education and his later unhappiness. George Spencer's irritability and depression had prevented him from demonstrating to his son that geniality of behaviour "which fosters the affections and brings out in children the higher traits of nature. There are many whose lives would have been happier, had their parents been more careful about themselves and less anxious to provide for others".[70] Herbert believed that his father's affectionate and energetic teaching of other people's children, before his own birth and during his infancy, had rendered him ill and inadequate as a parent. George could surely have foreseen the consequences this would have and have acted differently. Herbert blamed his father for losing control of his emotions, and leaving his son adrift in a world in which he was plagued with an abundance of paternal figures. Although it may not have been George's intention, Herbert's harsh education left him incapable of rebellion.

The impossibility of rebellion shows up clearly in Herbert's only major attempt to disobey. This incident became the centrepiece of the story he himself prized as a major fragment of his imaginary nonconformity; it was to illustrate his supposed rebellion from authority and order that linked him with a tradition of seventeenth-century independents. The story he told began with the information that, at the age of thirteen, he was taken from Derby to Hinton Charterhouse near Bath to be taught by his uncle Thomas, an Anglican clergyman who took in private pupils to prepare them for university. Herbert's parents intended to leave him in Hinton when they returned to Derby, but they kept their plan secret from Herbert. They pretended that they were paying an ordinary family visit. After a few days Herbert was told he was to study during the visit, but still nothing was said about the plan. Then, after four weeks, there occurred the "startling" revelation that they would be leaving him behind when they returned home.[71] Herbert felt betrayed, and this led him to take an uncharacteristic action. Ten days after his parents' departure, a young Herbert "was quite prepared to break out into a rebellious act".[72] He ran away, walking most of the long distance home in three days, covering 48 miles in the first day, 47 in the second, and 20 in the third. This got him close to Lichfield, from where he begged a ride on the coach to Derby.[73]

Thomas Spencer made no attempt to pursue him. Herbert's action in running away was such an enormity that any notice of it on his uncle's part was impossible. His wife, Anna, wrote a letter to another of Herbert's uncles, William, hoping to spare the parents the anxiety of worrying about Herbert.[74] She herself felt no anxiety, only anger: "It is my decided opinion that unless his Parents punish him severely, and return him again to us *immediately*, it will not only be *insulting* to us, but *ruinous* to the boy himself!"[75] The Reverend Thomas Spencer and his wife were successful in having their wounded feelings placated. Herbert's parents replied by saying that Herbert was sorry to have offended them by absconding, and then apologised for the boy's silence before his departure.[76] Herbert was returned to Hinton. He was then set to learn his mathematics, French, Latin and Greek. Thomas was a stern taskmaster whose reports to Derby were full of self-justifying comments on the need to impose strong discipline on the wayward youth. Gradually the uncle was victorious; he began to feel that "There was a quietness about [Herbert] and an evident desire to be satisfied".[77] The discipline was a highly personal one, a situation that no doubt satisfied the uncle that he was the essential tool in the process of breaking Herbert's spirit. He would repeat that he was now satisfied with Herbert's obedience to him, but suspected that he would not submit himself to any other person.[78] It was imperative to Thomas that the curb on the boy be enforced for a long period. He also strove to alter Herbert's character, which he saw as being deficient in the principle of *fear*, by which he meant "both that Fear of the Lord" which "is the beginning of wisdom", and "that fear of Parents, Tutors, etc".[79] Ominously, he recorded that – "after a few struggles" during which he *induced* this principle – Herbert "entirely surrendered himself to obey me".[80]

David Duncan, who printed some of the exultant letters in which Thomas Spencer revelled in his victory over his nephew, said that too much had been expected of a boy aged from thirteen to sixteen years. Herbert, he suggested, had been subject to too much exhortation by both his uncle and his father. Duncan believed that it was only the striking individuality of the youth that allowed him to "have held his own".[81] However, Duncan, as a trusted assistant and former secretary of Herbert's, had allowed compassionate optimism to soften his judgement on this point.[82] The young Herbert had not held his own at all; he had been broken. His letters from the period referred frequently to a lack of energy, which he was at a loss to account for.[83] A pathetic note survives, which thanks his father for his solicitude in placing him with his uncle. This should be read in the light of badgering letters from the father that complained about Herbert's reticence after he had been forced to return to the bullying and stentorian uncle. Herbert, crushed as he was, could only write that he attributed his father's solicitude towards him "to his sincere desire for my welfare of which I am convinced from good advice given, and hope with the help of the Almighty to follow it".[84] These sentiments, which were omitted from *An Autobiography*, should be connected, as they were by Spencer himself, with the lack of gratitude he felt towards his uncle Thomas and his aunt Anna. "It was", he remarked, "better to be under a control which I no doubt resented, but to which I had to conform, than to be under a control which prompted resistance because resistance was frequently successful".[85] This sobering reflection was a rejection of the fanciful nonconformity with which he occasionally masked himself. It suggests that, rather than Herbert Spencer's sense of independence being rooted in a dissenting rebellion, its source was a rueful recognition of his own defeat while a child. He had to deny that he possessed a puritan spirit. Instead, he tried to comfort himself with the admission that it was better to be forced into conformity than to be a successful rebel. To say otherwise would have been to deny the genuine roots of his own true individuality, which lay in a loss of personal independence, not in the assertion of it.

Herbert Spencer's half-hearted attempt to conceal his defeat by associating his individuality with the great cultural and religious forces of nonconformity was a kind of ironic playfulness. His autobiographical rewritings, which occupied his leisure time, were indulged in with a sense of genuine relief. They were a solace that made bearable the seemingly endless demands of the scholarly volumes of his "A System of Synthetic Philosophy". He knew well enough that his own life had been one of defeat. That is, he knew that his genuine acts of disobedience were not ones of open rebellion, but furtive ones, such as his secret reading of novels such as Horace Walpole's *The Castle of Otranto*.[86] Escape into an imaginary world might have provided some relief from oppression, but scarcely compensated for the passion that he felt had been stifled within him.

Spencer's political theory and his views on education were permanently shaped by his childhood experiences. In his early political writings he reaffirmed an individuality based on perfection rather than freedom. That is, for him the true individual was one who developed fully the potentials within: personal freedom was not the liberty to express emotions. This distinction is especially clear in his defence of the rights of children. Whereas a twentieth-century advocate of children's rights might support them so that youth could give free expression to their feelings, Spencer was adamant that the highest faculty that should be cultivated was self-control. He saw the perfections of the ideal man as: "[n]ot to be impulsive – not to be spurred hither and thither by each desire that in turn comes uppermost; but to be self-restrained, self-balanced, governed by the joint decision of the feelings in council assembled, before whom every action shall have been fully debated and calmly determined".[87] This perfect man who avoided impulsive desires, and who sought to be self-restrained and self-balanced, was the younger Spencer. His individuality had been constrained. It was a passionless self-control imposed by the strict education given him by his family. However, this was not to last. The mature Spencer regarded his inability to follow his feelings as a loveless tragedy, and wrote *An Autobiography* to warn others of his plight.

TWO

The longing for passion

Spencer judged that his own personality was a failure. He could boast that he had created himself, but the human core was absent. His inner life seemed barren and without passion. However, he was more successful when he constructed a social environment for himself. This exterior world was more responsive than his inner one. In the outside world he could address his two basic needs: humour and love. The first of these came in the form of a playful interest in the manipulation of symbols. Typical of this was his youthful insistence on wearing caps – rather than hats – both as a matter of politics and as a joke. The second need was more urgent and required him to be creative. It stemmed from his inability to experience physical love. By implication this meant marriage and children were beyond him. He felt nothing towards women, and he feared that if he forced himself to consummate a sexual relationship he would then replicate his father's systematic cruelty towards his mother. Spencer's solution to this dilemma was to create a substitute family that would provide love without an element of risk. This surrogate also had the advantage of being less likely to expose his emotional flaws to criticism. A new family would replace the dictatorial one that had crushed his youthful attempts to rebel and gain independence. It would also relieve his immediate distress. However, such relief was not a cure, but only a palliative. To be cured Spencer would have had to feel passions that were beyond him. All he could do was humour himself by jesting and by pretending to possess a family and children.

Spencer's personal life was often a matter of pretence and mimicry. In the realm of fancy he could imitate his father and his uncles. They were earnest dogmatic men who never spoke of frivolities such as royal personages, courts, bishops or lords.[1] It was not only the great who were beneath their notice. As Herbert mentioned at the beginning of *An Autobiography*, his family was not prone to gossip.[2] By this he did not mean that they failed to express unfavourable opinions of others. On the contrary, Herbert remembered distinctly that no one had a good word to say about his uncle John, a solicitor, who was thought to be selfish. The lack of gossip simply meant that no one in the family indulged in the kind of idle prattle about acquaintances or about the amusing foibles of relations. There were no human details to lighten the weight of the never-ceasing serious discussion and argument. It was not only people that were missing as subjects of conversation; no cultural topics were canvassed. The family omitted poetry, fiction, drama and history from their agenda. "Their conversation ever tended towards the impersonal."[3] It avoided what Herbert later called the "aesthetic" of life.[4] That is, they could not see gentle features in the visages of either real or imaginary persons, but concentrated on the skeleton underneath: the simple principles that they believed underlay and shaped both human behaviour and the material world. When the uncles brandished a symbol it was a political sign, not an artistic one. In his own later search for principles, and in his use of political symbols, Herbert was true to these family traits, but at other levels he passively rejected "nonconformity" as lacking in subtlety and beauty, or even as plainly false. There were political implications as well in that he could not espouse the extreme Protestant values of independence and individuality that have often been ascribed to him.

Since the examination of Spencer's principles of politics will be dealt with in later chapters, the philosophical interconnections between political independence and science will have to wait until then. However, the curious familial resemblance between the use of outward symbols and politics belongs here. The fact that Herbert Spencer partly imitated his father or uncles in his reactions against signs of status and wealth is part of his personal life. Although he adopted the family hostility to rank, his iconoclasm was frequently limited to the political sphere of life; he did not follow the family in its rejection of artistic and decorative values in favour of scientific and rational ones. This distinction can be most clearly seen in George's constant opposition to "aesthetic" values. George Spencer's resistance to custom was more a social and cultural imperative than it was a political one. George was a tall and handsome man whose passage along a street would turn heads in admiration. He always affected the same tight-fitting blue frockcoat no matter what changes in fashion occurred.[5] His simplicity, strikingly similar to that of Mr Barlow, the fictional hero in Thomas Day's *The History of Sandford and Merton*, led him to refuse to use the titles "Esquire" or "Reverend" or other forms of address: everyone was simply "Mr".[6] George Spencer objected to any façade or show through the use of clothing. In a time of increasingly complex funeral rites, he would not wear any mourning clothes to grieve the dead.[7] He also objected to the performance of plays.[8] To him any show, whether memorial or artful, displayed insincerity. The same lack of true feeling might emerge if one followed a particular form of worship.[9] George's behaviour does not appear to have been regarded as political by his contemporaries. Herbert noted that George Spencer was suave and sincere in his manners, and that no one ever seemed to take offence at the way he addressed them.[10] If anything, the stories that surround George tell of a deliberately contrived public appearance of calmness and frankness. He could, for example, stop in the middle of giving a child a lesson and ask the parent for the loan of £10. This smacks more of gentlemanly ease than political radicalism. Also, the fact that he was given to shaking his head in disapproval when Herbert, as a young man, uttered anti-monarchical views,[11] suggests that George's radicalism was of a mild variety.

While unconventionality in dress was primarily a moral and cultural matter for George Spencer, others in the family were more overtly political in their gestures. For example, Henry Spencer, one of George's brothers, denounced the Poor Laws and wore a white hat as a symbol of his radicalism. This was meant to be divisive as it exposed the wearer to insult, but it also symbolized an attachment to democracy and political radicalism.[12] In the matter of clothing Herbert took after his uncle Henry as well as his father, George. From the latter he drew a studied nonchalance about what he was wearing, and from his uncle he adopted sartorial politics. Herbert Spencer used his clothing as both a modest philosophical and a political statement throughout his life. From his youth to his old age he enjoyed making a point on the symbolism of dress. At the age of twenty-two, he could boast to a young woman that:

> Having patiently persisted in the practice of cap-wearing, notwithstanding the surprise exhibited by the good people of Derby at such an outrageous piece of independence and the danger of being mistaken for a Chartist leader, as I have frequently been, I have at last had the gratification of witnessing the result of my good example in the adoption of the cap as a head-dress by a good number of the young men of Derby. So that it appears that I may actually claim the high honour of *setting the fashion*.[13]

Many years later his young friend Beatrice noticed that Herbert was always clad in "primly neat but quaintly unconventional garments".[14] Here, as elsewhere, Spencer's attitude was that of a man who resisted authority, but did not rebel. His sartorial stances were noticeable, but not dramatic. For example, when he first became prominent he used to go to dinner in a frock coat, but later he went out in a swallow-tail jacket. However, he always refused to wear a white necktie,[15] and had some satisfaction in complaining about the social code that legislated this.[16] This slightly alternative mode of attire was reminiscent of his father's, but at times Herbert's views on clothing were more politically pointed. He was capable of indulging himself by sending a principled refusal to attend a Foreign Office "At Home" for the Emperor of Russia by stating that his insuperable objection to levee dress meant that he must decline the invitation.[17] When he was put on the spot by the Countess of Derby, who said that she would be sorry to be deprived of his company over a question of costume and that he should come in ordinary evening clothes, Spencer was forced to be explicit on how far he was willing to disregard conventions. He wrote that to make himself a solitary exception in so conspicuous a manner on such an occasion would be even more repugnant to him than conformity itself.[18] His nonconformity in dress was an occasional thing, and it was also a convenient excuse to escape from situations that he felt were politically embarrassing.

In the matter of clothing, Spencer felt he had the potential to rebel. Clothing was superficial; it was an outward symbol. This attitude was sanctioned by his family's refusal to obey the norms of social ritual. It was also an important aspect of radicalism because it allowed Spencer to deplore the falseness of his Tory opponents, who repellently used ritual to affirm their high status. The message carried by his unusual costume, and by his refusal to wear formal clothes, was to critique the pomp and circumstance that had crept over England during Queen Victoria's reign. His witnessing of the young Victoria's procession to Mansion House in 1837 shortly before her coronation was his sole participation in state ritual. He could later boast that, "It was the only royal procession, or display of allied kind, which I ever saw".[19]

The analysis of Spencer's rebellion leads to a much deeper level than clothing. While it is clear that he held to *outward* signs of insurgence, *inwardly* he remained excessively sensitive and passive. This dichotomy spread and ramified throughout his cultural perceptions and personal interactions. It also provided a framework that structured much of his autobiography, as well as his political views and his theories concerning the central tensions of social science. To understand why Spencer was not a rebel requires more sophistication than can be provided by exploring the ways in which he was superficially an insurgent while remaining essentially a conformist. This is a matter of sketching the appropriate historical context in which a display of passion represented failure rather than an admirable possession of sentiment. Spencer did not belong to a romantic sub-culture that put its faith in the cathartic and therapeutic effects of anger and rage. For him, and those of a like mind, the ideal was to avoid displaying passion. This restrained stance was best captured by his friend, Elizabeth Barrett Browning, in a poem on how to be trustworthy. When Spencer recorded the long-term consequences of tyranny, torture and slavery he could not simply describe them as absolutely bad or totally good. From his perspective they might possess both qualities yet still witness to the progressive evolution of humanity. Such a paradox of good and bad was painful to observe, and required "an almost passionless consciousness".[20] Spencer adopted Barrett Browning's painful depiction of grief:

> I tell you, hopeless grief is passionless;
> That only men incredulous of despair,
> Half-taught in anguish, through the midnight air
> Beat upward to God's throne in loud access
> Of shrieking and reproach. Full desertness,
> In souls as countries, lieth silent-bare
> Under the blanching, vertical eye-glare
> Of the absolute Heavens. Deep-hearted men, express
> Grief for thy Dead in silence like to death –
> Most like a monumental statue set
> In everlasting watch and moveless woe
> Till itself crumble to the dust beneath.
> Touch it; the marble eyelids are not wet:
> If it could weep, it could arise and go.[21]

To be passionless was not to manifest impotence; it was the calm expression of humanity when faced with cruelty or with the bleakness of an indifferent universe. The avoidance of shrieks of pain and anger signalled the courage and endurance that a person must possess when staring silently at the terrors of moral and scientific truth. To be passionless also meant that one possessed objectivity. When Spencer invoked the word *passionless* he simultaneously revealed two attitudes: first, he signalled his sense of emotional suspension and his inability to create strong emotional bonds; secondly, it was his brave claim to objectivity. Both these attitudes might now be regarded harmful and unnecessary, but in the context of Victorian culture objectivity was courageous and humane. The first was more problematic; it left him emotionally crippled, even by the standards of his contemporaries.

Spencer's picture of women was permanently affected by his bleak early interactions with his family. Both on an emotional level and an intellectual one, he was never able to develop the capacity for interacting with women in a way that he or his contemporaries perceived as normal. As a child he had been sequestered from women. Without a mother's protection he had been crushed by a male-dominated world. His reactions to women always remained the tentative, foreign and distant ones of a man who had only theories to rely on. He could either objectively claim that women were biologically inferior to men or, to re-cycle his borrowed French word "*brusquerie*", he could brusquely impose himself on young girls in a way he could never with adult women. When he could achieve some distance from them, women became symbols of an aesthetic that had been absent in his early years. He also grew to increasingly regret his family's condemnation of the arts as mere displays lacking sincerity and truth. He himself had a lifelong struggle to internalize the idea that beauty possessed a measure of truth and value. Ultimately Spencer found this truth and value in both the appearance of a woman and in an art object. In this way he cultivated a belated place for passion that his father and uncles had attempted to destroy in him. This late acceptance of the need for aesthetic values that, in themselves, possessed few scientific uses was a rejection of familial values. However, it came too late to be of any use to Spencer. He was never comfortable with women. The fact that his acceptance of aesthetic values was hard won also caused him to overvalue beauty. While many other Victorians also placed a great emphasis on beauty, Spencer's own contribution to this trend was perhaps more significant in that it emerged in the centre of his primary activities as a scientific publicist. That is, he was led to create a central place for a non-functional aesthetic in the midst of his scientific writings. His frequent evocations of the need for art and style, although often overlooked by recent scholars, signalled to his self-proclaimed role as a rival to Coleridge and Ruskin. From his perspective these were prophetic thinkers who decorated what they feared to be an increasingly godless universe with symbols or signs of the aesthetic.[22] Explanation and truth were not enough. Life had to be meaningful as well, and meaning was to be found in beauty.

At times, Herbert Spencer's disregard of passion worked as a simple affirmation of his family's belief in the certainties of science and radical politics. This feature can be seen in his early attempts during a lull in engineering work to become a professional writer. This was in 1844. His plan was to issue a periodical to be called the *Philosopher*, with a prospectus urging its readers on to that era of civilization when they will have shaken off the soul-debasing chains of prejudice. "The human race is not for ever to be misruled by the random dictates of unbridled passion. The long acknowledged rationality of man and the obvious corollary that he is to be guided by his reason than by his feelings, is at length obtaining a practical recognition."[23] He continued in this vein and condemned "mere antiquated authority", "the absolute dicta of the learned", "precedent" and "unerring custom". These were the words of a twenty-four-year-old political enthusiast, not of a mature philosopher. The old and young Spencers should be kept separate; while the striving for modernity and political equality never left him, he increasingly attempted to mollify the cruder ideological implications of his earlier goals. The most significant change was that the crude dichotomy between passion and rationality disappeared. Spencer began to see the distinction between the two as something that had prevented his own family from governing their relations with each other in a human way. His uncles, in particular, were able to use reason in such a way as to exploit their desires to be selfish and cruel, rather than to control them. Herbert Spencer, in order to avoid this family failing, attempted to restore emotion to an honoured place in the human psyche. The restoration of passion was not restricted to being a remedy for his grief; he generalized beyond this and erected a psychological theory that demonstrated that mental processes in general must be studied so that emotions were given their due importance. Although he is now known chiefly as an evolutionary theorist and as a sociologist, the genesis of Spencer's reputation as a serious writer lay in his *The Principles of Psychology*.[24] In this work he undermined the place of reason by attempting to prove that the commonly assumed line of demarcation between it and instinct had no existence.[25] For him reason was on the same level as reflex action, memory, feelings and the will.

From as early as 1853 Spencer began to see his family's emotions as seriously disturbed. Their heat in argument and their adhesion to doctrinaire beliefs was destructive. It allowed members of his family to mistreat their relatives. Speaking of his uncle Thomas, Herbert remarked that: "[h]is generosity however, had the peculiarity – family peculiarity it might in some degree be called – that it was more seen in large things than in small ones. The daily acts of domestic life did not exhibit that power of self-sacrifice which was called out on important occasions".[26] This, Herbert judged, was a moral deficiency. If one of the surrogate fathers was deeply flawed, the real father, George, was more damaged. According to his son, George was admirable in many respects, but he had "one great draw back": he was unkind, exacting and inconsiderate to his wife. Not only did he fail to display that sympathy that his son thought should characterize marital relations in general, but he punished his spouse every day of her life until, near the end of it, she became ill. Harriet was inarticulate while George held that either everyone should speak clearly, or they ought to suffer from the evil they brought upon themselves. "Hence, if he did not understand some question my mother put, he would remain silent; not asking what the question was, and letting it go unanswered." He seems to have begun this course of harsh corrective action after two or three years of marriage, and pursued it endlessly "notwithstanding its futility".[27] George made no allowances for his wife's lack of interest in academic subjects such as chemistry, mechanics and mathematics, nor for the fact that his wife had suffered from the deaths of six or seven infants and the loss of her only daughter, Louisa, at the age of two.[28]

Herbert reacted to his emotionally damaged family in a personal as well as in an intellectual way. He sought out substitute families to which he could attach himself. Also, he internalized his grief, which as he got older was often expressed as complaints about imaginary physical illnesses. As an intellectual reaction to his sense of loss he constructed an elaborate set of psychological and cultural perceptions of passion and its relationship with rationality.

Spencer attached himself to two surrogate families: the Moorsoms and the Potters. In both he discovered the laughter of children and a husband who did not hector his outspoken and successful wife. The women and children offered a kind of companionship that Spencer had largely done without when he had been a child. His youth had been a solitary time in which his major experiences had been listening to his uncles' heated debates about the principles of politics and morals and to scientific lectures at the Derby Philosophical Society.

Captain Moorsom, an engineer involved in the surveying and construction of railways, appointed the twenty-year-old Herbert to be his secretary. This was a promotion – Spencer had been a draftsman – and meant that he lived near the Moorsoms at Powick near Worcester. The Captain treated his young employee with as much kindness as he would have shown towards a member of his own family. Spencer went on excursions with the Captain, dined with the family regularly, and visited them on the weekend. Mrs Moorsom was Spencer's first "lady friend".[29] Spencer was thrilled to be a great favourite with the Moorsom children, and saw this good fortune as the way to "the mother's heart".[30] This idyll lasted approximately a year, and was broken by the sense of alienation that arose in Spencer when he heard a rumour that the Captain had underpaid one of Spencer's friends for work on a survey in Cornwall. As soon as Spencer entertained this suspicion he immediately forgot all the kindness shown him, and remembered only "that most abstract of the sentiments – the sentiment of justice".[31] As he remarked later, when he regretfully described his mood, he had allowed the abstract to obliterate all his other feelings. Spencer's ability to focus on a single aspect of a problem was a positive feature in his work but, as he himself remarked, it could ruin his relations with those close to him. When he was captured by "impersonal" feelings they destroyed his relations with his first substitute family. He had treated them harshly in the way his uncle Thomas and his father had treated their families. His unthinking imitation of ingrained familial behaviour allowed political radicalism and "passionlessness" to intrude on and distort personal emotions. "In later years I have never ceased to regret the error thus committed."[32] It was not to happen again: he was to retain the affection of his next substitute family, the Potters, for over fifty years, loving them loyally through two generations.

Spencer was treated as a member of the Potter family through much of his adult life. From meeting Richard Potter and Laurencina Heyworth a year before their marriage in 1845 until his own death in 1903, he enjoyed the comfort of frequent visits with Laurencina, and with her daughters after her death. The Potter family not only supplied him with warmth, but with a nearly endless supply of young girls on whom he could lavish his affection. In return, he gave the adult Potters access to the worlds of letters and science. The mother, Laurencina, in particular craved for this as she had wished to preside over a literary salon. This had been difficult to achieve since Richard was frequently absent on business, and kept her isolated in the countryside. However, Spencer's roots in the family went much deeper than this functional account of reciprocity indicates. It was with the Potters that he could feel truly human. In the security of their home he could joke, chaff, gossip and pretend to teach children, following the advanced educational lines he would have used if he had had children of his own. He believed he could save the Potter children from reliving his own childhood tragedies by repeating the axiom "No submission to authority" to the children at teatime. When he did this his normally placid features would be transformed into an attitude of tremulous exasperation. His eyes would darken, and his voice would become shrill and almost shrewish.[33] The sudden and uncomfortable emergence of a passion for the rights of weak creatures and children signified that this feeling ran deep. The presence of children relaxed him. He could indulge his engineering skills by building swings for the girls and little vivariums for small animals they had captured when he took them on rambles through the countryside.[34] With the Potters he shed his role as the cold philosopher, and abandoned his usual worries about his status. He did not care that Richard Potter had failed to read his books and dismissed his philosophy as impractical.[35] Nor did Spencer mind that his special protégée among the daughters, Beatrice, never read his books when she was an adolescent.[36] He would not succumb again to the "impersonal" judgements that had destroyed his relationship with the Moorsoms. With the Potters he was not an intellectual, but a person.

Spencer's most important relationship was with Laurencina. She was everything he could desire in a woman. Her marriage to Richard made her simultaneously unthreatening and deeply attractive. Her beauty was such that John Bright described her as one of the two or three women a man remembers to the end of his life as beautiful in expression and form. Her charm was so noticeable that the young anglophile Hyppolite Taine used it to sustain his hopes that England was a civilized country.[37] She was scholarly and gentle. Her detailed knowledge of that dismal science, economics, was leavened by her expertise in Greek. Most important, she shared Spencer's lambent, but still important, enthusiasm for radical politics. Finally, like many of Spencer's early male disciples, she needed him to talk about the "Unknown": his modern substitute for the deity. Spencer provided a calming metaphysical and scientific comfort as an antidote to her disquieting sense of religious doubt. At the heart of Victorian commercial smugness often lay a fear of the chasm of dark meaninglessness that Spencer alone could temporarily remove. For Laurencina this consolation was fleeting; after the death of her only son she withdrew from the cares of daily life and increasingly immersed herself in her studies. Even the running of the household was left to her older daughters. She neglected her ninth daughter, Beatrice, and to a lesser extent, her tenth, Rosalind.[38]

Richard Potter was a less fragile person than his wife. He needed Spencer less, and could even be dismissive of him. Yet Spencer was frequently his house guest, and repaid him with devotion. He described Richard Potter, with the kind of unselfconscious candour that was still possible in the nineteenth century, as "the most lovable being I have yet seen".[39] Potter was outwardly amiable to everyone, and was particularly devoted to the women in his family.[40] He seemed to have no close friends, and thus always found Spencer useful as a fishing companion. Potter's life as a railway baron meant that he was often absent from home, and he did not concern himself about Spencer's presence in his home during his absences. Presumably, since he thought that Spencer had no "instinct", Potter was unthreatened by jealousy. Even when he was at home Richard Potter was content to leave Spencer and Laurencina to their intricate conversations about politics and religion, and go up to bed by himself.[41] Richard Potter's own intellectual tastes were more conventional and muscular than those of his wife. He liked having senior ecclesiastics in his house, but among the acquaintances Spencer brought to him to stay he sometimes preferred the company of doughty manly figures, such as T. H. Huxley and John Tyndall, to Spencer himself.[42] Since Huxley and Tyndall were identified in the public mind as being as unorthodox on religion as Spencer was, Potter's admiration of intellectual figures seems, at first sight, to be based on whim. A modern chronicler of the family has suggested that Potter lacked moral depth and possibly a life of the mind.[43] This interpretation follows that of his daughter, Beatrice, who was baffled as to how Richard could simultaneously admire the figures of Edmund Burke, Thomas Carlyle and J. H. Newman, who she thought had nothing in common.[44] However, this was to miss the fact that all of Potter's favourites stressed authority, and the need for it in the modern age. It was unfair for Beatrice to have mocked Richard's intellectual choices as dictated more by emotion than by pure reason. She was also unkind and ironic about her father's role in the honorific processions that J. A. Froude organized to follow the aged seer Carlyle in his afternoon walks around Chelsea.[45] Beatrice was much amused by the thought of them all trailing the great man at a respectful distance. This devotion to Carlyle seemed to the daughter another sign of shallowness, but it was more likely a mark of genuine homage to a thinker whom Richard Potter admired, even though the rest of his staunchly Liberal family did not. It also suggests that while Potter was a personal friend of Spencer, he found Carlyle's praise of will and strength more to his taste than anything in the bloodless and altruistic "A System of Synthetic Philosophy". Certainly he paid no attention to this system. One of the few times Spencer became angry with him was while both were standing by a large pond, and Spencer was asked to explain his philosophy to Potter with brevity. Spencer, after complaining that his listener had not read any of the books he had been giving to the family for twenty years, began to speak. Potter seemed to be listening intently while he stared into the water, but suddenly exclaimed, "I say, Spencer, are those gudgeon?", and then rushed around the edge of the pond to see more clearly.[46]

Beatrice, and those who followed her lead, have not only underestimated the intellectualism in Richard Potter's choices, but have overlooked political divisions in the Potter family. If she had possessed greater political insight, Beatrice would have seen Richard as a devout conservative whose only brief liberal flirtation came over the issue of free trade. His son-in-law, Leonard Courtney, who, as an experienced Liberal cabinet minister, measured conventional politics with more care than Beatrice had done, knew Richard Potter not only as a financier, but as a politically active Conservative in Gloucester who once had almost won that seat.[47] Conservative politics would have been inexplicable to his family, which was one of the most entrenched Liberal clans in England. Richard Potter's father had been the first member for Wigan in the reformed parliament, and his oldest brother had been Mayor of Manchester, as had one of his nephews.[48] Another nephew, Thomas Bailey Spencer Potter, became MP for Rochdale in 1865 and was known in the House of Commons as "Principles Potter" for thirty years. Thomas Potter was also president of the Cobden Club until his death in 1897.[49] If these Liberal relationships are added to Laurencina's early radical politics, and the reforming politics of some of his daughters, then it is clear that, on the political level, it was Richard Potter who was the stranger in the house, not Herbert Spencer. The Potter daughters, true to their family traditions, took their father's political beliefs lightly or found them strange. Spencer's friendships with Liberals such as John Morley put him within the bounds of family politics and, therefore, made him an easier figure with whom to relate.

The daughters all reacted to Spencer as they would any ordinary member of the family. Even Margaret, who was clever and selfish and never made a friend outside the family,[50] would treat Spencer with a filial ruthlessness that gave them both vast enjoyment. Another sister, the kind and patient Kate, would protect Spencer from his own ill humours. He was often fractious. On an Egyptian holiday he was bored with tombs and would wait outside them muttering that the party had better go on because when anyone has seen one of this class of thing it was quite as efficacious, less fatiguing and more convenient to study the facts from books and pictures. Instead of being annoyed, Kate soothed him.[51] (Others of the party were angry at being rushed.) Beatrice especially, Spencer's favourite, treated him as a member of the family. She was not above silencing him in Cologne Cathedral during vespers, when he began to mumble "excessive monotony" and "superstitious awe".[52] While the young women found Spencer taxing on occasion, there is no indication that they treated him as badly as they occasionally did their parents. At times, the daughters had a group cohesion when dealing with their parents, and would openly rebel if Laurencina attempted to assert authority. They also resented the closeness of Laurencina's relationship with her husband, and his willingness to take business advice from her, but not from them.[53] After her death in 1882 there was a suggestion that they plotted to ensure that Richard did not remarry. Spencer escaped Potter traumas during times of open rebellion and filial jealousy. His own relationships with the Potter daughters stayed equable. He could even deal with the embarrassment Beatrice caused him when she decided to marry the socialist Sidney Webb. Even though her action may have been accompanied by an intention to irritate Spencer (then known for his anti-socialism), he greeted it with equipoise, not anger. That is, he did not perceive the breach with Liberal family politics as a personal betrayal. For Spencer it was an unfortunate action to be dealt with as a minor diplomatic incident. He told Beatrice that she could no longer be his literary executor, but he was secure in the knowledge that this would not rupture the bond between them. Beatrice continued to tend to his needs until his death, years after that of her real parents. The surrogate family that sheltered Spencer did so without the passionate dangers that would have accompanied a real one.

With the Potters, Spencer had only the kind and positive parts of family life. When the second daughter, Kate, was hoping to be wooed by Leonard Courtney, it was Spencer, not her parents, who helped entertain the prospective suitor on the first visit to the family home.[54] Spencer also thoughtfully invited Courtney to his annual picnic at St George's Hill so Kate could meet him again.[55] Later, when some coolness arose between Kate and Courtney, Spencer was useful in restoring Kate's spirits at the family gathering at Christmas in 1881. He willingly became an object of fun in order to raise everyone to good spirits. "How we teased Mr Spencer into kissing Beatrice under a bit of mistletoe and put a fool's cap out of a cracker on his philosophical head".[56]

At home with the Potters, Spencer too would arrange jests with headgear. With great foresight he would spend days carefully and slowly adding strips of paper to the inside brim of someone's hat. At last the victim could no longer fit his hat on his head, and exclaimed, "My hat's too small". Spencer, overcome with mirth, would gasp, "Your head's too big!"[57] In his adopted family, Spencer's almost anthropological fascination with headgear as a sign of status or of rebellion became humorous. The serious symbolic politics of his real family could be overthrown and safely transmuted into comfortable satire and self-parody. Only with the Potters would his sense of isolation diminish in the face of joviality.

Outside the warm ambience of his adopted family, Spencer lacked any way of establishing deep personal relationships. This affected him strongly. When he was young this weakness was not so damaging. Then his reactions had resembled those of a normal shy young man. Although he had lacked the courage to speak to a strange woman, he could furtively contemplate her beauty. Once, when observing a pair of remarkably fine eyes belonging to a young woman who shared a coach with him, he was greatly moved, not by sensuousness, but by the kind of passionless artistic appreciation he later theorized as a true aesthetic sense. Not that his purity freed him or allowed him to act; he could only observe. "There was a more special pleasure in contemplating the elegant curves of the eyelids".[58]

As a young man, when he dared not speak, his restraint was not outside the bounds of convention; it was a delicacy of manners, and an acceptance of convention. When in 1844 Spencer wrote to Edward Lott on the latter's announcement of his engagement he displayed the normal amount of congratulatory envy. He claimed that he had a considerable horror of making a declaration of love himself because of the ludicrous nature of proposals.[59] This, however, was well within the canons of normal behaviour for an overly sensitive young man. The difficulty was what lay underneath convention. Beneath his gaucherie, Spencer was conscious of being permanently damaged by the "impersonality" that his upbringing had imposed on him.

He felt himself to be incapable of forming a passionate bond with another human being. As Spencer grew older he kept most of his social contacts at the distant level of acquaintances of the kind he would meet at his club, the Athenaeum.[60] Anything closer was likely to be emotionally disturbing. Intimacy was to be avoided because of the probability that it would expose his inability to feel passion. He believed that his lack of passion meant that he could only feel the emotion of anger, not that of love. There was nothing he could do about his flaw "as unhappily I can testify from personal experience, consciousness of such lack does not exclude the evil or much mitigate it".[61] This was a painful and public revelation. However, it was not a confession of ill health; a revelation such as that would not have caused him pain. Like many Victorians, he talked about his health with an openness and frequency that would embarrass people of later generations. It was socially permissible to dwell on stress, migraines and sudden spells of weakness, regardless of whether or not these were perceived as having somatic causes. What was astonishing in Spencer's self-revelation was the admission of a grievous and apparently inescapable moral fault. From the pen of a man whose youthful friends had jovially called the "moral Euclid", this was an act of contrition. It was all the more remarkable because, in Spencer's worldview, expiation was not followed by forgiveness. His god was not a personal one; it was the "Unknown". However, if he could not make amends, he could at least avoid perpetuating the family "evil". He felt tainted by the moral failing of his father. The sin was in him as well. Therefore, it was his moral duty to suffer abnegation. He avoided love: to feel such a passion was to court the re-emergence of his past. His personal renunciation was an echo of his political rejection of society's past, but it was unaccompanied by the hopes he always had for the future of society. That is, his belief that he was duty bound to avoid close personal relationships was without the joyful fanfares that greeted social prospects such as the advent of altruistic social behaviour and the end of war. To lack passion within oneself was not as cheering as forecasting the end of violence and aristocratic privilege.

At times Spencer concealed his personal failings. The passages of the published version of *An Autobiography* that dealt with his only love affair, that with George Eliot, were skimpy and misleading. Even the fuller drafts of *An Autobiography*, which he circulated to friends during the early 1880s, were only partial admissions of the evil of being passionless. However, the chief effect of Spencer's recognition of his own flaw did not emerge in his reflection on himself; rather, it came in his late attempt to alter his philosophy from one of rationality to one that insisted that rationality had to share the stage with emotions. In Spencer's philosophy there was a distancing from both the enlightenment and romanticism. This alteration of his philosophy has been overlooked because it took place outside the formal structure of his multi-volume "A System of Synthetic Philosophy". Instead, he placed it in his much reworked and artful personal testament, *An Autobiography*. This was a significant act in itself: the insistence that the life of a person, even a flawed person such as himself, deserved a full account outside the laws of psychology and sociology. In the writing of his own life he could confront and overcome the philosophical laws that had crushed him under the weight of reason. It was late when he retrieved his humanity; certainly it was long after it could bring him any personal relief. However, his retrieval of humanity hinted at a kind of compassion that his philosophy could never develop. He could gift his own life to his reader, unencumbered by the slow work of social progress.

The effect of Spencer's shedding his early complete dependence on rationality, and his burgeoning belief that the avoidance of feelings was evil, affected more than his personal reflections. It spilled over into his thoughts about reason and emotion as these were spoken of by other cultural prophets. In his youth he had been willing to rely on Emerson's sentiment that love between men and women arose from each serving as a representation of the other's ideal. In this reciprocal vision there was to be a thorough recognition of equality. No amount of power should ever be acquired by a party that was greater than that claimed by the other.[62] The vehemence of his statement that "the present relationship" between men and women was nothing more than a remnant of slavery originated in his gloomy thoughts about his own family. He transformed this into a general account of love. In his early and Emersonian language he offered an idealized account of future relations between men and women in which privilege would be denied to the man. Despite his loss of status, the man would gain happiness by being united with another who, unlike women of the past, would not be a degraded inferior.[63]

Spencer's early theory of love did not survive his fervent encounter with George Eliot. He had relied on idealized equality as a basis for love, but the theory was useless in the face of the discovery that an equal, such as Eliot, did not fill the role of an idealized woman. The failure of his relationship with Eliot, and the consequential collapse of the shallow theory that had prevented him from experiencing intimate relationships, forced Spencer to generate a new idea in which passion was conceived of as more essential to love than reason had been. He became conscious of the disquieting discovery that a sense of equality did not raise a genuine commitment towards another human being. This thought took him beyond the analysis of a failed relationship into an intensely self-critical analysis. To him, this one failure meant he could not love. At the same time, his recognition of Eliot as an equal threw his general views on women into disarray. He attributed his failure to love Eliot, who was an equal, to the fact that she was male in intelligence. The idea of equality had confused his views on gender.

Spencer now saw Emerson's notion of love as a fraud. He realized that it had been a mistake to see a person of the opposite sex as the ideal representative of their gender. Spencer carefully replaced this error with a more modern account of passion, and in doing so became a critic of two of the great voyagers in the human soul: Goethe and Carlyle. Spencer's personal experience and his years of toilsome thinking about the development of human rationality in his metaphysics and psychology came together in this critique. It appeared to him that Goethe and Carlyle were simply replicating the falsehoods under which he himself had been raised. They had elevated the notion of the supremacy of the "will, which to him echoed the despotism of reason that had so plagued his own life. Both writers had laboured under the misleading impression that the feelings were under the sovereign control of "will".[64] Contrary to them, Spencer asserted that various feelings became the will in turn. His insight that Goethe's and Carlyle's notion of will carried a hidden and malevolent political premise transformed Spencer's personal psychology into a kind of cultural truth. His autobiography assumed a power that was never possessed in his academic work on psychology. By finding a democratic metaphor with which to oppose an autocratic one, he had become one with his time. The "will" was no longer a sovereign and, therefore, human beings were under the control of an egalitarian set of feelings.

In *An Autobiography* Spencer produced his mature theory of human relations, which would cure the lack of passion that had stifled his own life. When he was young he had believed that there was an inherent tendency towards good: "a patient self-rectification".[65] He believed that a natural medicine pervaded creation, and that the natural world and the human psyche showed evidence of this "essential beneficence". The smug optimism of his early essays and of the first edition of his *The Principles of Psychology* was built on science, reason and denial of the emotions. The light of the Crystal Palace illuminated his early hopes for the human race. It allowed him to synthesize his faith in science with his hopes for emigration, social improvement and psychological growth.

> Equally in the attainment of fitness for a new climate, or skill in a new occupation – in the diminution of a suppressed desire, and in the growing pleasure that attends the performance of a duty – in the gradual evanescence of grief, and in the callousness that follows long continued privations – [he] perceived this remedial action.[66]

As individuals evolved in this futuristic morality so their industrial culture would change to keep pace with these new hard-forged psyches. Spencer dreamt of succeeding stages of progress in which the human race acted with increasing prudence and foresight. Farsighted human beings would plant trees not bearing fruit for generations, would give their children elaborate educations, would build houses lasting for generations, would insure their lives, and would aspire to lives that would constantly be adjusting to future circumstances. In his 1855 *The Principles of Psychology* he could, without the bitter criticism that accompanied his later views of industry and commercial competition, admire the activity of his contemporaries. When the early Spencer remarked that the educated classes were mainly occupied by the struggle for wealth and fame, he was making a benign observation. In this early period he did not reject the idea that commercial struggle was analogous to "harsh biological struggle", nor was he yet alarmed by the thought that an excessive pursuit of wealth was psychologically damaging. He was simply and naively expressing the view that the middle classes would improve both their lot and that of the whole society if they practised thrift, self-denial and industry.[67]

In the early 1850s Spencer saw the commercial struggle as accompanied by an industrial transformation of the human form. Evolution worked to bring human beings into correspondence with their environment in such a way that they became partly machines. For him it did not matter if a man crushed an object with his teeth, with a vice or with a hydraulic press. Mechanical applications were simply extensions of the human organism. Measuring instruments such as scales, thermometers, microscopes and barometers were extensions of the senses, while levers, screws, hammers, wedges, wheels and lathes were extensions of the limbs.[68]

Spencer's early view of evolution as struggle was combined with the theory that organic progress for human beings incorporated symbiosis with mechanical senses and limbs. This struck his contemporaries simultaneously with both hope and terror. This unsettling reaction is typically displayed in Samuel Butler's *Erewhon*, which confused Spencer's views with those of Darwin in hopes of tarnishing both. Butler mischievously claimed to clear Darwin of responsibility of reducing the world to a materialistic nightmare by helping him imagine a utopia filled with mechanized human beings who were engaged in progressive and benign competition.[69] The novelist was not drawing on Darwin's notion of struggle, since that referred to organic interactions and was accompanied by a depressing Malthusian vision of the pain caused by the kind of change that was generated by excessive fertility. It was not Darwin but Spencer who created the conditions for the classic Victorian utopia. However, this futuristic image was a product only of Spencer's early writings, when his optimism for a peaceful industrial future was accompanied by his hope that duty and privation would overcome desire and grief. Emotions would be renounced and the human psyche and society in general would be improved by their absence. However, these ideas disappeared in his later writings; Spencer's mature views would have offered Samuel Butler no hope. Despite Spencer's lifelong insistence that the goal of human progress was an altruistic one, from the late 1850s he began to cast aside his philistine faith in the dreams of progress through hard work and the renunciation of pleasure. Increasingly, he no longer even saw happiness as an admirable goal. Spencer came to believe that life necessarily included doubt and grief, which human beings could only renounce if they were blind. By the time he came to construct his philosophical system in the late 1850s he abandoned such naivety and usually edited it out of his prose.[70] He also ensured that it would not appear in the collections of essays he republished. However, it was not until he came to consider his autobiography that he could clearly see the harshness and pathos that followed duty and callousness when they overcame desire and grief. His personal testament then became an act of contrition for his own part in furthering the doctrine of renunciation.

The testament was *An Autobiography*, where Spencer imagined that inside a person there was a presidential chair that would assume a control that was never absolute. The holder of the chair would be occasionally overruled, or even ejected, by other contenders for the position. Sometimes these contenders would be successful; at other times the existing "presidential chair" resisted their efforts.[71] In this promulgating idea Spencer, who was very reliant on Shelley's imagery, went beyond the poet, who stayed reliant on the will ruling the soul of man even though tyranny had been overthrown in the heavens. "Will" equalled a tyrannical emotion for Spencer in *An Autobiography*, and it was a tyrant that should be overthrown. What this signified is that the mature Spencer disapproved of the way in which the variety of emotions had been suppressed. This had been done through the expedient of arbitrarily selecting one of them, and naming it the "will". This seemed false to him, both emotionally and aesthetically. Emotions that arise naturally cannot, and should not, be suppressed. Instead, they should be allowed to preside in their turn. When he offered this image to the world he was not only being kind; he was espousing a truth. Spencer believed that it was worthless to advise the mind to assume a forced and rational calm. He wrote:

> Tell a mother who has just lost a child, or a lover whose tomorrow's bride has been drowned, that grief must be suppressed in conformity with the doctrine that pleasures are not to be counted upon, and that she or he must accept a lower standard of happiness. What result is there? None whatever.[72]

Grief, like other passions, must run its course, and only when "renunciation" has been spontaneously affected, could grief be put aside. Each faculty had a normal craving for action, and when the craving is strong, excluding it becomes "almost or quite impossible".[73] Spencer's own life and that of his family may have been ruined by the imposition of a rational will that autocratically ruled the passions, but he could warn against the same happening to others. The vantage point of his democratic radicalism gave him a sure insight into personal suffering and, while his own salvation was impossible, he could save others from false prophets who inhumanly upheld the dangerous falsehoods of ideal love, and who elevated the will in such a way as to subdue actual feelings.

To attack renunciation was to attack the central beliefs of the other great nineteenth-century prophets of modernity: Goethe and Carlyle. They had secularized Christianity but left much of its belief system in place. In this worldview, human beings still possessed a religiosity that compelled them to sacrifice themselves by giving up what was naturally attractive in the world. Goethe's religious views, as summoned up in Carlyle's translations of *Wilhelm Meister's Travels: or the Renunciants*, were expressed through "Three Reverences", the venerable spirits who reiterated the various ethical, philosophical and religious lessons that humanity had learned in the past. The "Third Reverence" espoused Christianity as a divine knowledge that lingered authoritatively in the human psyche. All the lessons were historical truths, and, although no longer plausible in the modern world, were somehow still spiritually present.[74] Goethe had imagined that the human spirit had evolved into its present form, and that it could still detect echoes of former stages of the soul.[75] When Carlyle rephrased these ideas in his work they became more astringent and grim than they had been with Goethe. Carlyle's version historicized spirituality and emphasized renunciation more than Goethe's urbane one had. The former portrayed the divine force within man and the cosmos as providing justice and salvation only if the individual worked and struggled.[76] Spencer's objections to this were manifold: he thought it was wrong to canonize historical knowledge – much of which was based on primitive errors and misunderstanding. In opposition to the past he believed that the spiritual needs of humanity could only be satisfied with current – not archaic – knowledge and belief.

Further, he thought that to identify the human spirit as the will was mistaken. Finally, he believed that to renounce the pleasures of the world, and abandon oneself to work and struggle, was cruel and useless. In brief, instead of seeing Goethe and Carlyle as fellow prophets of modernity, he saw them as turning their faces back towards a past of darkness, psychological tyranny and superstition. For Spencer modernity was kinder and more benevolent. Like a modern Prometheus he wished to give human beings some hope for their future.

There was a trace of both Shelley and Prometheus in Spencer.[77] He, like Shelley, wished to unseat the despot in the heavens, and leave the divine throne empty. Like Prometheus, he was sometimes passive in his feeling that nothing could be done to overthrow an evil power; one had to suffer until it toppled as a result of its own actions and the action of natural forces. Spencer could construct a version of independence in *An Autobiography* but, within himself, he was forced to reply on the distant and chilly hope that evolving nature would offer a kind and loving future for the human species. There was no triumphant sense of progress within him, only an increasing sense of loss and distrust of reason, will and knowledge itself. While the youthful Shelley had inspired optimism, the ageing Spencer could no longer see hope inside himself. He was not eyeless like Prometheus; worse, he was without emotion both when contemplating his own suffering and when reacting to that of others. When he responded to reports of cruelty from the edge of empire he could only shudder in his "almost passionless consciousness".[78]

Spencer's rejection of the doctrine of renunciation as set forth by Goethe and Carlyle was the bitter wisdom he distilled from his personal life.[79] He believed that in advocating a psychological passivity for the many varied emotions those prophets were substituting internal tyrants for heavenly ones that had already been overthrown. Spencer knew the costs of renunciation and of the control of passions by the will. His autobiography was a public expression of regret. It was also the presumption that his own life would alleviate the suffering of others. His doctrine of individualism was dependent on his rejection of renunciation. It did not rest on *laissez-faire* economics, nor did it rely on the choices that lay at the heart of J. S. Mill's *On Liberty*. Spencer's plea for individualism was even more internalized than Mill's. Spencer believed that all individuals should explore their own passions, not as a matter of establishing a sphere of individual interest, but as an imperative need. He also held that individuals should not subject themselves to an internal governor. Passions might lead to grief rather than to joy, but only by allowing the passions free play could human beings support the vicissitudes of life.[80] Spencer's feelings on this subject were especially poignant because his own lack of passion had prevented him experiencing love. He regretted he had never been able to resolve his problem with women.

THREE

The problem with women

Women were so strikingly absent from Spencer's early life that both he and his early biographer Duncan remarked that his first *contact* with them came at the age of twenty. Contact is emphasized here because it suggests a simple meeting with a stranger of the kind usually recorded by anthropologists; it is used because Spencer's interactions with women began with a shy awkwardness and seldom developed in such a way that the sexual connotations were clearly present. His relations with women stayed artificial and strained throughout his life, and his reflections on them struck a false and excessively ideological note. Women began as the "opposite sex" and remained so. Typical of this was his comment on the reproduction rate of upper-class girls in England. He thought it was lower than it should have been because they had over-taxed their brains in acquiring a "high-pressure" education. Curiously, this was "an overtaxing which produces a serious reaction on their physique". Frequently, he continued, such women were sterile, and so flat-chested as to be unable to breastfeed their own infants. Spencer contrasted this hypothesis with the suggestion that cerebral men did not possess the disadvantage of infertility, although, he added, it was difficult to find proof for this.[1] While these comments appeared in one of his scientific works, the only rationale they possessed was his personal bias against masculine-looking women, and his sturdy radicalism, which prejudiced him against upper-class luxuries such as the use of wet nurses to suckle the offspring of employers.

When Spencer was admired by women it always made him uncomfortable. If females enthused about his *The Study of Sociology*, he thought it "strange".[2] Their motives were never fathomable to him. While he frequently desired to be close to women – or at least to those who were safely married or who were pre-pubescent and seemingly non-threatening – he constantly theorized about them in a defensive and objective way. These theories were partly scientific, and partly an extended apology for the treatment that his father and his uncle Thomas had meted out to their wives. When his friends attempted to make a match between him and an early enthusiast for his first book, *Social Statics*, he felt frightened by the echoes of radicalism that he had provoked in this female admirer. His manhood was threatened even when he was merely facing a reflection of himself, and his own rueful comment on this weakness was that "one may say that as a rule no man is equal to his book".[3]

Duncan, referring to the years 1840 and 1841, remarked that Captain Moorsom's wife was Spencer's first woman friend.[4] Spencer himself also dated his discovery of women to this period, but focused on one of Mrs Moorsom's relations, a young single woman who flirted with him even though she was already engaged to be married. He was attracted enough by this flirtation to make her the subject of one of his artistic efforts: his only three-quarter portrait. He did not completely lose his heart even though, during the first two weeks of this embryonic relationship, he was "unguarded" by the knowledge that the young lady belonged to another. Despite the danger he passed safely through the only temptation of his youth. Emphatically, for Spencer, such encounters were perceived as threatening. He soberly recorded his escape as "no harm of any kind happened, notwithstanding the length of time we daily passed together".[5] As is often the way with unconsummated love, Spencer never recovered from it. He always remained distantly in love with the memory of this young woman, until in old age he exchanged photographs with her and was shocked to discover that she had lost her beauty. He clinically reported that the passing of half a century had left no trace of the prettiness and bloom she had had when a young woman.[6]

Spencer plaintively remarked that he had no sisters. Further, no girls had ever come to his parents' house, nor had there ever been a family visit to a house in which there were girls.[7] This being the case, he framed the hypothesis that his relations with members of the Moorsom family formed the pattern of his reactions to women for the remainder of his life. This theory was confirmed by his behaviour; Spencer always remained comfortable with married women and with the young girls who surrounded them. However, it concealed the salient detail that Spencer had possessed a close relationship with his own mother from whom he had been harshly separated at the age of thirteen. This was a bond with a woman who could protect neither herself nor her only surviving child.

His mother's brief appearance in *An Autobiography*, coupled with the comment that women were unknown to him until 1840, conceals an important fragment of Spencer's life. According to her son, Harriet was sweet, moderate in temper when irritated, and self-sacrificing to the point of exhaustion.[8] Since she had married a man who never yielded,[9] Herbert was left with a disaster to chronicle, although his filial sense of duty towards his father led him to do this in such a way as to make the latter appear less culpable. The same motivation drove Herbert to assign Harriet qualities that made her a natural victim. He noted that she lacked tact, that "her submissiveness invited aggression", and that "unlike women in general, she was too simple minded to think of manoeuvring, or if, exceptionally, she attempted it, she showed her cards in an absurd way".[10] In other words, he supposed that Harriet had encouraged the cruelty that George constantly visited upon her. Instead of pitying his mother, Herbert blamed her Christian passivity and her lack of cleverness. These negative qualities had elicited his father's almost animal-like assertion of dominance. He saw his parents as psychological opposites – one weak and one strong – locked together in a relationship that brought happiness to neither. There were overtones of biological science in Herbert's analysis. These had originated in commonly available medical and physiological works of the period, which attempted to demonstrate that women were essentially reproductive machines.[11] However, the master key to Herbert's analysis of the essential features of his mother and of other women was a product of psychological rather than physiological theorizing. That is, his sense of the determinants of personal behaviour tended to rely on mental rather than physical data. He felt that this kind of interpretation was less degrading for women than physiological perceptions because the evolutionary direction of his psychology portrayed women, rather than men, as the primary possessors of altruism. Since Spencer thought the future of the human species was one in which altruism would replace competition, he was assigning the highest evolutionary place to women.

For Herbert his parents were a contradictory puzzle. Both possessed qualities he admired. One was altruistic and the other independently minded, but together as a unit they simply harmed each other. Morally this seemed an impossible contradiction to him, so he took refuge in an interpretation that was heavily laden with science. He blamed his father, George, for being immoral while at the same time excusing him; George had been forced into this direction by attitudes and actions that were driven by a biological imperative he had failed to recognize. The father had not understood that women, such as his wife, were limited in their intelligence after they had been subsumed by their reproductive role: "he was not aware that intellectual activity in women is liable to be diminished after marriage by the antagonism between individualism and reproduction throughout the organic world".[12] This reflection was to affect Herbert Spencer deeply in his personal life and in his psychological and sociological writings. In his early writings he struggled to rectify the subordinate role of women despite being burdened with a deterministic biological explanation for the phenomenon. This changed near the end of his life when he abandoned both his pro-woman stance and his biological interpretation of gender politics. However, since he continued to reprint most of his essays and books without revising them seriously his positions on topical issues look confused. This is the reason why his public defence of women's political rights is accompanied by contradictions and reservations that have attracted the notice of some twentieth-century feminist scholars.[13]

His account of his personal life was unruffled by contradictions. Women were only represented by his mother, and she had become invisible for him. There was an element of anger and self-pity in the way Spencer dismissed his mother from his memory. She had failed to protect herself and her son and, as a consequence, she had no future in the record of his struggle to survive in an unpredictable world that was peopled with confident women. These were simultaneously attractive to him and completely foreign to his experience.

Outside the comfortable circle of his permanent surrogate family, the Potters, Spencer's attraction to, and reactions towards, women could result in explosive confrontation. Typical of this was his relationship with Henrietta Barnett, the unconventional wife of a clergyman. The two were forced into each other's company for many weeks while in a small family party travelling up the Nile river. The venture had originated with Kate Potter, who had become acquainted with the Barnetts in London when she joined them in attempting to bring charity and sympathy into the lives of poor tenants who inhabited Octavia Hill's Whitechapel lodgings.[14] Kate's parents, Richard and Laurencina Potter, decided to add further members to the Egyptian expedition by sending along one of their younger daughters, Margaret, and Herbert Spencer. Since Margaret was extremely high-spirited and Spencer was indisposed, it was probable that the older Potters were attempting to give themselves a holiday by sending two of their responsibilities away. Whatever their intentions, they unwittingly contributed to a social disaster. The group was an incompatible combination; Spencer was known to disapprove of charitable clergymen and Christianity, while the Reverend Barnett was intolerant of scientifically minded theists who had not studied ancient languages at Oxbridge. The clerical gentleman had not read a single book of Spencer's but, in any case, found his prose "alien and dull". His estimate of Spencer's conversation was that he constantly applied Darwin's principles to social life, and that he crudely grafted his ghost theory on to all religious phenomena.[15]

Henrietta Barnett had more tolerance and perspicacity than her husband and was able initially to enjoy Spencer's company, and to penetrate behind his façade of being uninterested in women. Early on in the trip she noted that Spencer "enjoyed the company of ladies, though he is much excited if he were suspected of so doing".[16] However, her early sympathy soon dampened. After several weeks of listening to Spencer explain the biological and sociological reasons for the subjection of women, Henrietta became so enraged that she physically attacked her husband – who unwillingly became a whipping boy – so she could prove that hers was the superior sex. Rushing at her placid clerical spouse and pulling him off his chair, ruffling his hair and calling him names was a reply to the philosopher.[17] Such incidents signalled that Spencer's presence in the party had become an embarrassment to the Barnetts, although the Potter sisters, who were used to him, were unmoved and calm. The key to the Barnetts' "difficulty" lay in the fact that while Spencer trumpeted his beliefs about sexual and psychological determinants of gender, their community work in London had uncovered sexual ambiguity and crossover. The Reverend Barnett was himself a prime example of late-Victorian gender reversal.[18] He was gentle and feminine, his wife the reverse. Spencer's didactic insistence on the irreducible quality of maleness may have been an attractive challenge to Mrs Barnett, who had penetratingly observed that he was fond of ladies. However, she lacked the sophistication to realize that she had been sexually provoked by Spencer's awkward assertion of masculinity. This resulted in a conflict because both the players in the scene were incapable of stifling ambivalent feelings. At some point the tension between Spencer and Mrs Barnett became so intense that he abandoned the trip early. As he left he exclaimed, "Good-bye, you are the most irrational woman I ever met".[19]

Spencer's response to Henrietta Barnett was unusually emphatic. More often his relations with women consisted of awkwardness and abruptness, without the passion she had drawn forth. As was mentioned above he referred to his own customary behaviour towards women under the heading of the now archaic term "brusquerie", which means rough playfulness. This was an ambivalent description. He did not admire this quality, and always regretted his uncouthness and lack of polish where women were concerned. Mrs Moorsom had attempted to improve his manners when he was young, but he himself felt incapable of adjusting his own behaviour. When Spencer was attached to the Potter family his fondness for the girls was demonstrated with rough play and humour. He tumbled them over when he took them for walks in the woods.[20] Repaying this favour the girls treated Spencer as boisterously as he treated them. On one occasion, when the philosopher was already middle-aged, they threw him to the ground and pelted him with rotten leaves.[21]

His attraction to girls was so marked, and his behaviour so peculiar, that his own hand-picked official biographer, Duncan, compared him to Lewis Carroll in his preference for the company of girls rather than boys.[22] One of the objects of Spencer's affections, Rosalind Potter, may have had a similar thought because she later declared that her affectionate mentor was the original of the "mad hatter" in *Alice in Wonderland*.[23] Spencer did not hide his predilection for little girls, and his contemporaries thought it a subject for humour, not repression. When he commented on his warm feelings he parodied his own psychological language and joked that they were aimed at the "vicarious objects of philo-progenitive instincts". These objects, who were girls under his control, were caressed frequently, and had to forfeit to him a kiss whenever he answered a question they put to him.[24] Spencer was oblivious to any suggestion that the objects of his affection might not want to kiss men. When the Potter sisters had grown up, he "borrowed" other children to accompany him for short periods of his summer holiday. He was still demanding kisses for answers late in the century. When he was queried about the practice during the 1880s, he pretended that he never imagined that it might not be as pleasant for his little companions as it was for him.[25] This comment was reported by his housekeepers when holding him up for ridicule. In reality he had *some* imagination, which was why he had warned posterity that, "Very commonly strangers begin to caress [children] forthwith without considering whether they may or may not like to have liberties taken".[26] This warning was issued because he wanted the individualities of the young respected, not because he was distressed by the thought of variant sexual desires. He thought of himself as having no physical desires, therefore he did not easily attribute them to others. Further, he would have believed it obvious that a surrogate parent, who knew particular children, would easily be able to see their separate individualities in a way a stranger could not. From his perspective there was nothing wrong with a "parent-like" caress.

There was always a supply of juveniles to keep the elderly bachelor from being bored with his own company. In 1884 it was the daughter and niece of his old friend, Edward Lott, who had the pleasure of accompanying Spencer on a tour of the West of Scotland.[27] In 1888, while in Dorking, he successfully implored yet another of the Potter sisters, Mrs Cripps, "Will you lend me some children?"[28] The parents of the borrowed youngsters saw Spencer's habits in a simpler fashion than the twentieth century would. His behaviour was regarded as naive or innocuous, and as just another of the novelties one would have expected from the author of the widely acclaimed *Education*. In this book, and in his private conversation, he deplored submission to authority. He also permitted freedoms, such as allowing the young to eat the same food as their elders. Worse, it was once remarked that he had even let some children consume supper as late as eight o'clock at night.[29] The presence of so many novelties, together with Spencer's insistence on teaching science to girls as well as to boys, meant that his behaviour was seen only as that which one might expect from a philosopher. When there was a reaction from the children, such as the time the Potter sisters "man-handled" him, pelted him and stole his hat, the other adults laughed and treated the incident as just deserts for a man who encouraged rebellion. However, the fact that some contemporaries could query his behaviour towards girls and identify it with that of Lewis Carroll suggests that at times Spencer was seen as unnerving and "fictional". He was regarded as a strange and wonderful creature who lacked humanness. This judgement catches one of the strands of Spencer's character.

He was too cautious to form a relationship with a woman, but he could shout at and play with the immature members of the female gender. Girls seemed to be non-threatening to a man who was self-consciously embarrassed both by his inability to love and by his fear that he could be capable of cruelty if he came close to an adult woman. At heart, he knew himself to be damaged in such a way that he fulfilled only one half of the evolutionary promise offered by his philosophical scheme. Spencer believed that the human species was evolving towards a future in which individuals would have a greater sense of individualism, and in which they would be altruistic or possessed of greater love for each other. However, for himself he could only claim the first of these. From his perspective he was a selfish individual who would have no posterity because he lacked the altruistic portion of the evolutionary gift. This meant that his style of individuality would not be replicated in the future, either by imitation or procreation. It would have been dangerous to attempt the latter. He sincerely believed that if he had produced offspring they would have been tainted by his father's nervous breakdown. He thought that George Spencer's distressed condition was an inheritable one, even though it had been caused by overwork. His understanding of contemporary biology led him to believe that morbid conditions that become habitual in parents could be transferred to children, where they would become hereditary.[30] This was not simply a distant scientific calculation that concerned the fortunes of a race or nation; it was a personal warning to him that he should avoid having children even though his need for a family life led him to treat housekeepers, friends and miscellaneous children as if they belonged to him.

Spencer's failure to reproduce himself spilled into his other thoughts. *First Principles*, the foundation text of his philosophical system, was his metaphorical child. When recalling the system's "highest truth" it was to remember that he was "a parent of the future; and that his thoughts are as children born to him, which he may not carelessly let die".[31] Each new book he referred to as a new child that made him feel more vulnerable.[32] This was a parental feeling that he desired to share. It was in this vein that he pathetically wrote to one of his women friends, Mrs Octavius Smith: "As with parents it ultimately becomes the chief object of life to rear their children and put them forward prosperously in the world, so, as an author's life advances, the almost exclusive object of anxiety becomes the fulfilment of his literary aims – the rearing of the progeny of the brain".[33] Spencer shared with his lover, the novelist George Eliot, the image that literary works were authors' children.[34] Both writers were childless and their use of this metaphor purloined the maternal roles of their old friends while, at the same time, fending off loneliness. Their claim to have given birth is especially parasitic and selfish when it is considered that both used it when writing to old acquaintants who lived in obscurity while they themselves basked in the warmth of public praise. The unpleasant part of Spencer's letter to Mrs Smith can only be savoured fully if it is recognized as an attempt to elicit sympathy from an elderly widow whose life had been bound up with her family, and who was grieving the death of her husband. Spencer's abiding sense of failure drove him to write such self-pitying letters. Conventional sentiments were excluded by his self-proclaimed Promethean duty of giving knowledge to the human race. He was occasionally left chilled by this task, and he was even sometimes doubtful about its importance. While he could joke about vicarious parenthood when referring to his "borrowed" children, his own written creations did not give him the parental satisfaction of which he boasted to his friends. For himself his writings did not mask the bitter taste of his failure to find a wife and produce children, although this sensation was carefully concealed from those of his friends who read the drafts of *An Autobiography* that he circulated.

The most important way in which *An Autobiography* was "doctored" by Spencer was in his concealment of his love affair with and courtship of Eliot. The encounter with Eliot forced him to recognize his own failure in masculinity by confronting him with a woman who seemingly possessed every one of the personalized and curious collection of feminine virtues on which he had mused when imagining his ideal woman. From his perspective she was without flaw, yet, when faced with perfection in a woman who truly loved him, he could feel no passion, but only emptiness and pain. During their affair he suffered for many months feeling uncomfortable with his own unresponsiveness; then he took his true love and fobbed her off on his friend, the notoriously sensual and libidinous G. H. Lewes.[35] After such a display of pusillanimity and sexual indecision, Spencer had no choice but to resign himself to being a recluse. His failure to love was the only adult tragedy of his life, and it was one that he was determined to hide. He had no choice in this as he lacked the literary genius that he believed was necessary to transfigure artistically either the passions or the coldness that, he knew, was their substitute inside him. If he had been a poet he might have made his failure to love into an aesthetically satisfying tragedy. However, in the absence of such talent, concealment was the only option.

In the surviving draft of *An Autobiography*, Spencer's account of his relationship with Eliot centres on a comparison between her and himself. The implication of this contrast was to charge her posthumously with backsliding from her early rebellious attitudes into a complacent respectability. Spencer believed that while he had remained true to his "nonconformity", she, possibly because of her need to feel in unison with the mass of human beings, had subordinated her personal convictions to convention.[36] This critical summation of Eliot's character was deleted from the published version of *An Autobiography*, presumably because it was seen by Spencer as excessively self-serving. However, the criticism, together with the other comment in the draft that was omitted in the published account, has a more subtle psychological message. The deleted comments stress Eliot's feminine qualities, while the published account focuses on her masculine ones. In the earlier version he perceived her gradual decline – from a radical woman into the grateful recipient of fame – as a change caused by her feminine moral nature. Even though he credited her with power and intellect greatly exceeding that of most men,[37] these qualities were still harnessed to femininity. This interpretation led him to the generalization that "it is a trait of the feminine nature to dislike to stand alone".[38] His draft reflections on Eliot generated other theories as well. For example, he thought that the ideal woman within her had led to her being modest, despite her great power.[39] Also, he noted that she was feminine in her gentleness when correcting others, and in this there was a predominance of domestic affection.[40] In the printed version, all of this was left out. He omitted the comment on how his own "nonconformity" contrasted with her increasing need for popular adulation; he also eschewed comments about her essential femininity. In their place was substituted a briefer, less comparative and more masculine account of Eliot. In the final version, while she was still portrayed as sympathetic and as lacking in self-confidence, her traits no longer occasioned him to dwell on the nature of womanhood. Instead, he mentioned only that her physique had "perhaps, a trace of that masculinity characterising her intellect".[41] He commented that she was strongly built, and had a head larger than most women. Her voice too was unusual; it was a contralto of a rather low pitch.[42] Her philosophic power was still mentioned as remarkable, and Spencer dealt with her by saying:

> I have known but few men with whom I could discuss a question in philosophy with more satisfaction. Capacity for abstract thinking is rarely found along with the capacity for concrete representation, even in men; and among women, such a union of the two as existed in her, has I should think never been parallelled.[43]

The overall impression given of Eliot by the published version of *An Autobiography* was of a strangely masculine figure whose variant sexuality was exuded in such a way that it was completely reasonable that a normal man not pursue his feelings of intimacy with her.[44] This differs sharply from the unpublished version, in which he had adorned her with feminine characteristics; then she had been the possessor of an ideal womanhood contrasting with his own masculine and individualist nature. In this vision, the two halves of human nature, the feminine and the masculine, had been equally matched in their love affair. The early account had to be discarded by Spencer when it came time to publish *An Autobiography* because it ventured too close to revealing the depth of his own emotional commitment to Eliot. There was also an additional complexity in that his actual relationship with Eliot after her involvement with Lewes had become partly that of a reader, not of a lover.

As a reader he could safely release his warm feelings towards her. When he read *Adam Bede* in 1859 he could write to her freely that he was moved to laughter and tears. He was able to tell her that finally his critical faculties were suspended.[45] When Eliot appeared on a printed page Spencer could cast away his fears and reservations and share her feelings. As a reader he surrendered his now useless idealization of her femininity and ideological womanhood, and regarded her achievements in the way she intended them to be read. He scanned her in the way her twentieth-century analysts have, as a person who was striving to evade identification with her own sex.[46] Spencer could now afford to be partly generous; Eliot no longer threatened the passionless "glacier" she had spied within him. In the distance his courage returned, and he could even risk making a fool of himself on her behalf.

In Spencer's private life, the mention of Eliot's name became a term of endearment. When he flattered his protégée, the young Beatrice Webb, he called her a "born metaphysician" and said she reminded him of Eliot.[47] In saying this, Spencer could hide his love behind a mask of familial affection. As the only adult person who had given Beatrice affection when she was a child, he was trying to honour her, and declare himself capable of love by comparing his feelings towards her with those he still had towards Eliot. His love could be enunciated safely when it masqueraded as admiration. While his final and published comments on Eliot showed some generosity in taking her complex and partially masculine self-evaluation at face value, part of his final "official" comment remained the selfish possessiveness of the foiled lover. The published version of *An Autobiography* saw Spencer boasting to have encouraged her to write the novels that had made her famous.[48] His claim was to have given her a divine gift: that of creativity. As with the original Aeschylean drama, Spencer saw himself as a creator who removed fears: he was a latter-day Prometheus who gave Eliot the courage to struggle with life and so produce her works.

Persistent rumours about his intimate relationship with Eliot circulated during his life, but he consistently claimed they were false. Spencer dealt with these by agreeing that he had indeed been acquainted with Eliot before she had become Mrs Lewes,[49] but this had been of no consequence. He noted that it had been suggested that "I was in love with her and that we were about to get married. But neither of these reports was true".[50] Gordon S. Haight dealt with this denial in a simple way, saying it showed a lack of chivalry on the part of Spencer.[51] However, this charge is simultaneously inappropriate and too kind. Spencer's consistent bias against the aristocracy meant that he cared nothing for imaginary chivalric qualities and would have been untroubled by their absence. More importantly, Haight's judgement fails to register that Spencer's denial was a monstrous betrayal of his affections and of his much vaunted truthfulness. It also should be noted that Spencer misled J. W. Cross when he claimed that his father, George, had had no knowledge of his relationship with Eliot. Herbert took his father to visit her about the time (late June 1852) he had agreed to marry.[52] This visit had to be concealed because it lent weight to the rumour about Herbert's close relationship with Eliot. His public disavowal of his love affair with Eliot contained two of the greatest falsehoods he ever uttered. In it Spencer abandoned the only feelings of love he ever experienced. His readers were also misled about his intention to marry Eliot. While Spencer could have correctly assured them that no bans had been read, he had in fact agreed to marry her despite his lack of enthusiasm about the match. Spencer's decision had been forced from him after months of hesitation on his part during which she beseeched him to allow her to fan his emotions into a flame. Eventually, with his resistance worn down he agreed to marry her, although, he complained, he could still feel nothing. At that point, belatedly showing some pride, Eliot rejected him.[53] Spencer had been completely routed; he was unmasked as a man incapable of open and passionate love, and as one who had consequently been rejected by the only woman who ever loved him. Eliot's love was transferred to another, leaving Spencer with a damaging form of self-knowledge and an unfading sense of devotion towards the woman he had not been man enough to possess.

The true story of Spencer's relationship with George Eliot is contained in a series of unpublished and partly published letters that were concealed by his official biographer Duncan and subsequently, in 1935, deposited in the British Museum by the Spencer Trustees, with instructions that they were not to be made public for fifty years. Together with these manuscripts is a lengthy letter written in 1881 to his American friend and publisher E. L. Youmans.[54] In this Spencer gave an *almost* complete and accurate account of his relationship with the novelist, while asking Youmans to deny it.[55] The letter was in response to rumours started by a Mrs Hartz, a distant acquaintance of Eliot from her early days in Coventry. The gossip had appeared in American newspapers and, subsequently, in England in the *Inquirer*.[56] Earlier, Spencer had disingenuously claimed to have kept all knowledge of this affair from his intimate friends and from his father but he was now forced to give details of his relationship with Eliot from 1851. It was not only pressure from newspapers that forced Spencer to clarify the situation. Respectable biographies of his early contemporaries were beginning to appear, and these threatened his carefully manicured account of himself. The most important of these works was *George Eliot's Life*. Spencer was concerned that its editor, J. W. Cross,[57] Eliot's husband, would spread truthful rumours about him. He therefore bullied Cross into inserting a disingenuous note denying them, but this historical falsehood was quashed by no less an authority than Lord Acton, who anticipated its potential to cause further controversy.[58] However, since Cross was as concerned as Spencer to protect Eliot's reputation, he needed no prompting to conceal the existence of her lovers. He dealt with the delicate subject of Spencer's intimacy with Eliot by restricting it to two innocuous passages. First, Cross coupled Spencer with G. H. Lewes, as if the former had come into her life as a companion of the latter.[59] Eliot's words were paraphrased to read as if she had never felt passionate about him. Spencer was reduced to "a good, delightful creature" who gave her a "deliciously calm *new* friendship".[60]

In essence, the former lover gained the complicity of Cross, the bereaved husband, in misleading the public. However, in seeking this collaboration Spencer's motives were not based on his normal desire for privacy. When it came to any mention of Eliot his anxiety about the exposure of his private life became more strident than usual. This was because he had more to lose; his early love had never completely vanished and he always remained deeply perplexed about their encounter. The truth was impossible. It had to be concealed because Spencer could not boldly face the pain that such a revelation would cause him.

This lack of courage makes the unpublished letters important addenda to *An Autobiography* and to the private account he gave his friends. The letters detail how Spencer's intimacy with Eliot began in 1851 and grew through frequent evenings at the opera, the theatre and concerts. The two were also frequently thrown together in discussions about editorial matters for the *Westminster Review*. After some time had elapsed, Spencer claimed to have begun to feel qualms about what might result from their constant meetings. While he admired Eliot morally and intellectually, and while he had decided feelings of friendship towards her, he wrote "I could not perceive [within myself] any indications of a warmer feeling, and it occurred to me that mischief could possibly follow if our relations continued".[61] He therefore took a "strange step – an absurd step in one sense". He wrote her a letter that delicately raised his fears that she might assume that their acquaintance would lead to love. However, he was incapable of such subtlety; the warning became confused because of his excessive sensibilities. That is, he began to worry more about the possible insult Eliot might read into such a message than the warning itself. How would she, he wondered, receive the inequitable suggestion that while he was in no danger of losing his heart, she was in the perilous state of falling in love with him?[62]

Spencer's obtuse innocence in penning such sentiments was unguarded by any suspicions on his part. He did not guess that Eliot had had a robust sexual life. He knew nothing of her affairs with Robert Brabant and John Chapman,[63] and assumed she was as pure and innocent as he himself was. Nothing else can account for the childish relief with which he greeted her easy acceptance of his letter. She had taken his attempt to distance himself "all smiling", quite understanding his motives and forgiving his rudeness.[64] The two resumed their intimacy as before, but on the understanding that Spencer was not emotionally committed. "And then, by and by, just that which I feared might take place, did take place. Her feelings became involved and mine did not. The lack of physical attraction was fatal. Strongly as my judgment prompted, my instincts would not respond."[65] This is unsurprising, as he was not given much to sexual desire; his friend Potter hinted at this when he said Spencer had no "instinct". Spencer could not force himself to lust after Eliot. His intellect told him this was the kind of woman for whom he had always yearned, yet his emotions would say nothing positive. He was confused because he did not find her beautiful enough to desire, yet he regarded his need to respond to physical charms as base or animal-like.

He fell into a despair of indecision about her. There was no escape, and even his wish to flee was ambivalent. He found it painful to be loved by her, yet the affair lasted through the summer of 1852, then through the autumn and into the beginning of 1853. As time passed, Spencer found himself in a unique situation; for once his emotions had become engaged, even though they were accompanied by negative connotations. All he could think was that during the brief period in which he was enthralled with Eliot he had "passed the most miserable time that has occurred in my experience". She would not agree that they should stop seeing one another. He was still paralysed with indecision when he surrendered and agreed to their union. This was the nadir of Spencer's adult life; it was the only time he was bullied into submission after he had left home as a youth. All other emotional commitments had been avoided or otherwise he would have been in crisis before this. His response to the demands was to give in – agreeing to marry George Eliot – even though he still felt nothing.[66] Yet even this surrender did not bring the desired release from unhappiness. That only came when he began insinuating Lewes into his private afternoon meetings with Eliot; a practice that began after Spencer met with Lewes by accident when on the way to one of his sad trysts. He asked Lewes to accompany him on his visit. Subsequently he did this repeatedly and, after the third or fourth joint visit, when Spencer rose to go, Lewes remained seated. When the former spied how matters were going it was a great relief to him;[67] he was free from the burden of being loved. Spencer's behaviour in this matter has been viewed as "cold calculation"; it has been suggested that he pursued his relationship with Eliot purely in order to enhance his intellectual life.[68] However, the reverse of this is true. Rather than being too cold and calculating, Spencer was an overly sensitive man who was filled with feelings. His failure was that he could neither express these nor dispose of them.

After his unsuccessful love affair had ended he could never completely escape from the bondage of his love for Eliot. Their relationship had deeper roots than he admitted to his friends; once roused neither his latent affections nor the public rumours about them would completely disappear. There was more truth in the gossip than his 1881 letter to Youmans and his other friends had admitted. The surviving manuscript letters from Eliot show that, in remembering the early 1880s, Spencer had reversed his role with that of his lover. His so-called warning to her had, in fact, been a love letter. In reality he had not been afraid that her feelings had become engaged; his fear had concentrated on preserving his own well-being. In truth, it was he who had declared his love to Eliot, and she who had displayed delicacy in reply. She had responded to the declaration with modesty: "How remote it is from my habitual state of mind to imagine that any one is falling in love with me".[69] However, despite Eliot's disclaimer, she had intuited not only that Spencer loved her, but that his affection towards her was flawed. He was capable of missing her, but something had been awry with his expressions of feeling. When she returned his affection, he was less responsive than she had liked. She complained about the "tremendous glacier" he carried inside him.[70] She attempted to make a virtue of his passionless love-making, treating it as a kind of honesty. It was, she guessed, based on his refusal to be fooled or importuned. This speculation allowed her initially to treat his incapacity to feel as a frivolous quality. She chided him for his coolness and advised that credulity had advantages. It was "an agreeable quality; let me tell you, that capacity of being humbugged. Don't pique yourself as not possessing it".[71] However, when ridicule seemed to have no effect she tried to admire and emulate his qualities. She wrote that his artificial calm and coldness made her feel "imperfect" and "unworthy of her better self". Eliot then tried to offer him the passionless affection she thought he could tolerate. "I can promise you such companionship as there is in me untroubled by painful emotions".[72] This attempt was doomed to failure; the power to be lukewarm in her feelings was not part of Eliot's character. Soon she was excitedly begging for reassurances that she would not be forsaken, and for the hope that he would always share his thoughts and feelings with her. She was troubled; jealousy, despair, humility and rejection all welling up in her breast. These were deep emotions that could not be echoed by Spencer; such feelings had been surgically removed from his psyche during his youth. He did not possess the attributes with which to reply to a lover who cried, "If you become attached to someone else, then I *must* die, but until then I could gather courage to work and make life valuable – if only I had you near me I do not ask you to sacrifice anything".[73] If Spencer could find the patience to bear the "concentration" of feeling that she focused on him, he would "not curse it long".[74] She too would learn calmness if she were delivered from the fear of losing the object of her love, even though he could offer little in the way of a return. "I suppose", she continued: "no woman ever before wrote such a letter as this – but I am not ashamed of it for I am conscious that in light of reason and true refinement I am worthy of your respect and tenderness, whatever great men or vulgar minded women might think of me".[75]

Alas for Spencer, his love was too feeble to respond to such warmth before he gave her to Lewes. Only when reading *Adam Bede* could he respond with laughter and tears. He was, unusually in his opinion, moved to "enthusiastic admiration".[76] He could dote on the novel because it was written in an almost private language that only he shared with the author. It spoke of affection in a way that conveyed poignant reminders of their liaison. The passages on love could be read as a rebuke to Spencer's simplistic fascination with physical beauty and elegant shapes. *Adam Bede* had included a defence of traditional love – with its Christian and Platonic overtones – against Spencer's simplistic biological view that affection was a response to instinctive drives.

Adam Bede's narrator admired the delicate feelings of "Seth Bede", whose love for a woman was scarcely distinguishable from religious feelings. His emotion had rushed beyond its object and lost itself in a sense of divine mystery.[77] Seth's brother, Adam, had narrower affections. He was attracted to the pretty but empty-headed Hetty, whose resplendent physical charms were no better than a peacock's.[78] Hetty's beauty did not mask hidden emotions; her feelings were unconnected with sentiment. This is a comment that men who thought like Spencer mistakenly assigned excessive importance to biological attraction. Eliot took the contrary position to convince her reader that evolutionary philosophy did not support the theory that the moral qualities of an individual woman were represented by physical signs such as the exquisite curve of an eyelash or an eye. These charms might only express the disposition of a beautiful person's ancestor: perhaps a grandmother.[79] In case this argument was unconvincing by itself, Eliot also assigned an irresistible quality to love. It did not have regard only for Apollo-like features; "it does not wait for beauty – it flows with resistless force and brings beauty with it".[80] This passage addresses more than Spencer's fixation with physical attractions. He was being chided for his reluctance to lose his individuality. Since he never allowed himself to surrender to the flow of emotion, he stood condemned as a loveless man. His tears, when reading the novel, could well have been caused by perplexity. Eliot's fiction was consistent with evangelical anti-sensualism while his own sexual aspirations and language could be described as pro-sensual.[81] However, in behavioural terms these roles were reversed; she was sensual and he was chaste. *Adam Bede* could strike him a blow at two levels; he had been a moral failure for wanting sex, and he was incapable of the physical love that he desired.

Although he had failed in love, Spencer could not extirpate his feelings. He remained so enamoured of Eliot that, for the rest of his life, he was given to issuing defensive remarks about her. When their mutual acquaintance Eliza Lynn Linton called Eliot a hypocrite for abandoning her early Bohemian lifestyle and adopting respectability at the expense of G. H. Lewes's real wife, it was Spencer who undertook the posthumous defence of Eliot's reputation.[82] He also amused the sardonic Frederic Harrison during a committee meeting of the London Library, by suddenly requesting that all novels be removed from the library's shelves except for those of George Eliot, which were serious works. Spencer's request was not as humorous as it seems to later generations. While Harrison enjoyed it, he too thought that Eliot's stature was greater than that of a novelist: instead, she should have been placed in the rank of moralists such as Ruskin and Mill.[83] Spencer's prejudice against fiction being deposited in libraries was common until the end of the nineteenth century;[84] hence his anxiety about the merits of his friend being overlooked. His outburst about Eliot was not merely a concern over her posthumous stature. His late thoughts on his only love were not simple acts of homage to a woman who had achieved greatness; they were the tormented effusions of an unfulfilled personal loss.

Herbert Spencer's self-frustrated love for Eliot kept echoing inside the homes he later made for himself with the Potter family and, briefly, with his housekeepers during the 1880s. His special protégée, Beatrice Potter, whose childhood depression he had lifted by comparing her with Eliot, grew up with a knowing and amusing picture of herself as a surrogate love object for the philosopher. For her as for her sisters, Spencer, with his unfulfilled love for Eliot, remained a permanent subject for mirth, a humourless bachelor forerunner of their own marriage prospects during the London season. When she was an adult, Beatrice found the story of Spencer's love affair – the "Miss Evans" episode – to be delicious.[85] Whenever dinner-table conversation flagged in the Potter household there was always the old philosopher to bait about Eliot. On such occasions he would declaim:

> It would have never done for me to marry.
> I would not have stood the monotony of married life and then I would have been too fastidious. I must have had a rational woman with great sympathy and considerable sense of humour.[86]

Beatrice would counter this by offering the opinion that "rational women are usually odiously dull and self-centred". In response Spencer would take this as an attack on Eliot and insist:

This is a very erroneous generalisation; George Eliot was highly rational and yet intensely sympathetic, but there the weak point (which appeared a very important one to me) was physique.

I would not have married a woman who had no great physical attraction.[87]

His housekeepers also found amusement in Spencer's fears that he might have married "a philosophic woman".[88] His biggest imaginary fear, however, was not the presence of brains, but the absence of beauty. He once refused to lunch with a young lady on the grounds that her plainness upset him. He would not sit opposite someone who was "so ugly". When his housekeepers reproached him with the adage about beauty being only "skin deep", he became very serious and read them a lecture on aesthetics. Their retorts had struck at something he held dear. He had always accepted this cliché, claiming it was literally true. That is, for him beautiful features were generally accompanied by the beauty of nature "so that it means a good deal more than it appears on the surface".[89] Beauty was on the surface but, for Spencer, it was the only human quality that was not completely governed by the human search for food, shelter and the need to reproduce.

Victorian narratives containing Spencer's ludicrous statements about his personal need for beauty in a woman have to be read on more than one level. On an obvious level they are simply amusing. They are reminiscences of how the beautiful young Potter sisters and others amused themselves teasing an elderly philosopher. The ironic moral point of their tales of this was to show Spencer speaking about the need for a person to possess physical beauty while his own chief physical attributes were bony limbs, small eyes and a long upper lip. However, Spencer's response to such teasing had an underlying seriousness. He was being pensive – that was why the baiting was fun – when he took refuge in his complex theories about the biological and aesthetic need for beauty. The ideas were a screen that hid his ancient failing to return George Eliot's love. His conversation here was no different than his earlier sentiments in his widely read textbook *Education*. That is, there too he used his own inability to love without the presence of physical attraction to make a generalization about the feelings of "everyman". All men were like him. Spencer insisted that women's intellectual abilities by themselves might be insufficient to attract males. Men were driven by nature's ends, and by the unconscious desire to consider the welfare of posterity.[90] This desire was a biological drive, which might insist on a good physique over other qualities. Without such a physique, he believed, intelligence could die out in a generation or two. Since bodily attributes were also perceived as hereditary, Spencer concluded that beauty should be preferred to intelligence. This biological insight was seen from the position of the male, who, Spencer believed, should surrender to impulse even though he himself never had.[91] *Education* portrayed women as the opposite of intellectual. They were impulsive.[92] Women were presently in the grip of passion, but would be improved if they struggled to avoid this, at least when they were dealing with children. On the basis of this advice Spencer leapt to his general warning to the whole species: to save human beings it was necessary to abandon old customs and childhood memories, and to reverse the present attitudes of both women and men. Only then could it be seen that women's passions were an inadequate excuse for bad decision-making on their part and that men's lack of passion did not give them true human qualities.[93] For Spencer, the progressive future of the human species depended on each gender adopting some of the qualities of the other.

The people who listened to the elderly man did not hear the muted personal tragedy in his words. They did not care about the deceased George Eliot. When Beatrice Potter searched through Eliot's correspondence in Cross's *George Eliot's Life* in order to flesh out the image of the female role model she had had held up to her as a child, all she could see was Eliot's selfishness. Since Beatrice was engaged in a desperate struggle with her own feelings of privileged guilt and with her desire for renunciation, mid-Victorian searches for pleasure left her cold. To her, Eliot seemed only a woman who self-indulgently made her friends among the gifted and fortunate.[94] Only to Spencer did the memory of Eliot remain the distant epitome of perfection, rendered more fascinating as she flickered back and forth between her dual male and female personae. His fading image of her avoided the biologically determined deficits that made so many women problematic for him; only Eliot seemed to have achieved the perfect balance between sympathy and rationality.[95]

Spencer's views on the biological differences between men and women, and his frustrated need for passionate love, caused him to adopt an ambivalent view on women. Their biological role may have, in his view, shaped them to be less "individuated", and therefore less developed, but on a broader evolutionary framework he saw them as superior. As the future brought less competitiveness, less violence and more altruism, there would increasingly be less need for feminine qualities to be subordinated to masculine ones. Later generations brought up on Havelock Ellis and Freud did not perceive this view as radical, but in Spencer's time it was progressive. A contemporary conservative would not have thought like Spencer; such an ideologue would have appealed to a much firmer basis for his beliefs than a shifting evolutionary trend. The poet Coventry Patmore, for example, was fixated on sex as the primary and essential relation between men and women. He thought that it was eternal and coextensive with every form of existence, natural and divine. This meant that while Patmore gave homage to "Our Lady", he was not prepared to countenance any tampering with women's due subordination to men here on earth. To do so would upset the ordered nature of relations in general.[96] In comparison with a fervent religious figure such as Patmore, Spencer perceived sex as a calm secular affair. Since he did not believe in the Creator, he did not see sex as the generative centre of human existence. It was merely one aspect of reproductive biology that, for Spencer, neither exhausted the meaning of women nor the beauty that they symbolized for him. Somewhere on the surface of beauty was the meaning of life, a delicate zest or efflorescence, the contemplation of which always gave him a sense of infinite joy.

To Spencer, beauty represented the only aspect of human life that was removed from the biological imperatives of reproduction and nutrition. He believed that most of one's physical and psychological behaviour could be explained by the same organic laws that governed animals and plants, but he kept alive the hope that there was a small area of human existence that was not wholly explicable by the biological and physical laws that had shaped so much of nature. While the form and functioning of a flower or a tree were the result of evolutionary responses to natural selection and to physical forces, there was an inexplicable residue of beauty that remained after the responses had faded away.

It was common to explain this superabundance of beauty by raising it above biological necessity to be a spirited and soulful experience. However, this sort of explanation was the kind that Spencer had rejected in *An Autobiography* as having done him so much harm. It was not his intention to replace scientific reductionism with a doctrine of renunciation and spiritual love. Instead, he desired the immediate enjoyment of sexual beauty and of the sensual gratifications that one could receive from art. He knew he himself had painfully failed to love, but the retelling of his experience might help others to reconcile themselves to their humanity by adopting a pleasurable form of life.

The opposing doctrine in love, as well as in science, was Platonism. Spencer was hostile to Platonic love for the same reasons he rejected the presence of transcendental physiology in the scientific work of Lorenz Oken and Goethe. Both Platonic love and Platonic archetypes in psychology relied on imaginary inner forces or spiritualism at the expense of true human feelings and desire. As a youth Spencer had absorbed the theory of Platonic love in the form that Emerson had popularized in the English-speaking world. For Emerson physical beauty was portrayed as a stepping-stone towards the appreciation of the soul. The transcendental view was that if a man in search of love lingered too long in the contemplation of the body, then he became gross and reaped nothing but sorrow.[97] Emerson's "beautiful soul" was the centrepiece to his caustic and dismissive appreciation of the sexual life of married couples who would eventually discover that what had drawn them together was a "deciduous" or temporary impulse: a cobweb. Only the evergreen mattered. What matured from the pairing of two people in a marriage was the purification of their intellects and hearts.[98]

Spencer, like many other young Englishmen of his generation, had uncritically accepted Emerson's deprecation of the attractions of physical love and beauty. They were imagined to be insubstantial in comparison with the spiritual bond created by the joining of intellects. Spencer's failure to love Eliot had destroyed his soulful imitation of Emerson's ideal. He was shocked at the discovery that he was on the edge of feeling physical attractions and revulsions that had nothing to do with his estimate of a woman's intellect or soul. This threw him into a lifelong desire to free his emotions, and his appreciation of art, from the idealistic shackles that imprisoned them. If, as he painfully reflected, he could not save himself then at least he could warn others that they must follow their physical desires and find their pleasure in beauty while it lasts.

Spencer not only opposed Emerson's distant and impersonalized view of love; he also rejected his view of art. Where Emerson had said that the laws of nature do not permit the division of beauty from use or function,[99] Spencer insisted on such a division. While Emerson believed that it was degrading to find solace and compensation in art, Spencer found such relief. Rather than feeling degraded by the search for beauty in artistic works, such as statues and buildings, that no longer possessed their original function, Spencer felt ennobled. Beauty was non-functional as well as superficial and ephemeral, but for him it had replaced the inner spirit. The substitution of beauty for spirit left Spencer with a pleasing sensation when he regarded women; they were aesthetic objects who also represented the future of humanity. However, this was a distant prospect, and offered no guidance to Spencer's radical politics. Accordingly, his reforming stance on the possible electoral role of women was a variable one and less constant than his other political ideas.

FOUR

Spencer's feminist politics

Spencer's feminist politics belong to his biography, not to his political theory. His thinking about women displayed neither consistency nor the detailed empirical basis that accompanied his other political discourse. In discussions about the other sex he was likely to vacillate and, paradoxically, issue extreme and biologically reductionist explanations of behaviour. This was in contrast to his usual nuanced prescriptions about the needs of liberal society. When warning of the dangers of aristocratic status and manners, or groaning with despair over the stupidity of electoral politics, or reiterating the idea that war was unnecessary and primitive, Spencer avoided contradiction and backsliding. Many of his political arguments were carefully articulated and closely supported with data. Further, they did not alter over the decades. When contemplating political inequities in his deliberate and passionless manner Spencer was seldom exhausted or flurried; instead he was proudly confident that he was more advanced than his contemporaries. When, however, he thought of feminist politics he became irritable and inconstant. His only consideration seems to have been the need to keep in harmony with conventional liberal beliefs: he did not concern himself with whether his utterances on women had implications for his own philosophy. On the subject of women he simply drifted from one position to another, motivated by dimly remembered personal fears and longings. He was aware of his lack of ability to deal with women personally, and this left him ill prepared to engage in political discussions about their future. This was awkward because, at the beginning of his writing career, he had accidentally taken an extremely progressive stance in the debate over women's suffrage.

Spencer's public utterances on women began with the recommendation that they should be granted full political and legal rights. This was a startlingly progressive posture to adopt in the middle of the nineteenth century, especially since it was several years before it was made marginally respectable by J. S. Mill and Harriet Taylor.[1] Spencer, at this time, thought of himself a radical feminist, but his views were not primarily motivated by the need to reform voting. That is, it would be a mistake to think of the feminism of Spencer's *Social Statics* as a precursor to the 1860s electoral activities of Mill and Taylor. Rather, Spencer's ideas should be regarded as late applause for the speech Frédéric Bastiat made to a great gathering of ladies at the Manchester Corn Exchange in October 1840. The French economist had argued that because muscular power had given way to moral energy, women should no longer be restricted to the rear of the movement towards social progress. Unless, Bastiat continued, the empire of women continued to grow at the same rate as that of modern civilization, there would an inexplicable void in the social harmony, and in the providential order of things.[2] This depiction of modern civilization transforming social relations and carrying them along in its wake was the visionary edge of English and French middle-class reform ideas. To mid-nineteenth-century reformers this image of social evolution was much more appealing than a mechanistic proposal on how to change the method of selecting parliamentary representatives. Spencer's feminism was lit by the pure flame of middle-class radicalism, a light that was modified by his subsequent work in psychology, biology and sociology but never extinguished. Even his later suspicion that socialism was a doctrine designed to attract female voters could not totally undermine his beliefs that women were important as the carriers of the torch of civilization. He always remained faithful to the idea that modern society should rely on moral energy rather than muscular power, and that women would ultimately take the leading role in social transformation.

In his *Social Statics* (1851), Spencer had boldly rejected gender as the basis on which one should assign rights. He argued that sexual differentiation could no longer be the basis on which political power was determined in either government or family. Power should be underpinned by the law of equal freedom. This law was not to be overturned by "differences in bodily organization, and those trifling mental variations which distinguish female from male".[3] Spencer believed that arguments based on sexual variations were insubstantial. He was especially hostile to the idea that the "woman's mission" was an exclusively domestic one that precluded her from taking a role in public affairs.[4] This attitude was simply political sentiment masquerading as a scientific fact: there was, according to Spencer, no universal truth determining woman's place in society. It was not necessarily subordinate: among the Pawnee or the Sioux, women were treated as beasts of burden, while in France they held clerkships, cashierships and other responsible positions.[5]

Spencer not only rejected the dogma of the mental inferiority of women as false, he also held that its absurd implications made it inadmissible in political argument. It suggested, by analogy, that men of lower than average ability were not entitled to rights.[6] To Spencer this was more than a *reductio ad absurdum;* it pointed to the very core of what was wrong with the male arguments that denied women's rights. These arguments were equivalent to saying that a woman who had weaker faculties and who, therefore, was in need of protection, should not be allowed to exercise even those necessary claims to security. This ran against the sensible common idea that rights served to constrain the power of the strong.

Without limitations on dominance Spencer believed that there would be an absence of justice, because social norms would encourage the mistreatment of women. In some societies this had been extreme: offending Circassian maidens were thrown into the Bosphorus by the Turk, and Athenian women were sold by their brothers or fathers.[7] Even in Spencer's Britain there was a statute allowing a man to beat his wife – with moderation – and to imprison her within any room in his house. Spencer believed that such abuses could be stopped if rights were assigned to women; if they possessed claims to equal consideration then their mistreatment would cease. Spencer's early views on the connection between the public powerlessness of women and their private subordination superficially resembled those of later Victorian feminists such as Arabella Shore,[8] but his views were not products of suffragette politics. He did not have a feminist political platform in 1851; his beliefs were merely a belated advocacy of the position of the National Democratic Union,[9] which saw equality as a magical potion that would cure all ills. Women's rights were simply a special application of the general egalitarian rights possessed by everyone. Spencer should not be classed as a pioneer of woman's suffrage as he was unconcerned with how women would actually achieve political power.

However, despite Spencer being undeserving of a place in a pantheon for suffragettes, his approach to women's rights is theoretically interesting in that he insisted that rights not be limited to the exercise of political suffrage. To him rights in the domestic sphere were as important as political rights. He was determined to rectify power imbalances in the home as well as in the public arena. Spencer's own experiences as a child watching his father hector his mother echoes through the prose of *Social Statics*. He had a profound personal interest in overturning arguments that claimed that there was a sound psychological basis for inequality. Opponents of women's rights had argued that if one put a husband and wife on equal footing, this would provoke antagonism as the two equals would then vie for supremacy in the household.[10] Nevertheless, watching the behaviour of his father and uncles, Spencer had concluded that the exercise of supremacy within a household was not caused by the struggle between equals, but by the inherent masculine desire to dominate the weak. He deprecated this as a primitive urge, and looked forward to a time when society had become sufficiently civilized to recognize equality of rights between the sexes: "when women shall have attained a clear perception of what is due to them, and men to a nobility of feeling which shall make them concede to women the freedom which they themselves claim".[11] The fact that Spencer, an inveterate opponent of the aristocracy, unconsciously appealed to *nobility* in this passage highlights his passion on this subject. His vision of a future sexual equality was not an idyll, but an unsettling and gripping insistence that social justice should be enforced in the home before it could be effective in governance of the nation. *Social Statics* argued that increased social justice was dependant on a prior improvement in domestic relations. Only then would the desire for supremacy slacken. Political life would remain in turmoil until the perpetual squabbling of married life ceased. As long as a husband asserted his claims without consideration of his wife's, there could be no self-sacrifice in domestic or in political relations. Both the domestic and the political spheres were to be reformed the same way by Spencer's advocacy of moral reform. In *Social Statics* Spencer expounded the belief that if individuals learned to avoid selfishness then the desire to exploit the weak would disappear. This was a speculation in moral psychology that avoided conventional discourse about politics: Spencer did not speak of the representation of interests, of the balance of power, or of duty, but only about a psychological drive for power that might be restrained by the institution of moral rights. He began, as did James Mill in "An Essay on Government", with a statement that despotism occurs when a person with power attempts to bend another to his will. He suggests that there is a desire within each individual to command others, and that this is the basis for political activity. Here the parallel between Spencer and James Mill is so close that one could accuse Spencer, as T. B. Macaulay did Mill, of writing about politics as if it entirely took place in a theoretical domain, and as if no actual state had ever existed. Or, to put it in another way, Macaulay thought it wrong to write about politics as if there were no actual examples of how it functioned. Spencer's politics in *Social Statics* is even more etiolated than that of James Mill in "An Essay on Government". The latter text at least hypothesizes that a form of political representation that closely reflects the electorate will curb the dominating behaviour of the state. Spencer's views here are even more abstracted from political life; he takes it as a datum that the human species consists of individuals who selfishly desire to control others. His solution to this dilemma is simply to assert that as soon as these individuals recognize that others are endowed with rights domination will cease.[12] Such recognition will end selfishness because it would deprive domestic and political tyrants of their motivations. When this theory is applied specifically to women it entails that, as they acquire the right to make decisions, they will automatically accumulate sufficient power to protect themselves.

Spencer's early views on women's rights were a product of a straightforward political radicalism, which saw equity as a force that removed all forms of injustice. However, his rights theory had an additional source on this subject; it was a projection of his personal fears that the domestic tyranny that had ruined his parents' marriage would also destroy any civic relationships contracted by himself or his contemporaries. Those feelings did not provide enough impetus to ensure that Spencer's advocacy of women's rights would outlast the 1850s. Both his political radicalism and his personal fears were moderated by the biological and sociological investigations that accompanied "A System of Synthetic Philosophy". The new objective knowledge he gained in these subjects undermined his earlier views. His personal motivations were overshadowed by calculations about the probable political impact of women's rights in England. He became concerned that giving the franchise to women would increase the amount of conservatism in politics.[13]

The most significant change was that his previously unfiltered gaze now restricted itself to English politics. The parochialism of his later views reveals that he had abandoned universal rights, for women as well as for everyman. Universalism had disappeared: he no longer equated political claims across varying cultures. Spencer's views became narrowly focused on a late Victorian political struggle between Liberals and Conservatives in England. The former needed to be rescued from Gladstone's dictatorship, while the latter had to be opposed as the voice of privilege and militarism.

Spencer himself would have said his diminution of women's rights did not represent an important shift in his views because he always derived his politics from the same source: that is, he had always emphasized the need to avoid despotism. He would have insisted that he had displayed consistency in his desire to see all human beings treated with gender-blind justice. However, this argument would be disingenuous; Spencer's abandonment of the advocacy of women's rights saw the biggest shift in his political theory. Almost any other controversial issue would have given him better grounds to claim consistency. In his progressive views of children's education, his rejection of the Protestant work ethic, his pacifism, his hostility to empire and his radical distrust of big government and big business, Spencer could, with some accuracy, depict himself as a troublesome thinker who was peculiarly resistant to fashionable changes. His change on the status of women was not part of a "drift to conservatism" as some scholars have claimed.[14] Instead, it was an extensive repudiation of a single cornerstone of his early radical faith: rights for women.

A warning of Spencer's withdrawal of support for women's rights came in 1867 in his response to J. S. Mill's invitation to join a society to promote women's suffrage.[15] Spencer declined the invitation because, as he noted, his views had changed since the publication of *Social Statics*. He now thought that such an extension of suffrage would be best left to the future, rather than being treated as an immediate matter. Spencer no longer favoured female suffrage. He gave three reasons for his new beliefs. First, he suggested that if women exercised the vote they would do so in such a way as to strengthen authority and reduce liberty. Secondly, he believed that women were more likely to be influenced by the prospect of benefits and by the immediate achievement of their goals. Thirdly, he denied that the minds of men and women were alike, either quantitatively or qualitatively. He explained, "I believe the difference to result from a physiological necessity, and that no amount of culture can obliterate it".[16] Spencer expanded on the "relative differences" of women by saying that they arose in those complex psychological faculties – the intellectual and moral ones – that have political actions for their sphere. Two years later in 1869, when Mill sent him a copy of "The Subjection of Women", Spencer ungraciously responded by saying that the book rested too heavily on the assumption that gender relations were determined only, or mainly, by law.[17] Mill retorted by suggesting that Spencer's fascination with the physiological differences between men and women was unhealthy. This is ironic; if there was a greater prude in Victorian England than Spencer, it was Mill. However, writing to Spencer, who was pathologically concerned about his own health, about the "*morbid* habit of dwelling upon sex" when deciding the destiny of women would have been a telling blow.[18]

The bizarre nature of this debate should not obscure the fact that neither of these great Victorian liberals managed to generate a useable rights language for later generations. They disagreed about rights for women, but not philosophically. The two sides of this debate are clear: Spencer abandoned indefeasible or absolute rights for women in order to see them in terms of sociobiological determinism, while Mill relied on a politics of choice and suggested that there was something unwholesome in Spencer's fascination with sex. Their dispute was nugatory; both possessed the same desire to defend liberty against despotism and masculine conservatism. Their differences about women were insufficient to drive them to amend their normal ideals or language. In any case a resolution would have been uninteresting because both disputants possessed a similar liberal disregard of rights. For example, Spencer and Mill avoided mentioning natural rights when speaking of legal rights. There were few alternative strategies available to Victorians.

As Stefan Collini has observed, Mill distrusted general talk about rights, especially of natural rights, but he considered legal ones to be of fundamental importance. Such rights were not only protective of women, but expressive of what society considered unacceptable types of behaviour.[19] Collini's observation could be extended to cover Spencer's discussion of rights in *The Principles of Ethics*. This leads to the problem of how Victorian liberals were to assess the political impact of gender rights: would these be conservative or progressive? Whether women would use the vote to buttress authority or would follow their own destiny towards greater freedom was unclear. A speculative disagreement on this subject could not have been solved by referring to practical politics, since no one in fact knew how women would vote. The disagreement between Spencer and Mill did not concern the fundamental nature of rights to be assigned to women. A principled argument about rights would have been impossible: Mill avoided rights language, and Spencer's later writings were so vague on the subject that, in *The Principles of Ethics*, rights dwindled from restraints that protected the helpless to the mere consequences of their own merits or demerits. Spencer's "rights" are so ambiguous that a recent investigation of them ends with the suggestion that perhaps they are only clear in *Social Statics*.[20] However, since Spencer repudiated *Social Statics* as early as 1868, his early rights language should not be used to clarify his later attitude on women.

Ultimately the dispute between Spencer and Mill on the political position of women became a disagreement about whether or not physiological differences between genders determine political outcomes. Although this discussion was not underpinned – on either side – by a philosophical position on indefeasible rights, the stakes were still high. It was a disagreement about the nature of politics itself, and in particular about the malign or despotic form of politics. Spencer's argument on the subject gets to the root of his general posture on the nature of politics. In 1851, when he published *Social Statics*, Spencer had believed that the source of despotism was an individual's lust for personal dominion. By the 1860s he had come to believe that despotism was actually caused by political actors' worship of power. His earlier views had suggested that despotism could be restrained by each individual internally adopting a moral stance in accordance with the belief that other individuals possessed rights that should not be infringed. However, his later ideas suggested that self-imposed limitations on actions were trifling in comparison to the social forces that would compel the masses to submit themselves voluntarily to any authority. The question of women's suffrage could no longer be solved by invoking a rights theory. For Spencer, women had ceased to be individuals; they had become part of an undifferentiated mass who would in reality submit themselves to any authority. It therefore seemed imperative to him that he oppose their acquisition of political power until evolution eradicated this tendency. There was hope for women because Spencer thought that, in the future, increased equality and education would cause political authority to cease to exist. When this process was complete women's suffrage would pose no threat and could then be safely promoted.

On non-voting rights the later Spencer was actually less conservative and more favourable to women than he had been. That is, he advocated that women should have equal property and personal rights with men before marriage, and that they should retain these rights after marriage. To the modern reader it might seem awkward that he wanted to ensure this retention of rights only in so far as they did not conflict with the bonds of marriage.[21] However, the perplexity here is political, not philosophical; Spencer did not seriously advance absolute rights for unmarried women any more than for married ones. He was speaking within a Victorian feminist debate and his use of rights language has few universal connotations. Moderate feminists of his era would have accepted the exclusion of these women from the vote, as their serious battle was for married women to secure control over their property and earnings.[22] Spencer's advocacy of other non-political rights would not have seemed controversial to a contemporary liberal. His advocacy of women's rights to physical integrity, the ownership of property, freely chosen beliefs and free speech was conventional.[23] In emphasizing that women possessed a restricted set of rights Spencer was not reinventing the "domestic mission" about which he had been so scathing in *Social Statics*. Instead, he was underscoring non-political rights in order to establish a narrow definition of citizenship that invoked a duty of bearing arms while, at the same time, signalling a biological basis for politics. When Spencer insisted that "citizenship does not only include the giving of votes, … It includes also certain serious responsibilities",[24] he was, of course, restricting it to males. The duty he invoked was military conscription, which was common among Britain's continental military rivals.[25] Spencer's pacifism did not cause him to feel any contradiction here because his desire for peace, like many of his other hopes, pertained to the future rather than to the present. When the archaic military aspects of current political organization had been superseded by peaceful industrial systems, military service and maleness would cease to be important criteria of citizenship. At that time, women would possess political power. In the present, there should be opportunities for women where political administration did not entail military or naval duties; on these Spencer was happy to promote the claims of women. For example, he wished to give them equal shares in local government.[26] Again, Spencer's position on this was a moderately progressive one (at least in Britain) as the municipal election reforms of 1869 and 1894 had not yet given women access to all areas of local government.[27] It would be a mistake to contrast Spencer's insistence that citizen militias be male with his advocacy of women having an equal share in local government as if this were a philosophical inconsistency.[28] In his comments on women in *The Principles of Ethics* Spencer is simply appealing to a number of diverse strands of Victorian public opinion. Since, when he wrote that work, he no longer possessed a rights theory, inconsistencies about political claims based on rights would not have troubled him.

Spencer's desire to exclude women from political power was not merely a reaction to the real or projected implications of their possessing it. It is possible that Spencer feared that women voters would imitate the lower-class males whose 1867 enfranchisement gave the Tories a lengthened grasp of governmental power. He was, however, definitely worried by the prospect that women, who generally loved the helpless, would promote socialist legislation in order to benefit inferior members of society.[29] An excessive emphasis on this fear would not illuminate Spencer's politics. Labelling Spencer a conservative who was resistant to change does not truthfully explain his position. It is confusing because it creates a dissonance with Spencer's political language, which resisted giving women political power because of their "relative conservatism".[30] Further, it does not account for Spencer's ability to keep as close personal friends people such as John Morley, Frederic Harrison and Leonard Courtney, whose political credentials were impeccably liberal.

Spencer's decision to reverse his stance, and to oppose the empowerment of women, does not belong in the domain of conservative Victorian reactions to women's power. Of course, such arguments are difficult to label as liberal or conservative. Further, since nineteenth-century women politicians were careful to keep a distance from mainstream politics, the reactions they provoked were not necessarily reflections of ideologies. Spencer did not believe that he was reacting; rather he thought that his views on women were non-political and when he discarded his support of political power for women this was not to mollify anyone. He gloried in holding opinions that could irritate the establishment, and believed that on one issue he could abandon his iconoclasm. On the subject of women he could not argue convincingly, which is why he needed only a single piece of evidence to prove that women were more enamoured of authority than men. His datum was that, among the Juangs, women continued to appear in the nude even after the men had taken to wearing loincloths. This was phrased with characteristic Victorian modesty; the women were said to appear in "something less than Eve's dress". To Spencer, this proved that women were authority-ridden because "instinctive feelings might have been expected to produce an opposite effect".[31] Leaving aside what a follower of Havelock Ellis or Sigmund Freud might say about the possible reversal here in Spencer's feelings on nakedness, the idiosyncratic nature of this single fact compared to his usual multitude of data suggests that Spencer's views on women's political power were a response to personal trauma, rather than a contribution to public discourse. Whatever else one might say about the women of the Juangs, Spencer's datum suggests that they were not helpless followers of the men; after all, the women stubbornly adhered to their traditional nakedness. At the level of personal and marital relations (and Spencer was writing about these women in the context of a chapter entitled "The Constitution of the State") the women were not accepting the supremacy of the men. Yet he is relying on their submission as an example of the fact that women are always swayed by authority more than men. Presumably this was because he also believed that women are more likely to adhere to customs than men.[32]

In referring to perverse nakedness among the Juangs, Spencer was hinting that women are instinctively modest, but that even this in-built bias is overcome by their worship of the authority of tradition. His argument only makes sense when seen as a projection of his personal fear of women. It justifies J. S. Mill's slighting phrase about the "*morbid* habit of dwelling upon sex". Spencer's fear does not arise from the fact that women were likely to be politically conservative, but is a response to their nakedness while the men were clothed. This could not be explained as an imitation of male custom. Although Spencer, as an armchair anthropologist, had not seen this for himself, it shocked him vicariously. This was not due to his sexual repression – Victorians were very knowing about the nakedness of indigenous women – but because of the existence of women who resisted male supremacy before they had been transformed by modernity, and without waiting for the process of evolution to end male dominance. The women of the Juangs were a dangerous anomaly that threatened the edifice of his philosophical system. Rather than providing evidence of inherent conservatism, the women's nakedness bore an uncomfortable resemblance to the rebellious attitude to authority that Spencer himself had always longed to possess. Spencer's rationale for his changing views on women can be found in his own reflections on his life. The basis of both Spencer's advocacy of rights for women, and of his later retraction of this, lies in his lifelong preoccupation with the power dynamics of his own family. His ideas are structured around the harsh supremacy of his father and the corresponding powerlessness of his mother. When he wrote *Social Statics* he saw her helplessness as caused by the father's despotic desires. Later, when he abandoned the idea that women should have political rights, he blamed their vulnerability as leading to despotism.[33] Female weakness not only elicited tyranny; it also encouraged the propagation of socialist ideals as an extension of maternal instincts. Women's love for the helpless flowed from the maternal function, which was at the very core of their nature. This sentiment appeared in Spencer's *The Principles of Ethics*, where it echoed *An Autobiography* in condemning his father for not making allowances for his mother's intellectual inferiority, and for bullying. When he could see a father's unkindness as a biological response to a mother's helplessness it freed him from assigning blame to men.

The highly personal roots of Spencer's views of gender dominance left him ill equipped to enter the arena of English feminist politics. Close to home women were disturbing: he was more at ease when it was a matter of praising the general superiority of women in other countries. He stopped the press when it suddenly occurred to him that his readers needed to be informed that the Lapps had the world's most favourable position for women despite having wretched huts, dirty faces, primitive clothing and no literature, art or science. This single quality implied to him that this Nordic culture ranked above the British in possessing the highest element of civilization.[34] Spencer could praise women easily if they were in the distance. Then, too, if women were far away they could often be absorbed in his aesthetic notions. Spencer was not content to speculate about women only in the political realm. They also underpinned his beliefs in culture and beauty. In these subjects Spencer's personal and sociological perceptions were much more closely intertwined than on any other area of speculation.

FIVE

Culture and beauty

Spencer's name is associated with culture because he is remembered in histories of sociology. While he is not given as much space as Max Weber and Émile Durkheim in such works, his presence means that he can safely be docketed as the most consistent evolutionary theorist among the founding fathers of modern social science. This, in turn, means that his comments on culture are treasured or condemned as they seem to cast light or obscurity on social development. However, Spencer's contribution to the analysis of civilization is not exhausted by references to his part in the development of sociology. For him the contemplation of culture – in the form of *beaux-arts*, music and literature – produced the most important evidence of development. In his earliest general exposition of evolutionary theory, "Progress: Its Law and Cause", Spencer devoted ten pages to illustrating artistic change while biology and government occupied only three pages each.[1] However, culture was not merely another way in which Spencer demonstrated the universality of his scientific views and evolution. His perceptions of this subject were more intrinsically interesting when they appeared outside the framework imposed by science. That is, while his cultural analysis was often external to the systematic evolutionary theorizing of *The Principles of Biology* and *The Principles of Sociology*, it was also more novel. Free from his philosophical system-building, Spencer wrote with originality, drawing on both his own life and his psychological theory. This subjective input emphasized immediate experience, and avoided the distancing or objectifying effects that had accompanied most Victorian attempts to construe culture as exclusively residing in either classical texts and artefacts or in the mores, languages and technologies of distant ethnic groups. As well as being subjective, Spencer's culture was close at hand. It was a response to the beauty he saw both in natural objects and in the eclectic assemblage of aesthetic impressions that a person received from the *beaux-arts* and architecture. For Spencer, culture was not primarily concerned with recapturing the meaning of the ancient civilizations of Greece and Rome, nor did it focus on anthropological excursions into the minds of primitive peoples; instead, it was about beauty.

Spencer's views on culture were most appealing when he was evoking an individual's response to something beautiful. The response was more aesthetically significant if the object in question no longer possessed a biological or social function. A thing had beauty when it no longer fulfilled the rationale for which it had been designed. Whether it was a statue that had once been a god, or a shell that had previously protected an organism from a predator did not matter; it was attractive if its proportions pleased the eye of the beholder. Spencer's aesthetic principle implied that if something was beautiful one should ignore its purpose. Culture could be a refuge from the terrifying and ruthless evolutionary forces that shaped all other aspects of life. Although he was, and is, easy to mock as a philistine, Spencer was, in fact, a true aesthete who loved culture because it demonstrated a necessary frivolity. *An Autobiography* was an elegy to someone who had sacrificed his own life in order to give this insight to others. He was, naturally, reluctant to let this be marginalized by the conventional – and almost ritualistic – Victorian discourses on the meaning of culture. These had been arguments between classically educated scholars and objectively minded social scientists. However, before exploring Spencer's culture, it would be useful to rehearse a fragment of the normal nineteenth-century debate. Without hearing this, Spencer's claims to significance risk falling on deaf ears.

It could be remarked that the modern English meaning of culture is the result of an argument between two Victorians: Matthew Arnold and E. B. Tylor. They were engaged in a struggle over a definition. The former's *Culture and Anarchy* and the latter's *Primitive Culture* can be seen as encapsulating the Victorian debate on the subject. The two authors were attempting to stamp their initials on to "culture", which has come to refer to both our own social identity and is an objective term with which we can describe all the salient characteristics of the "other", whether this "other" is an industrial nation or a small group of hunters and gatherers. The first use is humanistic and points to our internal aspirations while echoing aesthetic triumphs from the past. The second connotation is scientific and coolly appraises the social structures and the languages of ethnic groups in an objective fashion. The two meanings were, and are, in conflict. Their utility can be undermined by a clinical examination of the class bias of Arnold or by simply focusing on Tylor's idiosyncratic grasp of subjects such as marriage customs. However, such critical strategies leave the central antithesis of culture intact. After it is shown that a Victorian figure, such as Arnold or Tylor, used the concept in ways in which he was unaware, the analysis stops short because, in some sense, we now feel comfortable recycling both uses simultaneously. Or, if we do feel uneasy about praising our own cultural attainments or aspirations, we probably feel equally uneasy criticizing the cultural apparatus of someone who belongs to "another" group. Rather than the antithesis of culture shifting, part of it is weakened by changes. New-found sensitivity makes it awkward to speak of oneself as cultured or to describe objectively the culture of the Inuit or the Chinese. However, these shifts ensure that debates over the meaning of culture continue to appear fresh and significant. Culture continues to invoke both the subjective focus of the speaker, and the social object that he or she is examining.

However fascinating we might find this perennial debate over culture, the challenge that Herbert Spencer issued in the nineteenth century appeals to a more fundamental concern. The debate between Arnold and Tylor lacks immediacy: the former insisted on recycling images from the past and the latter focused on societies that were so geographically distant that they often lacked relevance to his readers. The difference between these writers and Spencer was caused by his focus on culture that was close at hand. He did not fasten the flowering of culture, the "efflorescence" as he called it, to intentions that came from the past. Nor, although Spencer wrote extensively on ethnography, did he restrict his comments to primitive or exotic societies. Instead, he focused on culture as an internalized possession about which each one of us could also be objective.

In the Victorian era the primary problem of culture was the need to ensure permanence. For Arnold, as for many other Victorians, the essential purpose of civilization was to ensure the transmission of traditional wisdom and classical art in the face of social and industrial forces that had little respect for them. The primary cultural task was to ensure that symbolic figures such as Socrates and the Venus de Milo did not fade from the collective memory. There was a strategy by which one could preserve culture. This consisted of questioning how images were preserved within the individual mind. To understand this problem one first has to grasp that for a Victorian culture and beauty were often to be found in an object that no longer served its original function. They were highly aware that their aesthetic response to a Greek god or a medieval castle was the admiration of a product of a civilization they regarded as backward and credulous. Therefore, it was obvious to them that the intentions of the artist or architect did not create the cultural impact or beauty. Instead, it was in the eye of the beholder. It was this gaze that detected the permanence of beauty when it was detached from its original function. In a phrase, the Victorian aesthetic was the insistence on the beauty of form over function. It was this admiration of form that was adopted by Spencer when he dealt with literary and aesthetic qualities that seemed to have no function, and whose defence, therefore, was all the more pressing. Most sensations were transitory; they were mainly the kind of perceptions and emotions that prompted and guided conduct, and then disappeared. Aesthetic sensations were the opposite. They were kept in the consciousness and dwelt upon.[2] Spencer, who saw himself as the most important psychologist of his period, had no hesitation in brushing aside the views of other aesthetes when analysing why the human species responded to beauty.

Arnold and other lovers of antiquity had mistaken the nature of permanence in the high cultures that they had evoked. It was a mistake, according to Spencer, to suppose that Greek sculpture was the pinnacle of human achievement, that epics and neoclassical literature possessed fineness, and that the paintings by the Old Masters were without serious flaws. In reality much of this art was "low" because of the emotions it expressed. Greek sculpture and epics, for example, simply appealed to egoistic and ego-altruistic feelings.[3] Spencer was chiefly concerned to oppose "classical" aestheticism, but he also rejected modern artists who had similarly adopted "low" emotional goals in art. For this reason he found the battle scenes of Horace Vernet's paintings repellent because they alternated between the sensual and the sanguinary.[4] "Low" feelings were not permanent sentiments; on the contrary, they were the kinds of emotions that were similar to stimuli, and that faded from the memory. They concerned passion, not beauty. Accordingly, Spencer believed, before we can see or feel beauty we have to be without a sense of urgency: "So long as there exist strong cravings arising from bodily wants and unsatisfied lower instincts, consciousness is not allowed to dwell on these states which accompany the actions of the higher faculties".[5] It was these states of consciousness that contained permanence, not an archaeological museum of ancient sculpture, paintings by the Old Masters and classical texts. We, Spencer believed, would have more aesthetic activity in the future because as evolution advances there will be less appeal to egotistical and ego-altruistic imagery.

For Spencer, culture became the cultivation of one's own sensibilities. It was not a repetition of external values borrowed from the Greeks and other sources of "classical" taste, nor was it an objective measurement of social institutions. Spencer's insistence on the predominance of the individual psyche's experience, rather than that of the class or nation, sidestepped older cultural claims that appealed to shared sensations or common intellectual values. Or, to put it another way, Spencer rejected both the legacy of Romanticism and that of the age of reason; his was the new voice of Victorian aestheticism.[6] For him, the aesthetic sense – that which apprehends beauty – was an organ that perceives without passion. This was a novelty in two ways. First, as has just been remarked, it was a rejection of Romanticism and, more restrictively, of the notion of the sublime that Burke had interpolated into the theory of beauty when he grounded aesthetics on awe and emotion. Secondly, Spencer not only removed passion, he overturned reason. Reason and its operations (such as contemplation, meditation or just the sheer presence of rationality) were not the defining qualities of culture; they were merely analysis or advice on future action. Whether Spencer was referring to the kind of reflection that led to good health or to utilitarian calculation, he thought rationality was not distinguishable from other stimuli or gratifications. Reason could not be detached from function because, like passion, it was one of the "life-maintaining" activities. Neither reason nor passion could be organs of aesthetic feeling. For Spencer beauty was independent of other human considerations, and it became the sole object of culture.

The second edition of Spencer's *The Principles of Psychology* had a concluding chapter on "aesthetic sentiments". This exploration of the human psyche was ultimately about beauty; his thought was that engagement in aesthetic activity was an action that was not to be directly life-serving. More than this, Spencer thought that any activity rose to an aesthetic form in proportion to its quantity and its being without pain.[7] Spencer was writing about music when he offered this definition, but he held similar beliefs about poetry and the fine arts. The important part of this idea was the way in which it challenged the prevailing belief that beauty originated in the satisfaction of sensual taste and in other forms of physical gratification. Spencer's cultural theories were designed to combat the belief that art was an expression of emotion or passion. As has already been stated,

Spencer thought that passion should be restricted to "life-maintaining activities". To be passionate was to be in the grip of a functional stimulus that responded to the need for food, shelter and warmth. Or, it was a stimulus allied with the need to reproduce. Passion could not be beautified, and for Spencer this meant that the pursuit of beauty was imperative. With Victorian advances in psychology and physiology, passion seemed increasingly to be only a collection of mechanistic responses. If beauty was to remain in the world it had to be disentangled from animal responses or separated from life. Beauty's refuge would be culture. However, there was also another idea at work here. For Spencer culture should reinforce an objective sense of life. This sense was not a matter of rationality. To him a display of logical prowess or mathematical proof did not point to something meaningful, but to the cold and dead physical process of the universe. For Spencer the antithesis was that fundamental: beauty was a sign of life, while the intellect could only organize knowledge about things that were dead.[8] At the basis of this antithesis was Spencer's despairing autobiographical analysis, but he was also catching up and bringing together the diverse strands of Victorian aestheticism. In the Victorian period culture and beauty were given the role of elevating and dignifying life. Science could not do this; it could only chronicle the passage of time. Without beauty to give meaning to time one saw only death. There was no god in the heavens, but, as Spencer observed, you could attach a gilt Apollo to your clock.[9] What had been a god to the Greeks had become a thing of beauty. It was no longer functional but, precisely because of that, it was beautiful. Whereas in his biological and sociological writings Spencer often relied on functional arguments, these were deliberately put aside when it came to the discussion of beauty. To say that an object or a person possessed beauty was, for Spencer, an exceptional judgement that belonged to the tiny class of non-functional, yet uniquely significant, statements. From a Victorian perspective he was an aesthete because of his belief that certain forms or shapes could be described in terms of simple curves and geometrical figures that were intrinsically beautiful.

The beginning of Spencer's anti-functional account of beauty is contained in the essays that he published during the 1850s. Here he advanced the suggestion that arts, such as music, seem to excite for their own sake. In a comment on the superfluous nature of joy he suggested that the delights of melody and harmony do not obviously minister to the welfare either of the individual or of society.[10] In these early writings he always proceeded as an association psychologist, as did David Masson, Bain and other psychologists of this period when writing about art. Consequently, they reflected an explanation of art that would find a place for the intellect as a power that played an active role in aiding and creating sensation. In Spencer's words, this would explain that "*cadence is the commentary of the emotions upon the propositions of the intellect*".[11] While Spencer's argument that the appreciation of aesthetic beauty is non-functional, it is still within the conventional rational realm of the technical "common-sense" philosophy of the period. The emotions are subordinate to reason. This belief was tied to a utopian hope that in the future the aesthetic sensation could be rendered more perfect so that we eventually may expect all men "vividly and completely to impress on each other all the emotions which they experience from moment to moment".[12] Spencer, no less than Ruskin, William Morris and Havelock Ellis, expected the future to be one in which beauty would cease to be the preserve of the aristocratic elite, and, instead, would become common property.

However, even though aesthetic experiences would be cleansed of class bias they would still be anchored in responses to a non-functional sensibility. Spencer did not want his lesser folk to limit their appreciation to humble sights such as a cottage. The aesthetic gaze should be free to respond to a castle, that commonplace illustration of beauty that had originated in aristocratic dominance. Artistically familiar from the work of eighteenth-century painters such as Richard Wilson, castles were the obvious "metamorphosis of the useful into the beautiful".[13] Spencer, like Ruskin, saw these medieval remnants as a necessary part of the landscape. While the origin of the castle had been a desire for greater security on the part of oppressive feudal lords, it was now a part of the aesthetic culture. The meaning of a castle was not reducible to origin here. The origin was functional; its later effect was not.

The military excesses of past social states had become the ornaments of our landscape, while the past habits, manners and arrangements served as ornamental elements in current culture.[14] The metamorphosis of the functional into the beautiful was so frequent that Spencer, in this early period, suspected that nearly every notable product of the past had assumed a decorative character.[15] It was for this reason that he defended the selection of historical subjects by contemporary painters.[16] Spencer's aesthetic theories were culturally specific. That is, he believed that function preceded beauty in European society. On the subject of non-European societies, such as those in the Orient and in Africa, he would note that ornamental predominance precedes use.[17] While Spencer was offering a conventional idea when he associated beauty with something that had lost its function, he was more radical when he reversed this thought and offered the opinion that whatever was performing some practical function, either now or in the recent past, did not possess the ornamental character and was "consequently inapplicable to any purpose of which beauty is the aim".[18] There is a status quo aspect to this view, which Spencer shared with other psychologists of the period. For example, Masson's reaction to the Pre-Raphaelite Brotherhood was based on the belief that people generally would find their use of cadaverous and deformed models to be disagreeable;[19] if beauty was an appeal to what was generally liked, then innovation was difficult to accept. Mental associations that summoned up disagreeable images were inappropriate for art.[20] Such a Victorian perspective is difficult to recapture because twentieth-century art has accustomed the viewer to novelty. Perhaps one way of assisting in the understanding of nineteenth-century art appreciation would be to consider that in the hands of Spencer, and a number of his contemporaries, it was an exercise in the popular imagination. What the untutored eye could see without discomfort was beauty.

In his early writings, Spencer's non-functional analysis was accompanied by a note of deprecation. Art was not so much a pinnacle of human achievement as a bloom on its surface. As he declared in his essays on education, "Architecture, sculpture, painting, music and poetry, may truly be called the efflorescence of civilised life".[21] In case anyone missed the image of light radiating from the surface of life, Spencer added that the production of a "healthy" civilized life must occupy the highest place, while the fine arts, *belle lettres* and the other efflorescent parts should be subordinate.[22] Beauty was associated with life, but it was outside it, and was condemned to be posthumous. This early analysis of beauty was closely tied to Spencer's radical critique of society. During the 1850s beauty was a remnant in the same way that the aristocracy was a legacy from the past. To be without a current function was a reminder that life had departed. The best summation of Spencer's early idea of beauty can be found in a metaphor contained in a novel by George Eliot. In this, a young woman's beauty is likened to the surface of a peach while her genetic significance is the unseen stone within it. The latter was her meaning to an evolutionary theorist. Her downy cheeks were irrelevant to this: "People who love downy peaches are apt not to think of the stone, and sometimes jar their teeth terribly against it".[23] Eliot noted that the external beauty of the woman refers to her ancestry, perhaps to a grandmother. When a male longs for beauty it is sexually functional for the species, but otherwise impersonal and meaningless. The individual will not find satisfaction even though he has been prompted to perpetuate himself in offspring.

Spencer's mature views on beauty, like his early ones, relied on an artistic production or appreciation having a lack of function, but the later ones had lost their political emphasis and became purely aesthetic criticisms. His mature views, as expressed in the expanded edition of *The Principles of Psychology* and in *An Autobiography*, defined lack of function as a sufficient, but not a necessary, cause of beauty, but no longer placed beauty in a subordinate position to life.[24] Rather than being a residue left behind when the proper function had ceased to operate, beauty became a purified quality that was above life. Spencer's increasing immersion in the technical aspects of biology and psychology drove him to see most human qualities (intellectual faculties, instincts, appetites, passions and even altruism) as concerned with maintaining organic equilibrium or maintenance of the species. Whether such activities were concerned with food, shelter or sex, they were equally life-preserving features.[25] Any activity or gratification that was connected with this primary activity was ulterior. That is, the only activities that were not motivated by basic needs were play and aesthetic activity. Those could not refer to ulterior benefits because their goals were proximate.[26] Art had assumed a more essential place in Spencer's world and was now an immediate goal of life, whereas before it had been a bloom on its surface. It had occurred to him that only inferior animals expended all their forces in fulfilling functions essential to the maintenance of life, such as searching for food, escaping from enemies, forming places of shelter and making preparations for progeny.[27] Human beings were superior: they had a place for a superfluous category called beauty.[28] Spencer's rationale here was to purify beauty from degrading notions of taste. Artistic canons that relied on the desire to gratify, or to acquire, were too similar to ordinary life-supporting functions: they could not adequately describe art.[29] His illustration of this idea came in "thought experiments". He imagined an aesthetic consciousness in which it was the action itself, not some further end, that was being contemplated. For example, one might receive an aesthetic feeling from considering the attributes and deeds of other persons, real or ideal.[30] During the 1850s Spencer's "laws of life" underlay all phenomena and he had then objected to the "gross utilitarianism which is content to come into the world and quit it again without knowing what kind of world was there, or what it contains".[31] For Spencer, at that time, the laws of life were the same throughout the whole organic creation.[32] Art was no different.

Spencer's last thoughts on aesthetics differed from his early views; he discovered that art *was* an exception to the rest of organic creation. He was no longer able to say that the sensations one acquires while gazing at beauty were merely superfluous; they were now in opposition to all other perceptions and emotions that one may have experienced. They were "no longer links in the chain of [mental] states which prompt and guide conduct. Instead of being allowed to disappear with merely passing recognition, they are kept in consciousness and dwelt upon: their nature being such that this continued presence in consciousness is agreeable".[33] Spencer's belief in the lasting quality of aesthetic sensation, together with his insistence on its immediacy, meant that he could no longer countenance theories that relied on innate notions of beauty; nor could he produce a theory that stated that something was beautiful because it carried signs or meanings. These would have been a product of some instinctual or passionate life-maintaining mechanism, or they would have reduced art to a bundle of impressions that carried ulterior messages. Unlike Ruskin, Spencer did not produce a theory of artistic symbolism.

Spencer's language of beauty was teased out of an intellectual world in which functional explanation was the stock in trade of economists, biologists and ethnographers. His insistence that beauty was non-functional can thus be seen as an act of defiance against the pervasiveness of scientific language. Spencer was claiming that psychology, the branch of philosophy that dealt with the laws of the mind, was not reducible to biological stimuli and economic maximization. His argument was that human culture and the human psyche recognized and kept in contemplation features of the external world in a non-reductionist way. These features were independent of the organic meaning that underlay them. This aesthetic dimension of human life could be accurately described, but not in terms of the kinds of *internal* functions that structured all other human activities. To see beauty was to respond to *external* impressions from nature. There was no inner meaning to beauty. For Spencer there were no ideal forms that beauty could mimic; nor was there an innate sense of what was sublime. The forms, and the appreciation, of beauty had to be found not in underlying laws of nature, but in its surfaces. When Spencer referred to beauty as the efflorescence of nature, he was adopting an anti-Romantic stance. While Romantics looked for the laws of nature, they also attempted to look underneath its surface; that is, when they looked for unities in nature they did not consult their senses.[34]

Like many of his contemporaries Spencer was dependent on architecture for the basis of his belief. His essay "The Sources of Architectural Types" (1852) was the first occasion of his insistence on the external origin of beauty. He visited the gallery of the Old Water Colour Society and wrote about the incongruity of placing regular architecture into irregular scenery.[35] He felt an unpleasant sensation caused by the clash of a symmetrical Greek edifice arising from an irregular background. There was a disjunctive impression caused by the eye's failure to find "unity" in a picture. If a building was introduced into a landscape it should seem part of that landscape. This was Spencer speaking psychologically; as a matter of association, a painting of a symmetrical building with a balanced number of wings and windows does not belong among rough hills and woods[36] because one set of ideas will tend to exclude the other. However, an irregular castle would fit visually among hills and woods. The portrayal of a symmetrical building in a town is allowable because there it would be surrounded by men, horses and vehicles, which also consist of regular features. Everything would then stand in bilateral symmetry.[37] Both regular shapes, such as classical temples, and irregular ones, such as Gothic buildings, mirrored what was in nature. Spencer believed that the Greeks derived their buildings from the vegetable world, whereas irregular buildings, such as castles, had inorganic forms for their basis.[38]

Spencer asked, "whence comes this notion of symmetry which we have, and which we attribute to [an architect]?"[39] His answer was that unless we believe in the doctrine of innate ideas, we must believe that the idea of bilateral symmetry is derived from "without". That is, we get it from looking at the higher animals.[40] Having taken care of classical models, Spencer turned to Gothic architecture, where he followed the often used analogy between organic forms such as trees and pillars and arches reaching for the sky. He stressed the predominance of vertical lines as the most prominent feature of Gothic.[41] Spencer believed that our pleasure in irregularity grew as it became greater, because it was increasingly a reminder of inorganic forms, and a vivid and agreeable impression of rugged and romantic scenery.[42] Spencer's hypothesis is not that architects intentionally give their buildings the leading characteristics of neighbouring objects; rather, he considered "that in their choice of forms, men are unconsciously influenced by the forms encircling them".[43] He thought it impossible for a mind to conceive of a design without an external model.[44] Whereas the Romantics thought of artistic senses as protean – actively creating beauty – Spencer portrayed aesthetic sensibility as a passive receptor for the material world. He followed his own theory, and this had the effect of making his aesthetic observations strikingly discordant. For example, one implication of refusing to acknowledge that an artist had an internal creative inspiration was a disinclination to share the general Victorian desire to recreate and empathize with the sensibilities of dead architects and artists. Historical intentions were irrelevant to Spencer's pleasure. Thus his enjoyment of Rome stemmed from "its harmonious colouring rather than its historical associations, of which he had no vivid perceptions". Venice with its antique palaces held no charm. The city displeased him because of its "barbaric" excess of decoration – "it is archeologically, but not aesthetically, precious".[45] In an attack on the sensibilities of Ruskin's *The Stones of Venice*, Spencer insisted that Venetian architecture was monotonous, and had made insufficient use of light and shade. In his opinion the architects who had designed the palaces along the Grand Canal were inadequate because they had worked only for admiration rather than being guided by thoughts of function or simple perceptions of beauty.[46]

Spencer believed that, as a matter of fact, grace is imprinted on our consciousness by the world. A graceful form is that kind of form which impressed us because of the minimal effort required for self-support, or the effortlessness of movement. Spencer thought that we consider those shapes as beautiful that are slight of build, and those forms as ungraceful that are cumbersome and have underdeveloped powers of locomotion.[47] An example of the first would be a greyhound, while the second would be a tortoise. Spencer's argument here is obviously independent of the argument he produced by ignoring functionalism. The graceful greyhound and the cumbersome tortoise are equally products of adaptation. The fact that one has a beautiful shape, and the other does not cannot be a simple product of one being functionless, which was the basis of Spencer's usual criterion for beauty. To explain this new argument Spencer drew on an example from his favourite physical activity: skating. Skating, he observed, was a graceful motion; it caused many muscles to work together in moderate harmonious action without strain. Turning to forms, he observed that we delight in flowing outlines rather than in those that are angular. This is "partly due to that more harmonious unstrained action of the ocular muscles, implied by perception of such outlines: there is no jar from sudden stoppage of motion and change of direction, such as results on carrying the eye along a zigzag line".[48] This is a conventional description in the sense that it accepts as a virtue one of the criticisms made of mere beauty by advocates of the sublime. They would argue that bold interruptions of line were sublime, while waving lines with easy transitions were merely beautiful.[49] What Spencer had done was to give a physiological basis to one side in an old debate on the nature of beauty. From the image of the harmonious action of the muscles we receive the feeling of a full activity without the pain that would accompany excess.[50] As a matter of fact, Spencer argued, we do find certain forms or curves beautiful. In addition to this argument, Spencer offered a nineteenth-century psychological argument that certain shapes, in both people and works of art, are connected with agreeable recollections. The occasions in which we have looked at graceful forms – for example while observing a dance – have been happy ones. This leads Spencer to offer a "common-sense" explanation of why some people are cultured and others are not. This is not from "good breeding", but from repeated experience. The difference between the uncultured and the cultured is that the latter have experienced more occasions of aesthetic pleasure.[51]

To return to the idea that certain shapes were beautiful because they resembled, or were reminiscent of, forms that were efficient, it is important to illustrate how hostile to sensuality this idea is. When Spencer admitted to once glancing at a young lady in a coach between Cheltenham and Cirencester, he claimed that there had been a "special pleasure in contemplating the elegant curves of the eyelids".[52] Leaving aside that he had been twenty-two years old, and that this was the first and last time he claimed to have looked at a woman, it is of interest to notice what was being said. He could have looked into her eye, which according to one art theorist was a spiritual rather than sensuous organ at this time,[53] or he could have glanced at some part of her that was overtly sensual. However, to Spencer, that would have been looking underneath that which was beautiful. The curves were beautiful, not the body. Spencer continued with this analysis in his criticism of the paintings of Dante Gabriel Rossetti, whom he accused of lacking in science. It was not the hidden laws that the painter had missed, but the surface. Rossetti, "catching sight of a peculiar iridescence displayed by certain hairy surfaces under particular lights (an iridescence caused by the diffraction of light in passing the hairs) commits the error of showing this iridescence on surfaces and in positions in which it could not occur".[54] This comment is remarkable because it combines a faith in naturalism and realistic art with a reaction against physical pleasures. Rossetti was the member of the Pre-Raphaelite Brotherhood who discarded realism in favour of sensuality,[55] and was an obvious target for such comment. Rossetti's failings as an artist stood for Spencer's condemnation of both himself and any other man who cared too little for erudition in women, while too much for her physical beauty, good nature and sound sense.[56] Spencer himself had rejected George Eliot for her lack of physical charms; he failed to love her despite her erudition, and felt cursed that his passions were the servants of gross features such as reproductive drives. He was determined to purify his aesthetic sensibilities so that they would not display the lower passions. Since he was a psychologist who also studied the evolution of culture, he was in a position to cleanse the past of mankind. Beauty was to be rescued from the "life-serving functions", and freed from the body.

What, then, was beauty? It was a quality free from strong cravings and, additionally, according to Spencer, removed from the idea of taste. Spencer had noticed that arguments based on taste were common in defences of artistic appreciation and art collecting, where they enjoyed the support of philosophical relativism. (The commonplace contention would be: this object may be vulgar to you, but to me it is beautiful.) For Spencer such a relativistic defence was inadequate, because he regarded it as a literal claim based on actual taste. He believed that such a claim was simply false because it could be refuted by observing that a pleasant taste, such as a sweet sensation in the mouth, did not commonly produce an idea of beauty. With a memorable phrase, he pronounced "gustatory gratifications are but rarely separated from life-serving functions".[57] Instead of beauty being based on gratification, he thought it was more usually found in sensations that are distant from immediate satisfaction. For example, a scent of a flower or an intense colour is more likely to be beautiful than a flavour, which he believed to be more immediate.[58] Similarly, he suspected that sounds that did not simply warn or guide behaviour were more likely to be beautiful than those that did. With the human form, too, Spencer was concerned to banish from the category "beautiful" perceptions that were close to animal instincts. He had noticed, both in himself and in others, that admiration of bodily perfections was paramount, but he rejected this as a source of beauty because it was based on a passionate drive to reproduce. In place of "low" sensations, Spencer erected a number of criteria that art and literature should fulfil if they were to be beautiful.

The most important criterion was that to be beautiful a work must not introduce a note of incongruity.[59] Both in form and colour there was not to be any jarring sensation. Aside from this, Spencer's notions were similar to those found among his contemporaries such as Ruskin or Masson, although in Spencer's case they were pursued with an intensity that make them appear distorted. That is, Spencer sounded quite normal when he offered the common mid-Victorian belief that something was beautiful when it truthfully transcribed an image from nature. He also conventionally borrowed from the early views of Ruskin to argue that progress in painting came from greater knowledge of how effects in nature were produced.[60] However, Spencer's views also diverged from those of other Victorian writers, and this led him to resist a few fashions. For example, instead of being captured by the increasing taste among his contemporaries for Chinoiserie, he pronounced Chinese art to be grotesque because of its disregard of true perspective. Chinese paintings, he argued, were similar to those of a child in which the absence of truth is caused by ignorance of the way in which the aspects of things vary with their condition.[61] Greek art too had failings: many admired ancient sculptors had not succeeded in their representation of nature. For instance, the way in which ancient Greeks depicted hair was not natural. They failed to notice that representation must be done by suggestion rather than reproduction. Their habit of cutting the intricacy of locks of hair to the depth of the scalp made the control of light and shade too strong to be like that in nature.[62] Even the great painter J. M. W. Turner was not exempt from Spencer's charge of failing to portray naturalness. He thought that Turner's prints often lacked a proper contrast between earth and sky. Instead, they showed the average tone of air as deep as that of something solid. Spencer argued that the greater darkness of the earth must be represented otherwise there is "an untruth which nothing can hide".[63]

Spencer's sources for his notion of beauty were many. He believed in the need for unity in a portrait because of his early attachment to phrenology, which stressed the importance of agreement between head and face. He also claimed to admire variety in a work of literature or a painting because of his "organic needs". In other words, he got bored easily. For example, he admired the mixture of imagery in Shelley's poems, but not its lack in Dante.[64] Of course, the demand for variety in art was common from the eighteenth century, and certainly not original to Spencer. Similarly, his complaint about excessive uniformity in music was a borrowing which he attributed to Charles Burney, the standard authority of the previous century.[65] As with many aspects of his aesthetic theory it was not so much their uniqueness that distinguished Spencer's ideas, but the use to which he put them. A conventional criterion, when transferred from art and music to literature, became philistine. In the face of overwhelming Victorian admiration for Homer, Spencer would announce that one could scarcely bear to read the *Iliad*, even as a source of ethnographic information. The narrative was inadequately varied.[66] Why, he asked rhetorically, should one admire a repetitious tale extolling brutal passions and instinctual behaviour?

Spencer's views on the naturalness of beauty had narrow boundaries. Beauty could not imitate brutal and instinctual behaviour even though these were natural. On the subject of imitation of nature Spencer distinguished between savage and civilized. In his very first published ethnographic comment he noticed that the savage only imitates nature, and therefore is necessarily crude. Later art imitates both nature and existing art, and therefore achieves sophistication.[67] This development has to do with transfiguring nature, which by itself is too repetitious and uniform to interest the observer. He could not tolerate a depiction of repetitive reality in ordinary life. Yet, he also eschewed an "awesome" account of beauty. If art diverges from naturalness, it detracts attention from the meaning or intention of a whole work.[68] Artistic beauty must be a varied and agreeable use of ingredients that brings all sensations into play. It should not overemphasize one sensation at the expense of others. Spencer's theory of beauty was a defence of the eclectic habits of Victorian art collectors. Like his contemporaries he admired an abundance of clearly visualized detail and variety that appealed to several emotions at once.

In his defence of naturalness, Spencer was in agreement with the South Kensington School of Art. To achieve a natural look, design could use a few simple ideas in combination. These would issue in multitudinous products.[69] He believed that one could bring science to bear on the production of artistic works. There was no great divide here between science and art, or between scientific knowledge and artistic insight.[70] This results in a paradox: the paradox of philistinism. Spencer insisted that there was no special category for artistic appreciation; it was simply an extension of science. Yet he also insisted that the product of art was high and pure. As consumers or appreciators of art we are listening to, viewing or reading something that is untainted by low functions. It, beauty, is not to be reduced to "gross utility". Our notion of beauty is not to be explained as a function of life because that would be a pursuit of a proximate end. Instead, beauty is an ultimate pursuit. To Spencer art had become a replacement for religion, but it was a substitute that was closely connected with the future and not the past, which had produced so many of the beautiful things and texts that one admires. In the future, culture would hold a greater abundance of beautiful sensations for more people; this would provide an evolutionary substitute for what had been pure and fine in old images of the gods. A golden Apollo would continue to grace the Victorian clock, but, in the future its equitable marking of time would be seen, and heard, by greater numbers of people. Spencer's admiration of naturalness was not restricted to art. He also allowed it through to behaviour, which meant that he would probably act in accordance with any desire to maintain his bodily health or to amuse himself. While disregard of restraints on aesthetic judgements is easy to tolerate, this is less the case when the rules concern behaviour. Spencer's flouting of social norms often led to his condemnation as an eccentric.

SIX

Eccentricities: health and the perils of recreation

When he thought he was out of the public gaze, Spencer's care for himself was so solicitous that it now seems absurd and unpleasant. Since the publication of Edith Sitwell's *The English Eccentrics* in 1933, Spencer's excessive concern for his health has provided a never-ending source of anecdote about Victorian oddity. A typical example of Sitwell humour was this:

> The taking of Mr Spencer's pulse was one of the great ceremonies of the day, and often, when out driving in his victoria, a cry of "Stop" would be heard by the coachman, and then, no matter where the equipage might find itself, in the middle of busiest traffic, in Piccadilly, or in Regent Street, the carriage would stop dead, disrupting the traffic in question; and silence would reign for some seconds, whilst Mr Spencer consulted the dictates of his pulse. If the oracle proved favourable, the drive was continued; if not, Mr Spencer was driven home.[1]

Spencer combined hypochondria with radical political opinions, irreligion and the frequently repeated desire to remedy the workaholic mass psychosis that afflicted his contemporaries. The presence of so many novelties in one person meant that Spencer became a target for later generations of satirists who found the Victorian period ridiculous. He came to typify quaint self-indulgence of a mildly objectionable kind. He was the quintessential molly-coddler. There is an irony here; Spencer is seen as representative of a society that he did not like and for which he was ill adjusted. There is also an error of judgement, because Spencer was not a true eccentric if by that one means a person whose remarkable actions were unselfconscious. Much of his unusual behaviour was a deliberate, a measured, attempt to challenge the behaviour of his contemporaries.

From his youth Spencer had believed that the great design of human progress required him, and others, to think outside conventional norms. Even at the age of twenty-two he was defending prejudices, mental idiosyncrasies, the spirit of opposition and "the tendencies to peculiar views".[2] These, he believed, all conspired to bring about the mental, moral and, eventually, social perfection of the human race. Varied characters, as well as differing opinions, were needed. Without these truth would not emerge. Spencer always remained faithful to his early vision that truth was "like a spiritual Venus, the impersonation of moral beauty, it is born from the foam of the clashing waves of public opinion".[3] He never grew into conventionality. As an old man he despised those who then called themselves liberals but who displayed "an attitude of subordination to the decisions of Mr Gladstone". In Spencer's eyes this was no different than the people of France submitting, by a plebiscite, to the personal rule of Louis Napoleon.[4] He was unperturbed that his own political beliefs, his character or his behaviour might be seen as eccentric; the great design for human progress prescribed variation in such matters.

Spencer's behaviour in private was unrestrained in its playfulness. He not only felt his pulse frequently, he followed his humours in a way that later generations would repress.[5] When he was "at home" with friends and companions he did what he wanted, regardless of the diaries kept by others. In any case, before his death this freedom was not dangerous; the tact and loyalty of his friends kept much of this information from the popular journals. Sometimes, he and his supporters used censorship. Strong measures were occasionally needed because accounts of his private life acquired a commercial value. For example, a stranger, the daughter of one his neighbours, wrote to one of his biographers, Macpherson, offering to sell him the journal she had secretly kept of Spencer's activities. The response was to purchase it for ten guineas so that it could be destroyed.[6] Since the offer had been accompanied by a threat to publish elsewhere, it could hardly have been refused. Stories that were repressed until his death were not tales of scandal or violence, but accounts of innocent amusement and disregard for appearances. Typical of this would be the windy day he arranged a seaside picnic for some young people. As they walked across the sand they found themselves rained on by envelopes, much in the way that playing cards had rained on Alice in Wonderland. Spencer, who was lagging behind, had carried large packets of envelopes. He waited until the gale was at its height, broke the bands that held the packets together and let the envelopes fly upwards so that they unexpectedly fell on his companions. Another amusement he discovered when sitting on his drawing-room roof in the spring. From there he spied on a group of young children at play in the next garden. He had his servants fetch branches from flowering trees in his own garden: he himself tied them into bundles and threw them down to the children to their great delight and his own.[7]

Some Spencer stories referred to his disregard of appearances rather than to his sense of play. For instance, when he travelled with an unfinished manuscript he would appear at the train station with a thick piece of string tied around his waist. Two or three yards of this issued from underneath the back of his coat, like a tail, and the end was in turn attached to a brown-paper package that contained whatever manuscript he was working on. This package he held in his hand. Although Spencer joined in the laughter when his companion told him he looked like the story-book dog with a saucepan tied to his tail, he continued to secure his manuscripts in this way.[8] Spencer was also a figure of fun because his clothing was sometimes unusual. When he stayed at home during the 1890s he would sometimes wear a curious woolly garment he had designed for himself. This was constructed in order to give him only minimal exertion while dressing. He would step into the garment, give one pull and he was fully clad in boots, trousers and coat.[9]

When he appeared in unusual dress he was identifying himself with the radicals of his youth. Those attending a Chartist demonstration, a lecture on socialism, or a soirée for the Friends of Italy showed a preponderance of unusual hairstyles, bare necks, Byronic shirt collars, peculiar waistcoats and wonderfully shaggy coats.[10] Further, like Spencer himself, those holding populist opinions wore caps or felt hats instead of top hats. There was a general truth here: "We believe that whoever will number up his reforming and rationalist acquaintances will find among them more than the usual proportion of those who in dress or behaviour exhibit some degree of what the world calls eccentricity".[11] Spencer's own behaviour and appearance often indicated that he subscribed to this belief. He was visually signalling his dislike of the establishment while calling for reform. He saw top hats as the black cylinders that symbolized tyranny, and in deposing the one you toppled the other. In his comments about hats he felt that he was taking part in the advance of freedom towards the future. He was pleased that the practice of removing one's hat as a mark of respect was falling into disuse; further, it was no longer swept out at arm's length, but simply lifted: "Hence the remark made upon us by foreigners, that we take off our hats less than any other nation in Europe – a remark that should be coupled with the other, that we are the freest nation in Europe".[12] Spencer's sartorial eccentricity was not just a personal foible; it was a political gesture that offered the enjoyment of offending the respectable, but socially backward, part of the population.

Spencer's enjoyment at breaking convention was accompanied by a serious purpose. He knew that some of his liberal and intelligent friends believed that rebellion in small matters, such as apparel or hats, destroyed one's power to help reform in greater matters. He heard them saying "If you show yourself eccentric in manners or in dress, the world will not listen to you. You will be considered as crotchety and impracticable ... by dissenting in trifles you disable yourself from spreading dissent in essentials".[13] Spencer had two replies to this. First, this outcome was caused by the cowardice of those who disapproved of convention while obeying it. These cowards made rebels look eccentric when, in fact, they were courageous. Secondly, he maintained that social restraints and matters of form are not small evils, but among the greatest. If one added up the trouble, costs, jealousies, misunderstandings and loss of time and pleasure caused by conforming to fashion, it would be found that "the tyranny of Mrs Grundy" is the worst from which we suffer.[14] Artificial intercourse of the kind encouraged by society was of low quality and drove away many of those who most needed its refining influence. Rather than respectable society ameliorating social habits, it was worsening them. Men had begun avoiding stately dinners and stiff evening parties, and would hide in billiard rooms. It was society that caused men to spend their evenings avoiding conversation and adopting injurious habits. Men's refuge in cigar smoke and brandy was a kind of rebellion.[15] Spencer was speaking for everyman here or, rather, for the bachelor who took refuge in his club. He was a notoriously keen billiard player and a hater of the social straitjacket of formal parties. London life could offer freedom from convention, and he was not about to be shackled in such a way as to suffer unnecessary pain.

Despite being a bachelor, and despite his complaints of isolation when he was old, Spencer was almost always surrounded by others during the last four decades of his life. As a result, his unusual habits were better recorded than those of most of his contemporaries. After his death in 1903, published stories of his private foibles could always be relied on to amuse the general reader. Typical of this sort of treatment were the stories passed on to Malcolm Muggeridge, the former editor of *Punch*. His wife, Kitty Muggeridge, was the granddaughter of Spencer's friend Laurencina Potter. Kitty combined family papers and published accounts to produce an archly written description of Spencer's self-diagnosed "cerebral congestion". Her Aunt Beatrice's papers provided recipes of Spencer's cures and an account of his habit of wetting his head with brine before retiring for the night. Reputedly he covered his wet hair with a flannel nightcap, and then over this put on a waterproof cap to prevent the moisture from evaporating.[16] The favourite bizarre titbit of Spencerian behaviour – found in Kitty Muggeridge's reminiscences and elsewhere – is usually extracted from the small book published after his death by his long-suffering women housekeepers, who wrote under the *nom de plume* "Two".[17] Details of this story vary but possibly, since it described a recurring phenomenon, the reality may have differed as well. In essence the account describes his arrival at a London railway station, usually on his way to stay in the country with his friends, the Potters. He would arrive at the station with a small entourage consisting of his amanuensis or secretary, a woman companion to read to him and to wave him off when the train left, and a couple of porters carrying the luggage. Before Spencer mounted the train he would proceed to a waiting room to have his temperature taken by the secretary. If this proved satisfactory the secretary would climb into a reserved first class compartment and string up a hammock so Spencer could begin his journey. If, however, his temperature was too high, he would have the train sent on, while he stayed behind having his head wrapped up with vinegar and brown paper.[18] That Spencer was extreme in his foibles and his self-cosseting could be viewed as either extremely funny or as terribly wrong. Friends tended to see it as amusing while dependents felt it as a grave fault. Certainly, his former secretaries, such as James Collier and Hudson, seethed with resentment against Spencer. Another, Walter Troughton, left an unpublished memoir detailing Spencer's irritating demands for unnecessary attendance by medical men, his insomnia and his habit of constantly feeling his pulse.[19] Other hangers-on were made uncomfortable by the way in which Spencer insisted on things being carried out exactly in accordance with his specific demands. For example, when someone played the songs he had loved in his youth on the piano he was insistent that this should be done in a way that was contrary to technique common in the late Victorian period.[20]

Spencer had strong views on most subjects, and he would air them whenever he felt comfortable or unguarded. For example, in the middle of a detailed scientific disquisition on the probable evolution of the internal organs of fish, he could insert a comment on the thoughtless cruelty of keeping goldfish in small aquariums.[21] When he was speaking informally he was even more decidedly opinionated. The matter could concern the true method of lighting a fire, of hanging a picture and of arranging flowers. He could also determine the preferred colours for a carpet and the best shape for an inkstand.[22] In areas in which Spencer claimed special expertise, he was not so much amusing as dictatorial. When he "borrowed" children and received a thank-you note from their mother mentioning that on return they seemed better than before, Spencer's reply would insist that their improvement was not due to the climate at his holiday home, but to the difference in regimen. They had had hot baths in the morning instead of tepid ones. With him children had been thickly covered with flannel next to the skin, avoiding the fashion of bare legs and exposed necks, which he blamed for so much of the current illness. While living with Spencer the children had animal products at meals, which were four times daily: breakfast was at 8.30am and included fish, egg and bacon together with bread and milk. Dinner was served at 1.00pm with as much meat and other food as they liked. This was followed by a slight meal at 4.00pm and "another substantial animal food meal" at 7.00pm. In addition to supplementing their normal diet, Spencer increased their exercise in order to reduce what he saw as their morbid craving for water. The children also stayed up late as he thought it a mistake to send them to bed before they were sleepy. If a child seemed excitable he would abolish bedtime stories and substitute a mechanical game to induce a quiet mood.[23] Since many of these suggestions were in opposition to Victorian practice they were startling, but, since they came from the reigning English expert on education, psychology and sociology, they had to be entertained seriously.

When he strayed into a domain that was not completely private, such as when he sat on the committee of his club, Spencer would venture into the discussion of minutiae with excessive confidence. On such occasions he might indulge himself with speeches on the art of tea-making and the philosophical principles of administration.[24] His club was a safe habitat so he was extremely free in his speech or actions. It was his sense of ease that made Spencer an easy target. Those who obtained a glimpse of Spencer behaving naturally were often surprised that he acted without restraint if he thought an action would amuse him. Unlike some other great Victorians he did not bother to keep his mask in place while in private. Respectability was put aside casually by Spencer, as long as it was not the kind of convention that was the opposite of coarseness, drunkenness and lasciviousness. He was so foreign to the first of these that his friend Tyndall caused much mirth in their circle of acquaintances by suggesting that Spencer would be a much nicer fellow if he had a good swear now and then.[25] Spencer never drank to excess, and was abstemious in other respects as well. If he played cards he preferred not to gamble and if forced to play for money he would pay his debts but refuse to collect his winnings. In his relations with women there was never scandal; there was nothing to report. If he were judged on moral grounds by his contemporaries his behaviour would not have excited punitive comment. Yet something about him was provoking to middle-class notions of propriety.

He was not entirely respectable because he believed in play. He was loyal to this belief even as an old man. It was a matter of absolute conviction to Spencer that an adult should play. After his breakdown from overwork in 1855 (he had just finished writing the first edition of *The Principles of Psychology*) he spent eighteen months travelling and consulting medical men about his insomnia and his sensation of overexcitement. In rapid succession he tried living at the seaside, visiting foreign cities and "taking the waters". This last remedy had the advantage during the mid-century of being both a "natural" remedy and a recognized cure for hypochondria.[26]

Spencer took up hydropathy after it had become so thoroughly respectable that conventional medical opinion in England had given up ridiculing it. By the late 1840s the "water cure" was no longer seen as one of the fake homeopathic cures, such as mesmerism and chronothermalism, that Victorians favoured.[27] Of course, there was always some scepticism, even towards acceptable remedies. Spencer himself did not notice any marked effects caused by his water treatment, but he suspected these would be few since, at the time of his first visit in 1854, he felt very well except for some heart palpitations. In his opinion, a large portion of visitors to such establishments were as healthy as he was, especially since patients often brought along younger relations to keep them company.[28] The establishment that Spencer attended was Umberslade Hall near Birmingham. Its regimen – which included amusements and exercise – seemed to have been less harsh than those of the famous Malvern establishments during the mid-century,[29] but, as with them, the cure consisted of the ingesting of large quantities of pure water, together with the external use of water. The latter included being wrapped in wet sheets, immersed in water of various temperatures, and doused and douched with jets of water.

Spencer's opinions on hydropathy were the same as those of the well-known contemporary medical authority, Sir John Forbes. That is, the actual remedy might not be water, but the patients' avoidance of drugs and their change from a sedentary, stressful and hard-working life to one of ease largely spent in the open air while they kept regular hours and ate wholesome food.[30] As early as 1842 Spencer was railing against the malpractice of doctors who over-prescribed medication. He had harsh things to say about pectoral pills, sudorifics, aperients, tonics, diets, vapour baths, diuretics, emetics, anodynes, cathartics, opiates and febrifuges.[31] Even topical or external remedies were not beneath his notice, and he expressed his doubts about plasters, "blisters", liniments and emollients. Spencer's experiment with hydropathy reinforced his belief that "natural" medicines were to be preferred over the Victorian pharmacopoeia. Since powerful drugs such as opiates, mercury, colchicum and the newly isolated alkaloids – including veratrine, emetine, quinine and aconite – were prescribed without a clear distinction between experimental and clinical use, it was plausible to concur with hydropathic doctors that much illness was medically induced. In reaction to the excessive use of powerful drugs, hydropathy sanctioned natural medical practices in which the patient was purified with water, plain food and exercise while being denied alcohol and tobacco. Since

Spencer's interest and involvement in hydropathy was uninfluential, and since his attitude towards the subject was perfectly conventional for the middle of the nineteenth century, it is clear that the only reason his home practice of hydropathy became notorious was that he practised it for too long. His longevity meant that he had outlived his friends. His isolation from the 1870s meant that Spencer no longer sensed what was fashionable. By then he no longer possessed the kind of information that would have told him that the water cure at Malvern and other places had lost its modishness, so he persisted in his natural cures.[32]

Spencer's personal views on health always remained similar to those propagated in hydropathic "natural" medicine in the mid-century. Until old age, he preferred non-interventionist remedies such as hydropathic packs and exercise over medicinal drugs. While he had been dubious about the claim that the water cure was preferable to drugs, he was sufficiently influenced by it to abandon, as an experiment, the use of tonics. When this experiment proved to be beneficial he gave up drugs altogether and did not resume taking them until the onset of old age. Some remedy was required because he was always convinced that something was wrong with his health, and that he needed to take steps to revive it. His favoured "remedies" when young were riding and the heavy physical exercise of grubbing out tree roots with an axe.[33]

Spencer's friends, worried about his well-being, gave advice. Laurencina Potter, George Eliot and Lewes suggested that Spencer marry or, less politely, engage in "the exercise of emotions". They blamed his ailments on the "lack of the domestic affections".[34] This was a hint that sexual intercourse or, less prosaically, love would cure his disabilities, but Spencer was unable to accept such counsel because he suspected that he had no "emotions". He replied to Laurencina that his feelings had been in a "dormant state" for so long that the exercise of "emotions" was unlikely to be an efficient remedy. As a young adult Spencer had not permitted himself any jollity or hedonistic behaviour that was not conducive to increasing the amount of social justice. While he had thought happiness was important and that one should avoid restraints on desire if they were painful, his emotions were framed in so earnest a fashion as to seem teleologically inspired duties. That is, he believed that it was one's duty to experience pleasure if individuals were to follow the "Divine Idea" of developing mankind towards perfection.[35] Later, when he threw this excessive seriousness aside, he began to seek out casual ways to entertain himself, both as a relief from frustrated sexual urges and as a matter of welcoming frivolity.

Spencer found a cure for his ailments in "disciplining" himself to amusement. He began to take exercise and to do physical work in 1855, but these had proved to be inadequate by themselves. As an addition to them he consciously inserted pleasure into his life. This was not a natural process; he forced it on himself. As he advised his friend Youmans in 1871, "though at first you may, in consequence of having wedded yourself to work, find amusement dreary and uninteresting, you will in course of time habituate yourself to it, and begin to find life more tolerable".[36] He believed that he needed to force himself into pleasurable pastimes as an antidote to the kind of extreme asceticism that he had been made to practise during his youth. Anti-hedonism had not been just a product of his familial background; it was, he felt, a malignant general feature of nineteenth-century society. Harmful asceticism and self-denial were not only to be found in religious groups, but in those engaged in business. Among both groups one found the mistaken assumption that life was to be subordinated to work or learning.[37] This, Spencer felt, was the error that had ruined the lives of his modern contemporaries. He generalized from his own need for amusement to a social remedy for his many contemporaries, who, he feared, were living a grim and shortened life under the twin governors of God and industry. In this way he was an early prophet for the twentieth-century desire for leisure and play among adults. While rival secular prophets, such as Marx, Carlyle, Ruskin and Morris, deified industry, duty or physical labour, Spencer attempted to remove these false moral qualities from the list of essential ingredients for human development.

Joy and games were not to be abandoned with childhood; they were to be retained by adults. This was Spencer's gospel, and he practised what he preached even when it provoked ridicule. Of course, when he was in public Spencer attempted to observe the norms of respectable behaviour, but if he thought he was among friends he would engage in "low" activities with no care for his image. For example, when he was at Philae on the River Nile with a small family party in 1880, he spied some large boulders on a summit of a hill overlooking the water. Unhindered by thoughts of his age – he was sixty – he conceived a desire to amuse himself by athletic effort. He instantly concocted a plan to roll the rocks down the hillside so they would splash into the Nile. He recruited the "Arab" sailors to help him in his task, which turned out to be a quite difficult one, so much so that he speculated that Arab boatmen on the Nile lacked the capacity to combine their efforts at given intervals.[38] When the party had ascended the hill, they found that the boulders were wedged tightly in the earth. Spencer formed the sailors into a column, with himself at the front, to push the first of the boulders. There he was, a tall elderly man in black clothes, held by the coat tails, silhouetted against the sky while ineffectually pushing the boulder. He could not push strongly because the sailors, fearing he would fall with the boulder, held him so tightly that he was immobilized. This scene, witnessed from below by the rest of his party, caused much merriment. The laughter of his friends was not repressed by Spencer's angry shouts at them; rather, it was encouraged.[39]

Spencer was among friends; those on the river below included two of the Potter sisters. In such a situation he did not believe in keeping up the façade that eminent Victorians were always supposed to hold in front of them. He was a devotee of neoteny: the preservation of juvenile playful characteristics into adult life. He played ordinary games, such as billiards, but he also delighted in spontaneous playful inventions. His "discipline" of amusement meant that he rejected, at least in his private life, the convention that men should display dignity or gravitas. For him dignity was an artificial rather than a natural mode of behaviour, and he believed that it should be challenged whenever the social costs were not too high. So while others sometimes perceived Spencer's behaviour as the awkward and forced adherence to a theory of amusement, he himself was engaged in protecting what he saw as natural behaviour in opposition to the artificial mores of his society. Some of his awkward habits were in fact part of his reform agenda.

However, not all of Spencer's "eccentricities" had their source in his rejection of artificial manners and behaviour. Some of the adverse comment on him, especially that which followed his death in 1903, had its roots in his own public comments about the state of his health. At the time he made them it was considered normal to recount the state of one's physical well-being in considerable detail and with repetition. However, to later generations this custom has often appeared to be self-indulgent and manipulative. In other words, the valetudinarian Victorian became an excellent subject for posthumous ridicule. Twentieth-century commentators, especially when they were reinforced by the private diaries of Beatrice Webb, which passed into the public domain after her death, did not recognize the distinction between what was intimate behaviour and public behaviour in Spencer. Everything was classed together as equally strange. Free from Spencer's attempts to control accounts of his life, subsequent generations took a gleeful interest in hitherto suppressed material. In addition, the interpretive weight he placed on his own characteristics was read differently by succeeding generations. The phenomenon was the same, but the explanation of it differed, even though recent commentators have often shared Spencer's insistence on the need for explanations to contain a moral. The explanations vary because they issue from dichotomous worldviews. On the one hand, Spencer claimed that the display of individuality was creditable in itself, and a source of philosophical insight. On the other hand, Spencer's posthumous commentators have seen his individuality as a self-indulgent idiosyncrasy that had no role in philosophical discourse. In part this difference reflects a change in intellectual interests between Spencer and his successors. Since he was an evolutionary philosopher of Life itself, it was plausible for him to regard his values as rooted in the specific features of his childhood development and his particular somatic functions. Later generations, whose ideas were generated by professional cadres of philosophers, sociologists, psychologists and scientists, tended to be impatient with explanations that were not drawn from disembodied professional languages. However, part of the later misunderstanding of Spencer was the fault of his own prose. The effect of his autobiographical accounts was never quite what he had intended. For example, remembering that when he and his friend Francis Galton had conversed about fingerprints, the whorls and loops of which were newly discovered to be unique to each individual, Spencer seized the opportunity to tell a well-rehearsed anecdote about himself. He had been dissatisfied with Galton's investigation into the causes of fingerprints because it seemed to produce no interesting results. The procedure had involved the dissection of the fingers of embryos, but this had only revealed that embryonic prints were folds in which sudorific glands were formed. There seemed to be no reason why the folds had particular shapes. Spencer, in an attempt to help, had offered an elaborate theory that began with the proposition that the opening ducts of such glands would have to be in the hollow between the ridges in order to be shielded from abrasion. When Galton destroyed this theory by replying that the ducts opened on the ridges, not the hollows, Spencer burst into good-humoured laughter, and retold an anecdote in which their mutual friend, T. H. Huxley, had said that Spencer's idea of a tragedy was a hypothesis killed by a fact.[40] Huxley had meant to ridicule his friend for being too serious, but Spencer's recycling of the jibe was a boast. It stressed that he had the unusual mental characteristic of producing hypotheses rapidly. However, the effect of retelling such stories was to make him appear ridiculous outside his circle of friends. When he retold them he blurred the boundaries of the private–public distinction with which he had shielded his private persona. He knew that barbed comments about himself would be repeated, yet he published them with relish.

He would often engineer his own exposure as an excessively individualistic person. For example, if he had truly wished to keep his deprecation of the monarch and aristocracy to his private life, its appearance as a theme in his books made this impossible. He would claim to be seeking privacy, but when he sought privacy he was a public figure taking a stand on a matter of general importance. His occasional refusal to accept invitations from the great aroused much gossip, but instead of accepting this as the price of fame Spencer became irritated. While reading over an unpublished and unwanted account of himself by a neighbour from the days when he lived in Queen's Gardens, he became particularly enraged by a passage that related how that he was "often (!) invited to dine at Marlborough House, but would never go". Spencer's anger was directed at the absurdity that the Prince of Wales would have "often" repeated an invitation after it had been declined,[41] but his expectations were unreasonable. He could not have reasonably relied on his publicly known stance always being taken as if it were a private matter. This misunderstanding was compounded by Spencer's occasional rudeness to people when they seemed to intrude on a venue that he arbitrarily had decided was an intimate one. His actions were sometimes seen as fostered by a peculiar and inflexible pride. For example, he might refuse to see an important personage even if the latter had brought a letter of introduction from someone he knew. The Reverend Barnett was appalled while travelling with Spencer to see him brush aside a French savant with a curt phrase. However, while Barnet told this story as a criticism, Spencer kept it alive as a boast.[42]

Spencer's inability to accept the consequences of fame led him to adopt some bizarre avoidance tactics, which were seen as unusual. His responses were also random. At times he accepted admiration, but at other times this repelled him. His harshness to unwanted devotees could be extreme. On one occasion, while they were travelling in Egypt, Kate and Margaret Potter took pity on one of Spencer's admirers, a Dutch judge in the consular courts in Cairo. He had begged the young women to arrange a meeting with Spencer. They agreed to do this and, as an expedient, they arranged for a party to take a midnight ride to the Tombs of the Prophets. In this way Spencer would be drawn out of his hotel into the open, where the Dutchman could address him. It was all arranged without informing Spencer. The plan went ahead and mounts were hired for the whole party, even though the judge was fat and had not ridden for years. (The Potters hoped that their guest would be safe enough since the mounts were only donkeys.) At the beginning of the journey, Spencer's judicial admirer sidled up to him on his donkey and began to give a pompous and prepared address. Unfortunately, at that moment the Egyptian donkey-boys started to yell and strike their beasts. The eulogy stopped suddenly, to be resumed several times during the journey. Spencer became so angered by these unwanted attentions that he absolutely refused to attend the supper that the judge had laid out against the return of the travellers. The evening ended with the young women attending in Spencer's place, and finding the great man's books spread out on the table forlornly awaiting the arrival of their author, who had once again protected his privacy.[43]

One of the curious changes at the end of the nineteenth century was the reversal of what was open with what was concealed or closed. Victorians had a fascination with their health, and thought of it as something that should be discussed by their friends, acquaintances and even strangers. Yet their sex lives were taboo. Subsequent generations have been covert about health while open about sex. Spencer is an extreme case of the Victorian combination. He was an example of a virginal and fantastical hypochondriac. It was the combination of these qualities that was unusual because, taken by themselves, Spencer's psychosomatic symptoms would not have been considered extreme during the mid-nineteenth century. Spencer could not have successfully engaged in a psychosomatic competition with contemporaries such as William Carpenter and Charles Darwin. In 1864, Carpenter, then Registrar of London University, was ill for a year with "a listless torpor of mind and body" and temporarily lost the will to live.[44] Darwin's symptoms were longer lasting than Carpenter's as well as more spectacular. During the 1840s, any reading that caused Darwin to think deeply also gave him headaches. Other stresses caused him to develop heart palpitations and gastric upsets including periodic vomiting. He also suffered from eczema, which cleared when he stopped thinking about evolution.[45] Victorians were indulgent towards those who suffered from imaginary ailments, while in the twentieth century hypochondriacs are sometimes seen as repellent. For an example of twentieth-century intolerance towards hypochondria one can turn to Sturrock's comments on how Charles Darwin used his queasy stomach to protect himself from intrusion and the demands of sociability, even to the extent of avoiding attendance at the funeral of his much loved father. Sturrock treats this "illness" as cowardice, and as a self-advancing strategy on the part of Darwin.[46] The source of Sturrock's information was Darwin himself,[47] although its use as an anachronistic interpolation is wholly at odds with contemporary morals. Darwin's family and his contemporaries would not have found his actions as morally flawed and repellent in the way Sturrock has. On the contrary, they were capable of tolerance and could feel amusement at psychosomatic illnesses and imaginary cures.[48] It is not that the later biographer has discovered a new fact, but that he has given an old fact a new gloss. The novelty is that he regards hypochondria as a deficit, while the Darwins and their friends did not. Like many other Victorians they accepted and condoned both true and false confessions of illness.

Spencer's hypochondria seems to be without much basis in physical illness. He never contracted serious diseases, nor did he suffer injuries as a child or an adult. He lived into his eighties with little diminution of activity or vitality.[49] While Spencer had a nervous breakdown from overwork in the mid-1850s, and subsequently suffered from insomnia, generally he enjoyed good health until the last two decades of his life, when he suffered from a variety of minor ageing debilities aggravated by the use of drugs, such as opium, which he took to induce sleep.[50] However, even when he was not physically sick he luxuriated in his illnesses. There were paradoxes here. For example, his trip to Egypt in 1879 had ended suddenly because he claimed that nervous fancies had taken hold of him. Yet in 1881 he credited the excursion with having been decidedly beneficial, especially since he was subsequently able to drink beer with impunity for the first time in fifteen years.[51] To his contemporaries such mysteries were acceptable. Few condemned his claims to be ill. Strangers were polite and his friends believed that suffering was the price he paid for valiant efforts to assemble and propagate knowledge.

An example of this comes in a contemporary work published to memorialize Richard Monckton Milnes, Lord Houghton. Monckton Milnes was a politician who, together with Lord Brougham, did much to promote the applied social sciences. He also hosted soirées where the literary and scientific elite of England could meet parliamentarians and other dispensers of patronage. While Monckton Milnes's country house, Fryston, occasionally saw Carlyle as a visitor, its usual denizens were attached to the liberal interest, and were regarded as mildly progressive in their views. When Spencer was visiting, he was comfortable enough with the company of Monckton Milnes to abandon his anti-monarchical prejudices and help entertain the King of the Belgians at a gathering in 1876.[52] He had stayed before, and his expression of gratitude is typical of his valetudinarian confessions. Dated at the Athenaeum Club on 6 April 1872 it read:

> Dear Lady Houghton,
> When I left town I was still suffering from a long fit of dyspepsia joined with uneasy bad nights. While at Fryston, both the original evil and its effect disappeared; my sleep, indeed, having been better than I remember for a long time past. And now that I have got back to my task, I find myself in very good working order. As I hold it to be clearly-proved fact that an agreeable emotional state is of all curative agents the most potent, it is clear to me that I must have derived much pleasure from my stay under your hospitable roof, and that I am indebted to you for an important benefit in the way of health and efficiency.
> I am very sincerely yours,
> Herbert Spencer[53]

The curiosity of this letter is not just that it was published thirteen years before Spencer's death with no thought that it might embarrass him, but that it was public long before this. Before 1890 it had been bound up in the Fryston visitors' book so that everyone could see that a great man had enjoyed his stay. There was no hint of irony or criticism in this double exposure of Spencer's ailments. To dwell on illness was simply part of a normal Victorian display of genuine feeling. How better to express gratitude than to reassure your hostess that her hospitality included curative properties?

Spencer mentioned his ill health constantly. Early readers and subscribers to the separately published parts of his *The Principles of Sociology* were greeted with a short gloomy preface warning about the lack of the author's physical well-being. He was concerned about a delay caused by illness during the early 1880s. Since he could only get through a small quantity of daily work, he feared this might cause further parts of the second volume of *The Principles of Sociology* to be even later than their predecessors. *Ecclesiastical Institutions* had taken him three-and-a-half years, and successive parts might, he suggested, never appear at all.[54]

Spencer's desire to share his health problems with his readers and listeners was most notoriously exposed in his only trip to America, the home of his greatest number of followers. He toured several cities, but would not speak in public. His disciples, who included his American publisher, Youmans, had hopes that he would address them, but he sternly resisted their pleas. Spencer gave only one interview in the United States, in which he corrected the misunderstanding that he was in favour of *laissez-faire* or free government in the sense of relaxed restraints on private activities. "There is", he said:

a persistent misunderstanding [by] my opponents. Everywhere, along with the reprobation of government intrusion into various spheres where private activities should be left to themselves, I have contended that, in its special sphere, the maintenance of equitable relations among citizens governmental action should be extended and elaborated.[55]

Spencer's refusal to lend his support to free enterprise at the expense of governmental activity upset the business community. Further, another of his American gaffes was even more disturbing. This was his stinginess with national compliments, something considered inexplicable by the new "Empire Republic", which had been accustomed to gratitude from those whom it had honoured. Lionized visitors usually gave lengthy speeches praising the grandeur of the country and the wealth and industry of its inhabitants. Spencer's silence on such matters caused his disciples to remonstrate with him. He finally agreed to speak in New York shortly before his departure. A gala farewell banquet was arranged at Delmonico's for 200 people. The dinner was elegant, and the decoration was in quiet good taste. No luxury was omitted; there was even a band.[56] Expectations were high when Spencer rose to his feet, but fell immediately because he delivered his speech in a low conversational tone without using gestures. The content was even more disappointing than the laconic style. Instead of extolling the prosperity and energy of his listeners Spencer told them that their lives were based on false goals. "The truth is, there needs [to be] a revised idea of life."[57] He extended his sermon with the thought that while modern human beings had replaced the ideal of fighting with that of work, this substitution would also fail. When the modern ideal had succeeded in conquering the earth and in subduing the powers of nature, the work ethos would disappear. He noted that J. S. Mill had been wrong when, in his speech as Rector of St Andrew's University, he had tacitly assumed that life was for learning and working. Spencer reversed Mill's phrase, and said that learning and working were for life:

Hereafter, when this age of active material progress has yielded mankind its benefits there will, I think, come a better adjustment of labour and enjoyment. Among reasons for thinking this, there is the reason that the progress of evolution throughout the organic world at large, brings an increasing surplus of energies that are not absorbed in fulfilling material needs ...[58]

Spencer told the assembled industrial and mercantile wealth of New York that human beings had had too much of "the gospel of work"; it was now time to preach the gospel of relaxation.[59] This was an offensive message at the time. It seemed unfamiliar to his audience, but it could have reminded them of the critical and oracular comments on the effects of hard work and industry that their countryman Henry D. Thoreau had prefixed to his *Walden*. This connection would have been unlikely because they had ignored Thoreau; Spencer, however, had their full attention. The significant difference between him and Thoreau was that he was famous and, therefore, could not be ignored. The elite of the city had gathered to hear reassuring words about progress and the shape of human evolution. They wanted an optimistic message about the future of the great industrial nation of the New World, comparing it to the decline of the old and aristocratically dominated Europe. Instead, they heard Spencer say that they themselves had not reached the final stage of social progress, and were forced to listen to his warning that they should not overwork but seek amusements and holidays.[60] He confessed to them that he had overworked, as had many of his friends, but this had been an imprudent expenditure of effort. The speech was a shock. Even though Spencer was more esteemed for his philosophy of altruism or benevolence than for the social Darwinism that the Yale sociologist W. G. Sumner had claimed for him, the evening was a disaster in terms of public relations. There was even a suggestion that the speech might be left unprinted. However, Spencer's remarks were not meant to shock; they were a mere extension of his general philosophy. Since he was addressing Americans, who he had mistakenly assumed liked to hear the truth, he had spoken more plainly than usual. His pithy comments represented no real departure for him; on the contrary, they were typical enough to be placed among his commonplaces, which one of his English followers put into a book of aphorisms: "There needs to be a revised ideal of life .... Life is not for learning, nor is life for working, but learning and working are for life".[61]

His New York outburst was not eccentric in the sense of being an involuntary and inexplicable utterance. Rather, Spencer was clearly enunciating his views on the importance of making people work less and enjoy themselves more. Such sentiments do not belong in the same domain as his preference for lying in a hammock while travelling on a train or his practice of attaching his current manuscript to his waist with a cord. While these latter practices are idiosyncratic, or "original" as his contemporaries might have said, his speech in New York was not, although it offended a greater number of people. His general views on the need for people to avoid excessive workloads were based on his objection to the industrialization of work and the excessive concentration on monotonous activities. His beliefs in play and relaxation were parts of his social philosophy: it was part of freedom to feel pleasure even at the expense of the industrial future.

Spencer's American speech was not an exceptional outburst detached from his philosophical beliefs. It was the expression of a belief supported by his intellectual ideals and his personal life. For him, overwork and ill health were causally linked, and he was gravely concerned that members of his family and other contemporaries had confused work with life. He believed that his uncle Thomas, seeing no alternatives other than work or *ennui*, constantly transgressed the laws of health, which had killed him.[62] His father, George, had worked himself into a breakdown.[63] He himself had advocated the gospel of work until his own nervous collapse in the mid-1850s. Significantly, he had phrased his early advocacy of the duty of labour in familial terms. His juvenile views had idealized parental duty when he had written about the role work had in improving civilization. The mother would practise foresight and economy, while the father's life was to be one of "laborious days and constant self denial". Only necessity, Spencer continued, could force man to submit to this discipline, and nothing but this discipline could produce continuous progress.[64] The necessity to which Spencer referred was Malthus's prediction that the increase of human population would always be greater than that of food production, and that this in turn would lead to improvements in intelligence and morals as people struggled to find the means to improve their efficiency and sociability. Later, in his mature views, he saw his early insistence on labour and self-denial as inhumane and self-defeating.

Herbert Spencer's message of the need for freedom from work was designed to save children from harsh task-masters like his father and uncle, but it did not apply exclusively to the young. Adults needed freedom too. In their case the threat was not a tyrannical parent, but life-threatening devotion to work. Spencer believed that in the process of industrialization, modern society had overemphasized the value of duty and labour. Human beings had come to be regarded as more praiseworthy the harder they toiled. To Spencer, this attitude was a mere superstition that ignored scientific truth. In accordance with that truth, he coined a slogan of freedom that proclaimed that "Life is not for work, but work is for life".[65] This was based on the evolutionary edict that the lower a life form, the more its energies were wholly absorbed in sustaining itself, or sustaining its procreative faculties.[66] Higher forms were freer from these simple life-sustaining demands. That is, they were relatively free from work. This was not only true about biological change, but of industrial progress. Each improvement in organization makes life easier. With evolutionary advance there is a surplus of energy. Spencer believed that this advance allowed all people to play. Human life would be liberated from mere toil, and become relaxed and pleasurable. It could then progress to a level at which an "aesthetic gratification" could take place.[67] Spencer thought that this highest source of pleasure was often ignored by his contemporaries, who were too engrossed in their work to see or to experience beauty.

There was a stridency to Spencer's message. This was caused by his occasional failures to obey his own injunction against work. His own attempts to reduce the amount of labour failed. "A System of Synthetic Philosophy" had ballooned into more volumes than his plan had envisaged. He constantly felt enfeebled by the magnitude of the research and writing that still lay ahead of him. However, even before he was trapped by his philosophical system, he felt overburdened. When he was still in his thirties he believed that he had been too focused on "mental absorption"; this had produced in him a sense of alienation that had not been present when he had been engaged in more varied work as an engineer.[68] Spencer blamed all his somatic ailments on work, although in doing this he was not making a distinction between ordinary disease and mental ill health. For him there was no distinction between real and imaginary illness; nor did they have different causes. His "cardiac enfeeblement" was caused by "an unhealthy, indoor, hard-worked, and often anxious" life. He continually cried that if only he had kept to a life of ease, novelty and amusement, and if only he had spent more time in the open air, then his life would have remained sustaining.[69] Like his heart condition, he believed that his insomnia had been caused by overwork since the mid-1850s. As has already been noted, he thought that he could only cure himself by recourse to hydropathic techniques, which have excited ridicule in the twentieth-century.[70]

Since Spencer was what later periods would call a hypochondriac it is, of course, difficult to distinguish in his account, and in those of contemporaries, the real from the imaginary. However, a true account of his health might run as follows. In 1856, at the age of 36 he suffered a physical "breakdown", while reading the works of Alexander Campbell Fraser, a Scottish psychologist and philosopher.[71] Spencer had agreed to write a testimonial on Fraser's behalf for a professorship at the University of Edinburgh.[72] The first edition of Spencer's own *The Principles of Psychology* had just been published and, like Fraser, he was deeply influenced by the work of Sir William Hamilton, the well-known Edinburgh scholar. Spencer was probably flattered by the invitation to write on Fraser's behalf, because, in his metaphysics and psychology, he owed much to the Scottish "common-sense" school of philosophy. Also, he would have taken the request as a sign that he was being offered the recognition by the establishment and the status that he craved for his written work even though he often attempted to distance himself from this establishment when it was exercising political power. Thus he loved consorting with Fellows of the Royal Society when he would never have consented to be an advocate for empire after the fashion of some of the more bellicose Fellows such as John Lubbock and Roderick Murchison. In the event, the extra reading imposed by the Fraser testimonial proved too much.[73] For the next quarter of a century, even though his general health, together with his appetite, digestion and strength seemed to have been at least average, he seldom achieved more than four or five hours sleep, and even that was in small segments seldom as long as two hours.[74] The time period mentioned, a "quarter of a century", took Spencer's life to about 1880. After that time he cured his insomnia with opium. This practice may have improved his sleep but, according to Beatrice Webb, who was in frequent communication with him at that time, drugs undermined his health. Writing in 1903, just after Spencer's death, Webb blamed what ill health he genuinely suffered over his last two decades on morphia and self-absorption.[75] However, she was ambivalent about how seriously to take Spencer's ill health during old age because, while he was vociferously angry about his feebleness, she thought he had above average vitality for a man in his seventies, and noted that as late as 1896 he was eating well, going to his club and working.[76]

Webb, like other members of the Potter family, had always stood by Spencer when he needed to be coaxed or teased back into good humour. In April 1856 Laurencina Potter put him up in a bedroom in which she claimed her brother had seen a ghost. As Spencer recognized at the time, she was relying on his scepticism and perversity to keep him amused. Her plan succeeded: he was intrigued into good spirits.[77] Spencer slept his normal short periods in the haunted room while using the wakeful interludes to wonder about the origins of superstition.[78] Such light and humorous methods of relieving Spencer's complaints only succeeded before the onset of old age. That is, when he was young his complaints about his health were valetudinarian pleas for relief from the self-appointed tasks of organizing all psychological, social and biological knowledge. He could rescue himself, or be rescued by others, from self-induced illness. Later, in old age, Spencer's illness was often real. During his last two decades his complaints were edged with the decrepitude of old age.

To later generations Spencer's health, his cures, his speeches about overwork, his obsession about minutiae and his novel ways of amusing himself have provided a source of entertainment. These accounts have entered the public domain as features of an amusing archetypical eccentric, and they should remain there. It is not the purpose of this book to remove these simple pleasures from the general reader; there is no more harmless a pastime than to summon up visions of those among the dead who were amiable and mildly idiosyncratic. Stories about Spencer were never completely concealed, and the ones he told about himself were no less remarkable than those told by others. He was, as Webb once remarked when she let her filial mask drop for a moment, thoroughly self-absorbed. However, Spencer's self-absorption was not self-indulgent or merely curious. He was a man who constantly examined his body and his mind as natural objects that were subject to natural processes. He examined his own ageing as a series of biological changes. For example, he saw his attempts to find pleasure in art and in nature as increasingly limited by his elderly body. This was why he noted his eventual loss of the sense of smell. By the 1890s he had reached the stage where he could no longer enjoy the beauty of spring, and sadly announced that one more "pleasurable sensation" had gone.[79] His scientific gaze made him hyper-aware that his morals, work habits and thoughts were only partly under his control. Many of his elaborate complaints about his condition, many of his jokes and many of his fierce defences of his privacy from curiosity seekers and journalists were attempts to create a small sphere of individuality around himself. This task seemed harder in a scientific age. He saw himself as a piece of "natural history", but he was also a self-created individual freeing himself from the routine of work and the tyranny of customary practices, so this task had to be achieved in such a way that he would not fall victim to the perils of industrial modernity.

The readers of his philosophical system would gain this salvation without bearing the costs of isolation and toil that had troubled its author. In a late prophetic utterance Spencer reminded the world:

There will remain a need for qualifying that too prosaic and material form of life which tends to result from absorption in daily work, and there will ever be a sphere for those who are able to impress their hearers with a due sense of the Mystery in which the origin and meaning of the Universe are shrouded. It may be anticipated, too, that musical expression to the sentiment accompanying this sense will not only survive but undergo further development.[80]

This thought had been prompted while reflecting on the future of ecclesiastical institutions; it made Spencer aware of his own ambiguity as a kind of agnostic priest of the "Unknown". At the end of his life he could only counsel his hearers to avoid excessive work, and to seek refuge in music.

II

The lost world of Spencer's metaphysics

SEVEN

The New Reformation

The gestation period for Spencer's philosophical system was the 1850s. His lifelong adhesion to the psychological and metaphysical doctrines that he shared with religious and scientifically minded contemporaries in the mid-century dictated the content of much of his serious writings for the remainder of his life. At times, his early attempt at coherence became an embarrassment to Spencer because it committed him to a position that he later distrusted. At other times his philosophical system wearied him by imposing endless burdens as he struggled through his last four decades with preordained tasks stretching endlessly ahead. However, there was one consolation that buoyed him up during his toil. This was a residual gift from the 1850s. His philosophical and scientific labours were never dry and empty; they were always a reservoir of religious meaning. To most of his readers, and to himself, his endeavours were always pregnant with spiritual vitality. Spencer's worship of the "Unknown" provided solace to those who feared that the universe was only a collection of lifeless material objects and physical laws.[1]

Even before he began writing his philosophical system, Spencer was keen to tell his readers that they should search for answers that transcended experience. Without such transcendence they would be atheists: people for whom the universe was empty. The choice was philosophically stark: one could follow J. S. Mill and rely on experience without knowledge of a first cause: alternatively, one could uphold the intuitionism of William Hamilton, which would also mean that one truly possessed no knowledge. The second choice was slightly less bleak because a person might feel meaningful responses that indicated there was something out there even if this could not be known. This was Spencer's solution; he believed that the responses he felt were significant, and on this basis chose to construct his own philosophy.

Spencer saw philosophy and religion as a coherent whole. His statements from the period linked his philosophy, which was based on intuition and which transcended experience, to the worship of the Unknown. His defence of his *The Principles of Psychology* in 1856 testified that:

I hold, in common with most men who have studied the matter to the bottom, that the existence of a Deity can neither be proved nor disproved. In the "Summary and Conclusion", which I was obliged to leave unwritten, I purposed showing that one of the corollaries deducible from the work as a whole, is, that the only things – subjective or objective – that are really cognisable by the human intellect, are *relations*; and that the things between which the relations subsist – in the one case sensations, and in the other the things which produce the sensations – transcend all analysis and can never be understood; in other words, that both in the external and internal worlds, science finally brings us down to a mystery that must for ever remain insoluble. This doctrine is in complete harmony with that held by the greatest of living orthodox metaphysicians – Sir William Hamilton; who, in his "Philosophy of the Unconditioned", shows that a knowledge of the absolute is impossible to man. Though it is an unavoidable corollary from this, that man can never know anything about the ultimate cause of things, yet Sir William Hamilton is not therefore charged with Atheism.[2]

My available leisure is much taken up; and I have presently been able to read but little of the Secularism. The only objection which I feel at present inclined to [make], is to the name. The term "Secularism" seems to me open to the criticism that it in a manner positively excludes the recognition of anything that transcends experience.[3]

In fact Spencer was not acquainted with atheistic literature. He could write truthfully to George Holyoake, the secularist, that he had read little of his beliefs, but that they seemed too definite in excluding a feeling of transcendence.

Spencer, who relished the idea that he was original, obscured the composite nature of his philosophical framework, which he had gleaned from many minds. Rather than being seen as a system, it should be primarily understood as a collection of texts that responded to, and inspired, the movement of mid-Victorian intellectuals that called itself the New Reformation or the spiritualists. Although Spencer's philosophy grew beyond its beginnings and acquired its own momentum, it never completely lost its rationale of creating a new morality and metaphysics with which to replace both orthodox Christianity and materialistic positivism. He saw himself as "the sincere man of science" defending true religion against sectarians.[4] Only a fearless investigator such as himself could find meaning in the perpetual changes in the internal world of the psyche and the external world of physical phenomena. Both worlds were terrifyingly protean, but the former alarmed Spencer the most. When looking inwards he perceived "that both terminations of the thread of consciousness are beyond his grasp: he cannot remember when or how consciousness commenced, and he cannot examine the consciousness that at any moment exists".[5] It says much about the quality of Spencer's optimism that when he lacked knowledge about both his mind and the physical universe he could still believe that the contemplation of the unknowable was intrinsically valuable.

Spencer's philosophical roots have been overlooked because he concealed his connection with the other radical intellectuals who wrote the briefly fashionable journal the *Leader*. In *An Autobiography* the only comment he made on this publication was the statement that he had not acknowledged authorship of the Haythorne papers that he had written for the *Leader* because he did not want to be identified with its socialistic views.[6] To some extent this claim reflects the anti-socialist and individualist bias of a later period when Spencer was writing *The Man "versus" the State* (1884).[7] It does not ring true for the 1850s, when the socialism advocated by Spencer's friends was not state socialism but voluntary reform that was disconnected from a strong theory of the state. Spencer's explanation of his desire for anonymity was disingenuous. Economic and political issues were secondary, both for his Haythorne papers and for the *Leader*. Even if they had not been, his reputation as the author of *Social Statics* and an exponent of land nationalization made him one of the more radical economic writers in the London of the 1850s. Finally, Spencer's anonymity was not a secret. He republished the Haythorne papers in his own name in 1857, with acknowledgements to the *Leader*, when that journal was still in existence.[8] Spencer's ambivalence about the goals of the *Leader* was a response to the journal's variable editorial policy. On the one hand the newspaper reflected Thornton Hunt's continuing Chartist bias and was, therefore, an organ of extreme radicalism. On the other hand, its financial backer, the Lincolnshire clergyman, Edmund Larkin, wanted an organ to promulgate "a larger Christian liberalism than then existed".[9] Spencer's sympathies lay closer to the goals of the latter; while he had been a fellow traveller of Chartists, his heart was with the goal of liberalizing religion. He was never enthusiastic about the merits of democracy, whereas liberalism in all its early manifestations was attractive to him.

Spencer's connections with the *Leader* and its editors were close. This ensured that his books and even his articles received favourable notice throughout the 1850s. Spencer was one of their authors, and no occasion was missed to give him a "puff". Even in 1851, when he was comparatively unknown, he received eleven mentions in the *Leader* comparing him favourably with established thinkers such as J. S. Mill, Auguste Comte, and Pierre-Joseph Proudhon.[10] He also appeared, completely irrelevantly, in discussions about literature and children's stories as "our admirable friend Herbert Spencer".[11] Of course, Spencer also curried favour with the *Leader* and its editors. In July 1853 he sent Lewes the proofs of his article on "The Universal Postulate"; Lewes read it with "immense interest", and thought Spencer had made "an irresistible case against Hamilton, Hume, Kant and Co."[12] When the article was published, the *Leader* duly praised its "anonymous" author in grossly flattering terms as:

> one of the profoundest and clearest metaphysical essays we have read. In it Common Sense is reconciled with Philosophy: a scientific basis is given for our Universal conceptions. We hope to discuss this paper more at length on a future occasion; meanwhile we call attention to it as a great contribution to philosophy.[13]

In the following month, the *Leader* devoted several columns to expounding this newest contribution to the advance of philosophy.[14]

An examination of Spencer's Haythorne papers shows that they were very close in style and content to the usual *Leader* fare. For example, in June 1853, when Spencer wrote on the value of evidence, he began by mentioning that gullible people believe in spirit-rapping, table-movings and the spontaneous combustion of human beings.[15] In this Spencer was simply recycling examples of credulousness used by G. H. Lewes and others in the *Leader*. The example of spontaneous combustion referred to Lewes's semi-serious critique of Charles Dickens's description of the phenomenon in *Bleak House*.

The development of Spencer's metaphysics – from the publication of "The Universal Postulate" in October 1853 to the end of the decade – was inextricably tied to the *Leader* and its readers. Spencer's statement that the ultimate psychological fact was "belief" attracted the attention of the *Leader*, which was attempting to find a foundation for its new religion. It was Spencer's reliance on belief as a psychological fact rather than "classical" accounts of feeling by eighteenth-century philosophers such as Hume that set the stage for the Victorian synthesis between the mind and the natural universe. Science was always present and intermingled with the desire of the human spirit.

Every title page of the *Leader* was dedicated with the following quotation drawn from Alexander von Humboldt's *Kosmos*:

> The one idea which history exhibits as evermore developing itself into greater distinctiveness is the idea of humanity – the noble endeavour to throw down all barriers erected between men by prejudice and one-sided views; and by setting aside the distinctions of religion, country and colour, to treat the whole human race as one brotherhood, having one great object – the free development of our spiritual nature.[16]

The founders of the *Leader* seized on this inchoate idea of the development of the spiritual nature of man as a weapon to use against their fellow radicals who still believed in a materialistic and Godless universe. Humboldt was an ideal source of spiritual inspiration. It was not simply his great popularity – *Kosmos* received three contemporary translations into English – but it was ambiguous in a useful way. Spencer and his friends already possessed an abundance of religious and philosophical ideas in their own language, and had no need to borrow these. They simply needed time in which to extend their ideas, and it suited them to use a short opaque quotation from Humboldt that did not commit them to a developed philosophy or religion. What they found empathetic was the sense of a quest, the search for unknown truth.

The first issue of the *Leader* contained the first chapter of an intellectual novel, *The Apprenticeship of Life*, by Lewes, which artfully pointed to the object of the new journal. The hero, a young sceptic named Armand de Fayol, had been raised on a diet of the Encyclopaedists:

teachers who dethroned God to put a phrase in his place; thinkers to whom the universe was no mystery at all, as everything could be explained by "Matter and Motion"; men who, disregarding the instincts of their souls, declared religion to be a *fraud* – the *invention* of crafty priests supported only by the terrors and prejudices of the credulous, not the spontaneous product of the human soul – the instinct imperiously moving the whole being of man.[17]

Armand was troubled; he manifested an irresistible craving for belief. Unfortunately, this craving was not enough to keep Armand alive, and, after a few more chapters, the young hero vanished.[18]

The *Leader*'s opinions and views evolved and matured over the first four or five years of its existence. During 1850, the two important developments were the recognition of F. W. Newman's two works, *The Soul* and *Phases of Faith*, and a series of papers entitled *Social Reform*, which were mostly written by Hunt and Lewes. These emphasized the pre-eminence of religion in social reform. The elevation of religious radicalism above politics and economics filled a vacuum and signalled a generational shift. Older radicals had usually avoided religion as a subject likely to cause division and hinder social progress. For example, J. S. Mill refrained from publishing his religious writings of this period.[19] It should also be observed that, despite their name, the Christian Socialists avoided religious discussion.[20]

The *Leader*'s labels that designated its unique combination of religion and science were "spiritualism" and the "New Reformation".[21] Its writers believed that "infidels" and "Atheists" were dated, and that they should aim their messages at the large and varied class of "spiritualists", who currently constituted a large proportion of the educated community.[22] To these: "religion is nothing, or it is a primary truth essential to man's nature – inborn, and the more fully developed in proportion as he is awakened to a knowledge of his own nature and a sense of his own relation to the universe".[23] This dependence of religious feelings on intuitive knowledge produced the guiding doctrine of Spencer's early psychology. This doctrine was not based, as pantheism was, on feelings about the universe or the "material" world. That would have left the mind as a mere passive recorder of external events. This new Victorian stance gave the mind an active role in interpreting the world. Further, it kindly suggested that there might be something in the universe that would satisfy the yearnings of the soul.

The author who most completely captured the spirit of this age was F. W. Newman, whose *Soul* and *Phases of Faith* were required reading for any self-conscious radical of the mid-century. It was a sign of the uniqueness of this period that Newman's progressive attitudes were restricted to religion – he eschewed large-scale national reforms such as universal suffrage and socialism[24] – but this was precisely the intellectual arena that was most exhilarating to Spencer and his friends.

Newman's rejection of pantheism was greatly admired as the work of "a spiritualist". His distinction between a pantheist and a spiritualist was that the latter sensed that the Infinite *did actually exist in the mind* – as an original and universal conception that all languages tried to express. With the former, the sense of the

Infinite was external; it related to something outside the mind. This distinction mattered because it allowed Newman, Spencer and Lewes all to discover a basis of morality that both drew support from the Infinite and existed within the individual. "Our own inherent spiritual instinct must be here a better guide than either dogmatic ethics or personal example: the spiritual, here as elsewhere, transcends the moral life."[25] These spiritualist writings followed Newman, who believed that the concept of immortality should not affect one's morality: "The idea of immortality, it is argued, is not *necessary* to stimulate our virtuous exertion, and that, if we rely on prudential motives of rewards and punishments, we ignore the power of conscience and the love of virtue for its own sake".[26] The argument that it was deeply corrupting to rely on prudential motives of rewards and punishments was not specifically directed at secular utility; that would have been seen by Newman as a relation of the more widely propagated Christian utility. To him William Paley was a greater enemy than Jeremy Bentham had been. His conflation of the two kinds of utility was followed by Spencer and others of their group. Of course, mid-century spiritualism was not solely an English moral theory with local targets; it echoed discourses taking place on the Continent. To an extent Newman was the same as D. F. Strauss in his *Soliloquies*, except that the former did not deny the future life, just a logical proof of it. Of course, since Newman depended entirely on man's spiritual instincts, rather than on logic, he could feel quite safe in his unorthodoxy on this point.

The aim of this was to condemn the orthodox Christians – many of whom had utilitarian leanings – and altruist utilitarians as equally immoral because of their reliance on rewards and punishments. The theological implications of this were unclear; Strauss was rebuked for denying the possibility of a future life. Spiritualism, like the agnosticism later developed by T. H. Huxley and Leslie Stephen, suggested that not only should everything be treated as possible, but that ontological matters should be treated with due solemnity. This ostentatious British display of tact contrasted with those continental writers such as Strauss, who had permitted themselves heavy sarcasms at the expense of the orthodox. Others, such as Renan, were exaggeratedly kind to their former fraternity. In the New Reformation neither hostility nor kindness would undermine the earnestness of the speculation that underpinned the harmony between morality and the spiritual instincts, and provided a psychological and scientific base for morality.

While Newman's *The Soul* was admired, his *Phases of Faith* was adored by the *Leader* group:

> No work in our experience has yet been published so capable of grasping the mind of the reader, and carrying him through the tortuous labyrinth of religious controversy; no work so energetically clearing the subject of all its ambiguities and sophistications; no work so capable of making a path for the New Reformation to tread securely on … Modern spiritualism has reason to be deeply grateful to Mr Newman…[27]

Phases of Faith was read by spiritualists as an emancipating work. It broke free from the liberal Anglican theology of Dr Arnold. The headmaster of Rugby had offended the young and credulous Newman by dismissing Noah's deluge as a myth and by reducing the story of Joseph to a beautiful poem. Newman had initially been staggered by what he saw as cynicism, but the long-term effect on him was to accept Arnold's statement that these stories did not matter to religion because our beliefs should have broader and deeper foundations than could be found in biblical statements on cosmology and physiology. However, Newman did not take traditional beliefs or myths as indifferent to the truth of Christianity. To do that would be to repeat Arnold's mistake and cause religion to be raised up against Christianity. Spiritualists avoided such disputes and claimed that, as a *system*, Christianity was dependent on its scriptural testimonies, but as a *sentiment*, or *doctrine*, it was not. In the latter sphere it had to be made acceptable to the modern mind.[28]

Newman's book became the personal testament for the New Reformation. The *Leader*, which had given it five reviews and eight mentions, certainly never devoted so much attention to a single work again. The most any author could usually hope for was two or three mentions. Newman's work served to launch the *Leader* into the propagation of its own religion and philosophy.[29] It was Newman who pioneered the key concepts that Spencer used later in his *First Principles*. Even the title was from Newman. When the *Leader* noticed the publication of *Soul* it had – according to the *Leader* – vindicated the claim of reason to be the arbiter of its own creed, witnessed to *First Principles* and refused to be subjugated by authority.[30] It is by the light of this three-point creed that Spencer's system of philosophy should be read. His *First Principles* originated in this anti-authoritarian attempt to coordinate religious enquiry with philosophical speculation.

Spencer's friends Hunt and Lewes had begun this search for formulations in August 1850 in a series of open letters called *Social Reform*. Hunt addressed the first letter to an anonymous friend who had reproached him with hostility towards the received order of society, and asked him to withhold his dissent from religion. Hunt's answer was that such advice had clogged religion with falsehood, and that religious dogma had hindered reform in every political and social area. It could not be avoided; religion was a word of fear and discord, only somewhat mitigated by indifference. In seeking remedies to the chronic disorders of society, Hunt put the reform of religion first, that is, he gave it priority over the reform of land tenure, labour, capital, taxation, poverty and even ignorance.[31]

The next letter on social reform was an open letter from Lewes to Masson.[32] "I have asked Thornton Hunt to postpone the subject of religion in this series until next week, that I might, if possible, broaden the basis by a rapid indication of the paramount necessity of including Religion in all Social Reform".[33] Lewes thereupon plunged into the subject in a bluff manner reminiscent of Hunt, beginning with a trumpet blast: "Religion is the heart of politics".[34] Religion was to be found everywhere, in politics, morals, art and education. Lewes also demanded that the Church reform itself to correspond to the religious truth as it was seen by the intellectual *leaders* of society. Churches were a necessity, but they were only acceptable when they had been reformed.

By religion, then, I do not mean the Church, for I believe the Church itself is in need of reforms as radical as any other portion of that remnant of feudalism we call the British Constitution. In *The Leader* more than one call has been made for the New Reformation, – or Church of the Future. Let no one idly deem it the vision of a few enthusiasts. It has become the practical, though often the unconscious aim of energetic thinkers, who see plainly enough the truth of what Machiavelli long ago declared, "I will never believe in a change of government until I see a change of religion"; men who see that if the ardent aspirations after Social Reform, which now so profoundly move leading men, are ever to become realised in an enduring form, they must be based upon a *Faith shared in common*, a conviction binding men together, not a Creed officially thrust upon them.

The New Reformation will start from a fuller development of Luther's great principle. He founded Protestantism on the liberty of private judgment: this liberty has scattered religion into sects. Its weakness lies in its restrictions; it is not *absolute* freedom, as persecution clearly shows. The New Reformation must make that liberty absolute, giving to every soul the sacred privilege of its *own* convictions, and by the illimitability of freedom in opinion making the *unity of sentiment* all the stronger.

Inasmuch, therefore, as the *religious sentiment* in man is universal, enduring, and his *religious opinions, or theories*, are *necessarily* wavering and changeable (a twofold demonstration afforded by all history), the Church of the Future should endeavour to found itself on what man has in common (sentiment), admitting all possible varieties – or heresies – in matters of opinion.[35]

The third open letter, by Hunt, concerned religion and anti-utilitarian ethics:

[T]here is no part of political or social activity in which we do not detect the want of a true religious influence as the motive to cooperate in advancement. The want of such an influence I believe is one main reason why the motive to activity has degenerated to the single imperfect and falsely-working motive of self-interest. Amongst some of our "practical" philosophers the predominance of that motive is boasted as a great sign of peculiar enlightenment in our day.[36]

Hunt was uncertain of how to express his dissatisfaction with the lack of true religious influence among utilitarian reformers.[37] He lacked the facility in philosophy possessed by Lewes and Spencer and his statement of belief was full of outdated and vague terms: "I believe that all forms of faith whatsoever have had a common origin, and have in them a common principle of truth".[38] From his perspective the laws of nature and God were obvious to "all human children of God", and, therefore, some credible interpretation of them would be found.[39]

Hunt elaborated his statement of belief as a creed. This would not have mattered except that it heralded the religion of the "Unknown" that was later developed by

Spencer in his *First Principles*. Hunt began by mentioning that the new spirit of religion had much support among younger members of the Church of England and among the dissenter sects. He then quickly sketched in the changes in radical religious thought in the first half of the nineteenth century.

> A still more remarkable change is taking place before our eyes among the scattered class who may be said to represent the free-thinkers of the last two or three generations. Their demeanour is characterised generally by two traits new and most important in their consequences. In place of speaking in that veiled language, which was usual among all but the vulgar and audacious, they now speak in direct and open terms. Heterodoxy, to use its antiquated name, no longer resorts to the bye-way of wit for its freest issues, but can speak in the simple language of common life. It does so without fear and with impunity. The other trait is, that in place of the blank scepticism which was prevalent among the intellectual classes, there is now a disposition to reunite under the common influence of the universal religion. For signs of these several approximations I might point you to the admirable books of Francis Newman on *The Soul* and the *Phases of Faith*; of Foxton, on the tendency of society towards a new kind of *popular Christianity*; of Leigh Hunt, my most beloved father and friend, who makes an explicit declaration of faith at the same time that he points to that "great revelation of the universe" of which Humboldt is the great prophet in our day; to the generous book of the Episcopalian Bushnell, on *God in Christ*; could I proclaim their authorship I might point to the admirable anonymous letters on the subject of Religion in the Open Council of my own paper. In short, the disposition to appeal from dogma to the universal conscience and faith of mankind is to be found in every class of society, and every class of public discussion.[40]

Thornton Hunt called for a universal religion, based on the ever-present religious impulse that he saw as a fixed trait of human nature. Universal religion was not to be confused with pantheism, which was an inferior and rudimentary impulse that had moved too hastily from conceiving the unity of the universe to considering that unity was God, and that this was All. Instead, Hunt again made an obeisance to the "Unknown":

> My slower mind will not move so fast. I still hear the voice of the instinctive revelation, which tells me that there is a great region of the unknown, greater than the known, but not dissevered from it, since I imagine its existence.
> My instincts tell me that there is a great influence ruling over the known and the unknown; men have called it God, and imagine for it every sort of attribute; I cannot follow its vastness, ... I worship it in the beauty and goodness that exist – so my instinct tells me – by its influence.[41]

119

The subject of religion was also taken up by Hunt's friend, R. H. Horne. Like Leigh Hunt, Horne had been editor of the *Monthly Repository* in the 1830s.[42] Horne had the same formulation of religion as the Hunts and, like them, floundered about using ill-defined psychological terms. Horne believed that religion was not a question of this or that faculty of the mind or the instincts, but of all the faculties: a sincere interpretation of the whole being. He thought that it was the free examination and public declaration of feeling and thoughts that would eventually work out "the Redemption of religion".[43] This was scarcely adequate as an account of contemporary psychology because Horne blurred terms such as faculty and instinct, which had distinct meanings. His only guideline was a vague liberal theory about the progress of the intellect, and his views here were suggestive rather than complete. It became Spencer's mission to clarify these confused ideas and give scientific guidance to the spirit in its search for a new religious faith.

Clarifying the vague ideas of Romantics such as Leigh Hunt and Horne was not the only task for spiritualists. They had a modern and scientific rival in the English Comteanism. The Comteans' most significant work in the mid-century was a book by H. G. Atkinson and Harriet Martineau, *Letters on the Laws of Man's Nature and Development*. The adherents of the New Reformation were alarmed by its open avowal of atheism[44] and renewed their attempts to define spiritualism by purloining terminology from Scottish philosophers such as Thomas Reid and Hamilton and combining this with the concept of the soul from Newman. The battle cry was: "*The soul is larger than Logic*".[45] This phrase signified that religious experience was outside the laws of ordinary thought and would always remain unknown.

> There are many things which we can truly be said to know, which, nevertheless, we can neither define nor prove. There is, so to speak, a logic of ideas ... a faculty which may be called altogether transcendent, the province of this faculty being precisely those ideas which the understanding or common logic of man has failed to grasp. Kant is the last great systematic psychologist who set this notion clearly forth. We are not Kantists [*sic*], but detect in his system the indistinct expression of that consciousness of a transcendent faculty we feel within ourselves, and which we see so powerfully operating on man.[46]

The spiritualists had no scholastic knowledge of Anselm and his proof of the existence of God. For them all statements about finite knowledge of the infinite were quite fresh, perhaps shared only with Kant. It seemed a novelty when they proclaimed that the existence of God could not be proved because the infinite must remain incomprehensible to the finite. This, of course, raised the question of whether a philosophic justification for a belief could transcend logic, but it was thought to leave the spiritualists no worse off than those orthodox who believed that the universe showed evidence of design, or the Comteans who believed that it was also controlled by law. Such beliefs were substitutes for true spiritualism, which did not pretend that God could ever be known. This almost foreshadowed Huxley's "invention", agnosticism, but avoids this outcome by abandoning logic for psychology. That is, for spiritualists, such as Spencer, important beliefs were not verifiable through the understanding but through instincts. In order to prove the existence of God and of the external world, one had only to appeal to what they called "the irresistible evidence of our instincts", which was never wrong.

Spiritualists were not only critical of English Comteans, but of Comte himself.[47] Significantly, the criticism was centred on religion. Lewes, who had been a Comtean, had distanced himself from the great Frenchman.[48] He now believed that while Comte's intellectual life had made an admirable exponent of scientific principles, his narrow perspective marred him "for that intense and enlarged conception of our moral or emotional life, with which Religion and Morality are inseparably connected".[49] Comte's positive philosophy failed to observe that the intellect was subordinate to the heart, and science was subservient to morality. The truth is:

> that man is moved by emotions, not by his ideas; he uses his Intellect only as an eye to *see the way*. In other words, the Intellect is the servant, not the lord of the Heart; and Science is a futile, frivolous pursuit, unworthy of greater respect than a game of chess, unless it subserve some grand religious aim, unless its issue be in some enlarged conception of man's life and destiny.[50]

It was this religion of the heart that underpinned Spencer's distrust of logic both in his psychology and in his own life. He always claimed that feelings were preferable to the intellect as guides to moral and practical decision-making. This reaction to Comtean views of science always grouped Spencer with his early friends, who had reacted against those nineteenth-century rationalists who were continuing the work of the Enlightenment.[51]

The parallel between the New Reformation and Scottish philosophy was deepened by Lewes's eighth article on Comte's positive philosophy, "Astronomy and Religion".[52] Lewes, in particular, was in an ambiguous position regarding Comte. On one hand, part of his own reputation as a philosopher rested on his extolment of Comte as the new Bacon. On the other hand, he could not conceal his growing rejection of Comte's philosophy. On 1 October 1853 he reviewed his own book *Comte's Philosophy of the Sciences*, stating that Comte needed popularizing. On 29 October 1853 Lewes reviewed the third volume of Comte's *Système de politique positive* by remarking that the preface must cause pain to his sincere friends since it was succeeded by two circulars, one addressed to the Emperor Nickolas, the other to Reschid Pasha: "both the naïve productions of a man who, living in hermit-like retirement, occupied in revolving his own thoughts, has lost the sense of ordinary affairs". Lewes was also critical of the fourth and last volume of this work. He suggested that Comte's view of the future was worthless.[53] This accusation was the harshest the New Reformation could make.

The moral earnestness[54] of the New Reformation elevated the "Heart"[55] above the "Intellect" in order to maximize spiritualist ethics. The emphasis here was on the moral, rather than the intellectual, progress of humanity. This split Victorian progressives into two camps.[56] Lewes and Spencer denied that there was progress through intellectual improvements, while friends of J. S. Mill such as Buckle denied that there was an evolution of morals, upholding only intellectual progress. This debate continued for some time. In the late-nineteenth century, the Spencerian philosopher John Fiske was still castigating Buckle for denying that there had been moral progress. Spiritualists, when rejecting the intellect as the basis of religion, developed a line of thought parallel to the more academic metaphysics developed by Hamilton and Henry Longueville Mansel when they attempted to establish religion on internal acceptance of faith, rather than on reason. The *Leader* intellectuals relied on Spencer. That is, they needed him to reinforce their beliefs with a philosophy, which he provided in the early editions of *The Principles of Psychology* and *First Principles*.

Rationalists, whether English Comteans or Buckle, were disliked by spiritualists because they worshipped the kind of abstraction that was concerned only with reason and intelligence. Such abstract thought was not only superficial, but anthropomorphic: "What they seek in the universe is not Life, but evidence of Design!"[57] The idea condemned here not only included attempts to prove the existence of a Designer, but any rationalistic comments on the universe, even those that deny the perfection of the universe. For the spiritualist, reason had no validity in religion. Only human emotion was constant and could appreciate the universe and God. "Anthropomorphism" had become a popular word among the *Leader* people. It was used rather loosely, and had only a tenuous connection with Ludwig Feuerbach's ideas. With Spencer, "anthropomorphism" was merely a derogatory term indicating a variety of primitive religions – including modern orthodox Christianity – that had well-defined beliefs. Spiritualism was seen as containing little anthropomorphism; this was why its deity had so few characteristics, and approximated a purely beneficent necessity.[58]

Rationalists were a minor irritant to the New Reformation; its major enemy was that great bastion of English orthodoxy, Joseph Butler's *Analogy of Religion*, which seemed to its numerous Victorian followers as impervious as it had been the previous century. Butler's synthesis between natural religion and science had to be overthrown before a modern one could be established. The attacks on the *Analogy of Religion* began in October and November 1852. These, like Sara Hennell's contemporary counterblast on the same subject, purported to show that the *Analogy of Religion* did not meet the requirements of modern religious speculation because its author had assumed the truth of revelation, and confirmed it by showing that it was no more difficult to accept than the religion of nature. This "natural theology" had founded its pretensions not on the true and devout interpretation of nature, but on mechanistic "contrivance" and "design". The metaphysics supporting this argument was as bad as Comte's assumption that "science permits us easily to conceive a happier arrangement".[59]

Butler's analogy was not only mechanistic; it kept science separate from religion when they were identical in being fastened by the same feelings. While the bulwark of the orthodox, Butler's *Analogy of Religion* was the butt of much criticism. Key spiritualists[60] such as F. W. Newman, Leigh Hunt, Lewes and Spencer regarded negative attacks as being in poor taste and slightly immoral. Their ideal was to accept all views as having some relative value and to reconcile them. Therefore Theodore Parker's pleasant *Ten Sermons of Religion* gave them an opportunity to escape from the negative work of denying and contradicting.[61] Few would have been able to quarrel with Parker's well-meant and vague platitudes such as the comment: "[t]hat Religion is a *binding together* of all our faculties – the keystone of our being's arch – no less than the binding together of all men into one humanity the keystone of the social arch … no man be his sect what it may will for a moment deny".[62] Parker was useful to the New Reformation because his sentiments such as "Science is the natural ally of Religion" and "the Religious intellect will above all things seek and welcome truth, believing all truth to be harmonious" were already gospel to spiritualists. There was nothing new in them, but it was pleasant to have one's beliefs confirmed by a famous American preacher.

> No man walks out on a starlight night without religious emotion, … It is an indestructible privilege of the stars to excite this emotion within us; and although this emotion will translate itself intellectually into various dialects and formulas, according to the various intellects of men, yet the emotion itself is constant; and the last Man, gazing upwards at the stars, will, in the depths of his reverent soul, echo the psalmist's burst of emotion – The Heavens declare the Glory of God![63]

This humble and emotive faith in natural religion was the creed of Spencer and his friends. God was unknown and indefinable, but infinitely greater than man. Those whose beliefs varied from this, whether they were orthodox Christians or materialists such as Comte, were castigated as anthropomorphic and irreligious. The impossibility of knowing the infinite should have made them *humble*. They knew nothing of God, and their ignorance should have restrained their assumption of knowledge of his existence, even if this only ascribed "certain attributes expressive of the relation in which he stands to us". "Do we *know* that relation when we call God a "jealous God", do we know the relation we express by jealousy … Do we not rather assume the relation of sin, and then argue anthropomorphically from that assumption?"[64]

If one had denied the validity of the rational explanation of religion, then – as the spiritualist discovered – it became difficult to write on the subject without being repetitive. By 1853 they had exhausted most of its religious ideas. That year saw the last major book of the New Reformation, the belated publication of Leigh Hunt's *The Religion of the Heart*. Hunt is usually remembered as a light versifier, and as a man of letters whose reputation was overshadowed by brilliant friends, such as Charles Lamb and Byron. In his own eyes, however, Hunt was a moralist and religious thinker. The anxious and painstaking works *Christianism* and *The Religion of the Heart* were dearer to him than the rest of his corpus,[65] so much so that he apologised for his vivacious style, which he could not shed.[66]

His first moralizing happened in 1832 when he persuaded John Foster to pay for the publication of *Christianism: Or Belief and Unbelief Reconciled: Being Exercises and Meditations*.[67] In this Hunt appealed to people of no religion on the basis that many of them had "a strong sense of religion at heart", and desired exalted notions of both the divine spirit of the universe and of the duties of beneficence.[68] To these potential believers he offered altruism. Religion and moral goodness were synonymous to Hunt,[69] and he only just refrained from intoning "Good" instead of God.[70] He offered his readers sample prayers to the "ineffable" God as the "Great Beneficience".[71]

These ideas were so obviously ephemeral that Hunt spent the next twenty years refining them so that they would resemble philosophy. *Christianism* changed into *The Religion of the Heart*,[72] while "exalted notions" became "moral sense".[73] The new version contained two important elements: the intuition of truth and its evolution (evolution here was in the sense of progressive change without biological overtones). The intuition of truth was achieved by the faculties, and verified by its consistency, or by its "harmony with itself".[74] A statement from the past – such as a biblical injunction – had authority if it was also intuited in the present. Without such confirmation traditional moral utterances, such as those contained in early Christianity, were merely barbarous misconceptions that human beings had outgrown. "Doctrines revolting to the heart are not made to endure, however mixed up they may be with lessons more divine. They contain the seeds of their dissolution."[75] In Hunt's mind all the barriers of knowledge were dissolved while moral and religious truth became indistinguishable from material truth or science:

> Let God's other Scripture, Science, again help us; let Divine Science, helping us through the stars, be heard in our pulpit; heard in a pulpit for the first time; a place, from which all the letter and almost all the spirit of this visible scripture of God have hitherto been excluded; though of all places none could become tidings of them better.[76]

Hunt's work was not intended as a simple exposition of the "church of the future" or the "New Reformation",[77] but as *the* Bible of the modern world. He gave careful selections from all the chief works of the "New Reformation", so that his readers would have "some manual of faith and duty, *in which the heart is never outraged*."[78] These selections were the new scriptures. He borrowed freely from F. W. Newman, Froude, Parker, James Martineau, Charles Hennell and the *Leader*. Even Chapman's slight work *Human Nature* was pressed into service.[79] Most significantly, Spencer's *Social Statics* was listed as one of the "scriptures" and its author praised as "a rising leader".[80]

At the beginning of a chapter titled "The Only Final Scriptures, Their Test and Teachers", Hunt defended his use of the word "scripture", while indicating which authors were divinely inspired teachers. This was a matter of establishing how they and their truths were to be tested. "For the word Scripture, in the sense to which it has been confined, means writing possessed of divine authority; and it is of the last authority to mankind, that nothing should be considered which goes counter to the first principles of good and true ...."[81] True knowledge had to spring from first principles, because only that would avoid the two extremes of "faiths which despise reason, and a reason exasperated or *mechanicalized* into no faith at all".[82] To reiterate Hunt's point, authoritative language could be neither traditional wisdom nor the kind of mechanistic reason admired by the Enlightenment. Instead, it had to be expressed in the form of first principles that could be intuited by the faculties of each human being. This reliance on first principles, which was adumbrated by Leigh Hunt and his followers in the *Leader*, produced both the New Reformation and Spencerian philosophy. It gave Spencer a means of distinguishing himself from utilitarian and anti-metaphysical philosophers such as J. S. Mill. For Spencer, empirical verification of truth did not lie in the objective world with its mechanistic laws, but in the subjective realm with its psychological principles.

The *Religion of the Heart*, although a romantic work, led its readers into those areas of Scottish psychology that supported the religion and ethics of the New Reformation. The "soul" and the "heart" were refined into philosophical statements about "faculties" and "necessary truths" by Spencer. That is, the first serious attempt to harness Scottish philosophy in this way was noted in November 1853 in a review of Spencer's "The Universal Postulate". The reviewer commented favourably on Spencer's reformulation of Reid's common sense as part of the psychological process.[83] There was a complaint, however, that Spencer had progressed beyond the *Leader*, and it was objected that he had gone beyond experience, and that his necessary truths or basic beliefs were "decomposable".[84]

After the publication of Spencer's "The Universal Postulate", his friend Lewes began to promote Scottish philosophy. They both had seized on an anonymous essay entitled "The Insoluble Problem" in the *North British Review* as a vehicle for expanding the relationship between religion and metaphysics.[85] The writing was reminiscent of Hamilton's early articles, but it was less orthodox than he had been and more in the style of a North British liberal theology. Its author, Alexander Campbell Fraser, had an eclectic philosophical position that hovered between the ideas of Hamilton, Victor Cousin, and Henry Calderwood. Hamilton's position was stressed as one that represented the finite as having absolutely no knowledge of infinity. Consequently, this implied that the natural religious sentiment in mankind was nothing more than an external pressure against an ever-resisting negative. Cousin's position was stated as giving man a certain positive knowledge of deity, while Calderwood was said to have controverted Hamilton's position, and revived Cousin's. Fraser himself attempted to find a middle way between the extremes of Hamilton and Cousin: "We believe, and therefore know, that the Infinite One exists; but whenever He is logically recognised as a term in thought or argument, either the object, like the argument becomes finite, or else runs into unnumerable objections".[86] This was the theology that powered the New Reformation. It was not sourced in Paley and the *Bridgewater Treatises* like so much English orthodox thought of the mid-nineteenth century.

This Caledonian legacy caused much difficulty in England. In the hands of Spencer and his friends, Scottish metaphysics and psychology undermined the empiricism that Anglican scientists had shared with Comte and J. S. Mill. That is, Scottish works such as Hamilton's edition of *The Works of Dugald Stewart* and James F. Ferrier's *Institutes of Metaphysics* were used to destroy a consensus between philosophical radicals such as Mill and Anglicans such as Whewell. Their empirical amalgam was vulnerable to a scientific study of human beings. It was due to Hamilton especially that "the old medley of odds and ends" that in Scotland had passed under the name of "science of the human mind" or "metaphysics" were partitioned off as metaphysics, logic and psychology. The impact of Hamilton's division of mental science on Spencer and his circle was to cause them to abandon their resistance to metaphysics. Their bias against that area of philosophy had existed because the subject had seemed incompatible with science. However, when it appeared that the study of the human psyche could be reconciled with metaphysics in such a way as to underpin religion, the new enemy became positivism, especially when that was in a Comtean form. The *Leader* observed that "Comte's doctrine of *Positivism* is that people ought to go on acquiring a knowledge of things in the ship without ever minding the ship's relations to the sea. But it *can't* be done!"[87]

The overthrow of Comte led to a revival of psychology. This development was caused by Hamilton, who, although he would have shared in the hostility to Comteans, would have been unhappy to have such allies. Nonetheless, his paternity cannot be denied; his progeny during the 1850s were Bain's *The Senses and the Intellect* and Spencer's *The Principles of Psychology*.[88] Both of these drew their inspiration from Hamilton or from his students, and both depended on the union between metaphysics and physiology, which the *Leader* had shown to be the basis of true psychology.[89] It was through Spencer and, to a lesser extent, Bain that Scottish metaphysics and psychology were absorbed by the religious and intellectual movement of the New Reformation and then, subsequently, the world in general.

This movement lost impetus after 1855. Now its chief significance is in having been the midwife of Spencer's philosophy. Its journal, the *Leader*, had been at the forefront of intellectual development, but its exclusivity was forbidding and it lost readers.[90] In addition, it lost its fashionable edge; there were no more contributions from prominent personalities such as Harriet Martineau, Robert Owen and Giuseppe Mazzini. One of the key figures, Lewes, slowly disengaged himself from editing and reviewing. Froude, F. W. Newman, Chapman and Spencer ceased to figure in the literature column. After 17 July 1858, the *Leader* was retitled the *Saturday Analyst and Leader*, and passed entirely into the hands of E. F. S. Pigott, whose instincts as a professional journalist had always been offended by the carelessness and extremism with which the paper was conducted.[91] Under his control the quotation from Humboldt's *Kosmos* disappeared from the title page.

Pigott, who was soured by the loss of his money, wrote to Holyoake that:

> The fact is the experiment of theological controversy has been fully tried in *The Leader*, and was found utterly incompatible with commercial success not to say with existence. Such reviews outrage or bore two thirds of our readers, and the other third is not sufficient to keep a paper alive, much less make it a property, and the "*Leader*" has I trust, lost enough. We cannot be a general newspaper and a controversial organ at once.[92]

Pigott's misguided efforts to save the *Leader* as a general newspaper failed; it died in 1860. It had only lived to outrage and stimulate, not to return a profit.

Previously the *Leader* had not been concerned with sales, but with the popularizing of radical religious and philosophical ideas. The minority of the readers that Pigott dismissed as commercially insignificant were those radicalized intellectuals who became attracted to Spencer's writings during the 1850s. The journal's success and influence can be measured by its impact on the thinking of two young men, Thomas Archer Hirst[93] and John Tyndall.[94] Unlike many of the *Leader* people, these two were not professional men of letters. Their attachment to the *Leader* was untainted by financial considerations. It was as enthusiastic amateurs that they began to think and write in the *Leader*'s own particular brand of spiritualism, and to preach the New Reformation. Hirst and Tyndall had become friends when they met in Halifax in 1845. Although there was a ten-year difference in their ages, they shared their intellectual experiences, whether this consisted of reading Emerson[95] and the *Leader*, or of travelling to Marburg for study.

The unpublished journals of Hirst show the effect the *Leader* could have on a reader's religious thinking. The first entry mentioning the newspaper in Hirst's journal is for 11 April 1850:

> Reading also the 2nd number (the first I have seen) of the new paper entitled the "Leader". From what I have been thus far able to judge, it promises to be a very able periodical. A tale called "The Apprenticeship of Life", by G. H. Lewes pleased me very much, especially some religious opinions therein, ...[96]

The next week again saw Hirst reading the *Leader*. He thought it worth recording that Froude had contributed "A Parable of the Bread Fruit Tree", which gave the history of religion very beautifully.[97]

In a short time Hirst was a devotee of the newspaper[98] in a way he had never been of the journals he had read during the late 1840s.[99] He began to read works reviewed in the *Leader*, while adopting the reviewers' opinions and language as his own. After Lewes's review of F. W. Newman's *The Soul*, Hirst read the work and made the following comment in his journal.

> Reading Newman on the Soul. The chief excellence and interest that I find in this work are, not only that it encourages one in many conscientious scruples against orthodoxy, and interprets many unconscious and yet deep feelings, but it concentrates and tends to strengthen the constructive part of spirituality.
>
> ... religion, as I said to Aunt, consists not in Doctrine, but in the *state of the heart towards its creator*. This is what is now before me, the prospect of a more personal relation to God, and I thank this book for this directive tendency.
>
> ... and yet I must confess to no vital belief in the personality and observance of God. I must think it intellectually, but it has as yet no practical vital effect on action – at least consciously so. Therefore when he [Newman] says on page 61, when showing the difference between remorse

and shame, that "Remorse is a convulsion of the soul, as it consciously stands under the eye of God", I must say I believe there is a remorse (quite apart and nobler than shame) perfectly unconnected with any conscious vital belief in God's personality or observance.[100]

In the last paragraph Hirst differed from Newman precisely where the *Leader* does. Whereas Newman posited his belief on feeling after the fashion of Hamilton and Mansel, Hirst found no such vital belief in himself, but began to generate a belief in the "Unknown". Hirst was also ready to discover his morality in psychology, in a way that was unsupported by God's personality.

When the concept became available Hirst was to adopt Spencer's "Unknown", but first his religious beliefs went through a form of vague pantheism common to the period. Such pantheism could be observed in a talk entitled "Education Societies, considered in their relation to Individual Culture" that Hirst read before the Halifax Franklin Society in October 1850. The introduction and conclusion of the paper consisted of phrases borrowed from Carlyle and Emerson, but its main body resembled passages from the *Leader* and from Humboldt's *Kosmos*. Hirst's mixture of the idea of progress, scientific adaptation, and pantheism was recorded in the following formula.

> I am so connected with the world, and everybody in it, that my own amelioration ameliorates the worth of man by so much. Nothing is ever destroyed or useless; and if not a particle of matter can be annihilated, neither can my knowledge; – it will exist and exert its due amount of influence, in spite of myself, both here and everywhere, now and forever.
>
> … Action, I repeat, is the unalterable law of the world. Nothing rests; not a particle of matter but must go through its endless courses and transformations.[101]

In the same address Hirst held forth against utilitarianism, in tones imitative of those Spencer used in *Social Statics*. Hirst remarked that utilitarianism did not tell one what is useful and that it weakened one's inner voice. By this he meant that it weakened the innate sense of morality, and was, therefore, low or base. In summarizing these views Hirst was consciously taking part in a contemporary reformation in morals, and this caused him to add a historical perspective to his peroration. "Time was, when an almost complete obtuseness on this subject was predominant, and a low animal-like utilitarian reigned supreme. Now, we are in a transition state; the claims of knowledge are to an extent admitted, but on faulty and partial grounds only."[102]

For Hirst, and for the others in Spencer's milieu, knowledge of human nature was a single subject encompassing all personal and social relationships, which could not be directed on the basis of a simple mechanical theory. Everything had to evolve at once, which is perhaps why Spencer's radical theory of land nationalization in *Social Statics* was a simple matter of progress unconnected to a specific plan of political action. A recommendation of a particular political action would have implied that change should be limited to one aspect of life. Hirst's comment when reading the *Leader*'s review of *Social Statics* is illuminating in this respect. Spencer's lack of a specific recommendation for political action was a sign of the book's sanity and truthfulness. For Hirst this feature signified that it rested on a firm foundation, and that, unlike works written by followers of Robert Owen and George Combe, it did not pigeonhole human nature too much into definitions and boxes.[103]

During the early 1850s much of Hirst's journal was filled with religious comment of the kind young men in Spencer's circles found inspiring. Like others of his kind, Hirst rejected any rational account of God or religion and, at the same time, insisted on the importance of a belief in the "indescribable, sacred *something*" that was recognized by the inner spiritual source.

> The most I can hope for is to see clearly that it is *un*accountable – this negative conviction should satisfy me. The highest height to which we can attain in this grand question is to be so fully imbued with its awful spiritual sanctity as to rest contended with the unaccountable fact, and to shrink from every hypothesis that would attempt to *explain* it.[104]

Hirst's friend Tyndall had similar views, also drawn from the *Leader*.[105] When reading Atkinson and Harriet Martineau's *Letters on the Laws of Man's Nature and Development*, Tyndall gave his reaction in language borrowed from Emerson and the *Leader*. She seemed an incomplete thinker to him, with only a partial view of the world. She amused for a while, like an exquisite piece of machinery, but Tyndall thought that her book would have been more satisfactory if she had possessed a life principle. In a remarkable admission for a man who was later the principal exponent of the merits of the physical sciences, he brandished the image of a living universe rather than a mechanical one.

> The contemplation of machinery causes a mere diversion of thought; while the recognition of life and of the alliance of the individual with this life, has a direct and sustaining effect upon the heart and morals. There are men who profess to detect this life principle in the universe – who protest against its being regarded as a machine, and the advantage that such men have over all others is that the grounds of the others' belief is perfectly comprehended by them, while the reverse is not the case. The transcendentalist understands the mechanic, but the mechanic cannot understand the transcendentalist.[106]

After this statement Tyndall quoted the *Leader*'s remark that, in her "deification of Law", Martineau herself had transgressed the limits that she had placed around others.[107] He refused to admit any of her dogmatic materialistic statements, on the same grounds as his refusal to allow dogmatic Christianity. Tyndall insisted on an open religious position. Since the Deity was unknown, it could not be defined or limited. Like Hirst and the rest of Spencer's set of friends, Tyndall had not yet adopted a final position, but already knew the direction in which he was tending. He was, as he immodestly described himself, "a daring soul" and had a premonition of his future religious beliefs. Writing rhetorically about himself in the third person he asked: "[w]hat if, in the pursuance of his studies, he arrives at the notion that matter and law are so to speak the sub-stratum on which a Deity inscribes himself, which is but an advanced premonition, and I think rests on grounds equally good with that of Miss Martineau"?[108] Tyndall's religious views of this period are best exemplified in a series of letters he exchanged in 1851 with a scientific colleague, Dr J. H. Gladstone.[109] An intimate conversation on board a steamboat between the two had moved Gladstone, and he wrote Tyndall an earnest letter in an attempt to Christianize him.[110] This was answered an hour after its receipt. In his reply Tyndall stressed the need for toleration and reverence, but rejected Gladstone's scheme of redemption. Tyndall had been studying the philosophy of the Hindus, and comparative religion had its usual secularizing effect; that is, it occurred to him that religion was not necessarily bound up with a Christian scheme. While Tyndall had no religious scheme of his own, he offered Gladstone a vision of expectant ignorance.

> But you will ask me what is *my* scheme. Anything I have to say upon the subject must be referred to a future day when my mind is more attuned to the subject than it is at present. Well I know the difficulty of an answer for the operations of God's spirit upon the heart of men are scarcely to be stated intellectually.
>
> I think our chief want is not a modification of our religious theories, but strength and courage to act up to the knowledge we possess. This I believe is the only way of access to still higher knowledge … A religion founded upon logic is for the curious and argumentative, but not for the man who is in earnest about the matter.[111]

Later, in response to another letter from Gladstone, Tyndall wrote a brilliant exposition of his current religious beliefs. In this exposition, Tyndall demonstrated his debt to Leigh Hunt and the spiritualists. This was particularly noticeable when he wrote of the religion of the heart as something distinct from religion of the head.[112] The former had impressed Tyndall because it appealed "to my own consciousness", and it broke "the rigid barrier which any mere intellectual scheme would draw around me". The religion of the heart concerned faith and love and was "beyond" logic:

> I grant the right of men to represent their religious experiences by symbols – nay I see the necessity of this. But I am careful of making my symbol immortal; I feel the possibility of substituting another for it equally as good and hence I should be sorry to attempt to force my particular mode of presenting religious matters to my mind upon others. I am afraid that the various phases of our present Christianity are so many symbols become

rigid; and that the quarrels and discussions of good men are to be traced to the unwilling substitution of the outward sign for the inward fact.[113]

Four years later, Tyndall wrote a similar letter to his American cousin Hector Tyndall. He warned the latter against the unchecked use of an analytic spirit that would strike at the very essence of faith, hope and love:

> All I feel is that there are sources of courage, strength, and consolation open to the human soul which are beyond my powers of analysis, they reveal themselves to me with an evidence equal to that of any physical fact, and I return to them sometimes with a certain wilful obstinacy in opposition to the analysis which would fritter them away.[114]

His letters to J. H. Gladstone and Hector Tyndall represented the summit of John Tyndall's early eloquence on the subject of religion.[115] Following the code of the New Reformation he was rejecting anthropomorphism, with passion and earnestness. Like Hirst, he anticipated much of the direction of Spencer's *First Principles*. When that work was published at the end of the decade both Tyndall and Hirst recognized themselves in it. It was natural that their letters of congratulation to Spencer were among the warmest he received.

The New Reformation, which had begun as an ideal "Church of the Future" propagating "spiritualism", eventually abandoned its attempts to organize a renewed national church and became an intellectual direction. The members of this movement possessed two common beliefs: the worship of an "Unknown" and the rejection of any attempt to formulate ethics or psychology on a rational basis, a critique that was particularly aimed at utilitarianism. Long after the decline of the "spiritualist" movement, its values were echoed in Spencer's *First Principles*. While the foundation stone of Spencer's philosophical system attracted varieties of readers as the nineteenth century wore on, these, like the "spiritualists" of the early 1850s, were drawn to it as to a beacon on the voyage between religion and science.[116] This passage was never completed; there was no progress from religion to science. The passenger always stopped in the middle.

Spencer had become the spirit of the New Reformation, and illuminated it by adopting the language of Scottish metaphysics. From late 1853, his "discovery" of Scottish metaphysics was shared by Lewes and his newer friends such as Tyndall and Huxley. At this point "spiritualism" came to be viewed as a psychological doctrine rather than a theological one, but it was still religious. The shift in emphasis did not redirect the appeal of Spencer's writings because – for spiritualists – religion and psychology were equally aimed at regenerating ethics. What they had in common was the complaint that orthodox Christianity and utilitarian psychology were both immoral. The first avoided the scientific truths on which true religion was based, while the second had wilfully and ignorantly failed to find any meaning in the universe when this offered so much evidence of the existence of the "Unknown". Spencer was first and foremost the prophet of a New Reformation that prided itself on the possession of a genuine morality lacking in normal Christianity and in positivism.

EIGHT

Intellectuals in the Strand

In addition to the intellectuals who clustered around the *Leader*, there was another group of London literati with whom Spencer associated during the late 1840s and the 1850s. This circle met at John Chapman's house in the Strand to discuss literary and scientific matters.[1] Of course, to say that British groups were intellectual could be seen to evoke a misleading continental association. It might be objected that reference to a class of intellectuals would be more apposite when recording the activities of the French, Germans or Italians. Emerson, writing at the time when Chapman was attempting to breathe intellectual excitement into London, said nothing came into English bookshops but politics, travels, statistics, tabulation and engineering, "and even what is called philosophy and letters is mechanical in its structure, as if inspiration had ceased, as if no vast hope, no religion, no song of joy, no wisdom, no analogy existed any more".[2] In general the American sage was accurate about the English; intellectual activity was too rare to form a national characteristic. The groups around the *Leader* and Chapman were too short-lived and as anomalous as the Pre-Raphaelites were to the traditions of nineteenth-century British painting. However, during the 1850s, the intellectuals of the Strand gave inspiration, hope, religion and joy to young intellectuals. These feelings also found a lasting life in the writings of Spencer, which carried what had been a parochial set of ideas to a vast international readership.

The *Leader* set and Chapman's friends did not form mutually exclusive groups. Lewes and Thornton Hunt would also appear at Chapman's during the evening.[3] However, Chapman's circle was markedly different in that it was less pretentious and self-consciously modern than the *Leader* set. While both affirmed the new truths in religion and science, Chapman's circle engaged in the repetition and popularization of these ideas to a wide audience. This was a contrast to the *Leader*'s fashionable dismissal of any idea more than a year old. The distinction between the two groups can be seen in their reaction to W. R. Greg.[4] His conversation was a mainstay of the evening gatherings at Chapman's house, and his *Creed of Christendom* was one of the publisher's more popular productions. However, for the *Leader* Greg's works were obvious and trivial; his work even compared unfavourably with the dated *Inquiry Concerning the Origin of Christianity* by Charles Hennell. Greg's essays were condemned by Lewes as the setting in permanent form of that which had only a transitory interest. The *Leader's* contempt was complete; Pigott refused to publish Greg because his articles lacked modernity.

Spencer never adopted the exclusive attitudes of other members of the *Leader* set. During the years 1845–60 he was often to be found at Chapman's soirées, sharing in the gossip as successive waves of crisis, scandal and new ideas rocked the small world of London literary radicalism.[5] At Chapman's house one could find all the latest celebrities and Spencer was thus able to keep abreast of each crisis as it developed. It was Spencer whose incautious speech allowed for the identification of George Eliot as the author of *Scenes of Clerical Life* and *Adam Bede*.[6] Spencer also helped Chapman in breaking up the monopolistic Bookseller's Association, and in drawing up the prospectus for the revived *Westminster Review*. His meeting with young Frederic Harrison at one of Chapman's Sunday gatherings gave him the knowledge to be able to identify the author of Harrison's article "Neo-Christianity", which created a storm of controversy around *Essays and Reviews*.[7] He was also able to introduce Harrison to Huxley and Tyndall,[8] who admired "Neo-Christianity" but had lacked Spencer's "inside" information.[9]

Usually the chief luminary at one of Chapman's evenings was someone like Greg, Froude, or F. W. Newman. Occasionally, however, Chapman could field a greater lion. For example, Leigh Hunt was drawn from his seclusion in Hammersmith to meet Emerson at Chapman's house.[10] Carlyle would also make appearances, drawn by the company of distinguished people. Of course, the discussion did not always produce intellectual sparks. One of Carlyle's followers duly recorded that, at one such soirée, the great man had spent the best part of an hour listening to Dr Elliotson on animal magnetism.[11] The ambition Chapman displayed in attracting distinguished guests was reflected by the quality of some of the books he published. In 1843, when he purchased the Unitarian publishing firm from John Green, his only major author was James Martineau. However, only a year later he was the English publisher of Emerson and Parker.[12] In 1845, while writing a preface to his edition of Schiller's letters and essays, he could proudly announce that he was publishing a second edition of Hennell's *Inquiry Concerning the Origin of Christianity* and the minor works of D. F. Strauss.[13] These successes continued. In 1851 Chapman's list included stellar publications such as: Herbert Spencer's *Social Statics*; Froude's *Nemesis of Faith*; Newman's *The Soul* and *Phases of Faith*; Atkinson and Harriet Martineau's *Letters on the Laws of Man's Nature and Development*; Greg's *Creed of Christendom*; William Smith's translations of Fichte; and George Eliot's translation of Strauss's *Life of Jesus*.[14] Also in 1851 Chapman expanded his empire by purchasing the *Westminster Review*, which had declined during a decade of weak editorship under W. E. Hickson. The *Westminster Review* again became a major quarterly because of Chapman's ability to persuade talented people to work for him. His editors during the 1850s included Eliot, Mark Pattison,[15] Huxley and Tyndall.[16] Chapman's 1851 list of publications would have been impressive for an established firm, such as Longmans or J. W. Parker, but, for a business recently started by a man of no experience,[17] it was astounding. This success was all the more impressive when it is considered that Chapman lacked both capital and great literary abilities.

Chapman was something of an enigma. Without being able to write, or to pay those who could, he initiated literary and intellectual turmoil. He was a catalyst, a man whose presence caused things to happen, while doing little himself. Part of his success was due to his personality and to his physical presence. He was noticed, and although contemporaries recorded his physical appearance quite disparately they all thought it worth mentioning. Eliza Lynn Linton remembered that Chapman was called the "Raffaelle bookseller".[18] In 1847 Carlyle mentioned that "Hedge came to me with tall, lank Chapman at his side, – an innocent flail of a creature, with considerable impetus in him".[19] In 1848 Leigh Hunt described Chapman as: "bookseller, writer, philosopher, succeeder, young, stout, good-looking, and scientific, all in one; so that authors will have no chance with him. He will ruin them, of course, as a bookseller; and then convince them that they ought to be content with being ruined, as a logician".[20] Chapman's charm and looks were accompanied by intelligence and vision. He had an idea of what he wanted written; his collection of authors was not random, but determined by his religious and philosophical goals.[21] The magnitude of Chapman's accomplishment was astonishing because his ideas were unpopular and not calculated to sell. There was ample justification for Carlyle's comment that the only book Chapman ever wrote himself proceeded from "the dust-hole of extinct Socinianism".[22] Initially, his political ideas were as awkward as his religious ones. When he sent the new prospectus for the *Westminster Review* to J. S. Mill – who had been the successful editor of the journal during the 1830s – the rude response was that Mill could not understand what Chapman meant by graduating reforms to the average moral and political growth of society.[23] Of course, its meaning was that the new editor was both a political trimmer and less radical than Mill. Chapman had developed his ideas with the help of Newman and Spencer, who stressed the importance of the natural change and evolution of society, while Mill was demanding forced change or revolution.[24]

Chapman's ideas went through two phases. In the early 1840s he was a disciple of the rational Unitarianism propounded by James Martineau. In the 1850s he rejected Martineau's views in favour of those expressed by the prophets of the New Reformation. Conveniently, three of these were authors that he published: Newman, Greg, and R. W. Mackay. In 1843, at the beginning of his career as a publisher, Chapman was a young unorthodox Christian who had just discovered a remarkable book by Martineau:[25] *Endeavours After the Christian Life*. This work, although often ignored by Unitarian commentators on Martineau,[26] was one of the most original he ever wrote. In it he offered the outline of an anti-Benthamite critique based on both moral and metaphysical grounds. Martineau drew a distinction between beneficent deeds and the sensibility of conscience in relation to right and wrong. The first appealed to prudential apprehensions, while the second illuminated "the secret image of right" that dwelt in every mind, and offered models of high faith and disinterested virtue that kindled "the reverence of the Heart".[27] This was an appeal to a morality based on intuition rather than on reason. In reacting against utilitarianism, Martineau was gesturing to the same inner inspiration as Leigh Hunt later did in his *Religion of the Heart*. The difference between the two was that Hunt worshipped a religious humanism based on the ideal of a happy man, whereas Martineau saw God as an ideal for man to emulate.

Martineau's appeal to intuition was not philosophical. He simply denigrated his contemporaries by suggesting that those who were anxious to "enlighten" religion had constructed a pretentious system in which the affections were not properly distinguished from one another, and in which they had no function. These utilitarians believed that moral qualities had been determined solely by *effects* of conduct in producing pleasure or pain, and that motives did not concern a moralist, provided they accompanied a large amount of pleasure.[28] For Martineau this was far from the spirit of Jesus.[29] Such a mild critique, hidden in the midst of an obscure volume of Unitarian sermons, would not have had any impact had it not been read by the young Chapman, who annotated and rewrote the *Endeavours*. Full of enthusiasm he rushed to Martineau's publisher, John Green and demanded that his emendations of Martineau be published. In response Green prudently sold Chapman his business to enable him to publish his own labours. Chapman's work bore a title that was remarkable – at least in the nineteenth century – for its length: *Human Nature. A Philosophical Exposition of the Divine Institution of Reward And Punishment, Which Obtains In the Physical, Intellectual, And Moral Constitution of Man; With An Introductory Essay. To Which Is Added A Series of Ethical Observations, Written During The Perusal Of The Rev. James Martineau's Recent Work, Entitled "Endeavours After The Christian Life"*. Chapman's book found few readers,[30] but it possessed the merits of combining a rich jumble of ideas, vigorous expression and fashionable quotations from Emerson. The book began with a criticism of Martineau's lack of scientific awareness. The Unitarian had failed to observe the development of man's nature or his psychical progression.[31] He was also ignorant of the way in which astronomy, chemistry and all other sciences, converge, and tend:

> to *One*, Unite, and Universality. All are progressing in the same path – onwards; uttering, echoing, and re-echoing the same great truths; and confirming the fixedness and immutability of those laws which have been established by him "in whom there is no variableness neither show of turning".[32]

Martineau had not demonstrated that all religious principles must have their foundation in nature, and should be discovered by reason.[33] This line of criticism was abandoned after several pages, when Chapman discovered that *Endeavours* offered a new foundation for ethics – including religion – that could be deduced in a scientific manner. The combination of ethics and science was important to Chapman. Ethics had to be sanctioned by both religion *and* science, even though most utilitarians had excluded the former and orthodox Christians the latter.

Chapman had evolved a view of religion similar to that enunciated a few years later by the *Leader*. While religion was the most important element in this radical philosophy, the result was not truly a modern theology. Rather, it was an all-encompassing framework of knowledge, akin to medieval theology. In this synthesis religion was to be the queen of the sciences.

> Religion, or the science of human culture and development, is the loftiest of every department of philosophy, and to which all others must yield subservience and contribution.[34]

> All discoveries in science and arts, are but other forms of revelation manifested in the more intellectual or material spheres.[35]

Chapman's beliefs were a transition between Unitarianism and a secular, scientific religion. Later, in the hands of Spencer, his "Great First Cause" became the "Unknown".

Chapman himself struggled when enunciating a scientific psychology that would fulfil religious criteria. He began by stating that the nature of right and wrong can be deduced by examining the sublime truths of our inner nature.[36] The purpose of religion was to cause the unfolding of man's being by the development of his faculties.[37] For him true philosophy flowed from the laws of this development, not from experience. Intuition was applied to experience and utilitarian ethics. The summit of human achievement would come when "all man's actions shall proceed from the fountain of spontaneity, or impulsive intuition. Then will his *entire nature* be evolved, all his faculties unfolded, and *genius* become the universal endowment of humanity".[38] Chapman failed to offer any plausible psychological reasoning that could support either his anti-utilitarian ethics or his religion; perhaps he was incapable of doing so. As Mark Rutherford had noted, Chapman's speculations on psychology demonstrated no deep understanding.[39] Nevertheless, despite his limited competence, Chapman's intentions were clear. He also knew how to ferment ideas and how to choose articulate authors. This is why he burdened the *Westminster Review* with the astringent psychology of Spencer. Spencer's goals resembled his own. In his diary entry for 27 April 1851, Chapman recorded that he saw in Herbert Spencer "in all his views, his zeal, his idealism, and confidence that there was no absolute evil, etc. the mirror of what I was in 1843–4 and 5".[40]

In addition to *Endeavours*, Martineau wrote another important book during this period: *The Rationale of Religious Enquiry*. This work also acted as a stepping stone for Chapman and the radical writers who were featured in *An Analytical Catalogue of Mr. Chapman's Publications* (1852). Most of the articles in this compendium were written by George Eliot, including an examination of the third edition of Martineau's *The Rationale*.[41] This, she wrote, "still indicates the right path and affords true though temporary guidance".[42] This was high praise considering that the work had already been abandoned by Martineau, and had been supplanted in the radical canon of spiritualism by more recent guides such as Newman's *The Soul*. *The Rationale* continued to be attractive because of Martineau's statement that reason was supreme in matters of religion, and that the special inspiration of New Testament writers did not guarantee freedom from error. They did not offer authority that could sustain a doctrine asking for the unconditional submission of the reason.[43] When Eliot paraphrased Martineau as saying that there was no escape from the final appeal to reason, she was advocating the idea that each person had an internal test for truth. This, in turn, meant that it was necessary to test "all doctrine and all precept by their harmony with those ideas of religion and morals which are gathered from the observation of nature, and of human life, and which constitute the primary revelation of God to man".[44] It was this internal test for truth that Spencer was to defend later in *The Principles of Psychology*.

The third edition of *The Rationale* – the only one published by Chapman – was more significant to Chapman's circle than it was to its author. The book represented an eccentric phase in this development of Martineau's theology.[45] It was only at the time of its publication that Martineau played an important role as an unorthodox thinker. Through his friendship with Joseph Blanco White,[46] he had temporarily come to see certain forms of "anti-supernaturalism" as benign, that is, as not hostile to Christianity.[47] This led him to be permissive about the rejection of the supernatural aspects of the "historical" Christ, providing that these were replaced with "philosophical" beliefs about the soul of Christ.

> [I]t rather shifts the ground than lessens the amount of supernatural belief, and transfers to the *soul* of Christ whatever has been lost from his outward *life*. Hence it is perfectly compatible with the acknowledgement of his divine authority to any required extent, and leaves the Christian characteristics wholly undisturbed.[48]

This statement was the key to the new spiritualism in Britain. Martineau's work helped to generate a form of theism that would be a safe refuge from the dangers of historically dubious proof. Religion was to be internalized into the soul, rather than relying on outward evidences. His lead was followed by two other Unitarian writers, Charles Hennell and F. W. Newman, who were also published by Chapman. Hennell's *Christian Theism* and Newman's *The Soul* were attempts to base religion on a doctrine of intuition, a vital task because the authors perceived the historical underpinnings of Christianity to be shattered. To them history's insights were unnecessary. A truth that is announced by heaven to one age may be discovered in another.

> A truth is a real and actual relation of things, subsisting somewhere, – either in the ideas within us, or in the objects without us, – and capable therefore of making itself clear to us by evidence either demonstrative or moral. We may not yet have advanced to the point of view from which it opens upon us; but a progressive knowledge must bring us to it; and we shall then see that which hitherto was sustained by authority, resting on its natural support. ... Thus revelation is an anticipation only of science; a forecast of future intellectual and moral achievements; a provisional authority for governing the human mind, till the regularly constituted powers can be organized.[49]

Martineau could dismiss historical proof because his theories were guided by two beliefs: that truth is progressive, and that the present possesses a higher form of truth than the past; and that truth exists inside man's psychological arrangements, "the ideas within us". Either of these notions would undermine a historically based religion. Martineau's work was filled with comments such as that the Christian writings were to be studied in adaptation to their own age and position.[50] Further, he believed that psychological development was more fundamental than religious beliefs. Human faculties were undergoing a development that would, for the first time, give them a true love of God. This transformation would come about because one's sympathies for others would unfold into an affection for God.[51] Martineau's psychological formulations were naive in comparison to Spencer's later ones, but they foreshadowed these in placing importance on mental development.

Martineau did as much as Newman and Leigh Hunt to invent spiritualism and the worship of the "Unknown". Although he was disappointed by the outcome of such ideas in Germany,[52] he nonetheless made a courageous attempt to synthesize his form of Christianity with philosophy. Together, he believed, they would form a progressive religion. "Refusing to dissociate philosophy and Christianity, [Rationalism's] genius has seized the glorious conception of a progressive religion, ever in advance of the understanding, and dilating the heart of individual man".[53] Radicals such as Chapman, Eliot and Spencer accepted this as a vision that stretched their minds to the limits of understanding when pursuing religious truth. It became a bedrock for them; long after Martineau had receded back into orthodox Unitarianism, they continued to pursue the "Unknown".

In the late 1840s, Martineau's friend Newman succeeded him as the chief spokesman of Spiritualism.[54] Newman, who was a constant visitor at Chapman's house, also attempted to bring order to Chapman's chaotic business affairs.[55] He was a stabilizing influence; it was Newman who guided the development of English theism through endless discussions. His young followers, such as Frederic Harrison, accepted his beliefs without reservations. To Eliot, he was "our blessed St Francis".[56] Newman's ideas extended on those of Martineau, but there was a difference. The latter had always professed a vague Christianity, whereas Newman insistently referred to himself as a theist or spiritualist. This gave Newman's rhetoric a more conscious radicalism. Newman's first exercise in radical religious thought was *A History of the Hebrew Monarchy* (1847), a book that was attacked by the *British Quarterly* as part of a great conspiracy.[57] Newman was not a serious scholar of biblical criticism; rather, he was a popularizer. However, his labours were intense enough to make him aware of the necessity of abandoning the Bible as the basis for religious belief. Accordingly, in his next work, *The Soul, Her Sorrows and Her Aspirations* (1849), Newman protested against relying on the Bible when searching for laws of men's understanding, or for the conscience or soul. These searches had stifled the spirit of God for seventeen centuries.[58] To end this labour Newman made three demands on religion. These epitomized the attempts of Chapman and the *Leader* people to capture the essence of spiritualism. First, historical proofs of religion were inadequate, and should be abandoned in favour of a direct appeal to the conscience and the soul.[59] Secondly, some *undefined*, but profound, change must take place immediately in order to regenerate morality. "If we continue to do as we are going, – *if no action of a totally new kind is set up*, – the present cause must go steadily forward".[60] Thirdly, there was a unity of science and religion, which could be discovered by the understanding. This synthesis was of primary importance, and could not be sacrificed for a belief in God.[61] Newman's conceptualization of the unity of science and religion is described in his insightful religious autobiography, *Phases of Faith*.[62]

Two of Newman's demands conflicted. He had asked that human understanding recognize the unity of religion and science, but had deliberately omitted the explanation of the nature of this unity, leaving it undefined and beyond the understanding. He had refused to answer his ultimate questions. To an extent this made him responsible for the curious arguments about metaphysical first principles that lie at the basis of Spencer's work. That is, thanks to Newman, Spencerian metaphysics became a question about understanding the unknowable. Newman displayed great pride in his deliberate ambiguity: "[y]ou appear to me to treat it as something bold and strange and unreliable in me (though frank and manly) to avow that I have no logical proof of my first principles. But this is a mere axiomatic truism. If a principle *had* a logical proof, it would be only secondary and tertiary, …".[63] However, despite this waiver, Newman did offer a *presumptive* proof for a first principle, which was: it is true if there is a uniformity of human opinion, and confirmed if there is harmony in distant results. Such a proof provided Newman with the basis of a *belief* in the existence of a *Mind* acting on the universe, which, to him, was the foundation of all religion.[64] While this presumptive proof might be unsatisfactory as a piece of logic, it did provide a convenient grouping of philosophical notions with which to bolster an ideal of morality. It also suggested that Mansel was wrong to attack Newman in his *The Limits of Religious Thought* because men relied on limited statements about religious and philosophical truth supported by presumptive axioms. Neither issued any statements about absolute truth; they only offered the *promise* of such statements. Their writings were equally pregnant with metaphysical hope in the mid-century. It was this promise that attracted the attention of radical philosophers in the 1850s, and that made Spencer an exponent of their metaphysics.

Following publication of *Phases of Faith*, Newman was intellectually exhausted: "I have written one or two books. If they are good ones, praise be to God. But I cannot carry my heart on my lips; and you might talk to me for a year and not hear from me truths so deep and spiritual as I have ventured to print".[65] This was an accurate prognosis; several years passed before he again made any detailed statement of his belief, and then he had nothing to say beyond reiterating his beliefs in poetical form. His poem was titled, *Theism, Doctrinal and Practical; or Didactic Religious Utterances*, and in it Newman attempted to tread a narrow line between orthodox Christianity and the scientific religions of pantheism and atheism. His *Theism, Doctrinal and Practical* fell between literature and philosophy; it was a populist creed expressing established beliefs in a simple fashion.

The gulf between Belief and Unbelief perpetually lessens,
Giving hope, that as Bigotry vanishes and Love strengthens,
All shall blend in one Faith, in religion as in other Science.
Hence both Theists and Atheists are now made wiser.
The Theist disowns all caprice in the Creator and Ruler
Believes in his imparted Wisdom, and discerns that it has Laws
The Atheist admits that chance is never a cause of things,
And that the harmony of the universe is a product of Law
That the universe is pervaded by Forces which obey fixed Laws,
And that by such Forces, the whole is entwined and guided,
All are agreed; and that these forces have organic harmony
Pure Intellect has no will, no desire of constructions,
No approval of Right, no living Force nor Motive
Nor does the Infinite One abide apart, but dwells in our bosoms,
Exciting men's affections and awaiting his cry.
Such Pantheism is but Atheism veiled in poetry,
And if it affect to be religious, is worse than Atheism;
For it leads straight to Paganism by deifying brute force.
Our knowledge of Goodness is prior to our knowledge of God; ...[66]

Newman's poetic muse failed him completely when he reached the arid philosophical subjects of verification and intuition; only then did he break into prose. After expressing his satisfaction with J. D. Morell's writings on the subject,[67] he offered a restatement of his earlier unpublished writings. His claim was that ultimate principles were verifiable by common agreement, not by logic. This was similar to nineteenth-century formulations of "common-sense" philosophy that influenced Mansel and Spencer.

> When we can once penetrate to the ultimate principles, which are real foundations of knowledge, no other verification is imaginable than agreement with other men and agreement with ourselves. We cannot *argue down* an insane man or a pertinacious criminal: we can do nothing but *overbear them* by our mere numbers and by our internal concord. Neither in physics nor in morals can the laws of proof be proved; they can only be enunciated and approved. The same is true of primary facts and primary judgements.[68]

For Newman, ultimate principles could be established, but not proved. The act of establishing an ultimate principle was the same as the "presumptive proof" he had offered in 1851. It was Newman's proof that attracted Spencer and he spent part of the 1850s demonstrating that proof by common agreement and internal concord was an actual, rather than a presumptive, exercise.

Newman's doctrine of intuition developed in the same way as the one in Spencer's *Social Statics*, because both originated as anti-utilitarian arguments that if the happiness principle rested on intuition, then it was secondary, and one should seek for a more fundamental principle of right.[69] If, on the other hand, the happiness principle did not rest on intuition, then it was immoral:

> It is surely mere imbecility in a moralist to betake himself to the "Happiness of the Greatest Number" as an argument which will avail against those who disown the intuition that "we ought to seek other men's happiness". Besides, it remains to discuss what is happiness; and of the several kinds of partial happiness, which is preferable. To abandon the decision to the variations of private taste, is, in one who makes Happiness the basis, to overthrow Morals fundamentally: if he will not do this, he must resort to *Intuition* to decide, of several tastes or impulses, which is nobler and better. Whatever atheistic or necessary logicians may wish or assert to the contrary, there is no more solid foundation for truth than Instinct and Intuition afford when they rest on the collective agreement of mankind …[70]

Newman's anti-utilitarianism, like his argument on verification, led to intuitionism and "common-sense" philosophy. The only difference between the two arguments was that when he wrote of morals he had no hesitation about treating intuition as a basis for first principles, whereas when he wrote of verification he hedged his views with formal qualifications.

The writings of Martineau and Newman inspired one of Chapman's most frequent visitors, Greg, to write "a manual of modern biblical criticism",[71] *The Creed of Christendom; Its Foundations and Superstructure.*[72] Greg had been exploring radical religious thought for many years, but had held back from expressing his own feelings because he was afraid that it would be injurious to mankind to speak openly about the truth.[73] He was also concerned about offending his family, which held orthodox Unitarian views.[74] When others, such as Newman, Charles Hennell and James Martineau, had blazed the way and made it safe, Greg ventured into print with the risible boast that his work was "a pioneering one".

The Creed of Christendom was tentative to the point of vagueness; it was also contradictory.[75] However, when Greg avoided the shoals of biblical scholarship and restricted himself to modern religion his words seemed to carry more authority. He constantly stressed the necessary limitation of both religious and philosophical knowledge. "Being finite, we *can* form no correct or adequate idea of the Infinite:- being material, we *can* form no clear conception of the Spiritual."[76] Greg also came close to adopting an idealist stance when, in an echo of German philosophy, he wrote that, "The consciousness of the individual (says Fichte) reveals itself alone". The individual's conception of other things and other beings are "*only his conceptions.*"[77] Logically, Greg should have developed this notion so that it resembled an idealist philosophy but, unconcerned with rigour, he affirmed another doctrine instead. This was borrowed from Sir William Hamilton, probably because Greg had been one of Hamilton's favourite students twenty years before.[78]

Greg wrote that, "The Deity is thus not an object of knowledge but of faith; – not to be approached by the understanding, but by the moral sense; – not to be conceived but to be felt".[79] This use of the moral sense as the antithesis of the understanding is pure Hamiltonian; although Greg had presumably forgotten most of the structure of Hamilton's philosophy, he had remembered its spirit. That is, he remembered that faith could not be defended by reason, but only by the moral sense.

Greg combined Scottish moral sense together with concepts such as the progressive development of religion that he had gleaned from Newman's *The Soul*. Absolute truth would be unattainable, yet, he believed, we should press forwards because the amount of error in our views would be progressively and perpetually diminished.[80] All that man could obtain was an idea of God that would satisfy the soul, or be subjectively true. In order to satisfy, the subjective truth must be the highest and noblest that man's mind is capable of forming. "Religious truth is therefore necessarily progressive, because our powers are progressive, – a position fatal to all positive dogma".[81] Greg postulated an ever-changing religion with no dogma, based on a progressive development of human powers or faculties. There was no God to be objectively examined; He was unknown.[82] While not original, Greg's *The Creed of Christendom* was significant in linking two intellectual movements, Scottish common-sense philosophy and spiritualism, both of which balanced between idealist and empiricist philosophies and espoused a limited intuitionism concerning "other objects and other beings". Greg even foreshadowed Spencer's adoption of intuition while supporting spiritualist religious beliefs. If Greg had spun his views with more panache he would have avoided being slighted by fashionably "scientific" writers such as Lewes. Rather, he would have been fêted like Spencer.

In 1854 a new star appeared in the firmament of radical religion. This was R.W. Mackay.[83] Mackay became Chapman's favourite and his name began to twinkle in Chapman's correspondence, and his writings in the *Westminster Review*.[84] Pattison, who was editing the theological section of the journal, was instructed to insert Mackay's notices without alteration: "I hope you may concur in the opinions they may express because I agree more nearly with the philosophical and theological views of Mr. Mackay than with those of any other English writer intimately known to me".[85] Mackay's two chief works, *The Progress of the Intellect* (1850) and *A Sketch of the Rise and Progress of Christianity* (1854), were published by Chapman. The first began with a section entitled "Intellectual Religion" that reiterated, in a careful manner, all the doctrines of Newman and the New Reformation. The kernel of Makay's argument was that former religious forms contained only partial truths and, during the development of thought, these were absorbed into a philosophical and natural system.[86] The ideas of traditional religion, such as "the hypothesis of miracles", had harmed true piety through promoting anarchy, rather than convincing believers in the merits of comprehensive beneficence and wisdom.[87] In the religion of the future, miracles and other superstitious beliefs would be discarded. Religion would be inseparable from science.[88]

In Mackay's vision religion was above science; it was the "unknown" that was beyond knowledge:

> In a more limited sense religion may be contrasted with science and
> something beyond and above it; and beginning where science ends, and

as a guide through the realms of the unknown. But the known and the unknown are intimately connected and correlative. ... Philosophy and religion have one common aim; they are but different forms of answer to the same great question, that of man and his destination.[89]

Mackay's utterances presage the "Unknown" of Spencer's *First Principles* even more completely than those of Martineau and Newman. The latter had possessed a residue of a belief in a personal God, but Mackay's God was the distant governor of the universe: an unintelligible force the outlines of which were faintly visible at the edge of the known and scientific.

Mackay also held the anti-utilitarian beliefs that were common in his circle. Pleasure and pain were rejected as inadequate moral teachers, and replaced by the sphere of benevolence, in which a neighbour's interests were regarded as one's own. An individual's benevolent susceptibilities were multiplied by his relations with other men.[90] Mackay did not explain this altruism by using the language of psychology, but he indicated that knowledge of morals and metaphysics was to be found through intuition by use of faculties or internal sentiments.[91]

Mackay was advising Chapman on theological matters as late as 1857.[92] He was able to expand on the subject with a precision and learning Chapman himself could not manage. The latter's helplessness can be seen in his reply to Pattison's request for an explanation of his editor's theological views, when Chapman was able to respond with an echo of Mackay's *The Progress of the Intellect*:

> It seems to me that this is pre-eminently an age of transition; and that though a few minds here and there over Europe may have completely risen above the mists of superstition ..., the work of destruction will have to be long persisted in before a constructive creed can grow up which shall be wide enough to embrace all the truths of modern science, and profound enough to minister to the deepest feelings and innermost yearnings of the human heart. For myself I live in suspense, and entertain a sort of dumb faith, inarticulate echoes of which I recognise now and then in music, but rarely elsewhere.[93]

This creed was simple, it ignored the complex problems of how one verified first principles and how one intuited truth but, combined with Chapman's enthusiasm and his gift for making others write, it was enough to make him one of the sources of an intellectual movement that revived intuitionist philosophy and bolstered the new foundations of progressive religious thought. Chapman was the impresario for a short-lived movement that had an enduring monument in Spencerian philosophy. After Chapman was driven from his publishing business in 1860,[94] the impetus and direction he had given to the movement could still be seen in Spencer's *First Principles* (1862), which bears the Chapman imprint, although not on the title page.[95]

NINE.

The genesis of a system

Before the publication of *First Principles* in 1862 Spencer was an obscure figure, uncertain of his abilities and his future. He was what nineteenth-century biographical dictionaries called a miscellaneous writer. During the 1850s his books sold slowly. The first edition of Spencer's *The Principles of Psychology* was small; only 251 copies were published on 29 October 1855. By June 1856, a mere 200 copies had been purchased, while sixteen complimentary copies went to Stationers' Hall. It took until June 1860 for the remaining 35 copies to be sold.[1] A similar fate greeted Spencer's *Essays*; 201 were published on 18 December 1857. By 18 June 1858, 144 copies had been sold, two had been given as presentation copies, and five went to Stationers' Hall. It took from August 1858 until June 1862 for the last 50 copies to sell.[2] The subscription to Spencer's "A System of Synthetic Philosophy" did not do much better. Spencer wrote to Holyoake, who was printing the prospectus:

> I shall be glad, when you can afford the time, to hear what is doing as to the "System of Philosophy" – and how the names are coming in? I have reached a little more than half the requisite numbers, and there is still a slow increase going on; but I do not think that I shall myself exceed two-thirds.[3]

Ultimately, the number of subscribers was under four hundred, and this meant the scheme was a financial failure. Spencer, however, concealed this in *An Autobiography*,[4] euphemistically suggesting that the plan had "succeeded fairly well".

The fact that Spencer's small publishing runs took several years to sell, and that "A System of Synthetic Philosophy" failed to receive an adequate number of subscribers, is an indication of his initial obscurity. This can be seen clearly if Spencer's sales are compared to the publishing successes of his contemporaries. J. S. Mill's *The Principles of Political Economy* had a first edition of 1,000 copies in April 1848, and by July 1857 required a fourth edition of 1,250 copies. Volumes

I and II of Froude's *History of England* had a first edition of 1,500 copies in April 1856, and a second edition of 1500 copies in May 1858. The first volume of Buckle's *The History of Civilization in England* had a second edition of 2,000 copies in July 1858, shortly after the first, while in May 1861 Buckle's second volume had an immediate sale of 2,956, that is, 3,000 minus copies for author, reviewers and universities.[5]

Spencer's articles were not eagerly sought after by publishers. He was forced to write them begging letters. Even then it seemed hopeless; the editor of *Blackwood's Magazine* was uninterested in a paper on the "Force of Expression" in 1851.[6] Spencer's supplication to the same journal six years later was also spurned:

> It has occurred to me that an article on the "Origin and Function of Music" for which I have recently been collecting materials would be well suited to the pages of *Blackwood's Magazine*; and I write to inquire whether the topic is one you would like to have treated. The course I thus take is somewhat informal. Had my friend Mr Lewes been in town I would have fulfilled the usages by asking him for an introduction to you, ...[7]

This was so desperate that *Blackwood's Magazine* did not bother to reply. The "Origin and Function of Music" was published in *Fraser's Magazine* instead, but Spencer courageously wrote again to *Blackwood's Magazine* offering further articles:

> Last spring I wrote to you proposing an article on "The Origin and Function of Music" for your magazine; but did not receive a reply.
>
> Happening recently to mention the fact to my friend, Lewes, he expressed the belief that your silence must have arisen from inadvertence; and thought it would be well to act on this supposition.
>
> Since these were written a number of other ideas, similarly admitting of brief and popular exposition, have been accumulating; and it occurs to me that they might form a suitable series for *Blackwood's Magazine*. Among other topics I may name are the following:
> 1. Dress
> 2. France and England
> 3. Love-Affinities
> 4. Selfishness
> 5. The future of English Art
> 6. Flavours and Odours
> 7. Lineage
> 8. Male and Female Character[8]

As usual his proposal was rejected; Spencer never wrote these articles, nor did he appear in *Blackwood's Magazine*. Faced with neglect, he even thought of applying for work in the civil service. Understandably, rebuffs and fear of poverty were downplayed in Spencer's *An Autobiography* and omitted from Duncan's *The Life and Letters of Herbert Spencer*. They present a picture of a young man whose interests were more trivial, and whose efforts were less successful, than the later Spencer could accept.[9]

He was saved from failure by the compelling integrity of his mind. Those around him began to believe in his prowess as a thinker on philosophy and science, before there was any basis for this. He had yet to write his *First Principles*, or the rest of his "A System of Synthetic Philosophy". Nor had he given evidence of his potential in verbal communication; he was not a brilliant speaker. Spencer simply seemed to be "a very remarkable man".[10] There was nothing to support this impression: it was just that many of his earlier acquaintances had faith in him and gave him encouragement. The most important of these were Huxley, Tyndall, Sara Hennell and Alfred Russel Wallace. They formed a small band of Spencerian philosophers. Outside this circle were some great Victorians, such as Charles Kingsley and J. S. Mill, who would not have sympathized with all the goals of Spencer's prospectus but nonetheless were willing to pay for it.

Huxley was one of the first to appreciate Spencer, and later played an important role in the development of *The Principles of Biology*. As an indication of Huxley's respect, it should be pointed out that Huxley, whose rudeness and aggressive energy were notorious, always went to extreme lengths to avoid quarrelling with Spencer.[11] Huxley first befriended Spencer in 1852,[12] but their friendship did not become close until 1858, when he moved to St John's Wood to be near Huxley. At this time it was only to Huxley that Spencer could speak freely of the vast metaphysical schemes that had begun to emerge in his head. Without Huxley's encouragement these would have probably come to nothing. Spencer's reliance on his friend can be seen in a letter written during 1858, which was remarkable for its emotion and humbleness. Usually, his letters were dry and formal.

> I had no idea that you had so far divined my aim; though you have given to it an expression that I had never thought of doing. I know that I have sometimes dropped hints; but my ambition has of late been growing so wide that I have not dared fully to utter it to any one. But having in some sort recognized it, you, who so well know my weak points, should still think I may do something towards achieving it, is I assure you an immense satisfaction, and will be to me a great encouragement to persevere.[13]

Huxley encouraged Spencer by reading the proofs for his *First Principles*.

> To my mind nothing can be better than these contents whether in matter or in manner, and as my wife arrived independently, at the same opinion – I think my judgment is not one sided.
>
> There is something calm and dignified about the tone of the whole – which eminently befits a philosophical work which means to live – and nothing can be more clear and forcible than the argument.
>
> I rejoice that you have made a beginning and such a beginning – for the more I think about it the more important it seems to me – that somebody

should think out with a connected system – the loose notions that are floating about more or less distinctly in all the best minds.

It seems as if all the thoughts in what you have written were my own and yet I am conscious of the enormous difference your presentation of them makes in my intellectual state. One is thought in the state of hemp yarn and the other in the state of rope. Work away thou excellent rope maker and make us more rope to hold on against the devil and the parsons.[14]

This kindly letter is more characteristic of the relationship between the two men than Huxley's oft quoted gibe that Spencer's idea of tragedy was a deduction killed by a fact.[15] In the 1850s and early 1860s Huxley and Spencer shared the radical ideas of religion and philosophy that were propagated by the *Leader* and by Chapman's writings. Huxley depended on Spencer to put such ideas in a clear systematic form that he could accept easily. Huxley had yet to develop his materialistic form of atheism, and still thought of religious speculation as an important activity. As he said to his fiancée, he believed that the tendency of a man's religious speculations should be seen as a key to his character.[16]

Although Huxley was not given to autobiographical comment,[17] he did make one important statement about his early religious and philosophical beliefs during a controversy with A. J. Balfour:

My relations with Mr Spencer's philosophy are of a totally different order. Thanks to Hamilton and Mill the fundamental principles of what is now understood as Agnosticism, were more clearly fixed in my mind when in 1850, I return[ed] to England with a well studied copy of Mill's *Logic*, which, along with Carlyle's essays [and] some volumes of Goethe and Dante, had shared my little cabin for four years.

Consequently, when I had the pleasure of making Mr Spencer's acquaintance in 1852, it was with much satisfaction that I found we stood on common ground, and no one could have welcomed *First Principles*, so far as its critical portions were concerned, more warmly than I did. But even then Mr Spencer appeared to me to be disposed to travel along the path – by which, as I conceive, Hamilton had been led astray – further than I was. And in the forty-three years which have elapsed the divergence of opinion then masked has unfortunately become greater and greater, until now we are speculatively (I hope in no other way) poles asunder. It is impossible for me to approve *a priori* method; to admit that Mr Spencer's form of the doctrine of Evolution is well founded; or to accept the ethical and political deductions which he makes from that doctrine.[18]

This comment demonstrates that it was in fact Huxley who diverged from Spencer. An evolution of Huxley's philosophical position can be seen in his published writings; it was a slow progress without the cataclysmic shift that took place in his views on natural selection.

Huxley's essay "On the Advisableness of Improving Natural Knowledge" (January 1866) imagined a religious symbol of the altar to the Unknown and Unknowable, which had been borrowed from Hamilton and Spencer.[19] This image depended on a postulate that human knowledge would always be limited, and that, therefore, man's instincts would be aware of something that his reason could not grasp. Since Huxley was always impatient of philosophical detail he did not feel compelled to be consistent about what reason could – or could not – grasp. He worked in metaphors, and when these no longer served his visionary purposes he shifted to a more optimistic utterance to the effect that the human race would mature, and that its spirit was fated "to extend itself into all departments of human thought, and to become co-extensive with the range of knowledge".[20] At this point there would have been no Unknown because humanity would be god-like.

Much later, with his mind sharpened by the study of Hume,[21] Huxley publicly rejected both Spencer's *a priori* philosophy and his form of spiritualistic religion. He was mildly embarrassed about his earlier adherence to these ideas. This seems to be the basis of the distance that came to separate the two close friends. Spencer always kept some faith in the Unknown even after Huxley had lost his. It was significant that Huxley's well-known "Agnosticism and Christianity" attempted to remove the religious content from Spencer's "Unknown".

The extent of the region of the uncertain … will vary according to the knowledge and the intellectual habits of the individual Agnostic. I do not very much care to speak of anything as "unknowable".

I confess that, long ago, I once or twice made this mistake; even to the waste of a capital "U".[22]

A crucial difficulty in interpreting Huxley is the integration of his early ideas, which were sympathetic to religion, with his later ones, which were not. One could accomplish this task biographically after the fashion of Adrian Desmond, by suggesting that Ernst Haeckel's conversion of Huxley extended far beyond the latter's new found fervour for the scientific merits of natural selection and applied to his philosophical ideas as well. That is, one could claim that the new rigour with which Huxley proclaimed the value of environmentally caused evolution left no room even for the possibility of an Unknown, and that he had become a thorough-going Teutonic materialist. However, this interpretation would credit Huxley with more intellectual coherency than he possessed. It is, nonetheless, significant to observe that Huxley's theological writings, as well as his scientific ones, underwent a signal change about 1868 even though premonitions of this were visible two or three years earlier. That is, at the same time that Huxley accepted a rigid evolutionary synthesis from a German scientist,[23] he began to reject the religion of nature that he had shared with Spencer.

Like Huxley,[24] Tyndall also encouraged Spencer to produce a system of philosophy that would satisfy the demands of the New Reformation. His sympathy for the movement was such that he propagated its values among the families of his scientific friends. He persuaded Frances H. Hooker, the wife of Joseph Dalton

Hooker, to read Newman's *The Soul*, one of the key texts of the movement.[25] During the late 1850s Tyndall debated with Spencer about the nature of the general laws of change and the nature of the "Equilibrium", topics that subsequently occupied much space in *First Principles*. Spencer hinted to his friend that his ideas were undergoing a change towards something vast and grand:

> Let me further say that you must not take the views put forth in the article on Progress and Philosophy as something more than rude, mis-shapen germs. During the past year they have been undergoing a development which I never anticipated; and the ideas at present published seem to me to stand towards a true theory much as an egg does to a bird, or an Amphioxus to a man. Indeed it is the wide and rapid evolution which these (and some connate ideas) have been undertaking, and the power which I see they have of absorbing and organizing all I had before thought, which renders me so anxious to get the opportunity of working them out.[26]

This conception of an overall philosophy of evolution generated *First Principles*, a work that should be seen as the climax of the New Reformation. The book was a testament to a secular modernism that was later prized by readers in every country on the globe, although they had no knowledge of its genesis among the earnest sensibilities of a small group of radical London journalists and scientists during the 1850s.

When *First Principles* was published, Tyndall welcomed it with enthusiasm because, of course, its religious principles were identical to his own:

> You get (I think) into mazes of words here and there through which I pass with a certain [thorny?] sensation. But taking it for all in all a better utterance I have never heard – so good and so clear one never. You may if you wish take an illustration of the "Absolute" from the wholeness with which I go along with you. My center of gravity is at its lowest point as I read these pages – on them I take my stand and defy the world, the flesh and the devil!
>
> My soul has already required this outlying region to breathe in, while theologians have filled it with miasma and the stench of rotten eggs.[27]

Tyndall's soul was overjoyed at the discovery of a systematic exposition of the "Unknown" that had cleansed religion of the impurities added to it by theologians. He felt flattered by the thought that Spencer had clarified the views Tyndall himself had outlined in his article "Physics and Metaphysics" (1860), although, of course, Spencer would never have admitted priority here. In reality, Tyndall's article, like Spencer's *First Principles*, was a common expression of the ideals espoused by many disciples of the New Reformation.

In "Physics and Metaphysics" Tyndall expressed his belief that philosophy in the future would assuredly take more account of the dependence of thought and feeling on psychical processes. Eventually, human faculties would develop to the point where they could easily grasp this kind of question. The exception to this prognosis was an area of thought that would remain inexplicable by such processes. That is, when we pass from the phenomena of physics to those of thought:

> we meet a problem which transcends any conceivable expansion of the powers which we now possess. We may think over the subject again and again, but it eludes all intellectual presentation. We stand at length face to face with the Incomprehensible. The territory of physics is wide, but it has limits from which we look with vacant eyes into the region beyond. Whence come we – whither go we? The question dies without an echo – upon the infinite shores of the Unknown.[28]

This "Unknown" or "mighty Mystery" was the basis of the scientific religion that Tyndall shared with Spencer, and it was from yarn such as this that both wove their metaphysics.[29]

Tyndall used to be depicted as a simple materialist. This was a pejorative tradition which harked back to Ruskin's *Deucalion*, and presented Tyndall as a caricature: a naturalist philosopher who was absorbed in the vulgar grossness of organs from vivisected animals, prehistoric beasts and huge primordial conflagrations of heat and gas.[30] This overlooked Tyndall's spiritualist side, which admired nature. Tyndall himself was responsible for much contemporary misperception. He provoked his opponents by using a combative language that depicted them as the forces of darkness while he appeared as the scientific force of light. He claimed that the struggle had begun with human history, when the physical operation of nature was ascribed to the work of gods, demons and the occult generally. He attacked his critics as primitives who avoided the light and whose overthrow had always been necessary for scientific progress.[31] Tyndall's polemical stance relied on this materialistic historiography as did his attacks on Christian theology, but it coexisted with his attempt to reconcile the claims of science and religion. This asserted that there was a real bond between mind and matter, of which science would always remain ignorant.[32] This limitation on science was a comment on the religious condition of mankind. Tyndall was paraphrasing Spencer to the effect that above, or behind, the phenomena of matter and force lay a mystery of the union that was incapable of solution, but that nonetheless offered solace to the spirit.

The human mind was seen by Tyndall as a pilgrim-like yearning for an ultimate home. The mind was an organ that sought to shape the universe so that it possessed unity of thought and faith. Tyndall believed that as long as the search for this unity was conducted without bigotry and dogmatism, then the restructuring of knowledge demanded by materialism could be put aside.[33] Tyndall's faith in Spencer's Unknown coexisted with his war on Christianity. The former was a vision of man as part of a living universe and offered the comfort of ending human isolation in the universe by endowing human beings, animals in general and even material objects with consciousness. For Tyndall the organic world was not disgraced by equating it with inorganic matter: on the contrary, inorganic matter was elevated by being full of life.[34] Every chemical experiment or astronomical observation became part of a religious cosmology, more meaningful because it was detached from the superstition of Christian theology. The new spiritualism had the advantage that it did not need to deduce uncertain propositions from revealed truths. Instead, one directly observed natural truth in the physical universe.

Unlike Huxley's religious beliefs, those of Tyndall remained relatively constant. That is, the views he expressed in "Physics and Metaphysics" were reiterated eight years later in an address in Norwich on "The Limit of the Imagination in Science",[35] where he defended scientific hypotheses, such as the theory of evolution, on the basis that they do not solve, nor profess to solve, the ultimate mystery of the universe.[36] The admission that science failed to answer ultimate questions did not matter because the role of scientists was to keep such questions open. They were the brave band that would not tolerate any unlawful limitation of the horizon of their souls. "They have as little fellowship with the atheist who says there is no god, as, with the theist who professes to know the mind of God."[37] Tyndall later wrote: "One of the objects I had in view in the composition of my Norwich address was to keep open for the faith and hope and pleasure of loving and religious men, a horizon the shutting up of which by science they dreaded".[38] He added that the great majority of his religious listeners felt a sense of relief when they heard his address. It seemed to please both churchmen and dissenters.[39]

In his famous Belfast Address of 1874 Tyndall expressed the same views. Although he was accused of "materialism" by the Presbytery of Belfast,[40] his religious views remained the spiritualist ones that he had acquired during the 1850s. In essence, his exposition of these views consisted of a summary of Spencer's psychology and philosophy.[41]

> With [Spencer], as with the uneducated man, there is no doubt our question as to the existence of an external world. But he differs from the uneducated, who think that the world really *is* what consciousness represents it to be. Our states of consciousness are mere *symbols* of an outside entity which produces them and determines the order of their succession, but the real nature of which we can never know. In fact, the whole process of evolution is the manifestation of a Power absolutely inscrutable to the intellect of man. As little in our day as in the days of Job can man by searching find this Power out. Considered fundamentally, then, it is by the operation of an insoluble mystery that life on earth is evolved, species differentiated, and mind unfolded from their prepotent elements in the immeasurable past. There is, you will observe, no very rank materialism here.[42]

Evolution, for Tyndall, was not a concept heavily laden with organic significance. After all, he was not a biologist.[43] Rather, for him evolution had primarily a philosophical and religious significance. A rough equation emerged in Tyndall's writings: moral evolution equals the mystery of religion. The vague idea that religion was evolution permeates all his works, and he gloried in "that wonderful plasticity of the Theistic Idea which enables it to maintain, through many changes, its hold upon superior minds".[44]

Aside from Huxley and Tyndall, one of the most enthusiastic early supporters of Spencer was Sara Hennell.[45] Even before Spencer wrote *First Principles* she was ready to become an apostle of Spencerian philosophy. They had only met a few times but his "Principle" had made a deep impression. She had already written on theology, and now she felt she had found a teacher in that discipline.[46] She wrote an emotional defence of Spencer to her friend Harriet Martineau:

> As to Mr Spencer's particularity I have *commanded* myself to set down precisely the impression it made upon me, and I can assure you the doing so required some moral courage when I do not know a single soul who is likely to agree with me: …. Since you have not read his book (and as my reader I am glad you have not) I am sorry you have had any personal impression about himself … .
>
> I have seen him several times, and, indeed, he was staying here for a few days, nearly two years ago. I was very desirous of entering upon these subjects, but the state of his health, that is, of his brain, forbade more than a very little and my own ideas were very unformed at that time. He is, as you say, extremely reserved and dogmatical and inaccessible (we have scarcely heard from him, or of him, since he was here) and because of the *thorough integrity* of his nature. Since you wrote he has sent the prospectus of his forthcoming work – and I must say, I cannot help rejoicing that he has found any way of bridging it over. What *is* to be done by a man who feels, as he does, that his mind is teeming with a vast scheme of thought, which he has no means of giving out because he has not the money to afford it! My own feeling is that of deep pity for the suffering it must have caused to a nature so proud and independent as his to be obliged to ask for that aid, – that is, to publish by subscription which after all is the commonest thing even with those who *are* able to do without it; – and, even of admiration that he is able to subdue his personal pride for the sake of his work. You see, I believe in him, – and perhaps nobody else does as yet! But I correct myself – it is not *himself* that I believe in; – he may be full of mistakes of all sorts: – what I do believe in is his *Principle*, – and even that I must qualify into his principle according to the light in which he appears to myself – for so far as I know he may disclaim the representation I have made of it. You may suppose I am very eager to compare notes with his forthcoming exposition.[47]

When she wrote the above letter in April 1860, Hennell was in the process of writing her first major book, *Thoughts in Aid of Faith*.[48] This gleaned ideas from most of the texts of the New Reformation such as those by Mackay and Greg. It also abounded with references to Unitarian writers, such as Parker, James Martineau and F. W. Newman. Her chief inspiration, however, was Spencer. She believed that those who sought after the real nature of "Religion" must seek aid in an "uncongenial alliance" with science.[49] Spencer's *The Principles of Psychology* was the very inspiration she needed in order to make the transition between theology and science.[50]

Spencer's definition of Life – "the continuous adjustment of internal relations to external relations" – attracted her attention, and she gave a précis of his argument of how higher forms of Life arose in consequence of more varied conditions in the environment. She became so excited by this logic that uncharacteristically she dared to say something herself, although it was not so much an argument, but a paean of praise:

> Here the reader feels himself on the verge of the Great Mystery, respecting which he cannot to himself maintain the cautious and conscientious reticence of the Author, by whom the idea is rather suggested than uttered, but which is so stupendous that having once entered the mind it becomes the all-in-all throughout the whole range of the subject.[51]

Spencer's work anticipated both religious and scientific certainty, and the fact that this certainty always remained in a promissory, rather than a revealed, form made it more acceptable to a mind such as Hennell's, which expected truth to be transitory. The concept of the development of the human mind, which she had drawn from Spencer, was neither precise nor academic. To her it came as a momentous generalization that the mind was henceforth to be considered as "a constituent part of the progress that belongs to the whole of Nature". Spencer's *The Principles of Psychology* was full of wonderful analogies,[52] which gave a vision to the soul of "a One magnificent harmony pervading the whole!"[53]

The first volume of Spencer's *Essays* (1858) had the same effect on Hennell as his work on psychology had. It was read as a revelation of constant plan in the mechanism of the heavens, as well as in human arts, and in social polity. Knowledge in all of these witnessed to a "new testimony" that pointed towards "the construction of a comprehensive theory that shall embrace the Universal Method of Nature's operation".[54] Like other spiritualists and disciples of the New Reformation, Hennell was an admirer of Humboldt's *Kosmos*, and Spencer appeared to her to be the philosopher who completed Humboldt's ideas on the unity and development of Nature. This had previously been a spiritual task; Spencer's writings furnished "a true basis for really scientific belief in the so-called spiritual world".[55]

Hennell's brother-in-law, Charles Bray, followed her in giving adherence to Spencer's religious and philosophical views. Bray, like Hennell, was a parasitic thinker; his writings reflected whatever he had recently read. The first edition of Bray's *Philosophy of Necessity* (1841) had been a series of stilted and mechanical borrowings from Bentham and Dr Thomas Brown.[56] The second edition (1863) was quite a different work and, without any apology adulterated his earlier sources with philosophical borrowings from Lewes and, more importantly, from Spencer. Bray did not delete his references to Brown, but tentatively hoped that Brown and Spencer might agree on psychological matters.[57] This was obtuse; Brown had died in the year Spencer was born, and while the latter was steeped in Scottish philosophy he took his cue from later generations of Caledonians, the writings of whom Bray had not discovered. However, Bray's ignorance did not matter; his reliance on Spencer was not a philosophical dependence, but an expression of solidarity with new ontological beliefs. Bray was more cogent on the subject of religion than on philosophy. To him Spencer's nebular hypothesis demonstrated that it was impossible to disconnect God from the material living universe, even though nothing was known "Positively", and the great mystery was still veiled. "The Nebular Hypothesis implies a First Cause, as much transcending 'the mechanical god of Paley', as this does the fetish of the savage".[58] To Bray this progress was illuminating; Spencer was the natural theologian for his generation in the way Paley had been for those who had come of age at the beginning of the nineteenth century.

Spencer's early followers such as Huxley, Hennell and Bray were almost an inner circle. Much more distant were figures such as Wallace,[59] who had spent most of the 1850s in the Malay Archipelago pursuing his studies in natural science. He also found time for religious speculation, and had developed heterodox beliefs. While in Timor he wrote his brother-in-law, Thomas Sims, a daring letter on this subject.[60] "In my solitude I have pondered much on the incomprehensible subjects of *space, eternity, life* and *death*! I think I have fairly heard and weighed the evidence on both sides, and I remain an *utter disbeliever* in almost all you consider the most sacred truths."[61] On his return to England in 1862 Wallace discovered Spencer's *First Principles*, which seemed to be "a truly great work which goes to the root of everything".[62] His admiration of Spencer was almost idolatrous. He was the only prominent Victorian to name his child Herbert Spencer.[63] *First Principles* seemed to hold out infinite promise, and Wallace visited Spencer in order to ask him about the "nature of things".[64] This was unfortunate; Spencer was always nonplussed by naive and casual enquiry from strangers and avoided answering. Wallace went away disappointed.

Wallace's admiration for Spencer was unaffected by his treatment and he began to read Spencer's other writing, particularly his essays. Like Sara Hennell, Wallace was particularly struck by the nebular hypothesis.[65] He wrote to Darwin that Spencer was "going to show that there is something else besides 'Natural Selection' at work in nature", and warned him to look out for a 'foeman worthy of your steel'!"[66] Reading Spencer caused Wallace eventually to introduce concepts of non-natural selection into his own work to explain the growth of intelligence,[67] and this can be viewed as an attempt to reintroduce religion into evolution. Contemporary scientists were sometimes distressed by Wallace's adherence to spiritualism,[68] a doctrine to which he stayed loyal long after it ceased to be fashionable. Wallace's ideas always displayed a Spencerian tinge, and it is mistaken to view him solely as a Darwinian because of the emphasis he placed on natural selection.[69] Wallace should be read in such a way as to encompass his late work *Man's Place in the Universe*, where he concluded that evolution led to the worship of an Infinite Being, which cannot be comprehended.[70] This was the last shot to be fired by one of Spencer's friends on behalf of the New Reformation and spiritualism.

Huxley's description of Spencer as a rope-maker was not an idle phrase. Spencer attracted his early followers precisely because his thought was an organized form of the loose yarn of ideas already familiar to them. His function was that of a systematizer of science together with radical religious and metaphysical thought. Some of the cogency of his arguments rested on a sense of familiarity in his reader; Spencer echoed and magnified their ideas, and they, in turn, regarded him as prophetic.

Most serious-minded Victorians read *First Principles* and, when they did, they felt that it had religious meaning. Those who took Spencer seriously regarded him as a prophet of a new religion. It was as a seer that Spencer aroused interest, admiration, emulation and fierce hostility. Discourses with and about Spencer invaded the privacy of drawing rooms, and disrupted the congeniality of clubs. Laurencina Potter was typical when she carried on interminable discussions with Spencer about the origin of religion.[71] Her daughter, Beatrice Webb, recalled her own admiration for Spencerian religion, for "that still reverent consciousness of the great mystery; that fearless conviction that no advance in science can take away the beautiful and elevating consciousness of something greater than humanity".[72] This inspiration was fashionable. The great and the good entertained an affection for Spencer because, as often as not, their views on religion were Spencerian. Typical was Lord Amberley, who believed in an Unknowable that could be regarded with religious veneration, without the trouble of a personal God.[73] Spencer was also useful in debates. The Metaphysical Society,[74] a remarkable club devoted to the discussion of the antagonism between religion and science, opened its proceedings in 1869 with an attack on Spencer's "Unknowable". The attack was delivered by R. H. Hutton – a prophetic journalist – who was angered at Spencer's failure to offer a theory of morals based either on the "Unknowable" or utilitarianism.[75] This was slightly unfair because Spencer's *The Principles of Ethics*, although yet unpublished, was to be the crowning glory of "A System of Synthetic Philosophy". However, it is unlikely that Hutton (a Unitarian whose religious odyssey landed him in the established church) would have appreciated it.

Outside his immediate circle Spencer was "lionized" after the publication of *First Principles* in 1862. It was only then that his company was desired by those who made it their business to meet "name" authors. The ubiquitous Sir Mountstuart Grant Duff, who knew everyone, mentioned meeting Spencer on 25 April 1863.[76] Sir Frederick Pollock, who also knew everyone of importance, made the following diary entry for 29 June 1864: "Met Herbert Spencer, Speding, R. Browning, G. H. Lewes, and Holman Hunt at breakfast at Houghton's".[77] The New Reformation had arrived socially. However, its force dwindled so that by the mid-1860s Spencer was its sole luminary. By then his philosophical origins scarcely mattered; he had found acceptance as a great Victorian.

Buoyed by his newfound fame Spencer could reach vast audiences far beyond the narrow confines of an intellectualized and radicalized English religious faith. He became the purveyor of modern science to the newly awakened intellectual masses of America, Russia, China, India and Japan. In these countries he became a secular icon with which to confront traditional learning. Whether he desired this global following is unclear. His writings are sometimes framed in the provincial boundaries of a mid-century radical, and lack the universalism that his distant readers credited to all his utterances. Also, his own sense of modernism was often accompanied by a kind of *angst*. Since he never revised his deep faith in the "Unknown" and since this overlaid his psychology, metaphysics and biology, his legacy to the great world was always an ambiguous and doubting spiritual one, not a triumphant form of materialism.

TEN

Common sense in the mid-nineteenth century

The religious radicalism of the New Reformation forced young English writers to search for a philosophical fortress to defend their sense that the physical universe was meaningful. Their belief in a natural religion caused them to suspect the existence of the "Unknown". However, they feared that their explanations of this moral and ontological truth were puny or, worse, literary. The need for a scientific and philosophical language caused young spiritualists to turn their eyes towards Scotland, where there was a school of intuitionist philosophy that was sympathetic to science and hostile to the ideas of Auguste Comte. This Caledonian philosophy – which had been previously adopted by the Oxford don Henry Longueville Mansel – created a safe haven, a refuge from the fear that the universe was a void. It was this fear of emptiness that led Spencer, the chief exponent of this New Reformation, to expound philosophy, a subject that had largely been abandoned by his contemporaries.

It was as a philosopher that Spencer took up speculative knowledge about living things. This was an unusual occupation because Victorian natural scientists and psychologists usually wrote without much regard for the philosophical underpinnings of their work. However, he should not be understood as an isolated genius. He was a competent mid-Victorian exponent of "common-sense" philosophy who happened to be possessed of a scientific vision of how the living universe functioned. If one sets this aside, and looks at the language in which he expressed his beliefs, he appears to be part of a school. To be a "common-sense" philosopher required the profession of great admiration for Sir William Hamilton, and for his edition of *The Works of Thomas Reid*.[1] It was also usual, although not necessary, to have studied under Hamilton. "Common-sense" philosophers all published works during the 1850s. The most prominent[2] were Thomas Spencer Baynes, John Veitch, A. C. Fraser and Mansel. Two of these, Fraser and Mansel, were critical in establishing the context of the metaphysical system that Spencer later developed in *First Principles*. Their ideas were useful in Spencer's reconciliation of science with religion, with which he provided a bulwark against utilitarianism. This last doctrine, utilitarianism, was the chief irritant for Spencer; it was to defeat it that he recruited the assistance of "common sense".

Except for Spencer's psychology and metaphysics mid-nineteenth-century British philosophy has left little legacy: as a result it has seldom been chronicled. Historians of nineteenth-century philosophy have usually focused on their rivals – such as the exponents of utilitarianism and Oxford idealism – at the expense of a group whose popular impact, through Spencer's writings, was unrivalled. It is important to record some of this lost history in order to measure Spencer's achievement. Without his milieu he cannot be adequately understood.

Baynes was the first of these philosophers. He had won Hamilton's favour in 1846 when he was awarded the prize for the best exposition of the new logical doctrine that Hamilton had been propounding in his lectures. Baynes was encouraged by Hamilton to publish his essay.[3] It appeared in 1850, the same year his *The Port-Royal Logic* was published.[4] This publication, like the prize essay, shows Hamilton's influence, and bears a dedication to him.[5] Baynes was at the centre of this new school of philosophy, as a contemporary noted: "Baynes was the assistant to Sir William Hamilton in 1858, and that the *Port-Royal Logic*, and *The New Analytic*, had already secured a reputation for him in Edinburgh then at its zenith as a centre of metaphysics".[6]

Baynes's publications rejected notions of utility, which he saw as crass, in favour of liberal humanism. This was an extension of Hamilton's idea that philosophy was the gymnasium of the mind. Baynes stressed the value of mental science as a crucial part of liberal education:

> The study of philosophy, in its various branches, tends directly to stimulate those higher powers of the mind, through whose awakened action we enter into the possession of principles, and receive the birthright of thought. Other branches of education quicken and train *directly* the facilities of perception, of memory, of imagination, and, *indirectly*, also that of reasoning; but the allied mental sciences are directly and pre-eminently the gymnastic of understanding. Logic, as one of these, addresses a special faculty, – that of comparison or reasoning.[7]

Baynes saw Hamilton and the common-sense school as reintroducing true philosophy into a country where it had been degraded by crude and degenerate utilitarianism that rendered logic and mental science obsolete. He believed that all that remained of philosophy were fragments embedded in Bridgewater Treatises, essays on population and political economy, together with occasional disquisitions on Bentham and his greatest-happiness principle.[8] It was this last comment, the common-sense attack on Bentham and utility, that made Scottish philosophy attractive to young Englishmen such as Spencer. Spencer's own anti-utilitarian ideas, developed during the 1840s, fitted comfortably with the more finely worked "North British" objections to the greatest-happiness principle.

Veitch, although a much younger man, began his studies in 1845, the same year as Baynes.[9] His philosophical training began under Fraser at the Free Church College, but in 1848 he enrolled in Hamilton's university class where he fell completely under his influence. Veitch's personal letters were dotted with his teacher's name. Hamilton was the "Glory of Scotland", and the "Dictator in Metaphysics".[10] Veitch was so overwhelmed that he took Hamilton's classes twice,[11] and helped to found yet another of the metaphysical societies that seemed to proliferate in Edinburgh in this period. Like Baynes, Veitch was put to work translating the classics. His translation of Descartes's *Discourse on Method* appeared in 1850, and a translation of Descartes's *Meditations on First Philosophy*, together with selections from his *Principia Philosophiae*, in 1853. Veitch's reward was to be made assistant in the university's logic class in 1856, a position he retained under Hamilton's successor, Fraser.[12]

Much of Veitch's career was taken up with teaching and with editing Hamilton's lectures.[13] However, he did leave one piece of independent philosophical work that may be taken to epitomize his own philosophical views. He saw metaphysics as dealing with the question of "the true character and meaning of human experience – of existence, finite, and infinite, in a word, with the ultimate reality of man, the world, and God".[14] This metaphysical conception had little to do with Presbyterianism – Veitch had abandoned that along with Jonathan Edwards after reading Shelley[15] – but it was concerned with rejecting the positivism of Comte.[16] Veitch could not accept that all human strivings after knowledge and science could end with the statement that there was no power in the universe higher than mechanical force, chemical affinity or an obscure vital agency.[17] He felt that a purely scientific philosophy was inadequate because it left no grounds for believing in human personality, freedom, responsibility to a moral law or reverence for a deity.[18] Veitch saw the central problem of metaphysics in the same way as Spencer did; it was necessary to reconcile science with the "Spirit", which represented freedom and responsibility to a moral law. Such a reconciliation meant that one had to construct a theistic doctrine.

> If man be, as he is, an intelligence free and moral, – if he stand erect out of the mechanism of physical law, a power above nature as well as a being in nature, – he is assuredly in relation to, and analogy with, a supreme moral Intelligence, of whom he is the dim and distant image, and in whom, not in any inferior reality, he must seek the satisfaction of his intellectual and moral nature. A rational theism is the only legitimate issue of a psychology which recognises the distinctive reality of mind in man; and it is possible only on such a basis.[19]

Veitch was sketching a moderate philosophical doctrine with which to oppose materialism. To combat its inherent godlessness he searched for meaning in metaphysics and rational theism. This was a philosophy that would defend humanity and freedom.

Baynes and Veitch are interesting because although they do not directly influence Spencer, they occupy a similar philosophical position to that of the New Reformation, without any shadow of its orthodoxy. Their views paralleled Spencer's as they too were reacting against utilitarianism and Comtean positivism by extending on Hamilton's ideas. Spencer's connection with the other members of the common-sense school Fraser and Mansel is more direct. He had studied them with enthusiasm and his views echoed theirs.

Fraser was the most influential and prolific of Hamilton's students. He had begun his studies under Hamilton in 1838 when that "Dictator in Metaphysics" had no rivals in Britain. Thomas Brown, Dugald Stewart, James Mackintosh and James Mill were deceased, and J. S. Mill had yet to publish *A System of Logic*. Hamilton courted followers; his lectures were frequently followed by evening meetings at his home where philosophical difficulties were debated.[20] These gatherings were more than student seminars because they were occasionally supplemented by the presence of Hamilton's older friends J. F. Ferrier and J.W. Semple (the translator of Kant).

After he had fallen under the sway of Hamilton, Fraser developed a similar line of thought based on the work of Thomas Reid. Hamilton had already begun working on his edition of Reid as early as the 1830s and Fraser, like most of Hamilton's group, favoured Reid's doctrines. At the same time, Hamilton encouraged those of his disciples, such as Fraser, who were captivated by the personality of Thomas Chalmers, to question the value of philosophical statements.[21] Hamilton, as an Anglican and a Whig, was a remnant of the small liberal intelligentsia of Scotland, and not responsive to the revival of Presbyterian divinity.[22] For Fraser, philosophy soon began to replace theology, and he answered his ultimate questions with ideas borrowed from Reid's common sense:

> It was now, too, that I began to see in our Common-Sense or Common Reason a reservoir which holds for us in a latent state the *rationale* upon which human action and knowledge at last depend, and which it is the work of the philosopher to interpret.[23]

> But it was to this Common-Sense, philosophically criticized, that I now began to look for relief in the ultimate uncertainties about the universe in which I formerly found myself a stranger.[24]

Fraser's first major work appeared in 1858, two years after he had succeeded to Hamilton's chair. This was *Rational Philosophy in History and in System*, a work that aroused Spencer's sympathies, but whose austere prose style helped initiate the nervous disability from which he suffered during the late-1850s. Scottish metaphysics was always quite stressful even though it was an elevated "search for Ultimate Truth, or that unity of Reason which is conceived to be the final reward of the philosophical impulse".[25] Fraser, like other members of the common-sense school of philosophy, found the question of how to relate the real with the theological the most difficult one. "But the Real – presented or presentable to our consciousness cannot without a contradiction, in any conceivable meaning of representation, be a representative of *Infinite* reality."[26] He believed that all finite analogies would fail to represent the Infinite or Inconceivable.[27] He himself found no clear solution to this dilemma, and, instead, offered "practical guidance".[28] This would have to serve when ultimate truth was beyond human faculties. The spontaneous feelings of human nature became the substitute for an intelligible theory of existence. "Human nature in its normal integrity is the rational light of man. The Practical Revelation of the Common Sense is the last tribunal of Reason on these high questions."[29] Fraser's practical solution, on first sight, appears similar to J. S. Mill's more secular "art of life", but Fraser was careful to offer his reader the possibility of metaphysical certainty that Mill, with his residual Comteanism, would have found primitive. Also, for Fraser, reason rested on common sense. This meant that reason was subject to spontaneous belief of men (i.e. a belief may be presumed true until disproved), and that it might be incomprehensible.

> [I]t is not to be modified and distorted, to gratify the Understanding, by factitious symmetry of a logical system. The Revelation contained in the unwritten Book of Common Sense, must be accepted, even although the propositions, which the Revelation implies, may be collectively irreducible into a logical system of dogmatic metaphysics.[30]

Late in his life Fraser wrote a popular book on Thomas Reid in which he asked: "How has Reid's protest of Reason in the name of common sense – a protest against sceptical paralysis of human intelligence, physical and moral – fared in the nineteenth century?"[31] Such questions can be answered from Fraser's own writings for he was the most orthodox, or the least extreme, of the nineteenth-century common-sense philosophers. The function of common sense in Fraser's arguments was always to force them into a moderate philosophical position. That is, he trusted reason unless it called for the false symmetry of a logical system. Metaphysics could be used to search for ultimate answers, but when it pointed to answers that were contradictory or inconceivable, human beings should take refuge in practical guidance.

It is now possible with the help of the analyses of the work of Baynes, Veitch and Fraser to define the doctrine of common sense at this period more clearly. First and foremost, it meant philosophical moderation. On crucial philosophical issues, such as the dispute between idealism (then often called transcendentalism) and empiricism, common sense dictated a position mid-way between the antagonists. Secondly, in reaction to utilitarianism, which was seen as a debased pleasure–pain principle, it meant the reintroduction of the study of pure philosophy, or of metaphysics. That is, it was the belief that philosophy should be studied for its own sake, and this was accompanied by a renewed interest in classical philosophers. Thirdly, common sense, in its search for meaning and truth in philosophy, reawakened elements of liberal humanism. Nineteenth-century common-sense philosophers followed Reid in reasserting trust in reason, and to this they added an attempt to modernize religious conceptions by developing forms of rational theism that united science with "spirit". This, in turn, produced admiration for all-encompassing systems of philosophy. As one of Hamilton's students remarked, the highest range of philosophy is "an all-comprehensive and soul-exciting Cosmology".[32]

It was Hamilton who led the way here. The usually bluff and humorous Edinburgh medical man John Brown lost his characteristic poise when he wrote with fevered admiration about his early mentor. There was no doubt in his mind about Hamilton's grandeur.

> In this respect, Hamilton is as grand as Pascal, and more simple; he exemplifies everywhere his own sublime adaptation of Scripture – unless a man became a little child, he cannot enter into the kingdom: he enters the temple stooping, but he presses on, intrepid and alone, to the inmost *adytum*, worshipping the more the nearer he gets to the inaccessible shrine, whose veil no mortal hand has ever rent in twain. And we name after him, the thoughtful, candid, impressive little volume of his pupil, his friend, and his successor, Professor Fraser.[33]

Fraser was praised for his "little volume" *Rational Philosophy* published in the same year as Brown's papers. From it was picked the following quotation: "A discovery, by means of reflection and mental experiment, of the *limits* of knowledge, is the highest and most universally applicable discovery of all; it is the one through which our intellectual life most strikingly blends with the moral and practical part of human nature".[34] Brown, with his usual eye for significant detail – contemporaries considered him to be a profound and scientific observer – had extracted the part of Fraser's book that stressed the *limits* of knowledge. The limitation was Fraser's contribution to British theism.

At the same time as Brown, another philosopher, far away in Oxford, was quoting Fraser. This was Mansel, who has been conscripted as a Kantian and as a source for modern agnosticism,[35] but whose reputation in the mid-century was as the most prominent English philosophical trimmer. He attempted to erect a defensible position between the scientifically known and the mystically experienced. This was not an agnostic position because Mansel was not denying the possibility of knowledge. Nor was he reliant solely on belief. Instead, he was searching for the shifting point of balance between our knowledge and religious experience. Mansel's Bampton Lectures of 1858 provoked an extraordinary interest in Oxford, filling the undergraduate gallery at the university church, St Mary's.[36] These lectures were a publishing sensation, going into four editions in two years, and being reprinted on the Continent and in America.[37] At Cambridge the Master of Trinity College, William Whewell, felt it his duty to preach a sermon warning that these Bampton Lectures would not only make natural theology impossible, but would do the same to revealed religion: "If we *cannot* know anything about God, revelation is in vain".[38] The dismay among the orthodox did nothing to harm the work's popularity. After the lectures, Mansel received special recognition in Edinburgh, and visited Fraser while he was there to receive an honorary degree.[39] The Athens of the North was a second home for the Oxford philosopher. Mansel's lectures were wholly dependent on contemporary common-sense philosophy. He began with a quotation from Fraser's *Rational Philosophy*, and developed his argument by peppering it with references to another Hamiltonian, James McCosh. Like Brown, Mansel concentrated on the religious implications of the limits of knowledge: "It has been observed by a thoughtful writer of the present day, that 'the theological struggle of this age in all its more important phases, turns upon the philosophical problem of the limits of knowledge and the true theory of human ignorance'".[40] Mansel couched his idea in the language of Hamilton's article, "The Philosophy of the Unconditioned", adding that this had suggested the principal enquiries pursued in his work. The borrowing rephrased Hamilton so that "the Unconditioned is incognisable and inconceivable; its notion being only negative of the Conditioned, which last can alone be positively known or conceived".[41] This statement of principles is deceptively commonplace; Mansel's use of them was more subtle and less consistent than the work of any other member of the common-sense school.

Mansel had seen the chief difficulty with religious thought as the conception of the infinite. He accepted that one was compelled to believe in an infinite by the constitution of the mind.[42] But "[i]f all thought is limitation; – if whatever we conceive is, by the very act of conception, regarded as finite, – *the infinite*, from a human point of view, is merely a name for the absence of those conditions under which thought is possible".[43] This contradiction is the chief difficulty with all thought. Mansel always stressed that philosophical and theological problems were the same; that is, one is compelled to believe in something as inconceivable that is merely an absence of thought.[44] From this Mansel drew the conclusion, not very logically, that:

> It is our duty, then, to think of God as personal; and it is our duty to believe that He is infinite. It is true that we cannot reconcile these two representations with each other; ... . But it does not follow that this contradiction exists anywhere but in our own minds: ... . The apparent contradiction, in this case, as in those previously noticed, is the necessary consequence of an attempt on the part of the human thinker to transcend the boundaries of his own consciousness.[45]

Mansel's thought was radically unstable here. He believed that "it is one's duty" to believe that god is *personal* and *infinite*, even though this is a *contradiction*. He then added that such an impossibility exists only in our minds; it is merely an apparent discrepancy. For him the important thing was the tacit reference to a reality "out there" that contained no contradiction.[46] This was the case even if one only knew the reality through a *belief*[47] that one was compelled to accept by one's mental constitution. This was the complement of human consciousness and of the relative and finite. It was an unknown: something that had no definition.[48] In other words, it meant that "belief" was unable to provide knowledge about the nature of God, which had been the original purpose of staring at reality.

Mansel later abandoned his attempt to approach reality through "belief", and left it, with its difficulties, in the hands of his self-proclaimed pupil, Spencer. Instead, Mansel adopted the face-saving comment that "belief" had only a practical and regulative application, not a speculative and theoretical one, and that it did not allow men to know whether or not their conception of divine nature resembled that nature in its absolute existence.[49] Mansel proclaimed, "We

163

must remain content with the belief that we have that knowledge of God which is best adapted to our wants and training".[50] His philosophical posture appears to be uncertain, especially if it is compared to the stringent attacks that he made on his opponents' lack of certainty. This weakness can only be explained by examining the central tenet of mid-nineteenth-century common-sense philosophy: "moderation". Mansel's basic position – and its flaws – was not unique to him, but shared by all common-sense philosophers.

Mansel had refused to give pre-eminence to reason, will, feeling or the moral faculty (intuition). All were only to be partially trusted. The conclusion that he offered in his eighth and final lecture was that "no one faculty of the human mind is competent to convey a direct knowledge of the Absolute and the Infinite, no one faculty is entitled to claim precedence over the rest, as furnishing especially *the criterion* of the truth or falsehood of a supposed Revelation".[51] Since all elements of the human mind were seen at the same level, it followed that any philosophical doctrine that relied exclusively on any one of them was illegitimate. The course out of the morass was to build on all of them simultaneously. This would naturally lead to a "moderate" philosophical position that was easily explicable to the common man. "Moderation", therefore, should be seen as having the same strength as "common sense".

In Reid's tradition, common-sense philosophers were to avoid excessive reliance on reason – because of Hume's successful attack on it – and they should make a direct and immediate appeal to the mind of the common man. No series of inferences leading from established premises were permissible. Causal links were untrustworthy because they depended on reason. So Mansel replaced all philosophical structures, which relied on inference, with a simple reference to belief, which had only a practical and regulative value.

Philosophical "moderation" led Mansel to balance between the contemporary extremes of scientifically based materialism and mysticism.

> In the material world, if it be true that the researchers of science *tend towards* (though who can say that they will ever reach?) the establishment of a system of fixed and orderly recurrence; in the mental world, we are no less confronted, at every instant, by the presence of contingency and free will. In the one we are conscious of a chain of phenomenal effects; in the other of *self*, as an acting and originating cause.[52]

To claim moderation was not, as it might be in many cases, a politic disinclination to become entangled in a dispute. Mansel adopted his stance because he found himself, along with the remnants of orthodox Christianity, on one side of a cataclysmic divide in British thought. On the other side were Unitarians, Comteans and – worst of all – excessively liberal Oxford dons, whose exclusive allegiance to science left no room for religion. His opponents were diverse and included Parker, H. G. Atkinson, and Baden Powell, but it was the last that he feared the most. Parker was an American and Atkinson was self-educated, whereas Powell was an Oxford professor. His apparent neglect of religion was more repellent to

Mansel because the unorthodox Parker had drawn support from the extremism of Powell's views.

> Professor Powell, in his latest work, though not absolutely rejecting miracles, yet adopts a tone which compared with such passages as the above, is at least painfully suggestive. "It is now perceived by all inquiring minds, that the advance of true scientific principles, and the grand inductive conclusions of universal and eternal law and order, are at once the basis of all rational theology, and give the death-blow to superstition".[53]

Mansel's resistance to the conflation of science and national religion was the basis of his attack on liberal theology. Most of the writers he mentioned, including Parker, Strauss, Atkinson and Greg,[54] were published by the notorious Chapman. Powell, later a contributor to the controversial *Essays and Reviews*, had already provided a theological article to its liberal precursor *Oxford Essays* (1857). Mansel's broad brush painted Powell the same colour as Unitarians and writers such as Strauss.[55] To Mansel these spiritualists all held truth, including religious truth, to be discoverable by laws in which cause and effect were indissolubly chained together.[56] In one extreme form, spiritualism elided into positivism; in another it became mysticism, which was also a form of the new spirituality. According to Mansel, writers such as F. W. Newman held doctrines that tied together elements of science and religion without basing themselves on common sense. This complication distressed Mansel, and he went to great lengths in disentangling the strands of Newman's thought.[57] After he had done this he replaced Newman's synthesis with a philosophy of the relative that placed the mind on an equal level with matter. It was from this point that Spencer began to develop his own philosophy. He had been vitally interested in Newman's bridge between scientific knowledge and religion, and now he accepted the necessity of rebuilding it on a common-sense foundation. Spencer had already accomplished this task in psychology; he now realized that he had to produce a metaphysically sound *First Principles*. His construction relied on the formulas used by Hamilton and Mansel when rejecting rationalism (or crude empiricism) and its sub-species, positivism. He also adopted their practice of leaving a philosophical niche for the unexplainable. In Spencer's hands this became the "Unknown".

In the 1850s British philosophers would not have regarded themselves as belonging to two traditions. That perception would come later. Victorian philosophers rarely read each other and even when they did it was carelessly. J. S. Mill was forced to make a damaging admission about Thomas Brown and Richard Whately, the two chief luminaries of his own philosophical tradition, when replying to Hamilton, the leader of the anti-empiricists:

> Of all persons, in modern times, entitled to the name of philosophers, the two, probably, whose reading on their own subjects was the scantiest, in proportion to their intellectual capacity, were Dr Thomas Brown and Archbishop Whately: accordingly they are the only two of whom Sir

William Hamilton, though acknowledging their abilities, habitually speaks with a certain tinge of superciliousness.[58]

Mill's own reading habits left something to be desired. It was only in 1861 – several years after Hamilton's death – that he began a serious examination of that Scot's works and discovered him to have been the chief pillar of an erroneous school of philosophy.[59] Until then Mill had thought that only minor disagreements existed between him and the chief of Scottish philosophy.[60] This erroneous belief was remarkable considering that, by his death in 1856, Hamilton was distinguished as the only British philosopher besides Mill himself who possessed an international reputation.

Mill chose his allies in the same haphazard way that he chose his enemies. Referring to an article in the *North British Review* for September 1865, which had been written anonymously by Fraser, Mill thought that he might have left the defence of his own philosophical doctrine to Fraser.[61] This was curious since Fraser was almost a clone of Hamilton and his work would not have aided Mill.[62] Fraser's important early work, *Rational Philosophy*, was a transcendental metaphysics, a search for ultimate truth. Mill's criticism of Hamilton for constructing a religious metaphysics would have been more appropriately addressed to Fraser, whose work contained more divine connotations than Hamilton's had done. Mill's acquaintance with other contemporaries was also superficial. He never got to grips with the modern renaissance of formal logic that had begun with Augustus de Morgan and George Boole. Nor did he understand the significance of the modern mathematics being developed by George Peacock and Karl Weierstrass.[63]

Hamilton was even more prone to ignore his contemporaries than Mill had been. He ended a heated debate on logic with De Morgan by returning, within a week, a gift copy of the latter's *Formal Logic* with most of the pages uncut. The flyleaf of the book bore the inscription "Sir William Hamilton, Bt. from the Author"; it was followed by "The Author, from Sir William Hamilton, Bt."[64] Comte's *Philosophie Positive* fared even worse; Hamilton treated it as unworthy of even a contemptuous allusion. In the 1840s, when one of Hamilton's followers mentioned the Frenchman, Hamilton asserted that he did not know that Comte had any claim to be numbered among philosophers.[65] Hamilton professed equal indifference to the merits of Mill's philosophy, presumably because he suspected that it depended on that of Comte.[66]

With the two doyens of British philosophy, Hamilton and Mill, ignorant of each other's philosophical position, it was unsurprising that lesser men confused them and eclectically borrowed from either when constructing their own ideas. The young Herbert Spencer was typical when he conflated the ideas of Hamilton's edition of Reid with Mill's *A System of Logic*: it seemed permissible to combine the former's intuitive metaphysics with the latter's empiricism. Although Spencer usually came down on the side of intuition rather than experience, he believed that he was striking a balance between the two compatible and parallel traditions of British philosophy. Similarly, James McCosh's first work, *The Method of Divine Government* (1850), equally praised Hamilton and Mill. The former had completed all past metaphysics on the subject of fundamental ideas by showing that the argument was based on common sense, yet was scientific and philosophic at the same time.[67] McCosh demonstrated to his own satisfaction that there was no important conflict between Mill's *A System of Logic* and Hamilton's philosophy, but this task was an easy one since no one thought differently.[68] McCosh's summation of the slight differences between the two, which appears naive in light of the later controversy, was that Mill had set an insufficient value on "original data or ultimate premises of our knowledge";[69] such a peculiar lack of emphasis must have been an oversight.

It was only after 1865 that British philosophers perceived a fundamental rift between Hamilton's philosophy and Mill's. This perception was the direct result of the publication of Mill's *An Examination of Sir William Hamilton's Philosophy*. Although Mill seemed to have exaggerated and, to some extent, invented philosophical differences, the effect of this book was immediate. It instantly stopped McCosh attempting to reconcile Mill with Hamilton. Instead, he used the warning that "By far the ablest opponent of intuitive truth in the country, in our day, is Mr John Stuart Mill".[70] Mill was accordingly reported as denying the possibility of any knowledge beyond phenomena while simultaneously maintaining that "we know body immediately", which implied that we did have intuitions.[71] McCosh's remarks were followed by "intuitional" attacks on empiricism by Veitch, Mansel and Spencer.

Suddenly there were two separate traditions of philosophers, one intuitionist and the other empiricist. The first school intuited truth from first principles, while the second discovered it by scientific experiment. This division of British philosophy does not pre-date Mill's 1865 work in which he unfairly condemned Hamilton's philosophy for contributing nothing to science.[72] The accusation troubled its victims, not only because they were all busily employed in extending the frontiers of science, but because they had taken much of their inspiration from Mill himself. Mill's philosophical reputation had rested on *A System of Logic*, which was originally published in 1843, before any member of Hamilton's revived tradition of common-sense philosophy had even published. Mill's *A System of Logic* was a key work for McCosh, Veitch, Baynes, Fraser, Ferrier, Mansel and Spencer, all of whom began coining their philosophical ideas during the 1850s. Hamilton himself was a successor to Mill. His edition of Reid was first published in 1846, and the first British edition of Hamilton's *Discussions on Philosophy and Literature Education and University Reform* appeared in 1852.[73] These dates are significant, although they have often been disregarded by commentators on Victorian philosophy who are chiefly familiar with Mill. It is important to note the twenty-three-year gap between the first edition of *A System of Logic*, a work aimed at overthrowing Whewell's philosophy, and Mill's *An Examination of Sir William Hamilton's Philosophy*. One should not accept the statement in Mill's *Autobiography* that his philosophy was unchanging.[74] A proper understanding of Spencer's ideas depends on the recognition of a temporal gap in Mill's philosophy. That is, an evaluation of Spencer's work is reliant on understanding that there can be a synthesis between Mill's early empirical work and the intuitionism of those whom he later declared to be his enemies.

It was the second of these doctrines, intuitionism, that most attracted Spencer. He loudly proclaimed his allegiance to the intuitionist school of common-sense philosophy in every writing from 1850 to 1865. It was the badge of philosophical respectability to which he clung even when ignored by other members of the school. They were a clannish lot and spent much of their time debating who was a "true" follower of Hamilton.[75] Despite his exclusion Spencer knew that he too was one of Hamilton's disciples because he had founded his first principle on intuition. McCosh ignored Spencer until he was spurred into action by Mill's *An Examination of Sir William Hamilton's Philosophy*. This had bracketed him with Spencer, who, when scrutinized closely, was revealed to have taken the doctrines of Hamilton and Mansel to their logical conclusions, which they, of course, would have disavowed.[76] McCosh's rejection of Spencer involved the reinstatement of "self" as a basic concept to replace Spencer's "belief". Once this had been done McCosh could set aside Spencer's universal postulate (the inconceivability of its negation is the test of a belief) as a mere secondary proposition. This emendation forced one to return to self-evidence as fundamental.[77] The rest of Spencer's philosophy was rejected without arguments being offered; to McCosh it was an ambitious but lamentable failure. The task Spencer had chosen was simply beyond human capacity.[78]

Mansel, although also viewed with suspicion by McCosh, was – at least in his own eyes – a faithful follower of Hamilton. Like McCosh, Mansel had ignored Spencer until the publication of Mill's *An Examination of Sir William Hamilton's Philosophy*, which had placed the author of *First Principles* in Hamilton's school.[79] Spencer suddenly became noteworthy as:

> a recent writer, who strangely enough, professes to be [Hamilton's] disciple, while rejecting all that is characteristic in his philosophy. Mr Herbert Spencer, in his work on *First Principles*, endeavours to press Sir W. Hamilton into the service of Pantheism and Positivism together, by adopting the negative portion only of his philosophy – in which, in common with many other writers, he declares the absolute to be inconceivable by the mere intellect ...[80]

However, it was too late for the Hamiltonians to expel Spencer from their midst. Spencer had been associated with them for years. This uncomfortable link had begun in 1853, when he proudly sent "The Universal Postulate" to Hamilton. The only acknowledgment was a terse note indicating that the recipient admired the talent with which the paper was written, although he did not agree with it, and that he was unable to answer at present.[81] Unabashed by this, Spencer continued to claim that his ideas were in harmony with those of Hamilton:

> Protagoras, Aristotle, St. Augustin, Melancthon, Scaliger, Bacon, Spinoza, Newton, and Kant, all teach "that men can know only the finite". Sir William Hamilton, who cites these and many other authorities, teaches the same; as does also his disciple Mr. Mansel. If then, for teaching "that man can know only the finite", I am chargeable with antagonism to sacred ideas, as too are all these.[82]

The boldest use of Hamilton's name came in the prospectus of Spencer's "A System of Synthetic Philosophy", which began with the statement that the unknown author would be:

> carrying a step further the doctrine put into shape by Hamilton and Mansel; pointing out the various directions in which science leads to the same conclusions; and showing that in this united belief in an Absolute that transcends not only human knowledge but human conception, lies the only possible reconciliation of Science and Religion.

Although it was originally printed for private circulation, the prospectus prefaced every edition of *First Principles*. It also accompanied volumes outside Spencer's "A System of Synthetic Philosophy", such as *Education*.

The constant brandishing of the names Hamilton and Mansel did not always reap approval. Some of his subscribers reacted with scepticism or hostility to such a banner. Froude confessed himself unable to understand it at all: "Mansel says the absolute is 'the unknowable'. How by following out his reasoning you are able to establish a belief in it I am curious to see".[83] He recommended that Spencer read Spinoza, which, given Spencer's distrust of rationality and his "spiritualist" yearnings after meaning in the universe, was rather harsh advice.

Another subscriber, Sir John Herschel, had different reservations. The names Hamilton and Mansel meant nothing to him, nor did their use of the word "Absolute". His concern was that Spencer had unwittingly adopted the shibboleths of the Hegel and Schelling schools of German philosophy. Spencer needed reminding that the Anglo-Saxon mind could not be led astray by a word:

> If by the "Absolute" it is intended to indicate that ultimate point (very different from $A \cong A$) in which all knowledge must centre and in which the Spiritual and Material worlds (already described as arithmetical) are connected and reconciled – that which "transcends not only human knowledge but human conception" and in which lies the only possible reconciliation of Science and Religion – well and good – but I wish it could be spoken of under some form of expression (and our language is not wanting in power) which would not recal [*sic*] the feeble peculiarities of the German authors I have named, who have done a crying injustice to a mighty theme.[84]

Almost as ethnocentric as Herschel's Anglo-Saxon bias was Mill's reaction. He had not confused Spencer with Hegel or with the positivists, but correctly classed him with the philosophers led by Hamilton:

> No Englishman who had read both you and Comte, can suppose that you had derived much from him. No thinker's conclusions bear more completely the marks of being arrived at by the progressive development of his own original conceptions; while if there is any previous thinker

to whom you owe much, it is evidently (as you yourself say) Sir W. Hamilton.[85]

In *An Examination of Sir William Hamilton's Philosophy*, Mill refashioned his remarks in a stronger manner:

> In this Sir W. Hamilton is at one with the whole of his own section of the philosophical world; with Reid, with Stewart, with Cousin, with Whewell, we may add, with Kant, and even with Mr. Herbert Spencer. The test by which they all decide a belief to be a part of our primitive consciousness – an original intuition of the mind – is the necessity of thinking it.[86]

This comment hurt Spencer's feelings in a way that it would not have done the previous decade when Mill had associated his philosophy with his own.[87] By the mid-1860s, Spencer was shedding some of this intuitionism because of his work on the philosophy of biology. He now felt closer intellectually to Mill because of his own disapproval of scientific Platonists such as Goethe and Richard Owen. Since Platonists resembled intuitionists in their scepticism about the possibility of discovering truth through empirical investigation, Spencer nimbly modified his intellectual alliance. This was difficult because he wished his metaphysics to remain unchanged while he gained support for his data-laden views on science. The moderate position, which had been uncontroversial during the 1850s, now seemed contradictory. Since he could not resolve this problem philosophically Spencer determined on a personal appeal. The friendly terms he was on with Mill allowed him to protest successfully about being classed with intuitionalists. Consequently, the third edition of Mill's *An Examination of Sir William Hamilton's Philosophy* obligingly omitted Spencer's name from the above quotation. However, despite his courtesy Mill indulged in a rare flash of wit and added a footnote explaining that he had not really changed his mind, and that Spencer could be refuted with the same arguments that were effective against similar philosophers.[88]

This jest was true; Spencer's psychology and metaphysics relied to some extent on intuitionism. However, it was also unkind; the contradiction between Spencer's philosophical foundations and the empiricism of his philosophy of science only existed because Mill's *An Examination of Sir William Hamilton's Philosophy* had declared it to be so after Spencer had begun his work with Mill's support. When Spencer designed his "A System of Synthetic Philosophy" it was still possible to reconcile one's intuition with science, and although troubled by the imposing difficulties this entailed, he proceeded as if it was still an attainable goal.

ELEVEN

From philosophy to psychology

Although common sense spoke clearly to the people, its philosophical advocates had reservations. These began with Thomas Reid's commentator, Dugald Stewart, who worried that common sense would draw on ignorance in such a way as to prevent free enquiry and uphold popular error.[1] His remedy for this laid down principles that had the additional advantage of causing the philosophy to develop into a science of psychology. Stewart stressed Reid's appeal to the constitution of human nature.[2] Deepening this emphasis his intellectual heirs focused on the laws and principles of the mental constitution. This led to Reid's philosophy being modified into a psychology by Hamilton in his supplementary dissertations and lectures,[3] by McCosh in *The Intuitions of the Mind*,[4] and by Spencer and Bain in the works they published during the 1850s. All of these publications framed the constitution of the mind as the workings of the brain.[5] Even Mansel, the common-sense philosopher who was most divorced from science, gave a powerful lecture entitled *Psychology, The Test of Moral and Metaphysical Philosophy*, which reduced virtues to psychological facts, discovered through investigations by the moral faculty.[6] Nineteen years later, one member of Mansel's audience still clearly remembered this lecture as "an assertion of the reality and necessity of psychology as a study and as a science".[7]

In the nineteenth century mental science was usually called association psychology. It had entered Reid's *An Inquiry into the Human Mind* as the answer to scepticism: fact was to reply where reason failed. The underlying premise was that human beings constantly infer there is a power or faculty of thinking and a mind that exercised that power. Yet it was impossible to demonstrate how these inferences could be made by logic or reason.[8] Connections between things, such as that between a sound and a passing vehicle, were established immediately, without reasoning. As Reid explained: "It is the effect of a principle of our nature common to us with the brutes".[9] These were first principles for which no further reason could be given other than that, by the constitution of our nature, we were compelled to assent to them.[10] An example of a first principle was "that similar effects proceed from the same or similar causes".[11]

Before Reid could wholeheartedly adopt association psychology as a defence against scepticism, it first had to be cleansed of godless elements. He had extensively criticized David Hartley's version of associationism because its doctrine of vibrations had a tendency to make any operation of the mind passive, a "mere mechanism, dependent on the laws of matter and motion".[12] Reid saw this as a tarnish, and to remove it he introduced moral senses as *active* faculties.[13] This marked the beginning of a split among the association psychologists: between the mechanists, such as Hartley, Condillac, James Mill and J. S. Mill and the intuitive or common-sense philosophers, such as Reid, Hamilton,[14] McCosh and Spencer. The only major exception to this division among mid-Victorian psychologists was Bain, who wavered uncomfortably between the two groups although usually siding with the former.

Reid's introduction of active faculties was the only significant change he made to association theory. For the rest he was content to repeat Hume's formulation of the three laws of association determining the order and succession of ideas, which were resemblance, contiguity and relation of place, cause and effect.[15] Hamilton too followed this formulation, although, in an idiosyncratic fashion, he gave new names to all the laws. Similar laws were accepted by everyone, even by James Mill and J. S. Mill.[16] This meant that when Hamilton adopted such a description of the laws of association he was merely reiterating a conventional formula. Adherence to this was the basic point of agreement between those who are usually described as association psychologists; it was incontestable. This is why Fraser, who was an intuitionist,[17] sounded like J. S. Mill.[18] While these philosophers disagreed with each other, and with Spencer on many issues, the laws of association were neutral conventions.[19]

After Reid had introduced the concept of *active* faculties, his definition of truth followed necessarily. Truths had to be divided between those that rested on the active faculties or the will, and eternal ones, which did not:

> The truths that fall within the compass of human knowledge, whether they be self-evident, or deduced from those that are self-evident, may be reduced to two classes. They are either necessary and immutable truths, whose contrary is impossible; or they are contingent and mutable, depending upon some effect of will and power, which had a beginning, and may have an end.[20]

To illustrate this distinction Reid chose an example from geometry: "That a cone is the third part of a cylinder of the same base and the same altitude, is a necessary truth. It depends not upon the will and power of any being. It is immutably true, and the contrary is impossible".[21] When Spencer came to borrow Reid's ideas he adopted all of them with little modification. However, Spencer did not keep company with Reid when the latter continued his analysis to include as a contingent truth the statement that the sun is the centre about which the earth and the other planets perform their revolutions. It was at this point that Reid's Christian formulation of science became unpalatable to Spencer. He could not accept Reid's belief that something was contingent because it was dependent on the will of the Being who made the sun and planets, and who gave them motions that best suited Him.[22] Their parting of the ways was not because Spencer was, at this point, secular; rather, it was because his kind of religiosity left no room for the unknown to intervene actively in the universe.

Reid had given two lists of first principles: one for contingent truths and one for necessary truths. He began with a dozen primary principles for contingent truths that were the foundations of association psychology and philosophy. The orthodox Scot felt the need to defend these against his opponents. They included: "*the existence of everything of which I am conscious*",[23] "*those things do really exist which we distinctly perceive by our senses, and are what we perceive them to be*"[24] and "*the natural faculties, by which we distinguish truth from error, are not fallacious*".[25] Reid's second list – first principles for necessary truths – had classes: which included grammar, logic, mathematics, taste, morals and metaphysics.[26] Confusingly, Reid's last class, metaphysics, included as necessary truths several first principles that had already been established as contingent truths. Such an overlap was not due to carelessness. Rather, it was caused by a fundamental split between Reid's Baconian empiricism and his anti-experiential rationalism. (The latter was later developed in an extreme form by J. F. Ferrier.)

Reid's distinction between necessary and contingent truths had never been very certain. In his first work, *An Inquiry into the Human Mind*, he ignored contingent truth, insisting that "All reasoning must be from first principles; and for first principles no other reason can be given but this, that, by the constitution of our nature, we are under a necessity of assenting to them".[27] In his last work, *Essays on the Active Powers of Man*, the need to distinguish between necessary and contingent was again put aside: all truths became necessary if experience gave no information about them.[28] Truths were intuitions or first principles that were not discovered by exploring what was contingent. The "sure mark of a first principle, [is] that the belief of it is absolutely necessary in the ordinary affairs of life".[29] This troublesome statement worried Stewart, who warned that obvious self-evident truths should be called fundamental laws of human belief – rather than the principles of common sense – so they would not be confused with mathematical axioms. However, despite this caution, Reid's two most popular nineteenth-century disciples, Hamilton[30] and Spencer, followed Reid in considering all truths to be necessary ones.

Hamilton emphasized that cognitions were uniformly second-hand. Furthermore, he believed that the original or primary cognitions necessarily manifested themselves as *facts* rather than as abstract thoughts.[31] Using Reid's language of necessary truths, Hamilton proposed that the cogency of such *facts* depended not on the immediate necessity of thinking them – for if carried out unconditionally they were themselves incogitable – but on the impossibility of thinking something to which they were directly opposed, and from which they have been immediately recoiled.[32] He concluded the speculation with the remark:

> I may here also observe, that though *the primary truths of fact*, and the *primary truths of intelligence* (the *contingent* and *necessary* truths of Reid)

form two very distinct classes of the original beliefs or intuitions of consciousness; there appears no sufficient ground to regard their sources as different, and therefore to be distinguished by different names. In this I regret I am unable to agree with Mr Stewart.[33]

This was a tenuous position, which Hamilton later attempted to fortify by introducing two kinds of necessary truth, one of which was similar to the contingent kind that Reid had left aside.[34] Consequently, Hamilton could *theoretically* deny a contingent or factual truth without being explicitly able to *think* this.

> Thus, for example, I can theoretically suppose that the external object I am conscious of in perception, may be, in reality, nothing but a mode of mind or self. I am unable however to think that it does not appear to me – that consciousness does not compel me to regard it – *as* external – *as* a mode of matter or not-self. And such being the case, I cannot practically believe the supposition I am able speculatively to maintain. For I cannot believe this supposition, without believing that the last ground of all belief is not to be believed; which is self-contradictory.[35]

According to Hamilton, a contingent truth was a vacuous necessary one; a truth that could be disputed only if it involved a self-contradiction. Thus Hamilton had effectively defined all truth as necessary; he would otherwise have been contemplating a vacuous category.

The disinclination of Reid and Hamilton to draw a useable distinction between necessary and contingent truths did not mean these were treated in the same way. Speculations were limited so as to apply only to the former in such a way that a necessary truth became one whose opposite was inconceivable.[36] This became the foundation of Spencer's psychology, metaphysics and, to a lesser extent, philosophy of science. It was this concept of "necessary truth" that underpinned his universal postulate. From this he was able to claim that all knowledge was of the same kind, and should therefore be considered in the same manner. To verify a piece of knowledge one simply established whether or not it was necessary. This use of reason to perform a mental experiment was so fundamental that it caused Spencer to cast aside the conventional distinction between scientific and ordinary knowledge. "The same faculties are employed in both cases; and in both cases their mode of operation is fundamentally the same."[37] However, Spencer was alert to the danger that a misuse of the term "faculties" could reinvent a division between kinds of knowledge. He wanted to prevent an outcome where some knowledge would be seen as the product of the will, some as the product of the feeling and some as the product of the intellect. So Spencer took the radical step of abolishing faculties as well as distinctions between kinds of knowledge. In *The Principles of Psychology* he held that, "Instinct, Reason, Perception, Conception, Memory, Imagination, Feeling, Will, etc., etc., can be nothing more than either conventional groupings of the correspondences; or subordinate divisions among the various operations which are instrumental in effecting the correspondences".[38]

Spencer partly owed this astonishing abandonment of the concept of faculties to his old friend, Thomas Hodgskin. In April 1855, while he was in Derby writing *The Principles of Psychology*,[39] Spencer had written to Hodgskin about the latter's review in *The Economist* of a new philosophical work by Samuel Bailey. Hodgskin had raised a point that Spencer had not previously considered:

> [Y]ou agree with [Bailey] in describing what are commonly known as *faculties* of certain kinds, as being in truth certain *classified mental operations*; and that our names for these faculties are merely names under which we group these classes of operations. Do you at all mean to imply in this that these various operations are *in themselves* processes of classing?
> ... it has a bearing on some of my inquiries.[40]

Spencer accepted Hodgskin's critique; these thoughts impacted heavily on his conceptualization of mental science. It especially altered his construction of faculties in *The Principles of Psychology*. As a result they became a mere set of conventions: a set of terms allowing him to classify mental phenomena. For Spencer classification became identical with the process of thinking. It was significantly more important to him than any real or imaginary division of the powers used when engaged in categorization. That is, it no longer mattered to Spencer if the process was performed by the intellect, the feelings or the will, because this had no impact on the process of thought.

Spencer postulated that there was one kind of knowledge, which had different gradations, rather than different kinds of knowledge based on separate faculties. This was his general theory of psychology, the purpose of which was to combine intuitions with experience. In this, Spencer's endeavours coincided with those of other nineteenth-century common-sense philosophers; they all advanced similar doctrines of philosophical moderation. However, Spencer's uniqueness was his combination of intuition and experience under the banner of the development hypothesis.

> [T]he simple universal law that the cohesion of psychical states is proportionate to the frequency with which they have followed one another in experience, requires but to be supplemented by the law that habitual psychical successions entail some hereditary tendency to such successions, which under persistent conditions, will become cumulative in generation after generation, to supply an explanation of all psychological phenomena; and, among others, of the so-called "forms of thought".[41]

Spencer introduced an evolutionary theory in order to prop up the intuitionist part of his common-sense philosophy. To him "forms of thought" or intuitions were necessary truths because it was impossible to dispose of them. Moreover, evolution had cemented them into the mind as necessary truths.

Spencer's earliest mention of Reid and common sense was in *Social Statics*.[42] This was before Spencer had begun his own investigation of psychology and at the time it was simply a reference to a useful moral philosopher. Reid was a member of the Shaftesbury school, which had assumed the existence of a faculty called "moral sense" or "common sense" that generated intuitions about right and wrong in human affairs. It was useful for Spencer to have traditional allies when he was preoccupied with supplanting the immoral doctrines of Benthamite utilitarianism, although at the time he accepted common-sense philosophy only with reservations. While it was not the "profoundest" doctrine "it is perfectly in harmony with that in its initial principle, and coincident with it in results".[43] Spencer was concerned that common sense had pretended to be infinite although it varied with each race, individual and time. However, he believed he could remedy this by constructing a purely synthetic morality.[44] This strategy might avoid the mistake of the Shaftesbury school, which was the assumption that instincts or intuitions generated by common sense were capable of solving all ethical problems that were submitted to them. "To suppose this, was to suppose that moral sense would supply the place of logic."[45]

While Spencer felt it unhelpful to compare geometrical and moral axioms he insisted on using such language himself. He believed that the moral sense needed a fundamental rule so that "reason may develope [sic] a systematic morality".[46] This idea did not hold; later, in Social Statics, Spencer abandoned the idea that one should reason from an axiom and decided that common sense was a purely selfish instinct. This led to difficulties; if one could not reason about one's relations with others then it was more difficult to assign rights to them. His answer to the question, "Whence comes our perception of the rights of others?"[47] directed readers to Adam Smith's The Theory of Moral Sentiments, where it was suggested that social conduct was determined by sympathy. Spencer amended Smith so that justice became only "a sympathetic affection of the instinct of personal rights – a sort of reflex function of it".[48] This idea was not argued in detail, and a close examination of Spencer's Social Statics would suggest that in 1851 he was only superficially acquainted with the work of Reid, and the rest of the Shaftesbury school and knew their work even less well than he did Smith's.[49]

In Social Statics Spencer had already begun to reject the notion of "faculties" when writing about moral sense. Although he could not express himself in the technical language of psychology until the mid-1850s, he started speaking about "the moral sense" in a way that sat uncomfortably with traditional common-sense philosophy. He assumed that moral sense was an evolutionary doctrine that would be incompatible with the traditional political theory that still guided European states:

> That moral sense whose supremacy will make society harmonious and government unnecessary, is the same moral sense which will then make each man assert his freedom even to the extent of ignoring the state – is the same moral sense which, by deterring the majority from coercing the minority, will eventually render government impossible.[50]

In this passage moral sense was no longer a faculty of the mind, but an unexplained force impelling people to act in a certain way.

Spencer's conception of moral sense in *Social Statics* was ambiguous. On the one hand, he saw it serving an active power that produced intuitions of good within the individual, but, on the other, he considered it as the cohesive force that governed the masses without need of earthly law.[51] There was no systematic development of either conception; they simply shared space in his political theory. As he himself admitted, up to the autumn of 1851, a year after *Social Statics* was written, he studied neither questions of philosophy, nor the phenomena of mind.[52]

At the end of 1852 Spencer began a systematic study of philosophy that was to produce its first fruits in October 1853. This was an article entitled "The Universal Postulate".[53] In preparation Spencer read Lewes's *The History of Philosophy* and J. S. Mill's *A System of Logic*, both of which followed Comte in rejecting metaphysics as a primitive form of knowledge. Spencer spurned these guides and stoutly attempted to re-establish metaphysics on the foundation of common sense. However, in doing this he jettisoned those views of common sense that he had espoused in *Social Statics*. Common sense no longer naively portrayed faculties that generated instincts of right and wrong. Instead, it had become an investigation of the truth of "universal beliefs" in the real world. These "beliefs" were universal because they were held by everyone; they were intuitions about metaphysical truths. Their purpose was to overturn the scepticism that had denied any possibility of real knowledge.[54]

The method used by Spencer in pursuit of these conclusions – which would seem mysterious if he is mistakenly classed with empiricists or positivists – was based on reason rather than experience. He could claim – with great ease – that experience was a subordinate means of acquiring knowledge. This, of course, left experience in an unsatisfactory position. Spencer had objected to one of Hamilton's main arguments by remarking that his premises would not be uniformly confirmed by people's experience, "but there would be no sufficient warrant for his conclusions" even if experience had invariably endorsed them.[55] This seemed to set Spencer squarely against the standard Baconian idea that we verify our scientific theories by checking them against the world. When he established his own main postulate that "universal belief" was the ultimate basis of thought, he demonstrated, through reason, why this must necessarily be so, remarking in passing, "That this division corresponds with experience hardly needs saying".[56] Significantly, Spencer almost neglected to include experience at all. At this stage of his thought, he held very similar views to another follower of Hamilton, Ferrier. This young Scottish philosopher had denied the importance of experience, and established certainty in knowledge, which was grasped by postulating premises and deducing principles of knowledge from these. The main difference between these men was that Ferrier more definitely rejected experience, using the philosophical language of nineteenth-century Germany, while Spencer took refuge in the eighteenth-century language of Reid.

Spencer's purpose in "The Universal Postulate" followed Reid's. He aimed to defeat the monster born in the year 1739.[57] This referred to Hume's *A Treatise of Human Nature*, which had originated in Descartes's ideas, and had marked the birth of modern scepticism. Spencer's reading of Hume and Reid revived this ancient controversy. There were costs to the choice; Spencer gave the work of the eighteenth-century Scots much more careful consideration and sympathy than he offered to later moral philosophers, writers such as Kant and Whewell.[58] In fact, Spencer was so carried away that he joined in the antique dispute; he boasted that he would demonstrate against scepticism what Reid had merely asserted.[59]

Spencer's refutation of scepticism relied on belief having priority over personal existence, consciousness and ideas. This was an attempt to establish a more basic principle than the cognition on which Descartes had based existence.[60] Spencer stated that "belief is the recognition of existence – is a knowing of the existent from the non-existent".[61] Therefore, he continued, our intellects find belief to be the fact prior to, and inclusive of, all other facts.[62] At this point, Spencer asked, "Why do we consider certain of our beliefs more trustworthy than others? What is the peculiarity of those beliefs which we can never question, and to which all the rest of our beliefs defer?"[63] By beginning with a Cartesian-like first principle Spencer was abandoning one of Reid's strongest points. That is, Reid was able to appeal to popular belief when verifying his arguments against philosophical opinion, and to show that both reason and experience could be trusted. Spencer, however, was unable to make such an appeal because people do not trace their opinions to one ultimate fact. In adopting this non-populist position Spencer became the most illegitimate offspring of Hamilton's revival of Reid and the philosophy of common sense.

Hamilton, in his supplementary dissertations to the works of Reid, "On the Philosophy of Common Sense; or Our Primary Beliefs as the Ultimate Criterion of Truth", asked the basic question that Spencer later formulated in his universal postulate. Hamilton's starting-point was that not all cognitions were second-hand. Any demonstration had to rest, at some point, on propositions that carried their own acceptability. There were primary propositions that were inexplicable and, therefore, incomprehensible. They manifested themselves not as cognitions, but as *facts*, of which consciousness gave assurance under the simple form of *feeling* or *belief*.[64] As a rider to this statement Hamilton had added that the analysis and classification of these facts, or primary elements of cognition, which he had passed over here, was:

> one of the most interesting and important problems of philosophy; and it is one in which much remains to be accomplished. Principles of cognition, which now stand as ultimate, may, I think, be reduced to simple elements; and some which are now viewed as direct and positive, may be shown to be merely indirect and negative; their cogency depending not on the immediate necessity of thinking them – for if carried unconditionally out they are themselves incogitable – but in the impossibility of thinking something to which they are directly opposed, and from which they are the immediate recoils. An exposition of the axiom – That positive thought lies in the limitation or conditioning of one or other of two opposite extremes, neither of which, as unconditioned, can be realized to the mind as possible, and yet of which, as contradictories, one or the other must, by the fundamental laws of thought, be recognized as necessary; – the

exposition of this great but unenounced axiom would show that some of the most illustrious principles are only its subordinate modifications, as applied to certain primary notions, intuitions, data, forms, or categories of intelligence ....[65]

In formulating his universal postulate Spencer was simply enouncing, as he would say, the great axiom for which Hamilton had called. That is, Spencer enounced axioms that were basic intuitions or beliefs that would establish "facts" (truths) about the real world. Spencer's axiom or universal postulate was as follows:

> Not only, however, is the invariable existence of a belief our sole warrant for every truth of immediate consciousness, and for every primary generalization of the truths of immediate consciousness – every axiom, but it is our sole warrant for every demonstration.[66]

That is, our cognition of logical necessity has no more certainty than anything else, and by reasoning we cannot transcend ordinary knowledge. Further, Spencer stated that "*a belief which is proved, by the inconceivableness of its negation, to invariably exist, is true*".[67]

The only important element of Spencer's explanation of belief that is not contained in Hamilton's dissertations on Reid is that the universal postulate allowed one to test immediately the truth of a *belief*, by showing its negation was inconceivable, without going through a chain of deduction. This emendation was adopted because Spencer considered a single use of reason to be more trustworthy than a series of inferences.[68] After establishing this additional point Spencer felt that he had reconciled philosophy with the kind of common sense that would oppose scepticism and be immediately understood. In any case, he believed that no other doctrine was sustainable:

> [T]he current belief in objects as external independent entities, has a higher guarantee than any other belief whatever – that our cognition of existence considered as noumenal has a certainty which no cognition of existence, considered as phenomenal, can ever approach; or, in other words – that, judged logically as well as instinctively, Realism is the only rational creed; and all adverse creeds are self-destructive.[69]

Spencer had delivered a trumpet blast against Hume and godlessness while reaffirming the philosophy of Reid. Also, he credited himself with protecting philosophy from the taint of German idealism that Hamilton had introduced by distinguishing between ego and non-ego.[70] After such a *tour de force* it is hardly surprising that Spencer's subsequent philosophical writings were almost all retreats or rearguard actions to defend the bold and complete philosophical premise he had laid down.

The changes Spencer made after the 1850s concerned the weakest point in his idea of the universal postulate: *truth*. That is, he needed to defend the statement that universally held beliefs are true. Spencer had already shown some ambiguity and hesitation about the word *true*. He was unsure whether it was an assumed correspondence between some objective fact and our subjective state, or only the continued existence of the belief to which it is applied. However, he felt that he could sidestep the issue by restricting his investigation to the contents of the intellect solely as a system of beliefs. In this way he hoped to avoid discussing the absolute validity of a system and focus solely on whether it was relatively valid.[71] In other words, Spencer claimed that our intuitions were true when they represented an assumed correspondence between object and subject, or if they had a continued existence. However, he did not pursue this idea further because he simply believed that some intuitions were truer than others.[72] There was an ambiguity here, and it is worth repeating in full his definition of the universal postulate: "Mean what we may by the word truth, we have no choice but to hold that *a belief which is proved, by the inconceivableness of its negation, to invariably exist, is true*".[73] This definition seems to favour the second of his two alternatives. However, in the conclusion to "The Universal Postulate" he rejected both options in order to establish a pure common-sense, or realist, argument on truth. This was as follows:

> It remains but to notice Scepticism's last refuge; namely, the position that we can never truly know that things are as they seem; and that whilst it may be impossible for us to think of them as otherwise, yet they may be otherwise. This position we shall find to be as logically inadmissible as it is practically unthinkable. For one of two things must be true of it: it must either admit of no justification by reason, or it must admit of some justification. If it admits of no justification by reason, then it amounts to a tacit negation of all reason. ... If, on the other hand, reasons in justification of the position be assigned – if it be alleged that we cannot know that things are as they seem *because* we cannot transcend consciousness – then there is at once taken for granted the validity of that test whose validity is called in question.[74]

Spencer's position was arguing that truth implied an *actual* correspondence between objective fact and subjective state, and that what the mind believed or intuited about reality was true. This statement was consistent with the rest of Spencer's common-sense philosophy and Reid would have approved of it. However, it was completely abandoned in Spencer's *First Principles* (1862), because it contradicted the doctrine of the "Unknown" that Spencer borrowed from Hamilton and Mansel.

In *First Principles*, *truth* became "simply the accurate correspondence of subjective to objective relations; while *error*, leading to failure and therefore towards death, is the absence of such accurate correspondence".[75] This was a radical departure from common-sense philosophy. Spencer's statement implied that this *truth* did not matter, which would have alarmed an orthodox follower of Reid. Spencer's new metaphysics and the foundation of his system of philosophy was in revolt against traditional philosophy. Spencer no longer concerned himself about what the mind might think about the external world because objective facts did not have to correspond to subjective states. A mysterious force called "Life" would ensure that the correlation between them would always be identical regardless of the state of the mind. This analysis concluded with the explicit denial of one of the fundamental canons of realism and common-sense philosophy. That is, he stated that "things in themselves cannot be known to us".[76] He was no longer in harmony with intuitionism as that was understood by his contemporaries. Moreover, he seemed to have destroyed the philosophical basis of psychological knowledge.

It would be convenient, and satisfying, at this point to suggest that Spencer had ceased to be a common-sense philosopher by the time he published the first volume of his "A System of Synthetic Philosophy" in 1862, and that he had adopted a semi-mystical evolutionary doctrine guided by an unknown life-force. Unfortunately, such a simple dichotomy is not a useful perspective when analysing Spencer's philosophical work as a whole, because he remained eclectic on metaphysics. He kept elements of both the universal postulate and common-sense philosophy in his system of philosophy after the publication of *First Principles*, although their value became increasingly obscure. Spencer's philosophy had always contained basic contradictions. He himself labelled it a synthetic product. This variety caused his work to attract empathy from a range of Victorian intellects. Someone like J. S. Mill could sense empirical echoes while Wallace or Tyndall responded to the spiritual harmonies.

Philosophical inconsistencies persisted from the early editions of Spencer's *First Principles*: the keystone of his philosophy. Spencer had begun this work with an appeal to common sense and realism: the truth of human beliefs could be assumed because of their long existence and their wide diffusion. Although not absolute, truth was perennial and nearly universal:

> The presumption that any current opinion is not wholly false, gains strength according to the number of its adherents. Admitting, as we must, that life is impossible unless through *a certain agreement between internal convictions and external circumstances*; admitting therefore that the probabilities are always in favour of the truth, or at least the partial truth, of a conviction; we must admit that the convictions entertained by many minds in common are the most likely to have some foundation.[77]

This seemed plausible, but when Spencer came to consider how internal convictions might agree with outside circumstances he rapidly departed from his common-sense principles and took refuge elsewhere.

His basic questions were how do we get sensations, and how do we come to have impressions of sounds, colours, tastes and so on in our consciousness? Spencer's answer was that we are bound to regard them as effects of some cause, because we cannot think of them in any other way.[78] He should have stopped there but, remarkably, he added, "and we are not only obliged to suppose some cause, but also a first cause".[79] He believed that any real cause could not have any cause behind it, but must be independent.[80] Spencer then quoted Mansel's *The*

Limits of Religious Thought for three and a half pages[81] to demonstrate that first causes were incomprehensible. From this he drew the conclusion that perennial human beliefs, or common sense, encompassed the existence of the world with all that it contains and all that surrounds it, and that this was in itself a mystery ever pressing for interpretation.[82] This was Spencer's first principle, and – rather than making good his initial claim to place common sense on a secure basis – it denied the fundamental premise of common sense that knowledge was a true correspondence between an internal conviction and an outside circumstance, and that popular beliefs directly intuited truth about the universe.

Spencer was slightly apprehensive about his first principle, and was aware of the conflict between it and common sense. His strategy had left him a hostage: borrowing a first principle from Hamilton and Mansel meant that he could not rationally affirm the positive existence of anything beyond phenomena.[83] This was perilously close to that scepticism which Spencer, like Reid before him, regarded as a chimera. To avoid this danger Spencer introduced a qualification. He made a distinction between the logical aspect of propositions and the more general, or psychological, aspect. Only in this psychological guise were propositions to be regarded as statements of truth.[84] In the second form propositions were recognized by the "indefinite consciousness" rather than by the "*definite* consciousness", of which logic formulated the laws.[85] Obscurely, Spencer did not extend on his distinction between *definite* and *indefinite* consciousness. This would have been difficult because the introduction of the second kind of consciousness – a form that recognized the truth behind phenomena – contradicted his first principle and, to a certain extent, re-established common sense.[86]

Spencer's habit of entertaining basic premises from more than one source made him particularly vulnerable to criticism when his great contemporary J. S. Mill decided to rid the nineteenth-century world of the philosophical ambiguity that was giving intuitionism a new lease of life. Mill's attack appeared as *An Examination of Sir William Hamilton's Philosophy* in 1865. In response to this, many common-sense philosophers whose thought derived from Hamilton published replies. Spencer's version, "Mill *versus* Hamilton – The Test of Truth", appeared immediately, in July 1865. He registered Mill's comment that Hamilton's doctrine on the ultimate fact of consciousness stated that we should not "reduce it (a fact of this class) to a generalisation from experience". This condition is realized by its possessing the "character of necessity".[87] To Mill this argument was a recipe for abandoning the guide that we take from experience. His criticism was taken badly by its target. It seemed harsh to Spencer; Mill was a subscriber to his published works, and to a limited extent had been his supporter until his growing popularity in the 1860s. In a sense Spencer was the only philosopher worth criticizing. By 1865, Hamilton had been dead for almost a decade, and Mansel was only read by dons, but Spencer had a fashionable following.

In *An Examination of Sir William Hamilton's Philosophy*, Mill classed Hamilton with Reid, Stewart, Cousin, Whewell, Kant and Spencer. This was an idiosyncratic list; contemporaries would have been surprised that Reid and Kant appeared together. Also, they might have thought that Cousin was similar to Mill himself in being eclectic. Spencer was "somewhat puzzled" at Mill's classification where it concerned him:

Considering that I have avowed a general agreement with Mr Mill in the doctrine that all knowledge is from experience, and have defended the test of inconceivableness on the very ground that it "expresses the net result of our experience up to the present time" (*Principles of Psychology*, pp. 22, 23) – considering that, so far from asserting the distinction quoted from Sir William Hamilton, I have aimed to abolish such distinction – considering that I have endeavoured to show how all our conceptions, even down to those of Space and Time, are "acquired" – considering that I have sought to interpret forms of thought (and by implication all intuitions) as products of organized and inherited experiences (*Principles of Psychology*, p. 579) – I am taken aback at finding myself classed as in the above paragraph.[88]

This was one of Spencer's many attempts to rewrite his own past.[89] It had been reasonable for Mill to criticize him because he had not in fact agreed with Mill in the two pages referred to.[90] Instead Spencer turned on Mill and, in a footnote that had originally been appended to his quotation from *The Principles of Psychology*, added:

To prevent misconception it may be well to remark that, though here apparently committing myself to the experience-hypothesis in its entirety, I do not hold it in its current acceptation, any more than I so hold the antagonist hypothesis of forms of thought, which, nevertheless, contains a truth. In a future stage of the inquiry I hope to show that both these hypotheses are right in a limited sense, and both wrong in a limited sense; that they admit of reconciliation; and that the truth is expressed by their union.[91]

Spencer did not truly alter anything in response to Mill's attack. To have done so would have obviated the whole rationale of his philosophical system, which was to mediate in the dispute between thought and experience by appealing to evolution.

Spencer's intention in 1865 was to demonstrate how his test of truth, which, despite the change of name still held the same formulation as it had when called the universal postulate, was assumed by empiricism. For Spencer empiricism ultimately rested on necessary truths.[92] His response to Mill had been mild because he also wished to show that necessary truths led to the same form of empiricism as Mill's. However, the pressure from his contemporary did have more effect than merely causing Spencer to reiterate his beliefs: it caused him to limit his test of truth so that it would apply only to simple precepts or concepts.[93] This change was important because, as was suggested earlier, Spencer's original contribution to Hamilton's doctrine, and the one that most strongly marked him as a common-sense philosopher, was the argument that his test of truth or his universal postulate would immediately apply to any concept, simple *or* complex, so that no chain of inference was necessary.

The final appearance of the universal postulate was in Spencer's belated second edition of *The Principles of Psychology*.[94] It was no longer among the opening chapters of his psychology; it was now hidden away near the back of the second volume, much truncated and modified in language. Whereas in 1853 his universal postulate had ended with a strong defence of philosophical realism,[95] by 1872 Spencer was anxious to show how it harmonized with Mill's "experience-hypothesis", and could be reconciled with its facts.[96] He had abandoned most of the common-sense elements of his early philosophy; he had even jettisoned his refusal to admit to any distinction between the two kinds of knowledge, that is, contingent and necessary truths. Instead, he came to rely on a weak distinction between unbelievable propositions and unconceivable propositions.[97] This was similar in function to the original except that the test of truth now applied only to logical propositions, and not to contingent and/or necessary truths about the objective world.

By the second edition of *The Principles of Psychology* Spencer's realism became "transfigured realism":

> which simply asserts objective existence as separate from, and independent of, subjective existence. But it affirms neither that any one mode of this objective existence is in reality that which it seems, nor that the connexions among its modes are objectively what they seem.[98]

In a similar way Spencer defended "transfigured realism" in "Replies to Criticisms" in 1873, writing that the acceptance of the "inexpugnable element" of common-sense judgement by no means involved the acceptance of other elements of this judgement.[99] This variant form of "realism" asserted that objective existence was separate from subjective existence, but that this did not happen in reality, and that the connections between the two kinds of existence were not what they seemed. Spencer's aim here does not resemble that of a modern philosopher whose chief goal is to achieve greater clarity and rigour. His contemporary readership was not composed of professional philosophers, but of questing and earnest amateur intellects who were seeking meaning in philosophy and psychology with the same intensity as they did in biology and ethnography. To such minds all of these fields were merely different facets of life whose meaning was self-evident regardless of whether it was subjective or objective. This search was not specifically religious or scientific; it was both together. The contradictions that stress and fracture Spencer's system of philosophy are only failings if one does not share the Victorian desire to see the living universe as personally significant.

It is difficult to share Spencer's belief that a science of psychology could invent a living universe with personal meaning. However, to his contemporaries, Spencer's discourse on this was the work of genius. In the nineteenth century he was one of the original minds who were "like men of another race, power depth and unity of thought …; they cannot touch any question without setting their mark upon it".[100] Spencer had made it possible to enlist the philosophy of mind under the banner of science. His thought had an evolutionary *frisson* that connected all knowledge of human existence, even those parts that had applied to first and final causes.[101] It was this insistence that the study of human mind was not merely the province of physiologists that set Spencer above his mid-Victorian rivals. He had taken the high road while his competitors, led by Bain, kept their gaze fixed firmly downwards on the limitations and function of human faculties.[102] To them faculties existed in a localized organ, the brain, and therefore would be thoroughly explained by reference to physical phenomena.[103] There was no need for the belief that a soul or spirit existed within individuals. Whether human existence as a whole had some spiritual meaning was also unimportant to Bain. Rather than reflecting on this himself, in his longer works he could only query the metaphysical reassurances of Spencer's *First Principles*. Bain was, in a sense, parasitic on Spencer's ability to furnish psychology with metaphysical meaning. Without this freight contemporaries would have ignored the discipline.

The difference between Spencer and Bain was partly a matter of method. The former had separated himself from the phrenological roots of experimental psychology while the latter had not.[104] At heart, the phrenologist's approach was a practical and therapeutic one that attempted to locate different emotions, aptitudes and functions in discrete areas of the brain. Spencer's tools were subjective; he *thought* about subjects such as emotions. Partly, however, Spencer's high road of psychological investigation was a product of his acceptance of the idea that the human mind had evolved so as to be able to intuit falsehood. That is, it was itself a scientific tool that could report on the nature of existence. As a method this was as contestable in the mid-nineteenth century as it is today. Comte's beliefs were known: he believed that there could be no science called psychology because it was an illusion that the mind could contemplate itself.[105] This, as Spencer remarked, omitted the subjective half of the discipline.[106] For Spencer scientific progress depended on this insight because the physical world paralleled the inner and psychological one. This allowed a sincere man of science to look inwards and get a glimpse of the ultimate mystery, which was denied to timid sectarians.[107]

Taken in isolation, many of Spencer's philosophical and religious statements can be criticized for their lack of originality or cohesion. Much was borrowed from various sources, and a contemporary, such as Mill, could rearrange Spencer's postulates so they would belong to antagonistic schools of philosophy. One would be mistaken, however, to dismiss Spencer's work as merely an eclectic compendium of common metaphysical and theological ideas of the mid-nineteenth century. His philosophy played the key role in combining two separate streams of thought so that the world of the spirit did seem, however briefly, to be anchored in reality. No other British writer had such a grasp of common-sense or intuitive philosophy combined with such close contact with the worlds of radical theology and science. In Spencer's published works the Victorian reader could discover the structured ideas of Scottish philosophers and the earnest desires of the English intellectuals whose metaphysical yearnings were still staunchly religious. One feature suggested that philosophical truth could be intuited through human faculties, and the other that religious truth should be understood by the soul. Both gave the mind an active role in discovering psychological and scientific truth. These two strands of Spencer's thought were intertwined. Common-sense philosophy offered a systematic exposition of the limited nature of religious and philosophical knowledge. The New Reformation expressed the need for new religious certainties to replace Christianity. The result was the Spencerian worship of the "Unknown": a prayer for certainty about the unknowable.

Spencer's evolutionary psychology is not available to be quarried by modern sociobiologists. This is because he remained loyal to self-reflection, one of the methods of traditional philosophy. He did not view the human mind as a developing biological organ, but as a self-reflecting spirit. Even in *The Principles of Biology* Spencer never felt comfortable with classifying human beings as just another form of Life, but in his psychology this hesitation became more pronounced because for him intelligent life was as much a matter for subjective reflection as it was for objective analysis. Spencer's *The Principles of Psychology* not only pre-dated his philosophical system: the work was its bedrock. In his later philosophical writings he never surrendered the idea that thinking was as much an empirical means of investigation as any other scientific procedure. To Spencer, it was always the case that a mental experiment was as valid as any other type, and while he had transformed philosophy into psychology he never abandoned the traditional reflective methods of the older discipline.

III

Spencer's biological writings and his philosophy of science

TWELVE

On goodness, perfection and the shape of living things

During the late-nineteenth century Spencer's ideas on evolution were often confused with Charles Darwin's theory of natural selection. Even John Lubbock, a scientist and a Fellow of the Royal Society, made such a mistake; quoting from Spencer on the subject of progress, he arrived at the conclusion that *Darwin* offered hope that man would soon be in harmony with nature.[1] Lubbock's conflation made Darwin appear in Spencer's garb. However, the opposite was also possible. When W. G. Sumner taught the first sociology course in the United States he gave it a social Darwinist bias towards struggle and competition, even though his textbook was Spencer's short *The Study of Sociology*, which actually emphasized that the human future was one of peace and benevolence. Sumner achieved his desired effect by omitting the last chapter of Spencer's *The Study of Sociology*, which contained the theories about altruism.[2]

It is important to examine the basis of the confusion of Spencer with Darwin. Seven years before Darwin published *On the Origin of Species* (1859) Spencer had used the development hypothesis to claim that, because of environmental changes, any existing species "*immediately begins to undergo certain changes of structure fitting it for new conditions*".[3] This claim could indeed be used to prove that Spencer, like Darwin, had focused on the way environmental changes had caused variation in species. However, as Spencer himself knew, it was a mistake to see his early work as generating the same theory as Darwin's. Spencer's theory, as stated in "The Filiation of Ideas", stipulated that a beneficial process caused inferior animals and human beings to disappear while leaving superior specimens to continue the race, but insisted that "there was no recognition of the consequences seen by Mr Darwin".[4]

Spencer was not modestly denying originality or admitting that his ideas lacked priority in the search for the origins of life. He would not have been humble enough to make such confessions and, in any case, they were unnecessary. He was not pursuing the same goals as Darwin. It was therefore painless for him to admit that he and Darwin had used evolution in different ways.[5] During the 1850s, Spencer did not consider sexual reproduction as more significant than parthenogenesis. This meant that, unlike Darwin, he did not see biological evolution as a theory whose function was to explain natural variation in species. Spencer was a man "on a crusade against the notion of 'species'".[6] However, later, during the 1860s, when he wrote *The Principles of Biology*, Spencer did employ Darwin's natural selection theories. In fact, Darwin's work on natural selection then became Spencer's most frequently cited source on the subjects of genesis and variation. Nevertheless, Spencer's focus remained different from Darwin's because he believed that biological science was not sufficiently advanced to provide an explanation of development based on heredity.[7] Although Spencer thought it important to work towards establishing a science of development, all that his contemporaries were able to accomplish was the promotion of speculative hypotheses on the role of genetic change and on the emergence or persistence of organic characteristics. It should nonetheless be emphasized that the object of Spencer's speculation in *The Principles of Biology* was *not* usually based on heredity, nor on embryology. Spencer was pursuing the facts, rather than the origins, of organic structure. His focus was on the shape and function of *mature* organisms.[8] He was not significantly interested in the question of whether species were permanent or modifications of earlier species. Nor did Spencer share Darwin's view that long periods of time were required to account for significant evolutionary changes.

Biological, social and psychological changes were all of equal importance to Spencer. He would not have conceptualized species change as more critical than other developments: to have done that would have prevented him from taking part in actual scientific debates of the mid-century. These discourses, like Spencer's earlier writings, emphasized the primacy of biological speculation in the understanding of the psyche or human society. One of his most protean notions, "the physiological division of labour", an idea that underpinned much of his comment about the importance of functions in adaptation, was not borrowed from economics or sociology, but from the great Belgian natural scientist Henri Milne-Edwards.[9] This was a theory that was also popularized by William Carpenter's standard textbooks on physiology.[10] Closer to his own family Spencer could have heard Thomas Rymer Jones, a former pupil of his father's, employing natural history to instruct legislators:

> With the advance of civilization among mankind, the division of labour, and the apportioning to each grade of society its proper field for exertion, becomes progressively more and more precise, and although few but the more enlightened legislators can understand the mutual bearings and relationships of all parts of the grand scheme, every one participates in the benefits that accrue from the harmonious adaptation of the whole.[11]

In adopting ideas such as these Spencer was merely accepting conventional scientific wisdom. That is, Spencer's partiality towards non-Darwinian evolutionary theories was consistent with his strong desire to conform with reputable mid-century science. An adoption of natural selection as the explanatory mechanism of evolutionary change would thus have been out of character. While Spencer could be bold, or even outrageous, in his views on politics or sociology, in the life sciences he was too cautious to take an extreme stance on a major controversy. He took his cue from the intellectual bellwethers Huxley and Joseph Hooker, who were his friends as well as Darwin's. During the late 1850s and 1860s they treated Darwin's ideas on natural selection with both admiration and scepticism. Spencer imitated this ambivalent stance, praising Darwin while adopting his ideas only selectively.

There was reason for caution. Spencer, along with other contemporaries of Darwin, was puzzled as to how the biological advantages offered by natural selection would be maintained from generation to generation. He thought it unreasonable to believe that spontaneous variation would occur simultaneously in many individuals in a species. Furthermore, if this did not happen, then back-crossing would cause naturally selected advantages to disappear in a few generations; novel characteristics would be swamped by breeding with individuals lacking the new characteristics.[12] These objections were not unique to Spencer: before the discovery of genetics such doubts about natural selection were commonplace.

In the two volumes of Spencer's *The Principles of Biology* (published in 1864 and 1867) there are dozens of references to Darwin's detailed scientific work. He also specifically gives Darwin priority in discovering natural selection.[13] It would be a mistake, however, to consider this as placing Spencer on one side of a subsequent debate on whether natural selection was the sole, or even the primary, mechanism for change. Historically such a description is bizarre; this debate would have been improbable in the 1860s because even Darwin had not dignified natural selection as being qualitatively different from other causes of biological change. Spencer's interest in Darwin's work meant that he could draw on the apparently functional causes of change (such as the use and disuse of organs or limbs and whether these are inheritable) as much as natural selection.[14] If Spencer did not distinguish between the two kinds of change this is perfectly understandable since Darwin himself did not make a clear division between phenotypes and genotypes.

Loren Eiseley and other mid-twentieth-century historians of science were incorrect to see non-Darwinian evolutionary theorists, such as Spencer, as Lamarckian simply because they were not Darwinian. Terms such as Lamarckian and Darwinian were seldom polar opposites or competing ideologies during the nineteenth century (except in the hands of *non*-scientists, such as Samuel Butler and Walter Bagehot).[15] Although it is not profitable to enter into a discussion of what Lamarckianism might have meant to Spencer had he considered it more carefully, it should be observed that Jean-Baptiste Lamarck, like Darwin, was focused on species change over long geological periods, even though his theories invoked the now heretical idea of the transmission of acquired characteristics from one generation to the next.[16] Spencer could not have been involved in a dispute between Lamarck and Darwin because, as he saw it, both of them urged natural scientists to consider species reproduction as the significant factor in biological change; whereas, for Spencer, such change had to be purposive for an individual organism. Reproduction could not be the purpose of life since the expenditure of sexual energy and the production of offspring often caused diminution of vital forces; it was thus a deduction from parental life, not its goal or consummation. The shape of life reached its perfection in the individual adult form, a form that was often damaged or destroyed by reproduction.[17] Rather than sexual reproduction being the key to survival, as it was for Darwin, Spencer's early views stressed that sex was a diminution of the parent that lessened its ability to survive.[18]

These differences in theory are a reflection of differences in the types of data. During the 1850s a chief variation between Spencer and Darwin was Spencer's lack of interest in information about fossil plants and animals. This was a matter of choice; Spencer could easily have become informed about such subjects. His father was a correspondent of Richard Owen's, and he was himself a friend of Gideon Mantell, and had spent many hours listening to the latter's views on the importance of palaeontology.[19] However, when Spencer used phrases such as the "lowest types of life",[20] he was not referring to the prehistoric giant hyenas and the dinosaurs popularized by William Buckland and Owen, but to simple organisms in the present. It was not the genesis of species that intrigued Spencer, but the *shape* of organic life forms *now* and their present responses to the environment. For him, the shape of an organic entity not only included what Darwin would have considered species variation, but also encompassed somatic differences between individuals and the varieties that could be observed in human psyches and societies. To him these were all equally the shapes and forms of life, and were all responses to their environments.

Before his ten-volume "A System of Synthetic Philosophy" began to appear in 1862, Spencer already enjoyed a reputation as a scientific writer and as a psychologist, based on his essays and *The Principles of Psychology*. The fact that he was already known as a thinker by the mid-1850s is often overlooked, as are the scientific ideas he generated in this early period. Spencer sometimes cannibalized his earlier works, and incorporated portions of them – only slightly edited – in the volumes of his "A System of Synthetic Philosophy". This has led some scholars to believe it pointless to investigate Spencer's ideas before they reached their final and mature form. However, such a teleological approach is wrong. The scientific ideas Spencer generated during the early 1850s were not embryos of his later thoughts, but fully developed theories of science and society that were, in fact, often at odds with the ones he produced later.

There was a cataclysmic shift in Spencer's ideas concurrent with his nervous breakdown in 1856. Prior to this his scientific work had been devoted to discovering the laws of progress. This programme was to be carried out by the use of psychological and logical tools; its goal was to provide a description of the shape of life. Spencer planned to analyse the directions of growth in all vital phenomena, from single-cell organisms to the human intellect and the growth of science, together with indicating the moral worth of civilization, and the means by which perfection would be reached. This optimistic investigation of progress was shattered by Spencer's loss of faith in his own abilities. He no longer believed in rationality in general, or in the moral quality of civilization. Illness and loss of strength soured his utopian dreams, and caused him to doubt the moral purpose of change. As a result his concept of moral progress began to transform into a scientific evolution, which offered little in the way of comfort. The certainties of his early science became the vague and distant coolness of the "Unknown". Spencer's opinions also changed on a number of technical subjects. In his scientific writing he reassessed the importance of species and the value of embryology. In the area of anthropology he became less optimistic and began to doubt that savages were morally inferior to their civilizers. He also became less confident about answers to the philosophical questions of whether human beings were capable of perfection and whether the mystery of the meaning of life could be unravelled. He began to suffer uncertainty. His early conviction that indigenous people were inferior to civilized ones was abandoned as a bias that, in itself, had become barbaric. Spencer had come to the sobering conclusion that moral qualities, such as respect for truth and gentleness, were possessed in greater quantities by savages than by the civilizers who had overwhelmed them. While Spencer could justify the destruction of non-European civilization in terms of a universal human progress, this rationale never alleviated the pain he felt in observing the process.[21]

Thus, Spencer's later views on evolution were accompanied by shadows of doubt that had not been present earlier. Up until 1856 his ideas had been bathed in the harmonious light given off by the Crystal Palace. To this point he had believed that mankind was moving towards perfection and would soon penetrate the mystery of "Life". When this movement was complete, perfection would be a condition of stasis held in equilibrium by nature's great tension: that between individuation and reproduction. Spencer believed that life would be in its highest form when the greatest possible numbers of individuals existed without suffering the pangs of reproduction or death.[22] These hopes were then discarded; his philosophical system seldom gestured towards human perfection, nor did it promise solutions to these mysteries. His later belief in human advance no longer pointed towards an ultimate goal; it had become an evolution from the past. What had been the mystery of life, waiting to be solved, was now the curtain before a distant horizon that had once concealed God, or even a whole pantheon of gods. The "Unknown" had become simply the edge of knowledge: an edge that retreated as soon as it was touched by science.

A constant refrain in Spencer's early scientific writings was that all the phenomena of the universe, from distant stellar objects to the microscopic activity of single-cell organisms, were subject to evolution. He thought the same to be true of human societies and the human psyche. Not only was everything evolving, but it was evolving in the same way: from simple homogeneity to complex differentiated structures. What linked all these separate processes together was more important than the processes themselves, and this common factor was "Life" itself.[23]

Spencer's initial conception of life was not cold and objective; he saw life as the general impulse towards goodness and perfection, evidenced everywhere one looked. It was the current of vital action that electrified the living part of the universe, and displayed itself in a hierarchical structure from lowly cells up to the pinnacle of human knowledge. The intelligence that perceived life was science, and Spencer saw this too as an organism. Science, in symbiosis with the human race, had reached the margins of the known world. In tropes worthy of a prophet, Spencer expressed his vision that observations on a star had to be "*digested by the organism of the sciences*" and that the environment for mankind now stretched into the surrounding sphere of infinity.[24] He believed that as scientific knowledge became increasingly accurate and specialized it would fall into equilibrium with its environment, which was the whole of the physical universe. Knowledge, and the species that employed it, would then have achieved perfection. In this early period, Spencer did not regard perfection as harmonious physical adjustment to nature; it was the attainment of all moral qualities, or of goodness itself.

Spencer's essay "A Theory of Population" (1852) is the key to his early views and, while he subsequently experienced a profound repugnance against his advocacy of the need to suppress desire, to vanquish grief and to increase work,[25] it motivated him strongly until his nervous collapse. Before this crisis in 1856, and its onset of self-doubt, he was inspired by the progressive view of natural science that he had borrowed from Carpenter, Louis Agassiz and Richard Owen. Behind these was the great font of scientific optimism John Herschel, who believed that, by cherishing an unbounded spirit of enquiry, science:

> unfetters the mind from prejudices of every kind, and leaves it open and free to every impression of a higher nature which it is susceptible of receiving, guarding only against enthusiasm and self-deception by a habit of strict investigation, but encouraging, rather than suppressing, every thing then can offer a prospect or a hope ...[26]

In these early halcyon days Spencer perceived his own experience, and that of nature generally as "the inherent tendency of things towards good – [and saw] going on universally a patient self-rectification".[27] He called this *vis medicatrix naturae*, and declared that it was not limited to the cure of wounds and diseases, "but pervaded creation" from the lowly fungus up to the tree. Reversing the direction, Spencer also noticed that from the highest human faculty down to the lowly polyp one found at work "an essential beneficence".[28] This progressive quality of nature even justified the creation of an empire and the subjugation of foreign peoples because what appeared on the surface to be suffering was, from a deeper perspective, necessary for benign progress. In any case, each conquered race or nation could acquire a liking for new modes of living, and "it is best for the world" that these societies should produce new commodities.[29] In this way, Spencer cheered on improvements such as free trade and the imposition of European government on distant countries; these kinds of progress would complement each other. In the future, Spencer foresaw further evolution of the same kind, when the human race would have sufficient intelligence and morals to offset diminished fertility and physical strength.[30] The futuristic balance achieved by such increases and decreases would maintain a perfect and long-lived existence for each individual.

Spencer was forever contemplating distant horizons. It was always in the long term that he predicted the abolition of pain and the postponement of death. However, Spencer doubted he could make this optimistic account of the future believable without answering to the gloomy parson and demographer, Thomas Malthus. Like Darwin and Alfred Russel Wallace, Spencer used the work of Malthus as his impetus for the investigation of Life, but, in contrast to their positive response, his reaction was resoundingly negative. Spencer believed it simply false to assume that fertility in nature was fated to cause misery. He argued that the mechanism of untrammelled fertility would *not* cause the population to grow until all available food was exhausted and starvation ensued. Rather, tomorrow would see a more efficient use of food. Moreover, technological innovation in the area of food production would provide the additional benefit of increased scientific knowledge. This would, in turn, facilitate further progress. The basis of Spencer's argument was the kind of definitional logic that he always applied to the phenomena of life. He argued that one could not correctly analyse the causes of the maintenance and multiplication of living organisms without first identifying the property that distinguished living organisms from other things.[31] This was the ability to maintain a correspondence between internal changes within the organism and the external environment. Spencer's focus on "organic correspondence" relied on the emphasis given to this by Spencer's friend, Gideon Mantell. Since Mantell, like Buckland, was indebted to Georges Cuvier it would be useful to quote the great Frenchman's "law of organic correspondence" at this point.

> Every organized individual forms an entire system of its own, all the parts of which mutually correspond, and concur to produce a certain definite purpose, by reciprocal reaction, or by combining towards the same end. Hence none of these separate parts can change their forms without a corresponding change on the other parts of the same animal, and consequently each of these parts taken separately indicates all the other parts in which it has belonged.[32]

Cuvier's original restrictive intentions did not hinder speculation in Britain, where *The Animal Kingdom* was accompanied by an editorial note on mammalia to the effect that "As a general, perhaps universal, rule obtaining in consecutive groups when sufficiently extensive, the summit of the inferior displays a higher organization than the terminal members of the superior …".[33]

Associating Cuvier's law with evolutionary change while discarding Mantell's interest in the reconstruction of the anatomy of extinct species, brings one close to Spencer's theoretical stance. That is, if one starts with the assumption that changes in the internal organization of living things always correspond with and produce a definite purpose, all that is then needed is the addition of the notion that internal changes correspond to external environmental forces: The result is Spencer's evolutionary theory. There is an end-point to this process. In *The Principles of Psychology*, where he fully developed the idea of organic correspondence, Spencer announced that life will be perfect only when the correspondence is perfect.[34]

The proclamation that organic correspondence would lead to perfection was a neologistic claim that much amused Victorian wits when they rearranged it in comic form. Spencer's solemn utterances about life were sometimes too bare-boned and ugly to be taken seriously. For example, Goldwin Smith poked fun at Spencer's late definition of evolution – "While an aggregate evolves, not only the matter composing it, but also the motions of that matter, passes from an indefinite incoherent homogeneity to a definite coherent heterogeneity" – by remarking that the universe may well have heaved a sigh of relief, when, through the cerebration of an eminent thinker, it had been delivered of this account of itself.[35]

Such humour should not mask the fact that Smith had himself once been a believer in the slogans of evolutionary truth. As a young Oxford don he had subscribed to Spencer's "A System of Synthetic Philosophy". Like J. A. Symonds, Grant Allen and other scientifically minded Oxford people he had found himself inspired by the modern possibility that philosophy could be employed to uncover knowledge of life itself. Instead of pursuing past desires recorded in classical texts one could touch the immediacy of the present.

For Spencer the evidence of progression towards perfection was ubiquitous. Whether it showed the development of amphibians or the history of the human race, the same general truths were displayed.[36] Increasing heterogeneities of flora and fauna had, in themselves, progressively complicated the environment of each species of organism. In phrasing his general truth in this way, Spencer indicated that evolution was often a cause rather than a consequence of the environment. His emphatic insistence that causes and consequences could be reversed was an attempt to avoid the type of flawed monocausal explanation that J. S. Mill had warned against in *A System of Logic*. Mill's modernism had partly rested on a rejection of philosophic doctrines that smacked of Platonism or intuitionism and he was, therefore, opposed to Whewell and Hamilton. However, Mill's hostility extended further than this, including any traditional philosophy in which long chains of inferences were dependent on a single cause, and where the result was claimed to be part of scientific knowledge. The adoption of Mill's argument gave Spencer a dilemma; he was attracted to the intuitionist philosophy of Whewell and Hamilton, but he also followed Mill in avoiding attempts to fit complex phenomena into simple monocausal explanations. When this last idea was applied to evolutionary thought, it fortified Spencer's belief that the environment could be simultaneously a cause and an effect. Conversely, this meant that the environment should not be regarded as a controlling factor. He could thus argue that the chief steps of civilization were made by people in regions exhibiting a complex physical geography, and "who, in the course of their progress, have been adding to their physical environment a social environment that has been growing even more involved".[37] Spencer's avoidance of monocausal and unidirectional explanations was complicated by his adherence to teleology in the early 1850s. He believed that civilized human beings were so constituted as to *realize* the Divine Idea as it operated in the universe.[38] Since these human beings were also part of nature, colonial rule over uncivilized people and use of their resources was necessarily directed to a higher goal. What was true in social theory was also the case in biology. In his earlier writings Spencer believed that the higher the life the more it was fulfilled. However, when this passage was recycled in the first volume of *The Principles of Biology* (1864), "higher" life was no longer accompanied by "fulfilment".[39] This textual change was significant because after his critical article on Robert Owen's scientific method in 1858,[40] he became suspicious of the use of teleological arguments in science.

During the early 1850s, Spencer saw Life as maintained at a high level because it was in direct proportion to the greatness of correspondence with its environment. Proportionality was scaled; perfect correspondence would be perfect life. "Were there no changes in the environment but such as the organism had adapted changes to meet; and were it never to fail in the efficacy with which it met them; there would be eternal existence and universal knowledge."[41] By offering this hope to his readers Spencer was suggesting to them that their future would contain two divine traits: immortality and omniscience. Spencer's pursuit of Life was the basis of a prophetic claim to knowledge of a true religion that would be free from the superstition and falsehoods of Christianity. His rivals were Thomas Carlyle with his "nescience" and James Hinton with his alternative religion of life. To his followers, Spencer's religion was part of a contemporary movement sometimes called spiritualism, or the "New Reformation", which was imagined to be a substitute for Christianity. However, it would be a mistake to suggest that this new religion constructed the world in the same way as its orthodox rival. Most significantly, humanity was part of nature instead of being above it. Spencer believed that, essentially, humanity was a vital force that responded to environmental changes. It was not to be distinguished from other forms of life except in the amount of force it possessed. Spencer foreshadowed the language of late-twentieth-century literary critics when he wrote of "an embodied Humanity"[42] that united the efforts of successive, and apparently unrelated, individuals as if they were components of a single organism.

Spencer's religion of life was designed to give solace to humanity at large, not to specific individuals. Even when it mimicked orthodox Christianity with promises of eternal existence and ultimate wisdom, these qualities were attributed to the human species as a whole; they were not aspects of individual redemption or personal salvation. Nor did Spencer believe that moral significance was essentially attached to national or racial groups. While his early evolutionary theory had suggested that indigenous peoples, such as Hawaiians and Australian Aborigines could respond neither to social complexity nor to the advanced numerical systems that accompanied science, this did not imply inferiority.[43] Ethical superiority was a myth because Spencer's religion of Life did not selectively bless a racially dominant group. Even those Europeans he pleonastically described as enjoying the "culture of civilization" were nothing more than a projection of general organic life. No individual, or group, was separated from any aspect of this Life by more than a matter of degree. Even the process of scientific reasoning, which Spencer regarded as the symbol of progress, could not be meaningfully distinguished from other vital workings of organic life. "The assimilative processes going on in a plant, and the reasoning by which a man of science makes a discovery, alike exhibit the adjustment of inner relations to outer relations."[44]

To elevate the place of reason in science would, in Spencer's eyes, have mistakenly emphasized a deductive method. His strategy was to focus on the collection of factual material and then to generalize and classify it. That is, he adopted induction, which was the common way of perusing the biological sciences during the nineteenth century; the patient collection of facts would be followed by the construction of generalizations based on them. When Spencer attached himself to this method of scientific procedure he was being perfectly conventional. According to the beliefs he shared with contemporaries, it was important to uphold Francis Bacon's dictum that knowledge proceeded from the "generalization of experiences". This was a truism; Spencer believed that "all educated men are in a sense Bacon's followers".[45] However, he was not content to leave this method as a commonplace because he was intrigued by J. S. Mill's claim that induction rested on a notion of causality that was not instinctive, but was itself based on experience. Mill had hinted that one could find empirical justification for his view of causality in psychology, but that he did not intend to investigate this basis.[46] Spencer perceived this suggestion as an open door allowing one to move backwards and forwards between the functioning of the psyche and the objective facts of science. Along one direction he was able to share Mill's belief that purely logical ideas, such as causality, were grounded in experience. Yet, if he moved in the other direction it was possible to proceed from induction, as a mere matter of collating experience, to a kind of proactive search for experimental knowledge.[47] To use Spencer's words, he wished to convert "aboriginal" inductive processes into premeditated ones. He believed that by studying the psychological procedures used by scientific thinkers one could probably predict which theories would subsequently be confirmed as true.

Spencer's philosophy of science was in the grip of his theory of "progressive intelligence." He viewed the original theory of induction as based on irregular or unconscious procedures.[48] In his essay "The Universal Postulate", which he later reused in *The Principles of Psychology*, he called those procedures "primordial" beliefs.[49] Since beliefs were all that we possessed when we grasped experience, it was clear that Spencer, by labelling them variously as aboriginal, unconscious or primordial, thought they pre-dated knowledge. This idea was gleaned from Mill's critique of Whewell, where he argued that Whewell's "necessary truths" were simply our earliest inductions from experience.[50] This was not an adequate rebuttal of Whewell's position since, like Mill, Whewell believed that, as a matter of historical fact, fundamental scientific principles were based on experience.[51] The difference between them was more subtle than Mill would allow because he had mistakenly thought that Whewell, as a historian of science, identified the origin of fundamental scientific principles with their philosophical grounding. Since this was not the case, Mill's critique left Whewell's defences undamaged.[52] Whewell's "necessary truths" were not necessarily ideal ones. For example, in the key passage where he proposed that "necessary truths" could be found in mechanics as well as in mathematics, he argued that the person who could see such truth was someone of "sound mechanical views".[53] That is, Whewell's necessary truth was not *a priori*. Given that both Mill and Whewell grounded scientific induction on experience, and that the latter offered a useful insight of fundamental laws as well, it was reasonable for Spencer to favour Whewell's formulation over Mill's. He also found Whewell's formulation attractive since he could extend it to suggest that inductions could be predicted.

This predictive process began when one classified phenomena together without being able to explain why some resembled others. The groups of phenomena classed together in this way were not subject to prior procedures. However, as both science and human faculties evolved, Spencer thought that one could use *prediction* (he coined the word "prevision" for this) to foresee which classification of data would occur. Whewell's classification had been incidental; but with Spencer it would now be intentional.[54] In this way, Spencer switched the emphasis of mid-nineteenth-century empiricism from the non-theoretical collection of data to a theory-laden logic of science. This would still be immediately inductive in the sense that it was based on the data of the human psyche, but not so if one was considering the data of the external world of the biological sciences. It was imperative to Spencer that his procedure continued to be seen as based on experience. He had taken a stance against the work of continental philosophers who had created philosophical categories that did not rest on experience. Spencer's criticism did not distinguish between Germans, such as Oken and Hegel, and the Frenchman Comte. The former founded scientific knowledge on *a priori* concepts about equality and identity, and the latter had arbitrarily pronounced that scientific knowledge moved from simple ideas to complex ones.[55] Comte in particular had erred because, as Whewell demonstrated, some science had, in fact, progressed from complexity to simplicity.[56]

For Spencer, only the modification of the original and non-theoretical notion of induction made it possible to reach true generalizations. In order to conceptualize this kind of induction Spencer proceeded by analogy with what had happened in the work of anatomists and physiologists when they had been confronted by embryology and palaeontology. They had found that "the real nature of organs and tissues" could be discovered by tracing the early evolution and development of organisms and of their constituent organs. In the early 1850s Spencer believed that the same considerations should be brought to bear on the growth of scientific knowledge.[57] He also borrowed an additional idea from his botanical studies, noting that in systems of plant classification the comparison of superficial resemblances between plants often misled the classifier. However, if one detected underlying parallels between plants that were seemingly unrelated this was a *prima facie* reason for assuming that the resemblance was a fundamental one: "A peculiarity observed to be common to cases that are widely distinct, is more likely to be a fundamental peculiarity, than one which is observed to be common to cases that are nearly related".[58] The predictive quality to this strategy was emphasized by Spencer saying that the classification of distant anomalies led to a "greater probability" of being able to establish generic categories than if one chose to class anomalies that seemed closely related.[59] When Spencer applied this strategy in his search for homologies in distant parts of the "vital" process as a whole it gave him a useful way to compare organic action with intelligent action. Since these two phenomena were extremely far apart, he considered that the ways in which they resembled each other signalled to basic features of Life itself.

The boldness of this predictive empiricism diminished when Spencer came to write the volumes of "A System of Synthetic Philosophy". He discreetly omitted his ideas about scientific method from likely venues such as his *First Principles* (1862) and the expanded edition of *The Principles of Psychology* (1870–72).[60] His early scientific method had sat more comfortably with the mathematical and physical sciences than it did with those biological and social sciences that increasingly occupied his attention from the late 1850s. During the 1840s and early 1850s the vast bulk of the scientific examples with which he illustrated his articles were taken from mathematics, physics and chemistry, in which a predictive model of explanation was obviously appropriate. Later, when he was engaged primarily with the explanation of organic and vital phenomena, he became concerned to expound the detailed findings of contemporary natural scientists, who were becoming increasingly professionalized. In this climate he found himself with less time and facility to predict which way their work would evolve. In addition, classification seemed less viable as a method because he had grown sceptical about "progressive intelligence". Predictive activity relied on reason, which, as he further explored metaphysics and psychology, seemed less trustworthy. His doubts about reason led him to adopt the more conventional empirical practice of letting the facts speak for themselves. Indeed, he eventually became quite extreme in this approach. For example, when publishing the elephant folio volumes of the *Descriptive Sociology*, he felt no need to apologize for issuing a data set. The assistants he employed made a simple tabulation of the social customs of both civilized and savage people without any accompanying explanation of the findings or the method. Spencer's name on the title page was a sign that the contents were facts.

Spencer's early advocacy of premeditated experience was an extension of the simple Baconian inductivism that Whewell, with the acquiescence of the British Association for the Advancement of Science, had popularized in the early-nineteenth century. In this sense of inductivism, the true path to scientific understanding was the avoidance of methods that did not immediately refer to experience. In England, even the most empirical of French scientists, Cuvier, could be viewed as an example of radical scientific theorizing. This was in contrast to the bulk of British work in botany, zoology, geology and palaeontology, which was pursued with few overt theoretical goals.[61] Spencer stayed loyal to the English tradition of basing knowledge on experience. Although he was indebted to Whewell for his anti-materialism and for some of his philosophical underpinnings, he did not follow him into Platonism when he insisted that knowledge contains an *ideal* element.[62] Spencer did not conceptualize a factual generalization as an ideal built on low-level data; to him it resembled an organic set theory. The procedure was simple; Spencer constructed a wide category under which one could fit living things that resembled each other. He was always in search of the widest net, which would ultimately capture not only butterflies, but also plants, vertebrates, crustaceans and even societies and mental phenomena. It was this classing procedure that eventually caused Spencer to abandon scientific induction as a philosophical method. He began to see the latter as relying too heavily on natural categories, which, in turn, led to Platonism. In reaction against this Spencer insisted on artificial, rather than natural, classification and procedures. The nature of life was not to be induced, but defined. After he posited his definition he would then decide where to classify information.

Initially, in 1852, Spencer defined Life as "the co-ordination of actions". He was never completely satisfied with this; even as the words left his pen he worried that the definition might be incomplete. His concern was that it failed to distinguish between organic and inorganic components. For example, the growth of a crystal might display the same actions as something alive.[63] However, he provisionally accepted this definition because he was heavily influenced by the language of physiologists, which was engaged in classifying "Life" under two headings: processing systems and working machinery. The first category was what physiologists called the "vegetative" system and referred to stomach, lungs, heart, liver and skin. These were organs and surfaces that had to work in concert with each other if life was to be preserved. The second referred to the working machinery, and was designated as the "animal" system. This consisted of limbs, senses, and instruments of attack and defence that had to perform their functions in the correct sequence if the organism was to survive.[64] Spencer's procedure here, like that of physiologists in general, was strongly influenced by the images from the mechanical sciences. However, this observation should not be taken to indicate that Spencer was attempting to unify the sciences by incorporating biological sciences under the laws of physics. Even though, in his early evolutionary theorizing, Spencer frequently used examples from the physical sciences and mathematics, it is important to note that in his general discussions of sciences during the early 1850s he was *not* attempting to incorporate the biological science into physics.[65] Indeed, his emphasis points to the independence of the former, as does his focus on multi-causal phenomena, and reflexive or self-maintaining processes. These ideas were integral aspects of Victorian physiological discourse and it is of critical importance to observe that this language was the source of many of Spencer's mature ideas in psychology and metaphysics as well as of those in his early science. Physiological distinctions were incorporated in his initial definitions of rationality and science. For example, rationality was defined as the power to combine or coordinate a great number and great variety of complex actions in order to achieve a desired result.[66]

Spencer fully articulated his definition of "Life" in 1855. At the time he had become increasingly frustrated by the problem that different aspects of the phenomenon were difficult to reduce to clear categories. While he still believed that he could see intelligence as closely resembling animal life in its lower forms, he became troubled that his procedure was an artificial one.[67] When he distinguished between types of mental life he was troubled that his categories seemed to be mere subdivisions that blurred into one another. For example, he could find no clear division between animal instinct and human intelligence.[68] Nor was he able to prevent higher and lower forms of reasoning from spilling into each other.[69] He recognized that his logical method was flawed; his aim had been to show that separate vital actions possessed a characteristic in common that he could then isolate as their defining quality, but this procedure would inevitably fail if his distinctions did not capture significant features:

> The further we carry our analysis of things, the more manifest does it become, that divisions and classifications are essentially human inventions which have no absolute demarcation in nature corresponding to them, but are simply subjective – are scientific artifices by which we limit and arrange the matter under investigation, and so facilitate our thinking.[70]

This doubt became a perpetual vexation. Spencer could never convince himself that his early "predictive" scientific method was anything more than subjective. Since he felt that qualitative procedures were inferior to quantitative ones, he found himself on weak ground. It is an apparent mystery that a philosopher who has been described as one of the three musketeers of positivism[71] should set about acquiring knowledge by definition when he viewed this method as lacking in objectivity. However, the paradox can be resolved on a superficial level: one can simply reject the suspected positivism because Spencer was sceptical about the acquisitions of direct knowledge of material reality, and because he did not believe that one would make scientific discoveries by consistently applying the same method. However, a better resolution would be to ignore the claim that Spencer was a positivist because to take it seriously would risk confusing the basis of his philosophy of science. His warning that knowledge was subjective was directed against positivists and anyone else who placed too much faith in rational processes regardless of whether those were built up from data, or were deductive knowledge based on principles. His definitions were never more than approximate generalizations based on classification. They did not have law-like pretensions. Instead, they were analogous to the way in which botanists identified a flower by combining their knowledge of fructification, the number of petals, colours, calyx, bracts and so on. These features were not considered separately, but as a *synthesis* of attributes.[72] Spencer's "A System of Synthetic Philosophy" was a result of this procedure: a procedure that lingered on in his later scientific thought, even when he had largely abandoned the search for Life that had driven it.

His survey of definitions of "life" in *The Principles of Psychology* (1855) began with Samuel Taylor Coleridge, who thought that life was the tendency to individuation. Spencer was sympathetic to this definition because it was comprehensive, but still thought it was flawed in being too inclusive. That is, the wording wrongly encompassed inorganic things such as crystals.[73] Coleridge's idea of life came first in Spencer's list of mistaken definitions because during the 1850s he felt it paramount to distinguish the growth of crystals from vital phenomena. This distinction between dead mechanical entities and living vital ones was significant even when talking about society.[74] Spencer's adoption of this distinction, and his later relinquishment of it, is critical to understanding how he construed philosophy to illustrate the meaning of life. Even when Spencer later abandoned his belief that organic entities were qualitatively different from inorganic ones he still seized on crystals as a visual synthesis with living things. The combination was a symbol emblazoned on the front covers of all the volumes of "A System of Synthetic Philosophy" (see plates), showing a plant carrying a symbol of a successive development – first as a chrysalis, then as a caterpillar and finally as a butterfly. The plant, like the animals, emerged from a cluster of crystals.

Another definition of life with which Spencer quarrelled was Henri Ducrotoy de Blainville's version: "Life is the two-fold external movement of composition and decomposition, at once general and continuous".[75] Spencer, in his search for the basis of life, had drawn on the ideas of continental scientists – especially those of de Blainville and Milne-Edwards – even though they had properly heeded

Cuvier's warnings and refused to consider the study of homologies as a key to a "philosophical" understanding of nature.[76] Since Spencer frequently relied on the use of homology it could appear as if he were invoking the wrong source when he mentioned de Blainville's name. However, when discussing Spencer it is important to discard the division between two schools of French natural scientists: the orthodox and their opponents, the radical transformationists, such as Etienne Geoffrey Saint-Hilaire.[77] From Spencer's perspective the division was unimportant; both sides were equally valuable sources, since he would have not, in the early 1850s, taken sides in a debate on the subject of the inheritable changes in species. That would have been pointless, since his own evolutionary theory did not depend on the transformation of species. Spencer's choice to discuss de Blainville's definition was caused by its importance in a German context rather than in a French one. In Spencer's circle, de Blainville was regarded as an important figure who had refused to recognize Goethe's eminence as a natural scientist,[78] and thus as an authority who opposed the transcendental approach to the natural sciences. In emphasizing de Blainville's significance, Spencer was signalling to his friends that his science of Life was not a variety of Teutonic vitalism.[79]

Nonetheless, Spencer rejected de Blainville's definition because it incorporated the processes that physiologists called vegetative, while excluding the functions of nervous and muscular systems, which formed the most conspicuous and distinctive class of vital phenomena. De Blainville's version had the additional flaw of being equally applicable to vital processes and to those that occurred in a galvanic cell or a battery.[80] This ambiguity was similar to that which had undermined Spencer's earlier attempt in 1852 to define life. On more mature reflection he decided that in focusing on coordination he had created a category that mistakenly included non-vital self-balancing systems such as the solar system.[81] His friend Lewes was also put aside. He had defined life as a series of successive changes,[82] which in Spencer's view excluded too much. First, it denied the more visible movements with which our idea of life is most associated and, secondly, it ignored the fact that many changes, such as nutrition and circulation, occur simultaneously.

Spencer's treatment of competing definitions of life was designed to demonstrate that a successful definition must be framed so as to be always distinct from both inorganic and mechanical processes. Further, he was insistent that Life could not be represented by a simple order of priority in which one factor is represented as causing another. Lewes, like Comte, from whom he had taken his direction, had erred in emphasizing mechanical processes and simple aspects of change, neither of which adequately accounted for the highest and most significant development, the growth of intelligence. Nor did Lewes's views account for the fact that the causes of many organic processes were interconnected and occurred simultaneously. At this time, the key idea for Spencer was to define life so he could draw an exact parallel between the assimilation of food by an organism and the assimilation of information by human intelligence. Spencer saw both as features of the process of growth. They were equally representations of Life itself because processes that digested nutriment were identical to those that absorbed information.

In Spencer's view, reasoning was similar to the assimilation of food because both processes were made up of *successive* changes, rather than instantaneous uniform ones. That is, transformation of food into organic tissue and the carrying on of a chain of reasoning were essentially the same; both involved a number of distinct interactive processes that followed each other in a way not usually imitated by inorganic bodies.[83] It was true that an inanimate object, such as a watch, could display a succession of quick changes, but this was an exception. In any case, such changes were greatly outnumbered by those going on in a living body.[84] The speed and frequency of alternatives was a gauge to indicate the amount of vitality present. Higher levels of life would display more rapid changes, or a lengthened series.[85] This last observation also served as a scientific recoding of Spencer's early idea of progress by providing analogies that would equally apply to individual psyches, social reason and the higher species of animals. Change in *higher* organisms included the following: the evolution of mammals, individuals acquiring civilized behaviour, and societies that were shedding their militarized aristocracies and rapidly undergoing industralization.

Spencer's recycling of these analogies in his popular essays produced quite startling effects, tempered neither by empirical observation nor an attempt at accuracy. Ordinarily an analogy must have several points of comparison to be acceptable, but his analogies with Life produced deductions that had no basis in scientific observation. For example, in his essay "The Social Organism" (1860), when he compared the organization of cells with the division of labour in human societies, he remarked that bushmen were equivalent to yeast plants, and that slightly higher groups of human beings were the same as hydras.[86] This did not mean that he had observed that the social organization of bushmen resembled yeast cells in multiplying rapidly, nor that he wished to offer such a hypothesis. His comment referred solely to what he saw as the simplicity of the organization. To refer to a society as "higher" only meant that it displayed more *heterogeneity* in its development.[87] He was simply suggesting that when something developed a higher plane, one expected that its component parts would display specialization or differentiation. Conversely, "primitive" life – as displayed in animal form, in mental phenomena or in social customs – was homogeneous. Traces of Spencer's early theory remained in his later sociological writings, hidden underneath his professional work on human societies. It continued to shape his ethnographic views in the 1870s and 1880s (for example, it was this theory that caused him to complain that the customs of primitive people were monotonously the same everywhere) even while it lacked intellectual coherency, because he had dispensed with the homology on which it rested. In his later comments, he modified his belief that there was a close parallel between a higher society and higher organism in such a way that it could not support deductions. While he continued to maintain that in both, "higher" was distinguished from "lower" by the display of increasing complexity, he had also developed the notions that, unlike individual organisms, societies had no particular "form", that they were not composed of a continuous mass, that they were more capable of movement and that they were not differentiated into parts some of which possessed sensitive tissue and some of which did not.[88]

Spencer's attempt to capture nature by definition had strongly anti-Romantic qualities. Images and forces that had so struck Romantics – clouds, volcanoes, the seas, mountains and glaciers – left Spencer unmoved. He believed them to be outside nature. This was because he perceived the qualities that had made them appear attractive as only mechanical alterations. For example, Spencer saw the changes that took place in the seas as merely mechanical.[89] There was a degree of ambivalence about Spencer's idea of "mechanics" in *The Principles of Psychology* (1855). On the one hand, if he considered that the internal processes of an entity were mechanical ones, then it was assigned to the inorganic world. On the other hand, if he were sure that something was an organism, then, when it displayed mechanical changes, it was a sign that "Life" had moved to a higher level. For example, he believed that sponges had progressed beyond single-cell organisms because of the mechanism they had developed to draw in and expel seawater.[90]

Inorganic phenomena, such as clouds or glaciers, only superficially resembled living entities. In his search for the essence of Life he had left the Romantics behind. Variations that took place in inorganic phenomena only simulated aliveness; they lacked the "definition" that accompanied heterogeneous changes in organisms. Unlike William Wordsworth, Spencer could not wander lonely as a cloud. Nor could he follow Ruskin in imagining that a fragment of a mountainside was alive. Spencer was not able to identify with the lifeless parts of the physical universe because he regarded these as lacking in the processes of mutual dependence that could be found in the leaf or root of a tree. For him, what defined an inorganic entity was its absence of features that could be seen to parallel assimilation, respiration and circulation. Arrangement was even more important. Things such as clouds lacked the "definition" that he had observed in the combinations of sequential and simultaneous processes and functions in organisms. In an organic entity it was not possible to vary any of these combinations without altering the whole ensemble.[91] Spencer found it more perplexing to exclude natural entities, such as clouds, than he did mechanical devices such as watches and steam engines.[92] However, he conceived of a reliable way to verify which of these resembled Life. That is, he gave a "proximate" definition that distinguished an organism from its inorganic look-alike by observing that only the first possessed variations and functions in a definite combination:

> [S]o dominant an element in our idea of life, is this definite combination, that even when an object is motionless, yet, if its parts be definitely combined, we conclude either it has had life, or has been made by something having life. *In its ultimate shape* therefore we read as our definition of Life – the definite combination of heterogeneous changes, both simultaneous and successive.[93]

This not only excluded mechanical devices and changes in the physical world, but it was also proof against dead organisms and chemical processes such as those that took place during crystallization or when a galvanic cell or battery was in operation.[94]

This "proximate" definition of life is remarkable because Spencer inserted the word "shape" into his formula. This, along with his use of "definitions" as referring to a combination of characteristics and functions, suggests that he believed that one could visually sight the qualities that composed Life. Taken together, the shape of Life and the definitions of its functions and processes signalled the presence of an irreducible form of Life, a form that underlay the common qualities of both human beings and other organisms and that propelled them towards some future goal.

Spencer's vitalism was designed to tame the wild portion of the soul, that part which Romantics had brought into close proximity with the human spirit. Although the first flush of Romanticism had died away, some of Spencer's contemporaries, such as Ruskin, Edward Forbes and John Tyndall, still saw life in a mountain or a glacier. In stressing that a large portion of the physical universe was lifeless, Spencer made it less awesomely attractive, but he also made it less threatening. In addition, he recreated a sense that human beings were unique even though they were not created by God. There was stability and homeliness to Spencer's vision; he did not agonize about the death of the God of nature like Tennyson in *Lucretius*. To see, as he did, human beings as the sentient peak of the only portion of the universe that could truly evolve was effectively a restoration of the image that Christian rational religion had unseated in the previous century. However, rational religion, as espoused by writers such as Paley, had not distinguished between mechanical forces and organic change, and it was this distinction that was at the very heart of Spencer's conception of Life. Spencer's goal was much greater than a simple desire to overturn Romantic definitions of nature; he also aimed to capture the ground occupied by orthodox natural theology. When exponents of natural theology claimed to find as much life in a watch as they did in organic functions, they were not merely pointing to evidence for God's existence; they were reducing the uniqueness and individuality of human achievements to the same level as the physical world. Natural theologians, unlike Spencer, did not promise human perfection. Human beings were merely one of the arches of creation, and while they were superior in mind to animals, they were inferior in many physical attributes.[95]

This vitalism was not Spencer's mature scientific belief; he had begun to relinquish the search for it by the time he came to write the first volume of *The Principles of Biology*. The transformation of his thought was hesitant; he never totally abandoned the task of using definitions to search for Life. He simply began to delete the word "life" from his prose, and reduce its significance. These changes were substantive, not editorial; occasionally they left him with a difficult residue of contradictions. Since he had reprinted or recycled much of his early material he often left an abandoned thought to intrude anachronistically into the volumes of his philosophical system. However, this confusion created an unexpected advantage. It reminded his readers that somewhere a philosopher of the "Unknown" had indeed known what the mystery was. The early Spencer had been able to see nature herself; this legacy remained in later writings, and gave a spiritual message to his reader, although his scientific philosophy was increasingly unable to support it.

When the young Spencer contemplated nature he borrowed the spectacles of both physiologists and psychologists. With the help of physiologists he could see phenomena such as the assimilation of food taking place inside a plant, and psychologists helped him to understand the reasoning that enabled a scientist to make a discovery.[96] There was no break in the chain of development from the lowest to the highest processes. Spencer celebrated this continuous progression with sanctified language. This progress, he intoned, "must ever remain, as it has been from the beginning, one and indivisible".[97] He had taken a mystery that had belonged to orthodox Christianity, and made it visible in the organic kingdom. He encouraged his readers to see the unity of life in the human mind and body and in human society. His early definition had captured the shape of life, and a faint image of this certainty always remained in his writings even though it was increasingly obscured behind the philosophical doubts and scientific sophistication of his systematic philosophy.

Of course, even in 1855, before he had begun seriously thinking about metaphysics, Spencer had had philosophical defences. These were largely in the form of self-criticism. While he relied on his vision of the shape of life to provide a sense of certainty, he believed that his procedure was flawed because it failed to convey to the mind a complete picture of the thing described.[98] That is, it had failed to summon before the mind an adequate conception of life. Spencer's self-criticism here shows hints of Platonism. He had carelessly adopted some of the neo-Platonist language of Whewell and Richard Owen, although he had not absorbed this thoroughly enough to accept Whewell's construction of ideal *types* in the organic world.[99] His borrowings caused him trouble because they were in conflict with the common-sense realism that underpinned both his psychology and his desire to perceive the world directly as a series of defined shapes.[100] Platonism was a doctrine of essences and underlying forms, not of outward shapes. The contradictions were confusing to Spencer: he began to see his conception of form as flawed because he did not have a clear idea of the shapes behind reality. Later, when he had a chance to reflect on Owen's scientific idealism, he severely restricted the application of Platonism to his biology, and reserved it for his metaphysics and psychology. However, in 1855 Spencer was still tempted to use it as a defence for his scientific position. He came to a provisional stance on this when he surveyed the philosophical battle between Whewell and J. S. Mill and took the side of the former. Whewell had suggested that necessary truths, such as mathematical axioms, were propositions of which one could not easily perceive the opposites. They were distinct from experimental truths, of which the opposites could easily be conceived. If an experimental truth was considered, it seemed easy to conceive of stars moving about the pole, or of the full moon appearing for a whole month. If, however, it was a necessary truth, we could not *distinctly* conceive unusual propositions such as two straight lines enclosing a space.[101] Spencer's novelty here was a consequence of his acceptance of Whewell's position. While the latter's view of necessary truths was a conventional one when applied to mathematics, it was more unsettling when extended to sciences such as mechanics and hydrostatics.[102]

Spencer's reading of the dispute between Mill and Whewell on necessary truths provided him with the notion that the mind could directly intuit when an idea did not match reality. That is, Spencer argued that even when one lacked a complete idea of life, one had a negative and instinctive mental picture that could be used to reject those propositions that were inconceivable.[103] He thought that this test of inconceivability would ensure that there was an accurate correspondence between subjective beliefs and objective facts.[104] He also considered that one could test the invariable existence of a belief in a given proposition by conducting a mental experiment to discover if its negation were inconceivable. The explanation for this was that one could not properly conceive of negatives for propositions if, in order to do so, one first had to change one's state of consciousness. If you could not imagine the inconceivability of a proposition without changing your state of consciousness, then the belief was persistent.[105]

Spencer's views on inconceivability can be described as a kind of mental environmentalism. If a belief could not be easily disposed of, without the subject having to adapt his or her mental environment, then the belief was likely to be closely related to objective facts. Analysing beliefs in this fashion caused the subject's mental picture to begin to resemble the environment. Any proposition that significantly disrupted the subject's mental picture could safely be rejected as inconceivable; for example, "An oak growing in the ocean, and a seaweed on top of a mountain, are incredible combinations of ideas".[106] Spencer supposed that there would be an invariable and necessary conformity between the vital functions of any organism, and the conditions in which it was placed.[107] The internal processes of life had to match up with the external processes occurring around it. This was the rationale behind his teleological statement that the actions going on inside a plant *presupposed* a certain environment.[108] Or, respecting a primitive animal cell, Spencer asked, "what are the changes in virtue of which it continues to live"?[109] The teleological language here is not Lamarckian,[110] because it does not concern the transmission of acquired characteristics within a species. Almost all of Spencer's early teleological and environmental ideas about evolution focus on single adult members of a species, not on a species as a whole. For example, he does not care how a medusa evolved from a primitive jellyfish, but only about the "general truth" demonstrated by the activity of a single living medusa. By showing that its capabilities, such as the stinging and contracting power of the tentacles, correspond to the reactions and strength of a living creature serving as its prey, he could judge if the medusa would be successful or unsuccessful. That is, unless the external contact between a medusa's tentacles and its prey were not instantly followed by the internal processes that resulted in its coiling and drawing up its tentacles, the former would die of inanition.[111] In Spencer's view the basic lesson of organic change was that an organism must stay in harmony with its environment. Otherwise, "the fundamental process of integration and disintegration within it, would get out of correspondence with the agencies and processes without it, and the life would cease".[112] For an organism to recognize objects and escape from predators presupposed that there was a counterbalancing between subjective changes and objective conditions of the kind specified in contemporary scientific work on animal instincts.[113] Spencer could see this process of adjustment in human activities as well as in general organic ones. It was exemplified in the actions of a farmer who rotated his crops, or a navigator who calculated his position at sea.

Internal responses were like natural calculations. Both were used to adjust relations with the environment.[114]

This overarching theory of how Life maintains itself was not a matter of adaptation. In his early period Spencer did not allow development of species to have a major place in his ultimate statement on how to classify life in relation to the environment. While it is true that Spencer believed that the development of an embryo suggested that environment could determine the structure of an organism,[115] this was a mere corollary of his general theory of correspondence. He was aware of the work of von Baer and other embryologists,[116] but this did not spark any novel evolutionary ideas for him. Instead of inducing innovation, the findings of embryology were simply more details that had to be fitted into his pre-existing evolutionary theory. This theory was not connected with change with species, but with Spencer's view that the "structural modification" of an organism would slowly alter until it resolved a "re-arrangement in the organic balance". Spencer's idea of growth involved the process by which the organism was brought into general fitness with its environment and, further, developed those "after processes of adaptation" through which it was specifically fitted for its special activities.[117] Unlike Darwin's theories, Spencer's ideas of growth and adaptation did not have *failure* as a primary focus. Instead, they were part of a general belief in *success* measured by the number of adult life forms successfully living in harmony with their surroundings. These theories were gleaned from such wide-ranging examples of Life that it scarcely mattered to him if a single species, a human individual or a specific nation failed. More of each would survive than would be lost, and these would represent the shape of Life just as adequately.

Spencer's correspondence theory of life did not resemble theories of biological evolution that were "pre-Darwinian." Nor did his theory of life spill over into his political views and cause them to be transformed. Although Spencer's politics were radical he did not follow the lead of many radicalized medical people of the period, and search for biological underpinnings in his hope for social change.[118] Spencer's work was not part of the thin British stream of pre-Darwinian technical work on species change. Neither in his use of data nor in his conclusions did Spencer's early ideas closely resemble theories of natural selection or of Lamarckian evolution. The adaptive aspects of Spencer's work concerned change within single animals, individual psyches and particular cultures. Spencer's scientific goal was to show that all these were self-regulating organisms that were successful if they stayed in close correspondence with their environment. Rather than prefiguring Darwin, Spencer's early work militated against theories that depended on struggle and failure. It was not until after Darwin's *On the Origin of Species* was published in 1859 that Spencer softened his emphasis on organic correspondence and biological stability. Later, by 1864, when the first volume of *The Principles of Biology* appeared, he had entirely abandoned his idea that the internal processes of an organism *must* be in correspondence with the external ones.[119] Thus, treating Spencer as a precursor of Darwin is a mistake on two counts. First, it is bad history because, like "precursor" historiography in general, it avoids or distorts the interpretation of intentions "not surprisingly since it is impossible, in principle, to intend to be a precursor".[120] Secondly, it obscures the novelty and interest of Spencer's early scientific writings, which were witnesses to an optimistic faith in a beneficial nature. While his later writings exhibit his increasing competence as a philosopher, this development occurred simultaneously with the gradual disappearance of his vision. His beneficent nature was replaced by a scientific theory of life. In "A System of Synthetic Philosophy" there was still comfort to be gleaned from Spencer's evolutionary theories, but it was cooler. It was the knowledge of what the human beings shared with the rest of the universe. What remained common was not an antique faith in destiny or fate, but a belief in membership in cosmic processes. This was a modern conception; by the 1860s Spencer had learned to avoid the archaic language of teleology.

1. Spencer's father, William George, after he suffered a nervous breakdown.

2. Spencer's mother, Harriet, who was bullied by her husband.

3. Herbert Spencer's family home in Derby after his father's failure as a manufacturer.

4. Spencer's uncle Thomas, an Anglican vicar and radical pamphleteer.

5. Spencer at nineteen.

6. Spencer at thirty-five, showing signs of overwork on his psychology.

7. Spencer looking insecure, as he planned his philosophical works.

8. Spencer at forty-six, displaying the passionless calm for which he strove.

9. Herbert Spencer felt that Mary Ann Evans (George Eliot) was his mental and spiritual equal, but the plainness of her looks meant that he could not love her.

ROMANCE AND REALITY.

Beautiful Being (who is all soul). "HOW GRAND, HOW SOLEMN, DEAR FREDERICK, THIS IS! I REALLY THINK THE OCEAN IS MORE BEAUTIFUL UNDER THIS ASPECT THAN UNDER ANY OTHER!"

Frederick (who has about as much poetry in him as a Codfish). "HM—AH! YES. PER-WAPS. BY THE WAY, BLANCHE—THERE'S A FELLA SWIMPING. S'POSE WE ASK HIM IF HE CAN GET US SOME PWAWNS FOR BWEAKFAST TO-MOWAW MORNING?"

10. According to John Leech, beauty in nature could not be seen by everyone (*Punch*, 1852).

DIFFERENT PEOPLE HAVE DIFFERENT OPINIONS.

Flunkey. "APOLLO? HAH! I DESSAY IT'S VERY CHEAP, BUT IT AIN'T MY IDEER OF A GOOD FIGGER!"

11. John Leech's classical ideals of beauty (*Punch*, *c.*1850).

12. The emblem of organic progress that appeared on the front cover of all volumes of Spencer's "A System of Synthetic Philosophy". Life evolved from inorganic crystals to a beautiful form, the butterfly. According to Julia Raymond Gingell, the editor of Spencer's *Aphorisms*, Spencer devised this emblem himself and had it drawn by J. R. Clayton.

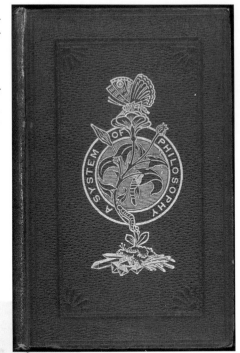

13. The frontispiece of Charles Kingsley's *Glaucus; or Wonders of the Shore* (1855). "Every well-educated person is eager to know something at least of the wonderful organic forms which surround him in every sunbeam and every pebble …"

DIAGRAM OF THE GEOLOGICAL RESTORATIONS AT THE CRYSTAL PALACE, BY B. W. HAWKINS, F.G.S., F.L.S.

Chalk.	Wealden.	Oolite. (Stonesfield Slate.)	Lias.	New Red Sandstone.
Pterodactyle.	Iguanodons.	Hylæosaurus.	Megalosaurus.	Teleosaurus. Pterodactyle.	Plesiosaurus.	Ichthyosaurus.	Labyrinthodon.
1.	2.		3.		4 & 5.	6.	7.

Mr. TENNANT, Geologist, 149, STRAND, London, is now able to supply the various Educational Institutions, Museums, and the Public, with these very instructive Models. Mr. WATERHOUSE HAWKINS has prepared for casting entire Restorations of those extraordinary British animals, carefully constructed to scale from the form and proportions of the fossil remains, and in strict accordance with the criticism and sanction of the highest scientific authorities. (See Mr. Hawkins's paper read before the Society of Arts, May 17, 1854, and published in *Journal*, No. 78.) Sold in sets of 7 Figures (from 1 to 7 as above), made to scale of one inch to a foot, illustrating the inhabitants of the Secondary or Reptilian Epoch, including the New Red, the Lias, Oolite, Wealden, and Chalk. Price of the set, £6 6s., or single figures at proportionate prices. Packing Case, 8s.; if wanted for the country, 8s. 6d.
149, *Strand, May*, 1857.

14. Dinosaur replicas for sale, advertisement (1854).

VISIT TO THE ANTEDILUVIAN REPTILES AT SYDENHAM—MASTER TOM STRONGLY OBJECTS TO HAVING HIS MIND IMPROVED.

15. John Leech's refusal to be frightened by science (*Punch*, 1854).

16. T. H. Huxley (1846).   17. J. D. Hooker (1855).

Spencer's scientific advisers

18. John Tyndall (1876).   19. John Lubbock (1876).

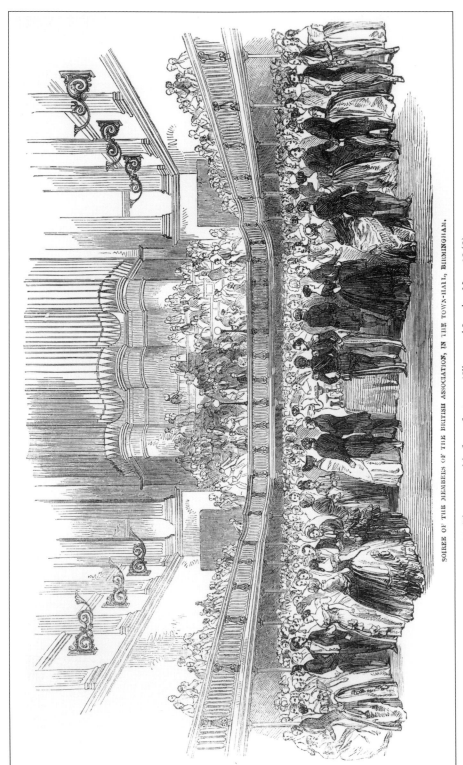

SOIREE OF THE MEMBERS OF THE BRITISH ASSOCIATION, IN THE TOWN-HALL, BIRMINGHAM.

20. The respectable face of science (*Illustrated London News*, 1849).

THE SOCIAL SCIENCE CONGRESS AT YORK: LORD BROUGHAM, THE PRESIDENT, DELIVERING HIS ADDRESS IN THE FESTIVAL CONCERT-ROOM—SEE PAGE 235.

21. Respectable social science (*Illustrated London News*, 1864).

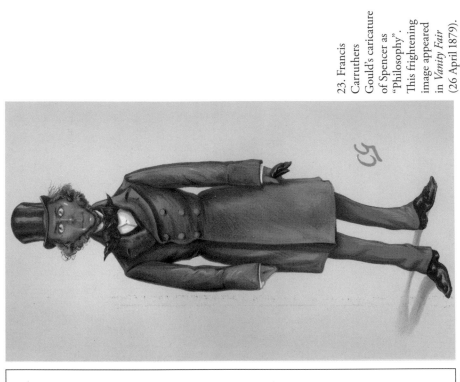

23. Francis Carruthers Gould's caricature of Spencer as "Philosophy". This frightening image appeared in *Vanity Fair* (26 April 1879).

MIND AND MATTER.

Navvy. "AH, BILL! IT SHOWS THE FORRARD MARCH OF THE AGE. FUST THE BRUTE FORCE, SUCH AS 'IM; AND THEN THE LIKES OF US TO DO IT SCIENTIFIC, AND SHOW THE MIGHT OF INTELLECT."

22. John Leech's amusement over scientifically driven progress (*Punch*, c.1850).

24. Max Beerbohm's retrospective view of eminent Victorians (*Manchester Guardian*, 13 March 1936). Spencer, like the others, was now a figure of fun, who should be unbearded.

THE RETURN OF "MAX": HE SHEARS THE VICTORIANS

"If they were flourishing in this our day"

Mr. Max Beerbohm's cartoon which we reproduce to-day is the first that this distinguished caricaturist has done since his return last year to England from Italy, where he has been living for the last twenty-five years, broken by occasional visits to London. He has chosen for his subject the Victorians, who are now so fashionable on the stage and in books, and shows us how they would fare if they had lived to-day and followed our barber customs. Their long hair, beards, whiskers, chintufts and heavy, curly moustaches would, of course, go. They would have Chaplin (or Hitler) moustaches; they would abandon pince-nez and monocle for horn-rimmed spectacles; they would brush their forelocks back instead of forward. One had not thought of all these things, but "Max" has done so. Shorn of all their hirsute decorations, they do not look such very grand figures now. Nor would our little heroes of to-day look so ordinary beside them! Anyway, here are the Victorians "if they were flourishing in this our day." How many of them could our friends—even the cleverest of them—identify?

They are, reading from left to right:

Back Row: Sir William Vernon Harcourt, Lord Leighton, Sir Henry Irving, Darwin, Whistler, Lord Randolph Churchill, Oscar Wilde.

Middle Row: Browning, Herbert Spencer, D. G. Rossetti, Swinburne.

Front Row: Gladstone, Ruskin, Disraeli, Carlyle, Tennyson.

25. Spencer's evolutionary diagram in which clusters of related organisms are depicted in such a way as to avoid the progressive scientific errors that necessarily accompany diagrams of evolutionary trees or cones (from *The Principles of Biology*, vol. I, 303).

26. Spencer's emphasis on form and symmetry in plants (from *The Principles of Biology*, vol. II, 143, 151).

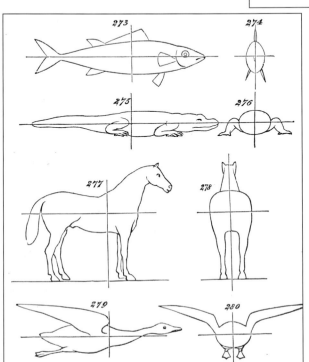

27. Spencer's emphasis on form and symmetry in animals (from *The Principles of Biology*, vol II, 182, 187).

THIRTEEN

The meaning of life

Spencer's *The Principles of Biology* was not a contribution to the growth of a science such as genetics or cell mechanics. Instead, it was a theoretical enquiry about the meaning of life. This quest was limited by the conventions of mid-nineteenth-century philosophical discourse, but nonetheless it contains interesting speculations about the nature of empirical investigation and the growth of knowledge. To Spencer, philosophy was not an abstract discourse; it was the pursuit of actual knowledge of nature in a way that could be seen as well as thought. Since nature was visible in forms and shapes, his questions about it were framed so as to account for these. It was the living part of nature that chiefly concerned Spencer in his mature biological writing, so he limited philosophical questions so as to exclude objects or systems that were not animate. That is, he concentrated on plants, animals and social systems. His answers were designed to combat the notion that living things possessed meaning because they were directed by some internal force. A proper search into the significance of life avoided probing between, and underneath, surface shapes for hidden clues. Such deep analysis would be guided by the rubric that what was casually visible in the universe was its actual face. "Superficial" knowledge was all that should be attempted; any attempt to investigate the underpinnings of nature would only uncover rational and functional laws, not answer questions about meaning.

Spencer adopted this philosophical stance because he wished to protect himself from the two currents of scientific thought that had originated on the Continent. The first of these was the philosophical idealism associated with Goethe and Oken. These German views were best known in English through Coleridge's renditions, although Spencer was more concerned about their presence in the work of Whewell. The second current was the materialistic one, expounded by the English followers of Comte. Spencer's basic enquiries were phrased in such a way as to avoid these two opposed extremes. He answered the question "What is the proper meaning of organic forms?" in such a way as to show that he had adopted an empirical stance that would radically exclude any knowledge or hypothesis based on transcendentalism. This was why, in *The*

211

Principles of Biology, he carefully avoided any hint of Christian Platonism, innate principles and teleological perfection, theories that had sometimes coloured his early scientific essays. Spencer was determined to excise these mistakes in a nuanced way without accepting Comteanism. That, he saw, was a repetition of the errors of eighteenth-century materialists who had crudely suggested that life was meaningless. Spencer, in *The Principles of Biology*, was a metaphysical trimmer who avoided twin perils: spiritually informed scientific knowledge and empty materialism.[1]

The first of these perils, which included the use of innate principles in biology, forced Spencer to adopt an artificial classification system for organizing knowledge in the life sciences. This system was so radical that it did not recognize any demarcations in nature.[2] That is, Spencer concluded that divisions such as those between species, genera or orders – and even the great gap between animals and plants – did not represent anything fundamental. He suspected that his contemporaries' search for a natural classification of living things with which to replace the artificial one of Linnaeus was doomed to failure. He also refused to put weight on the supposition that there were distinct boundaries between different species of plants and animals; these classifications could only be idealistic constructs that had little to do with scientific knowledge. Spencer also worried that those who saw species as the fundamental building blocks of life also espoused non-scientific notions, such as the theories that species were moving towards perfection and that they could be grouped in either progressive or linear series. Such errors would delay philosophical investigation into the meaning of nature. This was not to be found by interrogating hidden structures and groupings of animals and plants. Success would not come from searching for the origins of life, nor by speculating about the probable direction of organic development. The meaning of life would only be discovered in the detailed interpretation of the functions and structure of plants and animals, and in the examination of their relations with environment.

In *The Principles of Biology* Spencer distanced himself from the vitalism that had shaped his scientific writings during the 1850s. He also abandoned the idea that life was a separate force, the presence of which distinguished the organic from the inorganic. In its place he developed a science that demonstrated that living things were subject to the same laws as the rest of the physical universe. Nothing would be hidden; for Spencer the meaning of life was plainly visible in the forms and structures of the material world. The chief tasks of *The Principles of Biology* were to make the forms and structures apparent to the human eye and to explain them with the tools of science. These undertakings were spiritually significant because to see, and to understand, nature was to be part of it.

Spencer's *The Principles of Biology* was as much directed against the fear that the universe was empty and meaningless as it was against any traditional religious formulation of knowledge. For him, unlike Darwin, the criticism of orthodox beliefs, such as the special creation of species, was not a sign that he had plunged into the spiritual darkness of materialism. While he could echo with satisfaction the ironical comment of his botanical friend Joseph Hooker that the supposedly miraculous creation of "new" species habitually occurred in regions of the world remote from human observation, but disappeared when it was exposed to the gaze of zoologists and botanists,[3] this did not lead him to scepticism. Spencer's humour at credulous beliefs of theologically inspired scientists was tempered by his persistent attempts to provide the kind of solace usually given by Christianity. It was not that he offered a vision of God, or any hope of His existence; it was simply that Spencer's scientific writings mimicked the ameliorative qualities of theological writings. If the latter possessed cosmological or moral content, then so would his. This is why he framed his arguments in a scholastic language. For example, he defended his theory of evolution by using the theologically inspired argument that it implied a first cause. He claimed that like arguments for the special creation of species possessed the same rationale as those for evolution. Both implied that there was a cause.[4] From this, Spencer's reader could conclude that evolutionary theory provided an answer to the ontological question of why human beings exist. This answer, which was a reassurance that life was meaningful, had the merit of being backed by the latest findings of science.

If Spencer's scientific formulation of life was to be successful, then he had to demonstrate how it related to the stability or persistence of disparate organic forms as well as to their changes. From this perspective, it became more significant to examine how and why a biological individual, or a species, perpetuated itself without modification than it was to analyse development. That is, the static aspect of Life was more important to Spencer than its dynamic aspect. This focus explains why Spencer read Darwin differently from how a twentieth-century reader would. While Darwin's works were the most frequently cited sources in *The Principles of Biology*, they did not direct Spencer's discussion of species towards how these had been naturally selected, but towards how they had been able to perpetuate themselves.[5] For Spencer the most meaningful feature of a species was its continuation over time. This had occurred because of gamogenesis (sexual reproduction) and the neutralization of unfavourable genetic deviations. However, Spencer did not wish to single out species as the prime instruments of evolutionary change. He saw no critical difference between the perpetuation of a species and a single individual avoiding death. The species and the individual were similar because both exhibited the stability that was caused by the vital force.[6] However, this force was not progressive. In *The Principles of Biology* the vital impulse was not the agent of change, but of stasis. It was instability, not stability, that was the evolutionary force, which could jerk a species out of equilibrium and restore to it the developmental capacity that it had lost. In other words, some harsh novelty in the environment caused change, not the inner Life force, which by itself would always tend to prevent evolution. The argument of Spencer's *The Principles of Biology* was punctuated by repeated efforts to cleanse scientific thought of any connotation of innate drives, "natural" definitions and classifications of the divisions of life, or teleological theories that might suggest that the ultimate goal of evolution was perfection. Remnants of all of these ideas can be found in the scientific essays he wrote during the early 1850s but they were mostly expunged by the time he wrote the biological section of "A System of Synthetic Philosophy".

Aside from the abolition of teleology, the most important change to Spencer's biological writings was the decreasing utility of the guiding principle of individuation. When writing *The Principles of Biology*, he could not satisfactorily recycle his basic distinction that individualism in organisms was hostile to reproduction. His old attempt to analyse phenomena by using the logical tools of definition and classification now failed him. Try as he might, he could not discover a definition of individuality that was unobjectionable. "What", he plaintively asked, "constitutes a concrete living whole?"[7] He also began to define individuality in terms of a notion of life that would include groups or species. From this point of view, an individual might be an adult member of a population or the population in its entirety. His failure to defend the unique importance of individuals opened up a yawning gulf between his biological views and his political ones, which his general philosophical system did little to bridge. Political individualism no longer enjoyed support from a close analogy with the biological sciences. In *The Principles of Biology*, Spencer's progress was not a political futuristic ideal, but something that would diminish individualism and increase conformity.

Spencer's increasing difficulties with the use of definition in science gave him other problems besides a shifting notion of individualism. Definitional uncertainty also appeared in his account of genesis, the polar opposite of individualism. He could not discover an unambiguous definition of genesis (the basis of reproduction). His own definition was that genesis was disintegrative, not creative or progressive. He subdivided this disintegrative process into a negative and positive phase. This was an unsatisfactory outcome for a supporter of progress; if this definition of genesis was applied consistently it lacked progressive qualities. However, its threat to the overall anodyne quality of Spencer's philosophical system was limited because he failed to follow his own definition rigorously. For example, he believed that disintegration was less marked in higher organic forms,[8] a thought that allowed a hint of progress back into his view of life. Spencer's definitional approach contained a basic contradiction. The primary goal of his biology was to provide an analysis of evolution among human beings and the other vertebrates that he sometimes designated as "higher" animals, yet his findings suggested that genesis was *not* a progressive force, because it was often disintegrative, instead of creative. In essence, Spencer's biology was much more ambivalent about progress than Darwin's *On the Origin of Species*, which had focused on sexual reproduction as a force for change, not for stasis.

In his early scientific essays, Spencer had described the great tension of life as the antagonism between individualism and genesis. This tension no longer functioned as an explanatory mechanism for Life because it had been based on a notion of teleology, which was rejected in Spencer's *The Principles of Biology*. In that text there was nothing that directed the forces of Life towards a particular end. He had believed that sperm cells and germ cells possessed a mysterious power, but now he saw them as fundamentally similar to other cells.[9] The significance of this insight was that he could no longer causally anchor his notion of vital force to an individual organism nor, by implication, to the development of a species. If the cause of evolution was not seated within an individual organism or a species this implied that it was a response, or a series of responses, to the physical environment.

Spencer's account of science in *The Principles of Biology* was drawn from several disparate, and sometimes conflicting, sources. However, all of his sources began to give him the same message: that Life was inaccurately described if its evolution was portrayed as directed. From Darwin he took the suggestion that evolution was not proceeding towards specific goals.[10] From Richard Owen he absorbed the fact that more than one half of all species of animals were parasites, and that every animal had its own particular species of parasite.[11] This militated against teleology because it meant that, in nature, animals of inferior dignity harmed their betters. Nature was not progressive if by that one meant that "lower" animals without highly evolved faculties were subordinate to higher ones. On the contrary, an inferior creature could destroy its superiors and feed on them in a cruel way. This had always been true; the past had never been better than the present. His focus on Owen drew Spencer's attention to the time period occupied by the creatures that Owen had named dinosaurs, the terrible prehistoric lizards that had been equipped with extraordinary teeth and talons that could easily destroy early mammals. The fossil record could be read as one of "universal carnage".[12] What, Spencer asked, must we think of the elaborate appliances for securing the prosperity of organisms that were incapable of feeling, at the expense of misery to those that were capable of happiness?[13]

He drew on other sources besides Darwin and Owen. Cuvier and Agassiz taught him that the chain of being (the idea that species could be arranged in a single serial order) could be disregarded,[14] which meant that Spencer had yet another reason to reject teleological direction. His criticism of Lamarck reinforced this insight. Spencer was subtler than Darwin had been about the merits of the great French biologist. Spencer's chief criticism of Lamarck was not just that he argued that animals possessed the inherent tendency to develop into perfect forms, but that he mistakenly placed groups of animals into a uni-serial order that was substantially the same as an ascending order.[15] Spencer did not cite Lamarck when he theorized that modification in organic structure produced by changes in function was transmitted to offspring. He did not need to acknowledge the Frenchman because he could source the same idea to Darwin.[16] In any case, there was a more general problem: the scientific authorities on whom Spencer relied were having their doubts about progress. It was from Huxley and his interpretation of the German embryologist von Baer that Spencer learnt that a natural classification of groups of animals was not progressive.[17]

For the purposes of assembling his anti-progressive arguments, it did not trouble Spencer that the scientists on whom he relied were opponents of each other. It would not have mattered to him that Aggasiz was suspicious that Darwinism was a fad that would be as short-lived as Oken's "Physio-Philosophy". Also, it did not concern Spencer that his friend Huxley had devoted himself to destroying Richard Owen's reputation, or that Huxley also had reservations about the usefulness of Darwin's theory of natural selection.[18] From Spencer's perspective all of these scientists equally reinforced the notion that the appearance of progress was illusory when it was applied to animals and plants.[19] Spencer's reading of organic evolution as anti-teleological gave him a stance similar to that of a modern scientist. He was able to agree that environmental modifications of animals and plants had not come about because they were designed to secure these changes, but because natural selection had eliminated varieties that lacked such fitness.[20] It was only when Spencer discussed the human species that he partially relaxed his new-found anti-progressivism.

Spencer's doubts about progressive theories in biology were summed up in a blunt phrase that accused scientists of lacking empirical rigour: "the data are insufficient" and the theories based on them were weak.[21] He thought that one should not rely on the immense contrast between the small number of examples of "low" fauna in fossil records and the many examples of existing "high" fauna. The evidence was too variable in quantity to be reliable. Spencer believed that increased scientific knowledge had "made it manifest that remains of comparatively well-organised creatures really existed in strata long supposed to be devoid of them".[22] Further, it was a tenable hypothesis for Spencer that successively higher types had become fossilized in later geological deposits. This could indicate nothing more than that there had been successive migrations from pre-existing continents to those that had been gradually emerging from the ocean. Migrations might have begun with inferior orders of organisms, and would have continued with superior orders when the new land became more accessible and "better fitted to them".

Spencer's "tenable hypothesis" was not a serious suggestion; it was nothing more than a sceptical attack on what he saw as the remnants of transcendentalism. That is, Spencer believed that English geologists – especially Roderick Murchison and Charles Lyell – were unconscious apologists for a philosophy that displayed the tendencies of a Neptunist theory that they had already discarded. This theory had postulated the existence of uniform strata that had been laid down in the past during universal floods. To Spencer remnants of this could be found in Murchison's Silurian system because it was based on the idea that the whole world had uniform strata.[23] Similarly, Lyell's *A Manual of Elementary Geology* structured geological formations into "primary" and "secondary" categories as if these corresponded to distinct universal eras in the past.[24] Spencer's complaint was that while the Neptunist "onion-coat hypothesis" was dead its smell still lingered in the language of its antagonists.[25] This fragment from discarded scientific theory not only marred the ideas of Murchison and Lyell, it was also found in the work of palaeontologists, such as Buckland and Owen, whose lost worlds were respectively ruled by giant hyenas and dinosaurs. These writers, too, had foisted universal qualities on their subjects, imagining that their extinct creatures had been evenly distributed across the globe at the same time periods.[26] In this respect Spencer believed that there was little difference between the evolutionary arguments of more secular earth scientists and the theologically inspired geological work of Hugh Miller, which attempted to disprove evolutionary theory.[27] These scientific antagonists were conducting metaphysical enquiries when they should be restricting themselves to matters for which they had data.[28] This dismissal was not extreme: to Spencer's contemporaries there would have been nothing unsettling about his rejection of remnants of a systematic theory of geology. In any case the establishment itself was suspicious of the theories of earth scientists, a concern that was best captured in Whewell's quip that the favourite maxim of the Geological Society of London was that the time had not yet come for a general system of geology.[29] Spencer's statement that it was unscientific to base evolutionary hypotheses on unsupported statements was commonplace. Geology was unscientific because, in his opinion, there was no empirical basis for the supposition that uniform causes in the past must have had universal effects.[30]

Spencer suspected that the total amount of change in types of organisms that have existed from Palaeozoic and Mesozoic times down to the present day was not very great. Nor did the amount of change found in organisms necessarily point towards the development of higher ones over time. Although it was clear that nearly all living forms had prototypes in early geological formations, it was also true that ordinal peculiarities had been preserved to the present, and "we have no *visible* evidence of superiority in the existing genera of these orders".[31] The idea that evidence should be visible is, as will be shown later, an important aspect of Spencer's belief as to what counted as scientifically meaningful. If something could not be clearly envisaged, then it was probably what Spencer referred to as a "pseudo-idea",[32] an illegitimate symbolic conception.

Spencer's insistence on the anti-progressive nature of biology ran exactly counter to his own early ideas of progress in nature. It also left him in difficulty when he came to discuss the impact of biology on human beings because he felt obliged to make them an exception to his now non-progressive evolutionary theory. When most organisms achieved equilibrium with nature this led to a lack of change or death. However, when human beings achieved a balance with nature, there would be a more positive outcome. This unique result was the completion of the evolutionary process. Until human beings reached this goal Spencer thought that they would be progressive. His strategy here – of separating the human species from all other organisms – conflicted with some of his reform values. For example, Spencer's insistence that human beings should avoid artificial restraints on their natural emotions was based on what he felt was the imperative need that each human being should recognize the biological side of human nature. This idea becomes less attractive if one also adopts Spencer's argument that the evolutionary destiny of human beings is separate from the rest of nature. There is a fundamental confusion here; Spencer would not be able to answer the question why, if it is progressive to follow our animal-like emotions, our future should be so different from that of animals. This difficulty arose because Spencer did not completely shed his early progressive view of nature when he adopted the more neutral perspective of his scientific friends while he was writing *The Principles of Biology*.

Spencer's concept of "nature" was ambiguous. There was a tension between its insistence on an anti-progressive pattern for nature and its hope that human beings would have a benign future. This conflict had arisen because, while writing *The Principles of Biology*, Spencer's friends Huxley and Hooker had dragged him away from his early optimism about nature. During the late 1850s and early 1860s, he was partly under their scientific tutelage.[33] It was Hooker who drew his attention to that fact that some modern plants, such as the true pine tree, were scarcely distinguishable from their fossils. Huxley also gave him examples of "changeless" species – among corals, molluscs, insects, spiders, reptiles and sharks.[34] According to Huxley, even species of Triassic mammals did not differ from their modern descendents as much as they did from some current species or as these do from each other.[35] What these examples suggested to Spencer was not an absence of new types, or new persistent forms, but that organisms have no innate tendencies to assume higher forms. This meant that any hypothesis of progressive modification must be compatible with a hypothesis of persistence – without progression – through indefinite periods of time.[36]

Hooker and Huxley both guided Spencer, but Huxley was the *deus ex machina* of *The Principles of Biology*. His intention was probably to use Spencer to attack his opponent, Richard Owen. Spencer's role as a pawn caused him to echo Huxley's rude or overly robust language; the hyper-critical comment on Owen is uncharacteristic of Spencer, who was usually polite or oblique in his criticisms of scientists. When he remarked in *The Principles of Biology* that Owen's views resembled those of the *Vestiges of the Natural History of Creation* (the anonymous popular 1840s account of organic progress), and that Owen was possibly the last person to renounce the doctrine that animals have innate appetites that give birth to perfect structures,[37] he was writing as Huxley's proxy and imitating his mentor's bellicose tones. Huxley also had an unintended effect on Spencer: he had probably only wished to use his philosophical friend in a vicarious triumph over Owen, teaching him some academic caution about the progressive quality of Darwin's theory of natural selection in the process. However, Huxley was a catalyst for greater effects in Spencer's thought. He imagined the existence of something called protoplasm, which caused Spencer to reorganize his views on Life and its relation with the inorganic world. Having thoroughly absorbed the lesson that the appearance of progress might be an illusion, Spencer concluded that he should discover an alternative message in evolution. This reflection led him to the general hypothesis that evolution must be construed so as to include static as well as dynamic factors. In *The Principles of Biology* the "life force" was transmuted into a new substance, protoplasm, which exhibited stability rather than dynamic features, and which was believed to possess transformative qualities. The new substance lacked innate qualities or internalized forces that might have propelled change. Since protoplasm represented stability, its antithesis, change, was now seen as caused by environmental forces, such as gravity. Spencer's biological evolution had become a matter of responses to physical forces that shaped or formed organic entities in the same way they did inorganic ones.

While Spencer's early science had concentrated on defining Life in opposition to inorganic entities such as crystals, *The Principles of Biology* abandoned this distinction. Things such as crystals were now regarded as similar to organisms because they exhibited parallel structures. For example, the growth patterns of some species of algae seemed like the shapes of feathery crystallization.[38] To Spencer this suggested that both organisms and crystals displayed growth patterns that were responses to physical laws. What was true for simple shapes had a more general application to living forms. All of these were structures that had evolved in response to physical forces rather than to inner drives. If you wished to find out what Life was, it was no longer plausible to speculate on what happened inside an organism; instead, you described the exterior of an organism.[39] Such descriptive activity takes up over two hundred pages of text in *The Principles of Biology*, while the analysis of structure occupies a further one hundred and fifty. Page after page is devoted to the careful depiction of shapes of plants and animals together with the shapes of their components, such as branches, leaves and cells. The shape of cells in the interior of animals is also detailed, but this is a description of the outside of those cells – their surfaces – together with a comparison of how these differed from cells on the exterior. The morphological chapters have simple titles such as "The Shapes of Branches", "The Shapes of Leaves", "The Shapes of Flowers" and "The General Shapes of Animals", while the physiological chapters have titles such as "Differentiations Between the Outer and Inner Tissue of Plants". All the analyses accompanying these changes are deceptively simple. Their purpose seems merely descriptive, but the purpose of such exhaustive illustration of the shape of life was to deny the use of this material to those who still believed in the vitalist theory.[40]

Spencer's deliberate simplicity and reliance on description was a thesis in itself; it demonstrated that portrayal of structure and function exhausted the need for explanation; it was all of Life that could be examined. In lieu of inner drives or archetypes, the shapes of organisms became the sole evidence for the meaning of Life.[41] The numerous diagrams that accompanied his description were a visual demonstration of this. They were a display of surfaces laden with meaning, but one could not penetrate underneath this. This cover display was nature itself, and Spencer joined forces with the popular naturalists of his period in urging his readers to re-examine the commonplace plants and animals in the field or the hedgerow in order to witness life's meaning for themselves. Spencer's illustrations had a bias towards showing how evolution had affected homely things, such as the shapes of leaves of brambles and oaks and the structures of earthworms and buttercups. He drew his evolutionary examples from his country walks, and he expected his reader to sympathize with him in his botanizing when he found that he was in the wrong month and could not find any cow parsnip until July so that his text lacked an important illustration.[42] Often the mere mention of a familiar plant was enough to suggest appropriate connotations. He expected his readers to examine fragments of life on the wayside, rather than to imagine distant species in fabled places such as South America, the Galapagos Islands, and the Indonesian archipelago. This was a matter of taste: much of the mid-Victorian love of biology was entangled with the exotic sense of adventures that readers could share with Humboldt, H. W. Bates, Darwin and Wallace as they transported the imagination to distant tropical places. Spencer's examples formed a deliberate counterpoint to this. His personal botanizing – for example, he did most of his own work on fructification – was not a search for missing links or gaps between species, but a painstaking examination of everyday organisms. Since he was not considering teleological goals for plants and animals, and since he did not believe in serial progression, exotic species were no more or less important than local ones. Either could be used as examples to show that plants and animals were environmentally adapted.

Spencer's appeal to everyday organisms was a conscious attempt to illustrate the meaning of life while avoiding the scientific error of interpreting nature as a movement towards perfection. In Spencer's view this faulty interpretation had been caused by an excessive insistence on historical progression. However, Spencer also jettisoned the idea that analysis of a species required a historical dimension. To him species were simply persistent life forms. The analysis of such phenomena could be conducted in the same way as that of single individuals; neither required the consideration of historical evidence. Instead of seeing a species moving along a historical continuum, he viewed it as a moving equilibrium along a spatial continuum: "The moving equilibrium in a species, like the moving equilibrium in an individual, would rapidly end in complete equilibration, or death, were not its continually – dissipated forces continually re-supplied from without".[43] In Spencer's theory of life, perfection was replaced by completion. A species that had almost reached equilibrium was nearing "completion". He defines "completion" in terms of resistance to physical forces. According to Spencer, species would never actually reach the point of completion; they would always fall off this path before it ended. Equilibrium would never be achieved.[44] Since this final goal would never be reached, Spencer believed himself free of the teleological blemish that had marred so much evolutionary thought.

Spencer's concept of "completion" was a prosaic one because it did not imply that the movement of a species followed a plan. It was only a culmination of responses to physical force. An example of this was the probable evolution of the external shapes of species of lichen. Those lichen that possessed tubular shapes exhibited a reaction to the forces of gravity: tubular shapes were stronger than flat ones and had, therefore, an evolutionary advantage. Individuals with structures that lifted them above the rest were fittest for the conditions "and by the continual survival of the fittest, such structures must become established".[45] Spencer believed that, subsequent to an evolutionary process, survival belonged to those organisms that were adapted to resist external forces. It could be said that Spencer believed in the evolution of the *best type* of organism, but this would be misleading unless it is realized that Spencer defined type in a way that would have struck many Victorians, as well as many subsequent readers, as unusual. It is customary to see biological types, or archetypes,[46] as either species that were moving towards perfection, or groups of organisms that possessed special characteristics. For Spencer, however, the world "type" referred to the kind of general organic characteristics that are least likely to suffer pain or damage. Spencer was not applying a utilitarian calculation to nature: his evolutionary theory was the reverse of hedonism. That is, his evolutionary theory was anti-utilitarian.[47] He did not think that animals evolved by seeking pleasure and avoiding pain. Instead, he believed that Life would evolve to avoid cruelty. To him the present state of nature was morally intolerable; it included micro-organisms and parasites that lived at the expense of more highly developed organisms. There was no balance or harmony in nature because these parasites often did not possess organs that could feel pleasure yet their hosts felt pain. Thus nature, as it had functioned up to the present, could not be represented as a good or as generalized organic happiness.

Tomorrow was another matter. The future would not include species that would experience pleasure while causing pain to other organisms. Ultimately, evolution was progressive; in the future superior organisms would be less likely to suffer harm from inferior ones than they did in the present. There would be an overall improvement, not just progressive changes to individual species. Evolution would apply to "organic creation as a whole".[48] The process would not conclude with perfect species; instead there would be a high average benefit for all of Life. This benefit would not be an increased aggregate of pleasure, but the gradual abolition of evil, such as the practice of inferior organisms in preying on superior ones.[49] Evolution was a movement towards a kinder time in which the sight of victims marked with tooth and claw would be increasingly rare.[50] At heart, Spencer's evolutionary theory was a wish for Life to progress beyond cruelty. His concerns about suffering were broader than the traditional Christian ones of redemption and theodicy because for him pain and evil could not be explained by musings about the fate of human beings. The fossil record showed that pain and cruelty existed before human beings had evolved. This meant that in concentrating on human suffering Christianity had left unexplored the defects and miseries of lower creatures. For Spencer such ignorance was inexcusable; "what must we think of the countless different pain-inflicting appliances and instincts with which animals are endowed"?[51] This question would not have appeared laboured in Victorian Britain, where his contemporaries wept over the cruelties meted out to a fictional horse in Anna Sewell's *Black Beauty*, conducted parliamentary debates on the iniquities of dog carts, and founded the Royal Society for the Prevention of Cruelty to Animals. When cruelty was involved, Victorian moral consciousness did not distinguish between animal and human victims. Spencer was not eccentric on this point: his insistence that Life was not exclusively a human concern was a sign that he was in harmony with the ethical ideas of his time.

Spencer's sense of urgency, when combating cruelty in *The Principles of Biology*, was caused by the erosion of his early faith that nature was usually kind and creative. His abandonment of the vitalist doctrines that had accompanied his scientific essays during the 1850s, had also caused him to surrender the anodyne language that had accompanied vitalism. He had borrowed that language from medically informed physiologists such as Carpenter, who had portrayed nature as a benign physician. In his early scientific writings Spencer had imagined that nature possessed a force that would heal wounds and illnesses as well as cause the emergence of more efficient and less cruel behaviour. In contrast to this, Spencer's later scientific language was coolly disdainful of therapeutic progress. The disease of rickets offered a useful example with which to illustrate Spencer's shift. It had been "what, in teleological language, we call[ed] a remedial process".[52] In this process a deposit of matter had occurred on the concave side of the bone, which strengthened it. However, he now discovered that he had been misled when he had judged this process as a naturally curative one. The direct adaptation occurring in rickets, which was "seeming so like a special provision, and furnishing so remarkable an instance of what, in medical but unscientific language, is called the *vis-medicatrix naturae* is simply a result of the above-described mechanical actions and reactions, going on under the exceptional conditions".[53] There was no kindness or natural health in this present working of nature. It contained nothing to offset against the presence of cruelty. Spencer was forced to look to the future for hope, but he no longer sought the unity of Life, which had offered him comfort in his earlier writings. The future was narrower, too. It no longer included animals and plants, but only humankind. Spencer had been forced to divide nature. On the one hand there were animals and plants, which were scarcely distinguishable from inorganic things; on the other hand there was the human species, which alone could evolve beyond disease, stasis and death. This would be a triumphal progress, but one that would leave the human species alone in the universe.

Human beings – that species which Spencer referred to as "the most highly evolved of terrestrial beings" – were still subject to evolution, even though this would lead to a different outcome from that facing the rest of organic nature. To Spencer the arrow of time applied only to the human species; other species would revert to their primary structures when original conditions were restored.[54] Only human beings would achieve "completeness" through this process, while animals and plants would merely experience endless readjustments. When Spencer referred to the "incompleteness", or the evolutionary "failures", of the human species as it was presently constructed,[55] he had in mind a vision of human beings progressing beyond their current limitations and defects. They would do this by acquiring the ability to predict what changes would be needed to avoid pain and defeat. Spencer suspected that the human species might never achieve the perfection about which he had written so confidently in the early 1850s, but it could nonetheless edge closer and closer to it. In his analysis of domestic relations he expressed more optimism than he had when reflecting on affairs of state. Compared to the losses suffered by the parents of lower forms of life, Spencer thought that the lot of human beings had showed considerable improvement. As the social organism developed in structure and strength, parents were less likely to suffer because of reproduction, "and the implication is that in the highest type of man this sacrifice is reduced to a minimum".[56]

Nineteenth-century English thinkers possessed diametrically opposed views about progress. Some, such as W. E. H. Lecky and Lord Acton, saw it as an advance towards more freedom; others, such as Henry Hallam, James Mackintosh, T. B. Macaulay and J. S. Mill, thought that unrestrained liberty was a thing of the past, not a hope for the future. Spencer cast his lot with the second group. His biological studies caused him to reinforce contemporary regrets about the disappearance of freedom. However, Spencer's science directed him towards a more anodyne strategy than those adopted by his more pessimistic contemporaries. Instead of wistfully designing political and moral bulwarks to protect the remnants of freedom, Spencer took refuge in the hope that the new values gained with the help of increased scientific knowledge would more than compensate for those losses of individual personal and political freedom that his contemporaries would suffer.

With the increased understanding of our own natures and of the environment in which we live, Spencer thought that we would be able to anticipate, and therefore avoid, failures and painful struggles. Human beings will make progress by "ascertaining the conditions of existence to which we must conform, and in discovering means of conforming to them under all variations of seasons and circumstances".[57] This hope was not for the restoration of lost liberties or freedom. Spencer's progress would not be towards a future in which human beings will enjoy the luxury of unrestrained action. Instead, it would be movement towards conformity with the scientific laws of our own nature and those of society in general. In *The Principles of Biology* Spencer had broken away from the nascent liberal tradition, which endeavoured to find space for private desires within a public framework. As a philosopher of science he was not content to speculate about society as if the future of freedom could be defended by slogans such as that popularized by J. S. Mill: that every individual should be permitted to do as they wish so long as this did not interfere with the freedom of others.[58]

Spencer was uninterested in philosophical solutions that had to be imposed by an effort of political will. He did not believe that human beings could remedy their situation by exercising "deliberate choice to improve themselves". Progress for the human species would not be a consequence of moral effort. This was a factual matter to Spencer because he believed that a failure to be moral often occurred despite individuals making often repeated resolutions to be good.[59] Spencer saw the lack of desire, not the lack of insight, as the chief cause of faulty action.

> A further endowment of those feelings which civilization is developing in us – sentiments responding to the requirements of the social state – emotive faculties that find their gratifications in the duties devolving on us – must be acquired before the crimes, excesses, diseases, improvidences, dishonesties, and cruelties, that now so greatly diminish the duration of life, can cease. [60]

Human improvement would not happen in the way that it had been imagined by traditional philosophers. That is, it would not be a matter of the intellect determining which actions were good, and then enforcing them by an exercise of the will. Nor would progress happen spontaneously.[61] Individual human beings would need to develop the kind of emotions that would make them accept moral and social duties.

Spencer's stance here superficially resembles the one popularized in the twentieth century by B. F. Skinner. Both seem to view untrammelled human desires as an impediment to happiness. However, Skinner's solution to the problem of human misery was not the biological evolutionary one of Spencer, but a return to the even earlier tradition of Charles Fourier. That is, he believed that a new social reality could be constructed by political means. Spencer's approach was more passive than Skinner's. While Spencer regretted the lack of morals and the primitive emotionalism displayed by his contemporaries, he refused to sanction any artificial and political remedies for these evils. In part, Spencer's beliefs stemmed from his conviction that industrious behaviour was not a benign virtue, but psychologically damaging. This was true whether labour was forced on people by religious conviction or by state regulation.[62] However, his political passivity also grew from a central perception of both his science and his personal experience. For Spencer, to be in the grip of emotions was to carry a biological and primitive burden. When they intruded into one's personal life, or into social activity, feelings were likely to be destructive. The human tragedy was that this burden could not be escaped. One could not be safeguarded from destruction by exercising the will so as to keep emotions at bay. If one bound oneself to a rigid course of personal action, or if one politically coerced others, the result would be negative, and possibly worse than the harm caused by the primitive display of emotion and lack of morals. The lessons to be gleaned from *The Principles of Biology* were close to the personal reflections Spencer later mused on in *An Autobiography*. There, too, he was hostile to the suggestion that the will was the engine of social change. It was also in *An Autobiography* that he reminded his contemporaries that progress could not be achieved by idealizing tools such as fear and pain. Imagining that reform could be accomplished by the use of political power was an attempt to return to the cruel human past, not a glimpse of the future.

The anarchic freedom once dreamt about by rebels would gradually disappear as a desirable goal. The future advance of civilization would be at the cost of social and psychological individuation. This was a biological speculation: as societies grew more complicated there would be a concomitant increase in the mass, complexity and activity of nervous centres, which would result not in an increase of rationalism but of feelings. The larger quantities of emotion would be needed "as a fountain of energy for men" who have to keep their employment and raise their families under conditions of increasing competition.[63] For Spencer this increase in emotion meant that, unlike his fellow Victorians, human beings of the future would find enjoyment in emotional and intellectual activity. This would be a compensation, because when, in addition, the human population reached its maximum numbers, there would be less need to diminish the "procreative reserves" that were spent in the creation of new lives. Life would then be longer and more pleasant.[64] The idea that sex was opposed to pleasure, rather than part of it, might appear startling, but to Spencer, distressed by his inability to fulfil his own biological urges, it was comforting. The notion was the only remnant of his early optimistic belief that the future held a promise for human bliss: when there would be the final attainment of life in "a form in which the amount of life shall be the greatest possible, and the births and deaths the fewest possible".[65] Genesis, like death, would be infrequent because it detracted from the happiness of individuals.

Not only individual lives but the whole earth would reach a summit of maximum happiness. Spencer hinted that the world would realize a geopolitical stability in which all habitable parts would be brought into the highest state of culture. This benign transformation would be the result of the pressure of population. Instead of fertility being the cause of misery, as it had been with Malthus and Darwin, it would be the cause of improvement. Spencer believed that, in the past, pressure of population had caused the dispersal of species across the globe, the abandonment of predatory habits and their replacement by agriculture, the clearing of forests and wastelands, the creation of social organization and the development of social feelings. The pressure of increased human population was "daily thrusting us into closer contact and more mutually-dependent relationships".[66] Spencer's optimism did not stop with the suggestion that an increased number of people on the planet was progressive. It continued with his surmise that, after having improved the intellect and feelings, population pressure would gradually finish its work, and bring itself to an end.[67] This ultimate improvement was not the final development of Life, as it had been in his early scientific writings. Spencer had ceased to care about the uniqueness of living or organic entities: they were no longer distinct from inorganic ones. The final sentence of *The Principles of Biology* intoned the coda that man's stable and happy future was guaranteed by the same universal process that drove the simplest inorganic action.[68]

The great organic–inorganic divide that had run through Spencer's early scientific writings had altogether disappeared, along with vitalism and Life with a capital "L". This division was not incorporated into his philosophical system. Instead, organic bodies were now seen as welded to inorganic entities, and were subject to the same forces. The introductory volume of Spencer's "A System of Synthetic Philosophy", *First Principles*, traced "the metamorphosis of living bodies" to the derivative laws of force through a passage from infinite homogeneity to a definite heterogeneity.[69] What had been promised in the introduction was faithfully carried out in the *Principles of Biology*. Metamorphosis was no longer a spiritual change; living organisms did not transform themselves, but were transformed by external forces of the kind explained by the laws of physics and chemistry. In the *Principles of Biology* the uniqueness of living things was discarded. Like physical objects, they could be understood by chemical analysis.[70] He described the purpose of the opening chapter of his work on biology as giving a conception of the extreme modifiability of organic material in response to surrounding agencies.[71] This truth about organic material had its implications for the human species. Mankind no longer stood above other species, and at the head of *organic Life*, as it had in his earlier vitalist writings; it had become the leading edge of material processes as a whole. This change did not represent a demotion of human beings. Their species alone would evolve to find a fulfilment in a state of harmony with the cosmos. At the end of this process, competition, cruelty, pain and fear would only be faded memories of primitive days.[72]

FOURTEEN

Science and the classification of knowledge

Spencer's theory of scientific knowledge led him to focus on the external qualities of things – their shapes and surfaces – rather than on their internal structures and their essential or inner beings. His stance ranged him against the Platonism that was a feature of much contemporary philosophy of science. His concentration on externalities was also an integral part of his philosophical and cultural beliefs. Spencer believed that only while looking at surfaces could one perceive the beauty that was one of the scarce non-functional features of the universe. To see, or to touch, beauty was to sense something rare that had not emerged as a product of the normal human drives, which were rooted in reproduction or nourishment.[1] Yet, at the same time, a focus on beauty did not entirely ignore function. Feeling the outside shapes of physical things and organisms did not obscure the fact that they were fashioned by physical forces. For example, it was possible to admire the symmetry or shape of a plant even though its structure could be explained by reference to the most efficient way of bearing weight. Similarly, while the beauty of a rose or the shape of a human eye might be explained by reference to science, this did not exhaust the subject; Spencer believed that functional explanations always left an unexplored residue. There would be an extra tincture that was more beautiful and more meaningful than that required by function alone. According to Spencer the perception of beauty or meaning is itself a recognition of regularity that shows the object we perceive to be part of a universe is subject to the forces of evolution. We, as living things, share a common quality with crystals and other physical objects. Spencer took solace from the knowledge that human beings were one with the cosmos; it was a joyful realization that human beings did not have to evolve, or reach perfection, to feel at home in their universe. They were already part of it.

When Spencer abandoned the idea that Life was progressive it represented more than a change in biological thought; it pointed to a profound shift in his philosophy. To relinquish perfection, and the concomitant idea that Life would achieve archetypal or perfect forms, was to accept profound implications for the organization of knowledge. A biological philosophy was a task for which Spencer was uniquely prepared. Natural scientists had reached a point at which their attempts to erect "natural" categories within which to organize knowledge had become increasingly implausible. Arrangements of organisms in linear orders seemed increasingly dubious, and provided no assurance that there had been progressive change. Worse, the state of knowledge gave no indication that the direction of Life was overseen by God. The provision of such knowledge was the task that Herschel had famously set before scientists. This was the study of resemblances between classes of phenomena, and was the province of inductive generalizations. The very existence of natural groups was one of the riddles of the mosaic, and could not be unravelled without acquaintance with the highest laws.[2] Classification was not just a matter of formal taxonomy; it was the most dynamic form of science. Herschel's admiration of John Lindley's botany as emblematic of the necessary organization of scientific knowledge was mystifying. If this was the paradigm for science, what could it mean? Lindley himself offered no clue in how to decipher his unstable version of Baconian empiricism. His comfortable adherence to natural classifications theory made his ideas about the organization of knowledge ambiguous. His followers, such as William Jackson Hooker, who were equally practical, found it increasingly difficult to categorize plants at all. It was, they thought, surprisingly difficult to ascertain the limits of a species.[3] Facts were not what they seemed, and analogies could no longer be trusted to demonstrate affinities between organisms.[4] It was necessary to face up to the theoretical issues that had left earlier botanists unruffled. This philosophical task was, however, unlikely to provide reassurance that the universe was controlled by a divine intelligence. That is, the complexities of nature were unlikely to be an analogue for religion. To Agassiz it seemed that the classification of Life, whether it was arranged naturally or artificially, was only an expression of human understanding of natural objects "and not as divined by the Supreme Intelligence and manifested in these objects".[5] Organisms that we saw as grouped together in nature did not represent God-given realities, but were, Agassiz warned, possibly reflections of our very fallible human mental processes. For a psychologist such as Spencer this was a clarion call. He had the mental equipment to bring order to biology.

Spencer imagined a library in which the librarians had arranged the books according to their subject matter. As they grouped the books according to their subjects, librarians ignored superficial similarities and differences and proceeded by discovering primary, secondary and tertiary attributes. The librarians would act so that if readers later inspected any volume, it would be possible for them to infer the character of neighbouring volumes. Books would be classed together in divisions for history, biography, science, travel and so on. Within these categories would be subgroups, like those that emerged when literature was divided into fiction, poetry and drama.[6] There might also be sub-subgroups, for example, if the sciences were divided into subjects such as physics, and those, in turn, were divided into sections dealing with heat, light, electricity and magnetism.[7]

Spencer designated this kind of classification as classifying by "natural" groups. This differed markedly from a second method of classification: simple classification. In simple classification one arranges books according to one characteristic. Spencer compared this to the way in which a child might organize books according to a single conspicuous quality such as similarity in size or style of binding. However, simple classification did not have to be juvenile, as Spencer would have known when choosing this example. There were contemporary libraries, such as those of the Royal Institution of Great Britain and the University of Cambridge, in which volumes were arranged primarily in terms of a single criterion, such as their height or the date of their accession. The point of this library analogy was to illustrate that classification of knowledge based on single characteristics would possess a linear quality. The organization of books by either height or date of publication might result in an ascending physical or chronological succession. In contrast, if things were classified in groups taking many diverse characteristics into account, the result would not be a serial arrangement and the resulting order would be idiosyncratic. For instance, priority might prove an entirely useless criterion for classification. This problem can likewise be illustrated within the library, where it would be nonsensical to say that historical works must come *before* scientific ones or *vice versa*. Similarly, it would be bizarre to give priority to novels over poetry, or poetry over drama, when organizing literature. The purpose of Spencer's library analogy was to elucidate his beliefs about the organization of scientific knowledge in biology and in other fields. It rested on a distinction between natural and artificial categories in the knowledge possessed by scientists. Spencer, as a didactic philosopher, wanted his contemporaries to become more conscious of how they categorized knowledge, and to be more aware of the extent to which their concepts were arbitrary rather than validated by nature.[8] It was not that he espoused a doctrine of philosophical realism with which to replace the idealism of Whewell and Owen; rather he insisted on a poised neutrality on such metaphysical questions. He wished to avoid idealism as much as he wished to circumvent the simplistic mistakes exhibited in Comte's progressive classification, and bring his own philosophical classification close to the best scientific practice.

This search for the optimum scientific model of classification made Spencer seize on, and transform, the "natural" classification systems of the Jussieus, and of Lindley's *The Vegetable Kingdom*.[9] These writers had expounded a type of classification that did not rely on organisms possessing inner drives or teleological goals. In accepting such a model, Spencer was choosing an organization of knowledge not based on the linear classification of organisms. The groups and subgroups specified by these botanists were not categorized in a serial order, but only in divergent or re-divergent terms.[10] This botanical classification suggested a similar pattern to that found in the way in which the classification of animals was organized. In the animal kingdom as well as among plants, classification should not be based on a serial order in which there was an imaginary progressive ascent from lowly or undeveloped animals up to highly developed ones. With the help of the ideas borrowed from von Baer and Huxley, Spencer was able to demonstrate that the supposed linear organization of animal groups was an accident of typographical convenience. It was a practical and visually clear form of exposition, but was extremely divergent from what it was intended to represent. In actuality, there was no simple serial order to be discovered.[11] Spencer reconstructed what Huxley would have recommended when visually representing this problem as an alternative to artificial order. Instead of an inverted cone of increasing diversity of species, Spencer's diagram displayed the orders of living things as simple groups that had no linear relationships with one another; it was a picture of unconnected clusters of dots (see plates).[12]

His argument about classification led him to doubts about Darwin's overall picture of evolutionary progress. Spencer was ambivalent here; he could recycle Darwin's idea that change in organic forms could be represented as a branching tree while chiding him for believing in "the extreme significance" of the subordination of groups under other groups.[13] Perhaps, Spencer suggested, Darwin was more closely aligned with the discarded theory of the serial ordering of organic forms than was scientifically supportable. Underlying this criticism was Spencer's radical distrust of the natural value of a biological unit. Like Darwin, he too wished to theorize about the evolution of life, but he had become profoundly uneasy about the connotations carried by descriptive categories used by natural scientists.

Spencer did not himself believe that the various categories of living things, such as species, genera, orders and classes, were assemblages to which definite values could be assigned.[14] He had more confidence in arbitrary categories, such as groups, than in time-honoured ones, such as species. In a statement that would be surprising to someone who believed that nineteenth-century evolutionary theory was dependent on changes to one of these assemblages – species – Spencer argued that it was obvious that classifications "have no absolute demarcations in Nature corresponding to them".[15] If the boundary markers between groups of organisms were disturbingly uncertain, then there was worse to come. Other categories, such as the internal divisions within an organism, also had blurred edges. For example, Spencer stressed that inside a plant there was no absolute boundary between the function and structure of a leaf and the stem.[16] He was also intrigued by speculation on the sexual organs of plants, which seemed to demonstrate that "ovules" could graduate into anthers and subsequently produce pollen in their interiors.[17] This suggested to Spencer that the categories used by naturalists were artificial rather than natural. He believed that disputes among botanists and zoologists on questions such as whether a particular organism was generally distinct, or whether it possessed a feature that was of ordinal importance, meant that the categories they used were uncertain.[18] "The endeavour to thrust plants and animals into these definite partitions is of the same nature as the endeavour to thrust them into a linear series."[19] Spencer thought that the subordinate groups referred to in natural classification systems had some correspondence with realities, but were assigned more regularity than they possessed.[20] He believed there was something more important than the partially artificial establishment of groups and subgroups of organisms: the fact that the attributes possessed by the largest assemblies of forms were "the most *vitally-important attributes*".[21] Large and complex organic shapes were meaningful displays of Life. To study organic classification was not an exercise in taxonomic dexterity. It was the pursuit of meaning, without which the universe would be dead. For Spencer the study of organic shapes was akin to the contemplation of God for a Christian: it was a handle on a physical universe that was otherwise cold and empty.

Spencer's theory of classification moved outside the main boundaries of the conventional debate on organic development. In the early-nineteenth century this debate had been between the standard sources of wisdom in the natural sciences, such as the work of Cuvier, which had emphasized categories such as species, and "transcendental" figures, such as Goethe, Oken and Saint-Hilaire, who had searched for a more basic classification than species. For these thinkers a basic classification would be something like an organic type that was patterned on ideal qualities. The debate between the followers of Cuvier and the transcendentalists had lost much of its steam by the time Spencer had begun to think about the natural sciences. For his contemporaries, such as Owen and Huxley, the type (or archetype) had largely shed its metaphysical characteristics, and had become a heuristic device. A type had become an organic category that merely described a generalized subject. In this way it was claimed that there was a vertebrate type. A category of this kind was not thought to exist in itself. Rather, it was an artificial classification of which there might have been no actual example. This use of type in classification procedures undermined the idea that the reconstruction of prehistoric worlds was an important part of current biological sciences; it became unnecessary to search the buried past for earlier versions of an existing type. Any impetus to look to the past would be slowed if it were believed that the object of the search was a hypothetical entity, which might not have existed. Spencer was so taken with the suggestion that "type" referred to an artificial category that he extended this notion to "species". For this reason he insisted that species were primarily classifications of organic groups. He was on a self-assigned mission to prevent biologists from erecting species as metaphysical entities. Despite the fact that his stance placed him in opposition to authoritative scientists, such as Cuvier and Darwin, who treated species as natural categories, Spencer was determined to free the biological sciences from this archaic philosophical remnant. This was a task fraught with difficulties. Spencer was aware that the targets of his criticism would be offended by charges that they were tainted with philosophical idealism. Like himself, all English scientists claimed to be empirically minded followers of Francis Bacon. In addition, Spencer's position as a philosopher who lived by explaining the meaning of science was a delicate one. If he offended too many scientists, his role as their spokesman would be undermined.

When Spencer came to write *The Principles of Biology* he could not attack Darwin directly; he was so deeply indebted to Darwin's *On the Origin of Species* that open hostility would appear as churlish ingratitude. In addition, during the period he was writing *The Principles of Biology*, his scientific advisers were Hooker and Huxley, who kept their misgivings about the theory of natural selection away from the public forum. Spencer was, therefore, circumspect and mild in his criticism of Darwin. However, despite this, the point of Spencer's comments was not missed by his contemporaries. For example, Lewes called on Spencer for support in an article on Darwin where he referred to the biological-category species as a "metaphysical figment".[22] Similarly, Grant Allen, one of Spencer's familiars, suggested that Darwin's theories found acceptance because they appeared in a historical climate that had been prepared by Spencer, Lyell and the younger Saint-Hilaire. The last of these was credited with insisting that the environment had caused modifications to organisms so that the modified organism belonged not only to a different species, but even to a different genus or a higher kind of organism.[23] In the absence of modern knowledge about genetics, Spencer was able to press-gang this organic development theory into the service of an evolutionism that insisted that organic change was not insulated from the laws laid down by the physical sciences and mechanics. In *The Principles of Biology*, historical entities such as species – together with concepts such as organic growth – were not treated as unique. They were not seen as needing special explanations that differed from those employed in sciences that dealt with inorganic phenomena. However, Spencer did not wish to reduce the study of organic phenomena to a few simple physical or chemical principles.[24] Rather, he believed that, with the help of the kind of explanations available in mechanics, he could offer a more adequate explanation of both simple and complex organic phenomena. With the assistance of compression diagrams and the analysis of the radial and bilateral symmetries of organisms, he could detail physiological functions and morphological shapes without inventing special laws that were unique to explanations of organisms.

In *The Principles of Biology* the study of organisms was conducted in a language that would have been familiar to exponents of vitalism in the 1840s and early 1850s, although the goals of Spencer's science were quite different from this. His theory of classification, which had led him to overthrow the notion that there were types, species and other "natural" groups among organisms, meant that when he said that he was searching for "the most *vitally*-important attributes",[25] he was now categorizing phenomena artificially. This was done in such a way that those organisms possessing the largest number of common features would be grouped together. Of course, Spencer noted that common combinations of attributes could not have arisen fortuitously, but this did not signify that there were necessary combinations or that these had been guided by a "vital" principle. Other practical organic arrangements were possible besides those that he presently contemplated. For example, he could conceive of no reason why creatures covered with feathers should always have beaks instead of teeth, where teeth might have served them better.[26] For Spencer, common arrangements were neither fortuitous nor necessary, but were "essential attributes".[27] It is important to note here that in his use of the suggestively idealistic word "essential" Spencer was not adopting a Platonic thesis. By "essential" he did not mean that an organism or an organic assemblage had an inner essence, but that it possessed "the greatest number of attributes in common" with other objects.[28] The goal of a scientist was not to dwell on characteristics peculiar to a species or a genre, but to search for those common to plants and animals, and that might belong to groups that were distantly related, or not related at all. The shared characteristics in such cases would be the essential ones. Matters of shape intrigued Spencer above all else. To study the surfaces of an organism was to consider all meaning in the universe. He saw Life revealed in the tubular or radial shape of a plant, and in the way that this was shared with the shape of a wholly different organism: perhaps an animal. The curvatures and external modulations that were common to widely differing life forms all equally reflected the effect of environmental forces on evolutionary development. Over a period of time separate organisms would come to display similar shape characteristics, both in their overall external surface and in their component parts.

Initially, Spencer saw the task of a biologist as investigating and explaining the structure and functioning of particular organisms. However, it did not stop there. Drawing on the procedures of other sciences, Spencer noted that physicists and astronomers do not pause when they solve a particular problem of planetary orbits, but proceed until they have offered a complete explanation of complex planetary motions in their totality.[29] Similarly, the ultimate goal of a geologist is not to explain particular strata, but the entire structure of the earth's crust:

> If he studies separately the actions of rain, rivers, glaciers, icebergs, tides, waves, volcanoes, earthquakes, etc.; he does so that he may be better able to comprehend their joint actions as factors in geological phenomena; the object of his science being to generalize these phenomena in all their involved connections, as parts of one whole.[30]

Spencer thought that biologists should work in the same way as physicists and geologists. This should not be taken to mean that he was a precursor of the early-twentieth-century movement for the unity of sciences, believing that all sciences should be reduced to one general science based on physics. It was simply that Spencer felt that, in each of what he called the "concrete sciences", the goal should always be to explain all the phenomena considered by that science as parts of one whole. In this way the "concrete science" of biology was necessarily an elaboration of a complete theory of life:

> If different aspects of [biological] phenomena are investigated apart – if one observer busies himself in classing organisms, another in dissecting them, another in ascertaining their chemical compositions, another in studying functions, another in tracing laws of modification; they are all, consciously or unconsciously, helping to work out a solution of vital phenomena in their entirety, both as displayed by individual organisms and by organisms at large.[31]

This holistic plan for the empirical sciences was so fundamental to his thinking that Spencer began to wish that he had the time to rewrite his *First Principles* so that it would reflect the idea that the biological sciences should be structured to show that "ultimate" nature of life and death or, to put it more prosaically, of evolution and dissolution.[32] His thoughts about classification, and his efforts to discover and trace out the various structures and functions of living things, had left him with the impression that a true philosophical method was not based on metaphysics and psychology, but on an empirical demonstration of what could be directly and visually experienced. He no longer saw *vital* forces and their shapes as unique to living things, but regarded them as parallel to the inorganic features of the universe. A science of those vital forces and shapes that had been produced in evolutionary change was a demonstration of the mystery of Life in the universe as a whole. Evolution was now a response to energy and motion, rather than the upward spread of organic forces.[33]

The mystery of Life was encapsulated in its visual form, and this was infinitely more exciting than a verbal formula could be. While it could sound uninspired to say that the form of Life was determined by the distribution of forces as this moved towards equilibrium, it was fascinating to capture the various shapes of life in the hundreds of illustrations that accompanied *The Principles of Biology*. These illustrations bore witness to the fact that everywhere glanced at in nature there were organic shapes that directly revealed the truths of evolution. While Platonists and vitalists had dealt in indirect truths, such as inner drives and hidden principles, Spencer pointed to immediate examples found on the surfaces of organic things that one could see for oneself on a country ramble. For example, the common shape of a bramble leaf or a lichen provided pictorial evidence of evolutionary truth. The sight of any biological shape would have the same effect because Spencer believed that his law of organic form linked *all* organisms to their environment.

In order to understand why Spencer devoted so much effort to thinking about the classification of organisms, and why he believed that this was crucial to a proper understanding of Life, it is important to investigate the actual historical context in which he conceptualized science in the middle of the nineteenth century. Often, so much reference has been made to his personal relations with individual Victorian scientists such as Darwin, Huxley and Hooker that it has been overlooked that Spencer's role in the period was neither as a helpmate nor an impediment to the development of professional science, but as a philosopher of science. To explain Spencer's historical context it is necessary to grasp that his work was first and foremost an investigation into the state of philosophical knowledge, not into the technical details of early cell mechanics and genetics. If Spencer was useful to his contemporaries, or to his successors, it was because he thought systematically about the meaning of knowledge and of life itself, not that he himself made scientific discoveries. This historical context is not found amid a cluster of pre-Darwinian scientists or their religious opponents. Spencer was concerned about how we acquire knowledge, what this was and, in addition, what the meaning of life was. Since these are orthodox philosophic enquiries, his antecedents and contexts are obviously to be found in conventional orthodox philosophies of the period.

Most significantly, it should be observed that Spencer's scientific philosophy often followed the most orthodox of all philosophers, the Cambridge don Whewell.[34] This was to be expected, given that it was Whewell who had directed many young Britons along the nineteenth-century route towards progress in the sciences. In brief, Whewell's theory viewed scientific activity as a form of classification that postulated that there would be increase in scientific knowledge when its system of categorization was improved. When the discipline in question was a life science, Whewell's directions came attached to an unusual thesis. In his magisterial *History of the Inductive Sciences*, Whewell stressed that progress in a scientific subject, such as botany, would occur when its practitioners adopted a natural rather than an artificial system in classifying all the features of an object. This was perceived as a new doctrine because, as Whewell remarked, the discovery of a natural classification system had been an "insoluble" problem in the eighteenth century and before.[35] Whewell based his views on this natural system on the work of Cuvier, Robert Brown, John Lindley and Bernard and Antoine Laurent Jussieu: the standard sources for botany in the early Victorian period.[36] Spencer found such a juxtaposition of authorities too strong to ignore; he was forced to frame his scientific discourse in Whewell's language. Although he opposed both Whewell and the standard biological authorities Spencer was obliged to conduct his battle in terms they had specified. He agreed that scientific progress occurred when a system for classifying knowledge was constructed, but his own system was the exact opposite of Whewell's. Spencer emphasized classification as an *artificial* way of organizing knowledge rather than seeing it as a set of categories that referred to *natural* underlying features of the things being described.

For Whewell, botany was the premier science; he believed it was sufficiently typical of the order of progression in knowledge to represent how classification functioned in the other sciences. There was a convenience in seeing knowledge as progressive because it meant that one could understand sciences that had developed later more easily than the earlier ones. For example, botany had preceded zoology as a form of scientific progress, and therefore less effort was required to understand the latter discipline. Botanists, such as the two Jussieus, had constructed a system of natural classification that grouped plants according to their "natural" affinities.[37] This "natural" method directed botanists to the study of the physiology or structure of plants as the primary means by which they could attain knowledge of them.[38] Although it may seem surprising to mention physiology in connection with plants rather than animals, this was commonplace during the mid-nineteenth century. The study of the physiology of plants was the most theory-laden form of investigation into natural science.[39] However, even if the mid-century bias that favoured botany over zoology had been reversed, physiology would have retained its position at centre stage because the contemporary study of zoology relied heavily on Cuvier's *Règne Animal*, a physiological treatise dealing with natural classification.[40] This work had been so paradigmatic that Whewell, despite believing that the study of plants had priority over the study of animals, thought that the modern epoch began with it.[41] It was also from Cuvier that Whewell took his doctrine of vitalism,[42] a doctrine that heavily influenced Spencer's scientific writings during the early 1850s. This emphasis on physiology privileged structure and function, and these became the primary focuses of many works in Victorian natural science, including Spencer's *The Principles of Biology*.

Spencer was well versed in Whewell's sources, and was in total agreement with Whewell that scientific progression was dependent on being able to classify natural data in such a way as to reflect the physiological structure and function of an organism. However, Spencer resisted Whewell's thesis that classification was based on knowledge of "real" features. For Spencer, classification was both progressive *and* artificial. His views on classification resembled those of the French botanist Michel Adanson, whose views were in opposition to the natural system later developed by Cuvier and Whewell.[43] Adanson had spent years in Senegal gathering botanical material and formulating his system. His difficulties in describing exotic plants led him to adopt a procedure in which each organ or part of a plant was classified separately. He then arranged all known species according to the feature of that organ alone. Doing the same for each organ in turn, Adanson constructed a collection of systems of classification, each of which was artificial. When he matched the different artificial systems, those plants that came closest together in all systems were proximate species, while those that were separated in a few of the systems were slightly more distant species. A large amount of variation between systems indicated that plants should be classed further away from each other.[44] By using this method one obtained the means of estimating the degree of natural affinity of all species *without* considering their physiological basis. Cuvier's objection to this form of classification was that it presupposed a vast knowledge of descriptive natural history. All species and all their separate organs had to be known, and if only one of these were neglected then the classification of a particular organism might be faulty.[45] Spencer ignored this objection; his love of empiricism rendered him impervious to warnings of the difficulties in assembling a large collection of facts.[46]

Adanson's "universal method" had an influence on Spencer's basic metaphor for the classification of scientific knowledge. It created a librarian who artificially classified books on single features so that they could be easily arranged in a serial order. To Spencer this image of a serial progression of natural objects demonstrated that a "natural" system of classification (one based on *real* features of an organism) was just as artificial as the one that it purported to replace. Spencer focused on this flaw out of genuine interest in the discovery of what made knowledge progressive. From this perspective, Adanson's ideas undermined the "natural" ones of Cuvier and Whewell. Classifying by single characteristics could produce progressive results, but if one was searching for a large amount of natural affinity then the progressive quality disappeared.[47] The point carried by the metaphor reinforces the message of Spencer's later science, which was that the natural systems of classification are objectionable because their advocates arranged the facts of natural history so as to reflect inner physiological structures and vitalist drives instead of the external feature of organisms. Only these features were significant to Spencer: they were shapes that could be explained by the ordinary laws of the physical universe.

Whewell, whose philosophical ideas had at times guided Spencer in his thinking on psychology and intuitionism, became his main antagonist in *The Principles of Biology*. Where Whewell had argued – together with Cuvier, the Jussieus and Lindley – that scientific progress took place if one adopted a natural system for the classification of data, Spencer claimed the opposite. He insisted that scientific progress was dependent on the artificial arrangement of knowledge. Whewell's views on how scientific knowledge progressed had variable effects on Spencer. First, during the 1850s, Whewell's work pushed Spencer in the direction of vitalism. Secondly, during the 1860s, Spencer reacted against Whewell's formulation and discarded both vitalism and the distinction on which it rested: the difference between organic and inorganic entities. That distinction had allowed Whewell to analyse scientific classification in such a way as to differentiate between separate types of scientific knowledge. Whewell had divided types so that some would fall under the organic classification and others would be inorganic. For Whewell, this division was watertight even when it seemed, superficially, that physical laws were able to explain organic phenomena and that there was, therefore, no significant difference between them and inorganic phenomena:

> And it will be apparent, on reflection, and though *symmetry* is a notion which applies to inorganic as well as organic things, and is, in fact, a conception of certain relations of space and position, such *development* and *metamorphosis* as are here spoken of, are ideas entirely different from any of those to which the physical sciences have led us in our previous survey; and are, in short, genuine *organical* and *physiological* ideas; real elements of the philosophy of *life*.[48]

Whewell was faced with the problem that some of the botanists on whom he relied had been engaged in reducing plant data to a small number of simple organic structures, which began to resemble those of the physical world. Even more worrying were the ideas of A. P. de Candolle of Geneva, who believed that all plants could be reduced to perfect symmetry.[49] This theory was extended by the British botanists Robert Brown, John Stevens Henslow and Lindley.[50] The last of these, Lindley, thought that the true characters of all natural assemblages were very simple.[51] The result of this botanical theorizing was that, to Whewell, the life sciences began to resemble the physical sciences. This seemed a confusion, and to put an end to it Whewell insisted that natural scientists adopt the ideas of Goethe. These would be a bulwark against the kind of materialism that would result if one attempted a complete explanation of life on the basis of the laws of physical science.

Whewell believed that Goethe was not only the greatest poet of all time, but that he was the discoverer of laws that connected the forms of plants into one simple system of life.[52] He did not consider Goethe's poetical prowess to detract from his scientific ability; on the contrary, art had contributed to scientific truth.[53] The beauty that Goethe found in the segments of forms and shapes of plants had actually enhanced his scientific insight. Whewell's enthusiasm for Goethe was such that he abandoned his usual taste for ratiocination and asked his readers to repeat the poet's aesthetic experiment. Whewell instructed them to imagine that a flower, such as a wild rose or an apple blossom, consisted of a series of separate parts disposed in whorls, which were superimposed over each other on an axis. The lowest whorl was the calyx with its five sepals, above this was the corolla with its five petals. Even the stamens could be considered in groups of five. The imagination could reproduce this pleasing development in growth as a close repetition that echoed a variation in a musical score. Outside the imagination – in the morphology of plants – the whorls in flowers were identical, and were similar to whorls of ordinary leaves that had been brought together by the shortening of their common axis. To produce flowers, leaves were modified in form by successive elaboration. This was caused by the distribution of nutriment and by the exposure of the plant to the elements.[54]

Spencer was aggravated by Whewell's admiration of Goethe. His reaction was all the stronger because part of his own stance was taken from Whewell.

He had accepted Whewell's theory that the basis of scientific progress was the classification of data; he followed Whewell in saying that sciences moved from complex to simple;[55] and he frequently reiterated Whewell's mildly intuitionist idea that there was a basic psychological test for determining whether or not an idea was inconceivable, and therefore false. However, the debt to the Cambridge philosopher did not prevent Spencer from using some of Whewell's formulations as straw men whose function was to permit their critic to deny their usefulness. In particular, both Spencer's idea of beauty and his science of biology were forged in direct opposition to the image of Goethe as an artistically inspired scientist. In *The Principles of Biology* Spencer's own conceptions of beauty and science were forged in order to rescue the shape of life from subjection to laws that were uniquely organic.

According to Spencer, Whewell and Goethe were mistaken to believe that the "organical and physiological ideas" that ennobled life were taken from the physical universe. Spencer adopted Whewell's insight that beauty and science gave meaning to life through a display of symmetry, but insisted that this display was not restricted to the shapes of organic entities. Those shapes were not unique, they were only part of generalized meaning that was to be seen in the cosmos. To Spencer in *The Principles of Biology*, a display of astronomical symmetry in the physical universe was like the efflorescence or bloom on a piece of fruit.[56] Both were beautiful, and both could be seen as that surface of things that directly demonstrated the meaning of life in a way that was beyond functional analysis. They were equivalent to ontological statements that posited existence; instead of Spencer intoning "I am" in order to find meaning, he pointed to symmetry and beauty. All shapes were functional, even those that had seemed unique to plants and animals. These were governed by laws indistinguishable in kind from those that determined the shape of inorganic entities. In brief, the symmetries of life were no different than the symmetries of crystals and other non-living forms.

In his later science Spencer grouped organic with non-organic forms, thus disposing of what he saw as a specious and non-scientific Platonism that lingered at the heart of the English intellectual establishment. Whewell, the *éminence grise* of the British Association for the Advancement of Science and the mentor of Richard Owen, had proposed that science should proceed by abandoning definition altogether when classifying botanical and zoological information. Instead, Whewell argued, one should organize the material in natural groups. He claimed that, in the organization of knowledge, "a definition can no longer be of any use as a regulative principle".[57] He applied what he knew to be true of biological science to science in general. Whewell believed that the practice of unfolding our conceptions by means of definition had never been serviceable to science.[58] Instead, Whewell focused on *type* as his directing principle. A *type* was an example of any class – for example, a species – that possessed all the characteristics of the class.[59] If one classified by type then it was possible to use the resulting classification as the basis of general propositions founded on *positive* knowledge.[60] Through this use of classification, Whewell was able to claim that scientific knowledge possessed certainty in that it was not based on artificial or definitional categories.

Spencer, like Whewell, had left behind the traditional distinction between inductive and deductive knowledge, a distinction that permitted only the latter to possess certainty. Both were attempting to erect a framework within which science would produce *positive* knowledge. However, beyond this common devotion to *positive* knowledge, Spencer's scientific thought was forged in opposition to Whewell. This was why Spencer insisted on using definitional procedures even when he could not conceive of a plausible way of doing so. To use a definition was to adopt a badge that would distinguish its wearer from philosophers who reified types, species and other idealistic conceptions. This was a crucial matter to Spencer; he felt compelled to combat Whewell's formulation of scientific knowledge established on natural classifications. The fact that Whewell also claimed that this knowledge was as real or positive as any knowledge established on sense perception did not mollify Spencer. Despite the fact that Spencer was sometimes confused with Comte, he was not a positivist. For him positivism was as unattractive as idealism; both were extremes to be avoided. Balancing between them, Spencer forged a philosophy in which knowledge bridged the laws of psychology and a series of hypothetical or definitional propositions about Life and its forms.

Spencer's philosophy of science was in part an adaptation of, and partly a reaction against, establishment science of his period. Some origins of his philosophy can be found in the work of Oxbridge philosophers and scientists, rather than in that of empirical thinkers such as J. S. Mill and the English Comteans. However, many of the former were clerics who adhered to a belief in natural theology that Spencer did not share, although he did adopt their practice of seeing nature as a combination of truth and beauty. These were contemplative qualities and could be found in the curves or shapes of natural objects. Spencer saw external forms in nature as an aesthetic that gave meaning to life: a meaning that did not refer to the existence of God, but was immediate and complete in itself. His philosophy was designed to replace both natural theology and Platonism.

Spencer treated Goethe as a prophet for a failed religion. The German savant, he thought, had misleadingly relied on hidden spiritual drives as the basis of scientific explanations when these were nonexistent: errors that fed into a malign philosophy. In *An Autobiography* Spencer maligned Goethe for subjecting human emotions to the harsh, although fictional, notion of the will, while in *The Principles of Biology* he complained that Goethe personified nature, depicting it as wishing organic components, such as flowers, to take pre-ordained shapes. He portrayed Goethe as an irrational figure who had mistakenly melded medieval physiology with Platonic philosophy. One of his images in particular embodied all that Spencer saw as wrong with the German sage. Goethe believed that leaves were shaped in such a way as to dispose of waste products: "that as long as there are crude juices to be carried off, the plant must be provided with organs competent to effect the task".[61] Transformation, or organic change, was seen as a result of the action by nature that had provided the plant with purified juices. Behind this process Goethe saw nature achieving its end when a plant flowered. Spencer, in marked contrast, refused to allow a fictitious agency to be credited with the cause of a natural phenomenon such as the shape of a petal or a calyx. If an organic shape conformed to a basic pattern, reasons for it were not going to be found in teleological argument.

This rigorous consistency with which Spencer combated any vestiges of idealism or teleology means that his own use of words such as "species", "type" and "typical" should not be understood as containing these qualities. Of course, he did adopt such words when, for example, he dealt with an integrated group of cells by describing them as having a "typical form". He remarked that the group of plant cells of which a leaf was composed had a typical form due to the presence of veins, and to the special arrangement of cells around the mid-rib.[62] If the multiplication of cells exceeded a certain limit – at a time when a plant bud was taking on its main outlines – the cells began to arrange themselves around secondary centres, or lines of growth, repeating "the typical form".[63] This could be misunderstood; the use of typical forms might be taken to indicate that Spencer too harboured a lingering fondness for Platonism, for Richard Owen's "archetypes" and for teleological explanation. However, this was emphatically not the case; the mode of operation behind the seventeen chapters in *The Principles of Biology* on plant and animal morphology was to use "blind" natural selection to explain how organic units achieve their shapes without following in-built designs, or without moving towards predetermined goals. Spencer chose the case of annelids (segmented worms) to illustrate this; each segment represented a physiological whole. It was his opinion that even if one believed in a Creator who had prescribed an archetype of organic composition and then built up a variety of physiological wholes based on those units, this would still not produce a plausible explanation of organic variety. The significant features of combined physiological units were not easily explicable by reference to their component parts. In addition, since each class of organism contained so many exceptions to the supposed archetypal unit, the hypothesis possessed no explanatory power. Most teleological explanations possessed features that made them similarly implausible: "That certain organisms of nutrition and respiration and locomotion are repeated in each segment of dorsibranchiate annelid, may be regarded as functionally advantageous for a creature following its mode of life. But why should there be a hundred or even two hundred pairs of ovaries?"[64] The abundance of these internal arrangements seemed to conflict with Milne-Edwards's principle that the physiological division of labour was efficient. It was also less advantageous than it would have been if it had been designed.[65] Such apparent difficulties would, however, disappear if it were supposed that the annelids had arisen through the integration of simpler forms.[66] In short, the worm did not bear the fingerprint of God; its shape bore witness to a historical accumulation of responses to the environment.

For Spencer the problems that arose when explaining a worm were no different than those that had arisen when analysing any organic form or group of forms. As has already been observed, distinctions between organic groups were insignificant to him, regardless of whether they were apparently fundamental differences, such as those between species, genera and orders. He found even the great division between plants and animals insubstantial because it made divisions between organisms that were equally responses to the same evolutionary force. Plants, which had retained their current shapes for longer than animals, could be more easily explained in terms of the impact of physical forces on the structure and functioning of cells as well as on larger botanical components. In part, Spencer's deprecation of the distinction between animals and plants was caused by his rigorous application of the theory of natural selection to the explanation of shapes. His basic explanation was the same for both; the cause of the spiral shape of a plant and the elongated shape of an animal was to be found in the physical forces that surrounded it. As an example of animal change, Spencer noted that the evolution of a vertebrate skeleton implied the development of a strong internal fulcrum. The strength of such a fulcrum also meant that it became denser, and if this process occurred while an animal retained lateral flexibility, then it implied that such an animal structure would be divided into segments.[67] Spencer was proud of this style of evolutionary explanation because it was not reliant on hidden factors; it only assigned causes of already known kinds, which would produce effects such as they were known to produce. "It does not allege a Platonic *idea*, or fictitious entity, which explains the vertebrate skeleton by absorbing into itself all the inexplicability. On the contrary, it assumes nothing beyond agencies by which structures in general are moulded ...."[68]

While Spencer wanted to combat transcendental physiologists – Oken, Goethe and their followers – he took his arguments about the meaning of organic shapes far beyond what was needed to dispose of these antagonists. His own fascination with the forms of plants and animals, and with the shapes of their component parts, led him to produce an exposition of the meaning of living shapes that was as heavily laden ontologically as Platonism had been. He believed that what one saw in the external shape of an organism was not a reflection of some hidden truth, but the sight of truth itself. Meaning was in every natural object. Spencer was unafraid of appearing prosaic; this was not a danger because he glorified in the ordinary. Few other philosophers would have so rejoiced in the commonplace as to see verity in a centipede or a cow in the way Spencer had done.[69] Truth was the same as life: it was revealed by the search for a clear and exact description of external organic form. In this search, Spencer found the analysis of plant forms more appealing than that of animals. This may have been because he was more original in his botanical theory than in his zoological observations. Plant forms remained more significant to him because they were stable, and had maintained themselves unchanged over more geological periods. Plants also more clearly displayed basic symmetries in their shapes than animals did.

When Spencer rejected Platonism and teleology as hostile to proper scientific explanation, he held on to the question that Platonists and teleologists had asked. He still wished to know if, in general, the shapes of plants admitted of being expressed in "universal" terms.[70] By "universal" Spencer meant that the specified features would stay constant over all genera, orders and classes of organisms. He also wanted to know if organic shapes could be expressed as laws.[71] Since Spencer had rejected the notion that archetypes or teleological goals could provide a meaningful description of the evolutionary process, and since he still thought that shapes were universal and meaningful, these must, he concluded, refer to the external regularities possessed by plants and animals. A faint hint of the certainties of progress accompanied Spencer's optimistic words. This comforting quality did not depend on the origin of organic entities, or on their conforming to preordained shapes. Instead, progress took its meaning from a cosmological prediction that was awe-inspiring to its nineteenth-century audience, despite being formulated in a stiff scientific phraseology. Spencer's readers did not find Spencer's style at all awkward. It was seen as a necessarily earnest reference to the meaningful universe. They found encouragement in these words: "Direct equilibration in organisms, with all its accompanying structural alterations, is as certain as is that universal progress towards equilibrium of which it forms part".[72]

The significant part of existence or being was no longer in the hidden spirit, but in external living shapes. These globular, spiral or elongated figures could be expressed in terms of various symmetries, such as spherical symmetry or bilateral symmetry.[73] Familiar flowering plants such as oak trees and daisies displayed these comforting regularities.[74] Spencer's attempt to reach his reader did not stop at a few homely examples; he showered them with hundreds. His repetitive demonstrations exhausted all the regular organic shapes that he could find. There was no reason for him to be parsimonious by selecting a few organisms and their subordinate parts. His vision of the universe was uncontrolled by a natural hierarchy or by a linear succession of organisms. So every single example of a regular shape was equally important. Spencer only desisted from piling up yet more data when he was faced with organisms such as the "gregarinida" and the "infusoria", whose structures were so small they could not be specified by Victorian scientists.[75] When dealing with botany his main object was to present "under their simplest aspects, those general laws of morphological differentiation which are fulfilled by the component parts of each plant".[76] He offered similar morphological illustrations for animals. After pondering such examples, his readers could follow Spencer's logic and test the truth of his statements about the shape of life. They too could see how, for example, a radial shape could develop into a bilateral one. This would allow them to observe evolution with its display of symmetry at first hand.[77]

Spencers' belief that organic shapes were universal led him to enquire if there were laws for shapes of organisms. He saw this as the most important aspect of biological science, even insisting that he had priority in the discovery of this theory and dating it to the beginning of his scientific thinking. He refers to this theory as a "discovery" made during a country ramble with his friend Lewes in 1851:

> I happened to pick up the leaf of a buttercup, and drawing it by its footstalk through my fingers so as to thrust together its deeply-cleft divisions, observed that its palmate and almost radial form was changed into a bilateral one; and that were the divisions to grow together in this new position, an ordinary bilateral leaf would result. Joining this observation with the familiar fact that leaves, in common with the larger members of plants, habitually turn themselves to the light, it occurred to me that a natural change in the circumstances of the leaf might readily cause such a modification of form as that which I had produced artificially.[78]

Spencer recalled this innovation as the key to a general theory that explained the relation between the forms of leaves and the general character of the plants to which they belonged. This was an evolutionary theory of an environmental kind, which, in 1851, had competed with Spencer's vitalist theories of organic change. When he later discarded vitalism, his environmental evolutionary theory was already present, and thus able to occupy centre stage. Spencer could disingenuously argue that he had always focused on the importance of the external shapes of organisms, ignoring the fact that he had concurrently proposed vitalist theories that were teleological.

When Spencer's environmental evolutionary theory was separated from its accompanying vitalism, it had a strong bias against the search for the genetic underpinnings of species change. He considered that such a search would place undue emphasis on evolutionary ephemera and the construction of falsely "natural" categories in nature. What, he queried, would be the purpose of such a study when the actual shapes of organisms was similar even though they belonged to different species or genera?[79] Of course, Spencer's decision to base his evolutionary theory on the analysis of organic shapes embedded it firmly in a mid-Victorian milieu, which perceived the role of natural science as explaining the harmonies of existence. Spencer's discovery of repeated patterns in nature was felt to be as soothing as harmonics in music or the demonstration of a mathematical proof. Later, as the Victorian era drew to a close, *The Principles of Biology* became a monument to a lost conception of a meaningful and benign universe. The study of the shapes of things became a subject of such minor interest that a brilliant twentieth-century monograph on this subject by D'arcy Wentworth Thompson has often been regarded as a curiosity without an intellectual home.[80]

Progress served a dual function in Spencer's *The Principles of Biology*. First, it was the actual picture of biological change: a picture that had been cleansed of any Platonic or teleological taint. Secondly, it was the concept that linked the world of animals and plants to that of man. In the animal and plant worlds the final outcome of progress was an equilibrium unaccompanied by moral connotations; it might refer to stasis or even to death. As they evolved, individual organisms were inevitably killed.[81] Spencer's early hope that animals would find an evolutionary relief from pain and death had faded away. In its place was a notion of progress that had no more of good and evil about it than had fluctuations in the ecological balance between carnivores and their prey or in an animal's fertility levels.[82] Spencer's second kind of progress – which referred to the world of humankind – was different. In this world, the process of equilibration was altogether more hopeful.[83] Progression in the human world had a moral quality lacking in the ordinary organic sphere. According to Spencer, the difference between human beings and the rest of organic nature was caused by the fact that human beings avoided the antagonism between genesis and individuation that he saw as so prominent a part of the organic world. For ordinary organisms there was an inverse correlation between the life of an individual and the perpetuation of the species: reproduction often caused the destruction or diminution of individual life forms. This, however, was not necessarily so; in a "small departure" from this inverse correlation Spencer declared: "we shall find an admirable self-acting tendency to further the supremacy of the most-developed types".[84] This *small* departure was, in fact, a giant futuristic claim that individual human beings would escape the twin dangers of stasis or death that confronted other organic beings. There was a subtlety to Spencer's views that can only be appreciated when seen in contrast with those of his scientific contemporaries who had not fully absorbed the lessons of evolution when considering the human future. If, on the one hand, current human beings were placed outside the evolutionary process – a hypothesis offered by Wallace and Huxley – then their future would be one of stagnation.[85] On the other hand, if modern human beings continued to evolve like other organisms, then it would always be the case that individual numbers of the human species would be sacrificed in the process of biological change. Spencer's solution was to say that with progress drawing them forwards, future human beings would remain part of the natural world (and thus experience evolutionary change); yet, at the same time, they would be above it and thus able to avoid its perils. His vision had humanity ultimately evolving to the point where individuals avoided the cruelty and destruction that the demands of hunger and reproduction had imposed on other organisms.

IV

Politics and ethical sociology

FIFTEEN

Spencer's politics and the foundations of liberalism

Spencer's writings on politics sit uncomfortably with his philosophical system. With the exception of *Social Statics* (1851) and *Political Institutions* (1882) his political works were a series of rhetorical or tactical statements written in response to the momentary concerns of his contemporaries. Some of his political values, such as his anti-aristocratic bias, remain constant, while others, such as the need to promote women's suffrage, are reflections of positions briefly held by liberals in the early Victorian era. Compared to the sober and careful analyses that accompanied his scholarly books on psychology, metaphysics, biology and sociology, some of Spencer's political statements have an ephemeral air. Like newsprint, they provide a record of momentary concerns, that no one would dignify with the label "universal truths". However, despite the transient nature of his more polemical utterances, his more serious endeavours had an overall direction. As early as the publication of "The Proper Sphere of Government" in 1842 he possessed the principles for which his technical writings later provided an empirical basis. That these technical writings sometimes failed to support his ideals reveals one of Spencer's virtues as an intellectual. He never completely controlled his data so as to fit his pre-ordained writing plans. There was a stubborn scientific honesty to his writings that often caused facts to win over coherency.

The brilliance of his original conception was often compromised or deflated when it was fleshed out: his motivations were a few improbably tall and sunlit peaks and pinnacles, while the voluminous technical works were mist-covered valleys and low ranges whose topography cannot be easily delineated. The works have a certain glory in their vastness, and in their scrupulous inclusion of all the details that an energetic Victorian could glean from an exhaustive survey of contemporary biology, psychology and morals. Nevertheless, these "truths" of nature seldom allowed themselves to be completely subordinated to theory. The result is that his technical philosophical work failed to display the clarity of his popular political writings. For example, Spencer began his intellectual career by consistently and passionately arguing that the human species was progressive, and that Jeremy Bentham's principle of utility was mistaken. There is, however, no logical connection between these initial beliefs and the philosophy that subsequently developed from them. Instead, his support of progress and his opposition to utility were so fickle that they are hard to use in the support of a systematic ethics or public policy. An attempt to demonstrate how these early principles underpinned his system would serve only to emphasize the enormity of the contradictions that Spencer's philosophy contained. For instance, Spencer's early writings claim that all living things were progressing towards an unspecified future, while his later works suggest that such improvements apply only to the human species, severed from the remainder of the animal kingdom. To ignore this disjunction, and insist that Spencer consistently upheld the doctrine of progress would make it very vague. Likewise, it would not be helpful to emphasize Spencer's anti-utilitarianism as a coherent doctrine. While he was always hostile to the Benthamite hedonistic calculus, few other statements about his views on utility would be consistently true. In "The Proper Sphere of Government" he wrote as a Christian utilitarian opposed to individualism and thus was hostile to those who construed happiness as if the collective did not matter. Later, when he became imbued with a tincture of individualism himself, he argued against Christianity and was less concerned with the happiness of the collective. At all points in his career his notion of happiness was broadly conceived; it reached beyond the short-term joy felt by an individual, a civilization or even a generation. For Spencer, it was a long-term satisfaction, not an immediate one, which offered moral direction to society. There were, of course, limits to this, and, he later recommended that you should prudently refrain from sacrificing yourself when promoting social improvements.

When he eventually freed himself from Christian doctrine, Spencer's thought remained marked by its origins. To a limited extent he shared William Paley's scepticism over the value of dogma in moral and political argument. Instead of relying on the holy word, Paley had rested his arguments on nature, where God had revealed the essential happiness of life. This kind of Christian naturalism provided Spencer with a conception of natural morals to the effect that life is, in general, happy. The result was to place his liberalism in opposition to the fundamental beliefs of the other great nineteenth-century liberal J. S. Mill, for whom the individual was divorced from nature. It was Mill's opinion that "conformity to nature has no connection whatever with right and wrong".[1] If it had, he continued, why would the acquisition of virtue always have been accounted to be a matter of labour and difficulty?[2] However, this was not a conventional Victorian stance; Mill's distaste for natural morals would not have resonated favourably with most of his contemporaries. Spencer was speaking with the majority when he countered that nature was the source of both harmony and happiness. Each of us, he believed, sought conformity with nature because it was a cornucopia of ethical and political norms. It was also an ideal, providing guidelines for the regulation of state activities. Citizens and legislators had only to ask themselves the question of whether or not their activities impeded the operations of nature. If the answer was affirmative, then the offending action or regulation would be rejected even if it had briefly produced pleasure. It was a paramount principle for Spencer that long-term contentment could only be achieved if one acted in accordance with nature.

The principle has been obscured because most of his political writings were not republished in modern times, and what did appear was untypical. In the mid-twentieth century only one of Spencer's political works was in circulation: *The Man "versus" The State* (1884). The best-known edition was published by Penguin. Its front cover bore a picture of an antique lighthouse, which resembled a giant factory. The grossness and inhuman scale of the architecture suggested that individual endeavour was dwarfed by harsh, and largely militarized, initiatives. The image could suggest that Spencer's book is a libertarian tract that prevented the individual from being totally absorbed by society. However, the editor of this edition, Donald Macrae, was too scholarly to say something so obtuse about Spencer's political theory; the picture was merely to reinforce the argument that Spencer was a lone oppositional voice protesting against the high tide of bureaucratic momentum that was refloating the coercive hierarchies of the medieval state. Macrae's point was that Spencer's individualism and his talk of "the coming slavery" gave him the status of a conservative liberal – or perhaps a scientifically minded Burkean – who regretted the rationalist destruction of the organic contrivances that had kept harmony in modern societies.

In a political sense, Spencer was not an isolated figure, and did not become a conservatively minded libertarian in his last years. *The Man "versus" the State* should be read as a popular fragment of the social and political liberalism that occupied Spencer's mind during his whole adult life. While he consistently opposed socialism and communism, he remained faithful to his early progressive vision for over half a century. He should not be understood solely on the basis of a few essays published in 1884; he was the philosopher and social scientist whose technical volumes – such as *Social Statics*, *First Principles*, *The Principles of Psychology* and *The Study of Sociology* – validated Victorian social change and aspirations. He militated against war, agitated against the coercive treatment of children and animals, grieved about the displacement of aboriginal peoples and provided the politically confident bourgeoisie a classless vision of the future. It was Spencer's blend of sentimentality and optimism that created a liberal ideology in the Victorian period, not his later reaction to socialism. From the 1840s Spencer's political thought was not only an extension of his system of philosophy, but also provided its inspiration. His early writings such as "The Proper Sphere of Government" (1842) and *Social Statics* (1851) were not ephemeral, but central to Spencer's efforts to reshape the values of his society. Paradoxically these texts have been overlooked by modern philosophers writing about the foundations of liberal ideology.

These scholars assume that liberalism is an ideology that stresses a strong notion of the individual. Sometimes, indeed, liberalism is even conflated with individualism.[3] When this is given a historical gloss the practice is to show liberalism passing through a collectivist phase at the end of the nineteenth century, but then returning to its roots in the mid- and late-twentieth century, when philosophers began to revive Victorian philosophical tenets under titles such as "classical liberalism" and "liberal utilitarianism". This gives too much analytical coherence to liberal foundations, and has also restricted the kinds of statements philosophers permit themselves when defending, or criticizing, liberalism. It is often proposed that there is an essential connection between liberalism as a reigning and dominant ideology and liberalism as a coherent moral doctrine. Modern philosophers imagine that these operate in tandem to defend the morally autonomous individual whose choices structure society. They also believe that both give legitimacy to the state when it is restrained within proper boundaries (with the proviso that the state's actions foster individuality).

Margaret Moore provides an example of one of these currents of philosophical literature. She argues that thinkers such as John Rawls were mistaken in encumbering liberalism with communitarian additions because the former has an original understanding of itself that does not appeal to shared values in a community. She equates such original understanding with a fundamental liberalism that rests on the autonomy of the person.[4] Moore believes that if the basic doctrine is to survive, liberalism must prevent actions that cause harm or infringe the autonomy of others whatever the justification for such actions.[5] In a similar vein, Richard Arneson mentions Victorian thinkers as defenders of a liberal political order that delivered perfectionist values: values that could not be obtained except via autonomous choice on the part of each individual.[6] Although there are others who modify this foundation doctrine and warn that Victorian writers such as Spencer did not possess a systematic theory of individualism that would always require us to ignore the welfare of society,[7] the consensus seems to be that liberalism contains some essential individualist quality inherited from the past. Perhaps, it is speculated, citizens of modern liberal democratic states necessarily follow a path marked out by Victorian pioneers: a way that is still clearly visible despite the false trails of communitarianism, republicanism and extreme libertarianism.

It is not my task here to warn that this well-worn path leads nowhere interesting, although I feel bound to remark that it seems insulting to the memory of reforming and modernizing Victorians to recycle them as "classical" and traditional icons. I do, however, intend to cast doubt on the apparently well established belief that liberalism was fundamentally concerned with limiting social or political control over the individual. This is a shibboleth that an examination of Spencer's beliefs will overthrow. Mine is a hard task because, at first sight, the weight of scholarship to the effect that Spencer believed in the supremacy of the individual seems unquestionable. This is why James Meadowcroft referred to Spencer's ideas as the archetypical form of individualism that was out of step with the 1890s and, therefore, signalled that the doctrine had all but disappeared.[8] One might deduce from this that the late Victorian distaste for individualism was a reaction to Spencer's enthusiasm for the ideals of unrestrained competition. He would be described as a proponent of an atomistic individualism that was an archaic remnant of the mid-century.[9] To put it in other terms, Spencer has been scripted as a "classical" liberal who continued to elevate the individual in a time when most other reforming ideologues favoured socialism.

Such suppositions are erroneous; Spencer's occasional expression of sympathy for individualism was a product of the last two decades of his life, when he had already completed most of his philosophical system. Rather than being an extension of "A System of Synthetic Philosophy", or of his early political essays, the odd aside in favour of individualism was as unrepresentative of his own work as it was of late Victorian discourse. Both Spencer's philosophy and his political journalism during his first four decades had emphasized the primacy of justice over the individual. As early as the 1840s Spencer had adopted an extreme stance on justice and equity, giving these a prominent place in his theory that required the state to eliminate any inequalities in the administration of law, and to provide this free of charge to those who were incapable of defending their interests. His proposal was a cross between legal aid and roving policemen equipped with executive legal powers.[10] The value he placed on justice was extreme and, if adopted, would have led to very redistributive policies. For example, he advocated a progressive death tax on grounds of equity.[11] With the same enthusiasm he recommended that direct taxation be extended to the lower classes as fast as these were granted the franchise.[12] Equity was an ideological wild card; it could fit on progressive or conservative platforms.

Spencer knew as soon as he enunciated his doctrine of justice that it would require an enhanced state, not a minimalist one. This became a kind of bedrock; he was faithful to the same idea thirty years later when he was provoked by his friend Huxley, who had accused him of being an anarchist.[13] The charge had been particularly hurtful because it was accompanied by the insinuation that Spencer possessed *laissez-faire* views that were so riddled with contradictions that they could not draw support from biological evolution. Huxley had suggested that Spencer was unable to buttress his analogical bridge between the body physiological and the body politic. Obviously nettled, Spencer fell back on his old defence that it had never been his wish to undermine the restraining power that a state should exercise over individuals. Rather, he wanted this carried out more effectively, and much further, than it was at present.[14] He also rejected the idea that his statements on representative government were designed to limit the power of the legislature by ensuring that its composition reflected the interests of the electorate. His defence was that when he had referred to a national representative assembly as averaging interests of society and of the individual, he had not meant to specify these as two sets of interests that were opposed to each other.[15] He was merely suggesting that a representative assembly should reconcile conflicting class interests, a goal that Huxley approved of. This was a convenient argument; it would have been difficult for Spencer to make a primary political distinction between a person and a society because, in the peaceful societies, which he favoured as models for the future, individual interests referred to interior psychological actions, not to those exterior ones that were the domain of society.[16] This unorthodox notion was based on Spencer's psychological investigations and it empowered him to brush aside the conventional view, common to Kant and J. S. Mill, that social interests should be distinguished from personal ones.

For Spencer it was not that the individual and society operated in different spheres, as they had for J. S. Mill. That distinction would have allowed for a principled discussion of when interference with the former was justified. Spencer's conceptualization of the individual and society places them on separate planes, making it illegitimate to permit some restrictions on freedom while forbidding others. This feature of Spencer's thought should be read as a disinclination to engage in traditional philosophical discourse. It also suggests that the foundations of liberalism are not as entirely Kantian as many modern philosophers have supposed. Instead of anchoring liberal theory into a Kantian ocean floor along with Wilhelm von Humboldt and Mill, one might instead imagine that liberalism was a doctrine new to the world in the mid-nineteenth century. In this world the science of psychology had proclaimed laws of human behaviour that functioned independently of rights theories and the languages of consent and duty. Rather than recycling classical accounts of politics from Plato and Hobbes, Spencer avoided them. In their writings the body politic had been analogous to the human body. Instead, Spencer used contemporary mental science to argue that *representative* – as distinct from the original or *presentative* – consciousness was similar to the representative apparatus of a political body.[17] The implication was that the masses took no part in executive decisions made by that body; these were carried out by the executive division of government.[18] This does not suggest that Spencer paid special consideration to the individual as an important consideration when contemplating the functioning of political representation or voting. A combination of individual wills did not legitimize executive decisions. Spencer did not follow Rousseau and James Mill in recommending that the general will should be equated with a number of separate wills acting as one. Spencer's liberal doctrine was not dependent on classical discourses; nor was it a physiological metaphor with the human body. In any case, the latter alternative was no longer available because it had been repossessed by professional scientists such as Huxley, who were contemptuous of antiquated writers such as Rousseau for being parasitic on the antiquated natural science of Georges-Louis, Comte de Buffon. Spencer's own argument was invigorated by the belief that human behaviour and properly restrained executive government worked together in propelling society towards social reform. He did not see this process as caused by rational decision-making because any social change was a complex process that often took place despite, as well as because of, individual volition.

If one were to discard the notion that the individual is central to the foundation of liberalism and instead focus on the strong anti-individualistic streak of Victorian liberalism, this would not only illuminate accounts of Spencer but also perhaps cast light on the whole liberal tradition in the nineteenth century. Most notably it would cause the most troublesome contradictory Spencerian doctrine – the advocacy of land nationalization – to disappear as an anomaly and emerge as an exposition of liberalism. In the nineteenth century his enthusiasm for this reform gave him a radical chic that was lacking in J. S. Mill, T. H. Green and other philosophical Victorians, and this helped promote his reputation as a reformer in the new worlds of America and Australia. It was an integral part of Spencer's *Social Statics* (1851), which was reprinted in an unrevised form for British and American readers at least until 1868, and not edited into an individualist format until 1890. Until this last date *Social Statics* continued to adumbrate that the nationalization of all land was necessary before a person could justly possess *any* property. This radical stance on property was not extraneous to his political theory in the sense of being a mere by-product of youthful exuberance. It was part of his coherent philosophical opposition to the misguided rationalist attempts to explain politics as the expression of individual volition or as the satisfaction of individual pleasure.

From the 1840s Spencer had been opposed to Benthamite utility, which he regarded as a false moral philosophy that considered politics as a matter of satisfying the interests, or improving the happiness, of individuals. In *Social Statics* he broadened his attack to include Paley's Christian utility. It was not that Spencer had abandoned his Christianity; that was still to come. Rather it was because his liberal ideology had shifted so that it began to oppose principles relying on a language that contained references to non-specific qualities. In parallel to Bentham's own refusal to analyse politics in terms of power because that would evoke theological connotations, Spencer avoided the use of the non-specific term "happiness" because it seemed more than human. It mistakenly suggested that each individual had an identical mind or soul that could correctly answer a single abstract question about happiness. Spencer opposed this with a general proposition that human sentiments, and the organs that experienced them, should be considered as having several causes. In addition, he had begun to resist any supposition premised on the proposition that politics was a matter of satisfying a collection of individuals by addressing each person's happiness, reason or volition. To him these qualities hinted at an invisible or spiritual world, whereas he was determined to base government on the realities of psychology and administration.

Spencer began his early politics in an anti-individualistic fashion by stating that rules were needed for "a true philosophy of national life".[19] Government was a check not on individual wrongdoing, but on "national wickedness".[20] While general misbehaviour, together with the need for an administrative remedy, would lessen as civilization progressed, Spencer did not see such improvements as necessarily caused by the pleasure experienced by individuals. Although he could sound like an individualist when speaking of man as a social atom,[21] this was not a reference to moral autonomy. It was merely an insistence that his contemporaries should employ modern scientific analogies drawn from subjects such as chemistry. He wanted to avoid homeopathic medical language, which suggested that happiness would follow from individuals knowing their own cures. The liberalism of Spencer's *Social Statics* does not prefigure J. S. Mill's *On Liberty*, with its reliance on individual choice; rather, it refers back to Mill's earlier *A System of Logic*, which had contained sober warnings against a non-scientific and primitive reliance on remedying unhappiness as if this were a disease with a single cause. It also echoes the earlier Mill in its scepticism that factors such as speculation, intellectual activity and the pursuit of truth were powerful propensities in human nature that would cause social progression.[22]

The concept of human nature seemed too abstract to Spencer. Instead, he defined man as a visible, tangible entity with properties.[23] This was a focus on an outward corporeal being, not an inner spirit acting under its own volition. For Spencer this meant that, when we considered happiness, we should not locate it within an individual consciousness, but outside it. That is, we should examine happiness in the "social state".[24] Individuals who wished to obtain happiness must be conditioned to fit within an existing society, not an ideal one. This signified to Spencer that, for any given epoch, it was our social conditions that were constant, not our characters.[25] Individuals were not independent entities who could search successfully for contentment; they had to be moulded so they could adjust to a context. Sometimes this meant that suffering had been required or, at least, that pain had been experienced when the civilizing process was in conflict with recalcitrant individuals.[26] Spencer believed that it was fortunate for Britons that such compulsion had lessened as their nation had become more advanced.

This compulsive feature of liberalism is stressed here in order to demonstrate that – unlike J. S. Mill in *On Liberty* – Spencer did not rely on individual choice as the motivation for progress. Nor did he see the individual and the state as antinomies in the way libertarians have done. Instead, Spencer viewed the state as simply one of a variety of institutions that might coerce people into making progress. Rather than viewing the state as the sole organization with a monopoly on force, Spencer suspected that it might not be able to socialize people effectively, and that equivalent reforming effects could be achieved by other, and more efficient, social organizations. However, no matter which organization was responsible for directing change, Spencer's theory of progress did not depend on the reason or feelings operating inside single individuals.

Spencer had swept away much of the apparatus of traditional philosophy, which relies on each separate mind elucidating essential meanings. What remained were two of the most formal of juridical doctrines: administrative injunctions stripped of any morals that might have appealed to reason or participation by citizens. That is, justice and equity were the only norms that remained compatible with Spencer's early political theory.

When analysing Spencer's land nationalization it is important to remember that it was an integral part of a systematic discourse on politics. The idea should not be seen as anomalous, but as a projection of his core liberal values. Driven by his lifelong anti-aristocratic bias, and fuelled by his sympathy for the revolutions of 1848, Spencer's *Social Statics* proclaimed that equity could not legitimize property in land.[27] Humanity should not be degraded by begging the lords of the soil for "room for the soles of their feet". If people were made subordinate by landlords they might even be expelled from the earth altogether.[28] This warning caused Spencer to avoid that reservoir of individualistic values, Lockean property theory, in which a person acquired a right to land by mixing his or her labour with it. This labour theory of value seemed absurd to Spencer; he could not comprehend why the extermination of a set of plants or the ploughing of the soil to the depth of a few inches proved a claim to the earth beneath.[29] Instead, he believed that all land should be treated as unclaimed and as belonging to everyone. This egalitarian proposition was not restricted to undermining the validity of *original* possession because his justice theory did not permit the legitimate transfer of a possession that had been wrongfully acquired.[30] Spencer's was an extremely radical doctrine: nothing could be lawfully owned because the human race had been, and was still, the original possessor since "the world is God's bequest to mankind".[31]

Since Lockean theory had failed to legitimize individual claims to the soil, society could expel owners from their lands without behaving unjustly. It did this already when space was needed for the construction of railways and public utilities. Spencer speculated that if a similar procedure were applied to land in general this would not unsettle anyone because he was not advocating socialism. Unlike Fourier and Louis Blanc he did not promote the "community of goods".[32] He was simply arguing that society, rather than individual owners, would possess the title to lands. This proposition seemed non-threatening to Spencer because it entailed no difference in work practices: the current situation was that farmers worked privately owned land, while in the future they would till land owned by "the great corporate body – Society".[33] Land would be leased and the agent of "Sir John" or "His Grace" would be replaced by a community official. Stewards would be public dignitaries instead of private ones.[34] Spencer's adoption of business language here did not incline him to libertarianism, but to a communitarianism where everyone would be free to bid for vacant farms. This was a simple matter of justice to Spencer who worried that, "In our tender regard for the vested interest of the few, let us not forget that the rights of the many are in abeyance; and must remain so as long as the earth is monopolized by individuals".[35]

When Spencer published *Social Statics* in 1851 it competed with individualistic accounts of man and society produced by other radicals. This rivalry was not his intention, even though, given how rebarbative he was, he would have reacted against competitors if he had noticed they existed. As it was, he could truthfully say that he had never read Humboldt's *The Limits of State Action*,[36] which was the chief source of libertarianism during the 1850s. Nor, although he tried, could Spencer ever puzzle his way through Kant's writings,[37] which were the source of many of Humboldt's ideas. Later, at the end of the nineteenth century, when there was a vogue for neo-Kantianism, Spencer's admissions of ignorance were used in academic circles to paint him as a demotic writer, too unrefined to be worth refuting. Alternatively, he was glossed as a philosopher who would have been a Kantian had he been better schooled. These comments were worse than patronizing; they were misguided reactions to a philosophy that was diametrically opposed to Kant's. Spencer was troubling to neo-Kantians because of his lack of individualism, not because he had failed to read their mentor.

It was, of course, true that even in the mid-century Spencer could have read more Kant. His acquaintances – such as J. S. Mill, Pattison, Eliot and Lewes – could, and did, write fluently about German philosophy. However, Spencer's failure to imitate his friends was not caused by vulgar populism. The most profuse mid-Victorian writer on individualism, William Maccall, who was much more eager to appeal to the populace than Spencer ever was, knew all about the Germans.[38] When Maccall regretted that Spencer knew nothing – except at second- or third-hand – of the metaphysical developments during the previous hundred years,[39] he was not being a linguistic snob. Instead, he was expressing regret that, unfortunately, Spencer belonged to an intellectual tradition that was rooted in science. This was in contrast to his own thought, where Kantianism had to jostle for room in a space crowded with connotations of Goethe, Fichte, Humboldt and even Feuerbach.

This last figure's voice echoed in Maccall's definition of individualism, which he called "my Gospel, my substitute for a dying or a dead Christianity".[40]

The presence of Maccall in Spencer's milieu draws attention to the religious origins of much of the mid-century individualism, but it is also a reminder that orthodox faith was slipping away; the English future was to be secular. Maccall's individualism, like that of many of Spencer's acquaintances, placed "Man" in a historical, not a scientific, tableau. At its centre was a vision of the human spirit moving through time from a primitive past. This was a romantic ideal of the growth of the individual as a free and strong spirit, which had renounced Christian dualism in order to reintegrate humanity into nature.[41] As a political creed this was individualistic because it directed Maccall to oppose the despotism and priestcraft that had elevated the nation at the expense of the individual.[42] He believed that despots and priests had advocated equality so that their subjects would remain at a debased level while learning lessons that would prevent idiosyncrasy and rebellion.[43] Maccall's account of the rationales that support governments was neither philosophical nor juridical; it was a defiant and progressive bohemianism that would have appealed to a twentieth-century hippie rather than to an Ayn Rand. Its promise was that the oppressive and monotonous state would fall and be replaced by a reformed one in which each "man" would be free to yield to his unique instincts without being rebuked for rebellious utterances or actions.[44]

Maccall's simplistic ideology was the pronouncement, in what he imagined to be a Teutonic philosophical form, of the *idea* of the individual that was opposed to the nation. The latter was not benign like society; the nation was a tyrannical organization that enslaved individuals. In the future, however, there was to be no sphere for legitimate national actions because only "Man" was to remain an "actual" part of Maccall's pantheistic universe.

Maccall despised his contemporary J. S. Mill as a cold didactic person whose only decent intellectual garments were stolen from a passionate predecessor, W. J. Fox.[45] However, it was not Mill's theft of English garb that chiefly irritated Maccall; it was that the author of *On Liberty* was also clothed with the same Germanic and romantic clothing as himself. Mill might have been dry and excessively rational in comparison with Maccall, but he too had borrowed Humboldt's description of the true end of Man.[46] For Humboldt, as for Mill, human progress was prescribed by reason, "not suggested by vague and transient desires".[47] This development culminated at "the highest and most harmonious development of his powers to a complete and consistent whole".[48] In order to encourage this Humboldt had made freedom the first and indispensable condition of progress. Such a defence of freedom was problematic to Victorian liberals, and it split them into two camps, which effectively is a dispute between the Millites and the Spencerians.

Mill's *On Liberty* is in the first camp because of its proposal that human freedom can be acquired by the mind rather than being achieved through the pursuit of blind desire. Mill's individuals are closely aligned to Humboldt's depiction of human beings as isolated and rational beings controlling their own destinies and in harmony with nature.[49] To both Mill and Humboldt individuals possess inherent greatness because of their potential to cause social change. This, in turn, means that the individual's will and creative life have to be fostered. For both philosophers it was individuals who ultimately would achieve perfection – evolving from a rudimentary form in the classical world – and who were able to reach this goal by participating in the rich collective resources of a social context provided by their fellow citizens.[50] State activity was harmful when it attempts to expedite this process because it reduces variety.[51] Faced with this problem the correct strategy would be to leave social participation as a choice for the inner impulse within each unique being. This would then lead naturally to the unity of man.[52] While Mill was not as lyrical about this unity as Humboldt had been, it was clear that their ideas had the same goals.[53] The similarity in their beliefs is hard to ignore if one reads the chapter on individuality in Mill's *On Liberty* or glances at the proem from Humboldt at the beginning of the 1867 people's edition of that work.[54]

The value of Man for Humboldt and Mill was the intrinsic worth of each individual soul, which would achieve uniqueness as it moves towards unity with nature. The cultivation of the soul was a combination of cultural progress within a society and of psychological development within a person. Both processes were directed by the intellect. This description of human development would be unacceptable to the Spencerian camp. From that perspective there were a number of things wrong with the Millite synthesis. First, Spencerians would always see human perfection as something that would be achieved in the future, not as something that had already been attained through cultural evolution. Secondly, Spencerians would not have agreed that Humboldt and Mill possessed a principled reason for the limitation of state interference. Practically, of course, Spencer too thought that the state ought to restrict its management of human affairs to areas where it could effect positive change,[55] but on matters of justice or social equity it could be as intrusive as it wished. Since Spencer could not see individuals as capable of achieving perfection now, and since he believed that social progress would often be temporarily harmful, there was little rationale for him to erect a theory of individual rights which would always limit state action. Individual perfection was the end goal for Spencer. However, this would not be achieved by living in harmony with nature, but rather by the species evolving beyond it. His ideal was not the classical one of philosophically transformed individuals living compatibly with others, and with nature, in an intimate way, as prophesized by Humboldt and Mill, but the modern scientific view of the individual as a biological unit embedded in nature, which developed from a simple to a specialized form. Individuals lost independence in this process, but this was counterbalanced by freedom from superstition, poverty and violence. In the future Spencer's individuality would accompany the increasing differentiation of social roles, and the reduction of political domination, but these outcomes would not have their roots in a doctrine of individualism.

The current understanding of the foundation of liberalism needs relaying so that the dominant voices of Humboldt and Mill do not cause us to build solely on the division between the individual and society as if that antithesis had exclusively produced the innovative ways in which liberals have valued progress, liberty and restriction on government activity. One could go even further here and suggest with Patrick Joyce that distinction between the "self" and the "social" is not a productive way of recycling ideology from the nineteenth century because these opposites were always mutually self-defining, and belonged to false realities that were, in actuality, equally "bourgeois" and "working class".[56] Spencer's liberalism, in particular, is not usefully glossed as a "bourgeois" individualistic ideology that was forged in opposition to the collective. Spencer's early writings were more representative of liberalism than Mill's *On Liberty*, or even than his own late *The Man "versus" the State*, precisely because they did not echo the individualism of German Romanticism. To ignore Spencer's early writings when considering the foundation of liberalism would be like interpreting Friedrich Hayek as if he had never written *The Road to Serfdom*, but had contented himself with publishing on constitutional reform from the 1960s.

Spencer's political ideas seem problematic to modern scholars, who would prefer to have a sparse and coherent set of principles such as those that are often associated with Kantianism and individualism. Spencer's habit of thinking within a thick sheaf of psychological and ethnographic data means his views are less often co-opted into political debates. As one of his recent critics put it, his "methodologically holistic" values assume that the collective good and the welfare of individuals will coincide in the future, but offer no rules on how society should force the pace of change or, alternatively, how individuals should be protected from the collectivizing process.[57] This objection would be salient if Spencer's value to liberalism consisted of his ability to bridge a gap between negative and positive liberty. However, since he mostly avoided describing himself as a negative liberal, and never promoted collectivism or positive liberty, this should not be seen as his contribution to liberalism. In any case, if Spencer was correct in his assumption that individualism was a doctrine that was only relevant in primitive societies that were also coercive, then, from his perspective, it would have not been sensible to commend it as a way of overcoming the evils of war and domination, or as a way to promote social justice. If it were restored, political individualism would not encourage the humane outcomes that Spencer hoped would result from social evolution. Since, in its modern political form, individualism supported the exercise of will by a democratically elected sovereign, Spencer saw this as a regressive phenomenon. He could not posit intrinsic value in a particular constitutional form, such as democracy, and he was uninterested in whether the intentions of voters were taken into account. Instead, he focused on whether social changes had the effect of decreasing personal domination. His concern would not make an inspiring revolutionary or reform slogan, but it is, at least, a clearly expressed and non-contradictory criterion for the measurement of the social progress that Spencer hoped would inform liberal values.

To reiterate, Spencer's liberalism had more of a corporate than an individualistic focus. When the autonomous individual did make an occasional appearance in Spencer's writings it was as a biological unit of the human species that was necessarily sacrificed during evolution. If the protection, or the enhancement, of the individual had been the purpose of Spencer's liberalism, then this would have left his political theory in confusion. However, Spencer was unperplexed because he took his ideological direction from justice and equity. To determine whether a particular action was justified, or whether it would lead to greater social justice, he simply examined whether it met these formal criteria. Of course, Spencer's foundation of liberalism has its flaws. For example, it lacked the intimacy with nature promised by Humboldt or the spontaneity admired by Mill and, therefore, it did not explain why the individual needed to be involved with his fellow creatures. However, Spencerian liberalism had the advantage of possessing a political cutting-edge that was absent from the ideals of Humboldt and Mill. While liberalism in its Spencerian form did not particularly encourage participation, it was open to assimilation by democracy. This was not because Spencer advocated government by the people. On the contrary, his reservations about the value of popular political control were as great as Mill's. It is simply because Spencer's non-individualistic and evolutionary account of justice offers less resistance to democracy than the ideas of many Victorian liberals because, unlike them, he did not see any valid reason to perpetuate inequalities that had lingered on from earlier political and social structures.

Spencer's early writings free the foundation of liberalism from the constrictions that philosophy has often placed on it when assigning legitimacy. Primarily his ideas have this effect because they do not assign rights to either the state or the individual. While this stance was not unusual in the middle of the nineteenth century, Spencer took it further than his contemporaries. He did not share the belief that the state and society should be limited so as not to intrude on those aspects of private life that are imagined to be essential. For Spencer there was no domain where individuals possess a unique way of evaluating their own happiness. If one believes contentment to be the result of an evolutionary process, rather than of individual calculation, it would be a mistake to echo Millite liberals who offer principled reasons for protecting a person's moral autonomy from intrusive corporate initiatives. The implication of this is that while Spencerism is frequently *laissez-faire* or non-interventionist on practical grounds, it does not support a defence of individuals that privileges their possession of knowledge.

Analyses of liberalism that focus heavily on limiting political or social interference with the individual might see Spencer's neglect of this as a sign that he was not truly liberal at all. However, this would be a misunderstanding of the foundation of liberalism. In the past, liberals formed a chorus in which Spencer's voice was even louder than Mill's, because the Spencerian advocacy of liberalism as a doctrine of increased justice was easier to reconcile with a programme of political reform than moral autonomy was. To ignore Spencerian liberalism is to ignore a dramatically radical aspect of liberalism that did not recognize the authority of the state, or of any other institution, even if this were restrained within proper boundaries. That is, for Spencerian liberalism there was no justification for authority even when the state was functioning properly. There was no sphere within which a government possessed the authority to exercise lordship or dominion. Philip Pettit is quite prescient when, in his republican writings, he draws on Spencer as a source of anti-domination.[58] Spencer and Pettit resemble each other in declining to source justice in democracy. Neither cares deeply about the conditions under which political representation justifies domination, because such oppression can never be justified. For Spencer it would not have mattered a great deal if a constitution had a democratic shape because he was interested solely in whether or not the executive operation of a government was just and equitable. He was not speaking entirely in jest when he expressed preference for governance by an oriental potentate over a parliamentary system, if it functioned more effectively. Spencer's liberalism was not attached to particular institutions; instead it was a repeated insistence that social and political activity should be judged only on the grounds of whether it led to increased justice and equity. Institutional frameworks were irrelevant to the operation of justice; Spencer could not have advocated republicanism – even in the attenuated Pettit form – any more than he could have defended the monarchy. To do either would have committed him to the belief that political and social change could be achieved by adopting idealized constitutional designs. There were no constitutional boundaries that protected the person: if corporate activity ran counter to individual interests and happiness then this could be a cause for regret, but not for limitation or prohibition of such activity.

SIXTEEN

The 1840s: Spencer's early radicalism

In the 1840s Spencer's political writings displayed few philosophical features that would have struck his contemporaries as novel. His views resembled those held by other mildly unorthodox Christian radicals in the Midlands and the north of England. There was no hint of the secularizing metaphysics he developed later; nor was there, as yet, any thought that human beings were subject to instinctive impulses. At this early point in his intellectual career, Spencer perceived both individuals and societies as influenced by psychological or biological motors, but not as *driven* by them.[1] He imagined that human beings progressed by following invariable laws given by God. These laws were natural because they were based on the balanced and organic forces of the universe. In brief, "nature will be obeyed".[2] Mankind was simply one part of a harmonious living creation. As a political consequence of this Spencer insisted that existing constitutional arrangements be scrutinized to see whether they were necessary, rather than artificial, contrivances. If they were the latter, they would interfere with that beautiful self-adjusting principle that traded off competing social interests.[3]

Under Spencer's theory, a community, a nation or an individual was a natural projection of humanity. Even a government was a natural edifice. None of these entities had priority over the others; in terms of either importance or time they were equal. This proposition placed many constraints on "The Proper Sphere of Government", Spencer's first serious writing on political theory. For example, it meant that he could not successfully employ a social contract argument, because he believed that since governments were natural, they had not been artificially formed as a result of each individual's rational calculations. Spencer's treatment of individuals, communities and institutions as equal products of social evolution makes his ideas opaque to political theorists whose language has been shaped by classical discourse. Their questions would not be apposite. If, for example, they were to enquire about the relationship between Spencer's notion of community and his idea of the state, they would find that this could not be answered because the latter part of the question refers to an abstract identity, the state, about which he had little to say. Spencer did make many comments about the proper functions of *government*, but he tended to view these as efficient features of an organization rather than as factors that would make authority more acceptable to the governed. In addition, Spencer did not perceive progress as a consequence of citizens sacrificing themselves to the public good but, eventually, concluded that this would prevent progress. This comment, however, anticipates Spencer's mature political theory, when evolution suggested to him that individuals should not be regarded as citizens, but as unintended beneficiaries of improvements in the quality of life. These results occurred when political institutions operated effectively because they had developed conflict-avoidant procedures and functioned using legal and administrative procedures. This was not a lesson for citizens as the individual and the government would not operate in harmony, and benefits achieved by the former were not in response, or reaction, to changes in the latter. This was an anti-republican stance, and was not fully articulated until Spencer wrote his political sociology a quarter of a century in the future. In his early writings he could still propound the traditional notion that authority was legitimate because it rested on the submission of the populace.

For Spencer, in "The Proper Sphere of Government", governments were yet another benign aspect of the natural fabric of the universe. They were only malignant when they ignored their true functions and ventured on subsidiary activities. Provided, however, that public administration functioned properly, Spencer did not permit any justified resistance to it; his formula here was the traditional Christian one of passive obedience. His view was that – to an unbiased arbitrator of the relationship – the people had *submitted* to their government, and that, as a result, they must tolerate the exercise of coercive authority over them.[4]

Spencer's thoughts about the meaning of submission reflected the political language adopted by Jonathan Dymond. Dymond was a Quaker apologist who was much admired by Spencer's father and by Edward Miall, the editor of the *Nonconformist* newspaper.[5] Dymond, although clearly what contemporaries called "a friend of civil liberty", believed that, "Submission to Government is involved in the very idea of the institution".[6] From this he concluded that the duty of submission was necessary and that it was right, both as a matter of expediency and as a matter of acquiescence to the will of God.[7] The young Spencer's admiration of such ideas was encouraged by Miall, who was his employer as well as being a voluntaryist and a spokesman for the Complete Suffrage Union, the organization for which Spencer was the Derby organizer.[8] Miall's version of voluntaryism stated that, in the age of civil law, it was a matter of "perfect indifference" as to whether men consented to their government, as long as that did not meddle with religion. As far as the behaviour of citizens was concerned, the government simply drew around citizens a circle of prohibition, forbidding them to step over this on pain of its displeasure.[9] So long as they kept within that circle it did not matter whether their obedience was freely given. Since the institutions of the state did not have "the responsibility of winning man's consent to the attainment of their ultimate object, [they] need pay no regard to his nature, in securing their own support".[10] From this voluntaryist and liberal perspective an appeal to the human understanding was only significant in the area of religion. Obedience to the state should not rest on such an appeal; it was a lesser matter requiring submission rather than persuasion. Such a dismissal of the need to address the intellect might repel a later liberal democrat to whom individual consent was the necessary foundation of government. However, this repugnance would have seldom arisen in the heart of an English radical of the 1840s.

Spencer's early liberal doctrine was consistent with that of the *Nonconformist* in lacking a basis in consent. His politics also contained little evidence of the systematic and extreme vision of the morally autonomous self that so prominently features in his often invented posthumous reputation. In Spencer's early writings individuals lacked a real existence. Even their attributes were missing. Rather than appearing as the will of a single person, human volition appeared in containers labelled the "community" and the "national character". These group identities were the basic political units to which Spencer appealed in "The Proper Sphere of Government". The need to support these units was a moral matter that outweighed efficiency. For example, Spencer was willing, for the sake of argument, to admit that the public provision of welfare might alleviate the distressed condition of the working classes, but he remained opposed to it on the grounds that it would lower the standard of the national character.[11] He made a similar judgement in his first article on the railways, where he argued that even if state administration of such transport were more efficient, more profitable and of greater public advantage than that managed by individual or joint enterprise, it would have a deteriorating influence on "national character".[12] This negative effect would more than neutralize the direct benefits that had been obtained from public ownership. As an example of what might happen if this warning were ignored, Spencer cited the sluggish degeneracy that could be found in Sweden.[13]

The young Spencer was certain that the community – not the individual – possessed moral standing. This conviction rested on the distinction between what was permanent and what was contingent: individuals might escape the consequences of evil doing, but a community always remained to carry such burdens:

> The individual who breaks through the bounds of justice, and trespasses upon the rights of a fellow-creature, may perchance escape all civil reaction. Not so, however, with the different sections of a community. The life and health of society are the life and health of one creature. The same vitality exists throughout the whole mass.[14]

Such political logic was underpinned by a simple organic metaphor. Spencer believed that one part of an organic body could not suffer without the remaining parts being ultimately injured. Further, he thought that the body could not be disordered without this weakening the limbs, which in turn would reduce the nourishment delivered to the body.[15] If the social organism acted without harmony it would degenerate. The reverse was also true; an increase in harmonious social interaction would be progressive.

In "The Proper Sphere of Government" the grand and irresistible law of human existence was "progressive improvement".[16] All social phenomena, even negative ones such as the tyrannical behaviour of the aristocracy, witnessed to this advance. Spencer did not care which laws these phenomena were subject to: those of the organic world, those governing human society and those that governed the mind were equally important. All were signs that pointed to the ultimate goal towards which the Almighty had directed humanity.[17] This goal, as well as the smooth path towards it, could not coexist with pain or grief. Spencer's vision of a future without suffering would have contained anomalies if he had been a Benthamite utilitarian, but was perfectly compatible with the bland Christian utility that he then advocated. In his view, the Almighty did not rule the world through suffering; indeed, the presence of suffering in the human condition was a signal that humanity needed to make more progress.

Social distress also meant that those natural arrangements that should have been dominant had been set aside. In their place were a set of partly artificial constitutional and administrative contrivances that interfered with harmony. At this early stage of his intellectual development, Spencer was oblivious to Malthus's warnings that compliance with natural desires led to unhappiness. Even later in 1852, when Spencer showed signs of awareness of that gloomy parson, it was only as a historical figure in the history of population studies.[18] Aside from the fact that Spencer regarded Malthus as dated, he did not share the latter's view that sexual activity was always immediately pleasurable. Spencer's distaste for hedonism was not caused by his deprecation of the importance of the reproductive impulse; on the contrary, he suspected that this was the force that aided in the preservation of the species. Rather, his caution sprang from the belief that the maintenance of a species was often at the expense of the parent's survival and well-being.[19] Since, for Spencer, natural desires – at least in the human species – were supposed to lead to happiness, this meant that they should not be directed towards sexual gratification. Such an argument made sense if one considered happiness as a possession of adult numbers of the species, rather than of the species as a whole. Instead of childishly pursuing happiness as the kind of immediate gratification that could be found in sexual activity or eating, Spencer hoped that adults would devote themselves to the vanquishing of mental darkness. When they had escaped this gloom they would experience a kind of happiness that would be unalloyed with the pain associated with sexual reproduction and labouring for subsistence. This vision was based on a simple reversal of hedonism: civilization was to be subsequent to the fulfilment of bodily needs, not subservient to them.[20] It was not, as it became for Lubbock and many other Victorian commentators on culture, the possession of a high level of technological superiority and national prosperity. Instead, Spencer believed that as the human species made progress – or was civilized – it moved further from being subordinated to necessity. On this subject, he had laid down a thesis to which he clung long after jettisoning most of the transitory ideas of "The Proper Sphere of Government". His constant public refrain became the doctrine that the future of mankind would be unaccompanied by immediate physical gratification. Although privately, in *An Autobiography*, he doubted that his anti-sensuality had served him well as an individual, it continued to appear in the pages of "A System of Synthetic Philosophy". At heart, Spencer's theory of civilization was a matter of higher sensibility. Both his ethics and his aesthetic tastes were combined in such a way as to make him yearn for a time when the human species was not trapped in a cycle of sexual reproduction, pain and death. To be civilized meant the avoidance of these while exercising higher sensibilities. The converse was also true. To remain in darkness, Spencer intoned, was to be insensible to the highest pleasures of which "the Creator" had made human nature capable.[21]

For Spencer, civilization was not only the prospect that lay before the human species; it was also the foundation of his early theory of rights.[22] This theory should not be confused with the one that recent political philosophers have used when they discuss normative principles as if these had originated with rights-bearing individuals coming together at the beginning of a political society. In the work of these theorists rights are assigned to a state or a sovereign in such a way that this artificially formed ruler is dependent on the individuals who compose it, and is compelled to preserve the safety and well-being of the initiating members. This is similar to his ideas of social contractarians of the seventeenth century, which reinforced natural law as well as describing the process through which governments were lawfully established. At times this process was imagined to be a historical one to the effect that, at some point in the past, citizens had assembled to create a sovereign government, usually a monarchy. In erecting a sovereign the citizens were said to have vested their rights in this entity by means of a binding contract with each other. However, it was not necessary that a philosopher, when imagining such a compact, should provide it with a historical dimension. On the contrary, he might, like Spencer, be eager to avoid reinvesting government with the panoply of traditional authority. While, in his later writings, Spencer advocated respect for tradition, his earlier thoughts were that majestic connotations of the past should be ignored. When Spencer imagined society in a primitive condition,[23] this was solely so that he could speculate about which requirements would be most necessary when thinking about governance.[24] He did not admire early societies, nor did he put faith in any binding agreements of the kind which were supposedly formulated in a founding convention. Spencer – in contrast to other philosophically informed contemporaries such as James Mackintosh, W. E. H. Lecky and H. S. Maine – was not guided by historical research into constitutional origins. He had read the standard works by authors such as Sismondi, Hallam, and Stubbs but they did not convince him of the political merits of medieval liberty or Italian republicanism.[25] The task of breathing a semblance of life into long-dead liberties and political practices left him cold. Spencer recommended that his fellow citizens cease to trouble themselves with the useless petty details of history; these only denoted the local workings of human wills and passions.[26] Such details were only the "superficial tissue" of the past. Instead we should "dive beneath" this to discover the universal laws that determine the fate of nations.[27] Although Spencer's phrasing here was pregnant with anticipation of a future sociology, it also slammed the door on any empirical investigation on how social contracts might actually have been forged. All pieces of historical data were "mere insignificant details of outward show" rather than spiritual truths.

At the risk of discomforting some modern Spencer enthusiasts, I feel obliged to warn that the early Spencer cannot be an advocate for modern libertarianism. He did not see rights as flowing from the nature of the individual, nor did he give the individual a pivotal place in politics. For Spencer, nature spoke to the community, not to the individual. When he introduced rights into his early political discourse this was not to promote an autonomous sphere for individual activity, but merely to adumbrate his conviction that the *original* function of government – the giving of justice – should be furthered while other attributes should be jettisoned.[28] The prime function of government was original in the sense that it was, in Spencer's eyes, always rationally defensible to administer as if one were in agreement with it. This was true even when the government was not democratic. "Original" did not mean temporally prior to other functions; it was simply part of a definition that specified it as the antithesis of secondary functions.[29] Spencer guessed that his contemporaries would mistakenly list extra duties for the state such as the provision of religion, education, welfare, the means of communication (postal service, roads and railways), and defence. He also thought that his countrymen might require that the government regulate domestic relationships.[30] However, Spencer's own preference was to omit the first five of these requirements (namely, religion, education, welfare, communication and defence). He was sceptical about the motives of those governments that wished to provide for religion and education (it seemed obvious to him that their real intentions here must be to teach veneration of authority). He was also suspicious about state involvement in welfare, communication and defence on a variety of grounds, although late in his life he recanted on the last of these and said that he had erred in not emphasizing the need for a society to protect itself against external aggression.[31]

In the 1840s Spencer did not see defence as a proper function of the state because he believed that war was an unnatural activity. To spend money on the army and navy was analogous to drinking wine and becoming intoxicated.[32] War and drunkenness were equally unnatural and, therefore, wrong. This anti-Dionysian sentiment led him to say that conflict was against the spirit of Christianity and civilization, and that it encouraged "animal passions".[33] Spencer's strictures against military expenditure went to the centre of his early political theory. His stance on politics was accompanied by a greater speculative novelty than would usually accompany a liberal stance. This is because he had temporarily thrown aside the strong protective rationale that lay at the basis of much Victorian political speculation.[34] To appreciate this, it is important to emphasize that Spencer's definition of the original functions of the state included policing, but excluded external defence. This was provoking in a historical period in which internal security was not usually regarded as a state's duty whereas external security was. His views would have seemed bizarre and have required explanation to make them comprehensible.

Spencer's early scepticism about the need for the state to defend its citizens was partly based on his aesthetic sensibilities. That is, he felt that conflict involved a vulgar display of animal passions, which belonged to humanity's primitive past and ought to have been despised by the civilized. However, part of Spencer's objection to military activities rested on his understanding of international law. He began, as would most commentators, by stating that, "Viewed philosophically, a community is a body of men associated together for mutual defence".[35] The subsequent step in this argument would usually have been a justification of war, but Spencer did not explore this. Instead, he busied himself with limiting the "protection" activities of a state. This, in turn, weakened the capacity of social contract theory to support individualism because, by implication, Spencer's limitation of the "protective" quality of a government was that it might not be obliged to defend its individual members. His rationale for this meant that, when it was properly considered, communal membership was exclusive to occupants of a particular territory. From this it could be "fairly assumed" that the privilege conferred by membership would be enjoyed when citizens were residing within that territory.[36] Spencer, therefore, concluded that a nation could not be expected to give extra protection to its citizens when they may have chanced to wander outside territorial boundaries. Further, he suggested that when people left their community they forfeited their membership, and could be said to have foregone all claim to assistance from the state.[37] Spencer's argument invoked, without any explanation, "the natural laws of society", a phrase that he perhaps borrowed from Volume 3 of Adam Smith's *Wealth of Nations*. With these laws in mind Spencer then observed that colonists in territories such as Australia should not be assisted by the military because this would not teach them to adopt better behaviour, and to treat the native owners of the soil with justice.[38] This last argument provided only marginal support to Spencer's main point that individuals forfeited rights when they left their territory and suggests that, to him, it was the community, rather than the individual, that possessed the right to self-defence. His basic position seems to be that whatever motive individuals had had originally for joining a society, they did not possess a guarantee to later protection. Therefore, for Spencer, individual claims advanced by members of a developed state did not depend on the primitive origins of the community as a protective association. Further, Spencer's argument stressed that citizens should not claim that the right to protection existed independently from the community. A civil society was a natural body that could assign whatever privileges it wished to its citizens. This being the case, Spencer insisted that one of those privileges – the right to call on protection – was a limited one.[39]

Spencer's doubts about the value of social contract arguments were consistent with his rejection of the idea that it was the business of a state to promote the welfare of its citizens. His views were an extension of a popular reaction to the idea that governments should promote the happiness of their subjects. To govern in this way was to succumb to the Benthamite definition of utility, a weakness that was regarded as alarming in Spencer's circle. "Is it", asked Miall rhetorically, "to be considered true that governments must undertake happiness for man?"[40] Continuing in this vein, he rhetorically demanded answers to such questions as: were governments to be made responsible for citizens' beliefs and actions? "[A citizen] has only to become the passive raw material out of which his rulers may spin their purposes and weave their plans, and the result will be 'the greatest *happiness* of the greatest number'."[41] This quotation is a reference to Bentham's utilitarian criterion that the sole test for legislative acts should be whether or not these promoted the greatest happiness. Spencer was anticipating this tirade when he announced himself to be horrified by a parliamentary grant for the promotion of singing. "Truly, it would be a lucky thing for the aristocracy, if the people could be persuaded to cultivate their voices instead of their understandings." [42] The point of such irony was not to deprecate the joy of singing (Spencer himself was a member of a glee club in Derby), but to insist that happiness should not be a consideration when specifying which functions of government were legitimate.

The Benthamite test for the usefulness of a particular state function was not the only concept Spencer wished to combat. He was also suspicious of arguments for the public good when these were used to further the state's activities into areas other than the administration of justice. For example, when his contemporaries defended taxpayer-funded health legislation on the grounds of the general good, Spencer opposed this as a vague rationale that would too easily be extended into further areas. If the government were involved in promoting good health, then it might as well prescribe the amount of exercise, regulate the number of meals and publicly adjust personal intakes of meat, vegetables and fluids.[43] It could be extended so that all similar regulations would improve health. To Spencer this was an example of a *reductio ad absurdum*: such public good arguments also worked for medical qualifications and rules on the acceptability of various drugs, and implied that all were invalid. The crux of Spencer's objection to any public good argument was that it would assign a government duties, such as guarding people against dangerous pharmaceuticals or unqualified doctors, in areas where individuals ought to protect their own interests without state assistance.

At first sight this looks as if Spencer was defending the individual in the way recent libertarians might do. Such defenders have often possessed philosophical anarchist or Kantian beliefs that cause them to uphold the moral autonomy of each person regardless of any, or most, social or political consequences. Spencer's liberalism, however, could consider such consequences as pertinent criteria for policy-making provided these raised issues of justice. In extreme cases, Spencer's notion of the original function of the state might even reduce the autonomy of a person, not increase it. This can be seen if one contrasts Spencer's views with those of a modern individualist, who might be hostile to the state extending its coercive power when banning the sale of cigarettes while remaining neutral if a government merely offered advice in the form of a health warning on a packet. In this case the modern individualist would hold that coercion reduces choice, while regarding a warning as increasing the possibility for informed decision-making. In opposition to this, a Spencerian argument would be equally opposed to state coercion and to warnings in the area of health. This is because both operate in a domain that should be considered as outside the original function of government. The implication of this is that Spencer's liberalism is quite distinct from that of the followers of J. S. Mill, who have articulated their ideology in such a way as to make the increase of individual choice paramount when promoting social progress. Millite arguments contrast with Spencer's liberalism, where the motor that drove civilization forwards was increased justice, even when this did not further individual choice.[44]

If one compares Spencer's conceptualization of the individual with that of later liberals who wished to develop a person's potential by improved education and welfare, then it is clear that, despite its biological overtones, his early social thought did not foreshadow the "social organism" of new liberalism and liberal democracy of the late-nineteenth century. That use of organic metaphors advocated increased *social* justice, whereas Spencer contented himself with the rectification of injustice in such a way as not to include an increased benefit for the whole community. Spencer's argument against the Corn Laws is an example of this. He viewed these edicts as unjust because they specified that a person must purchase his staple food from a specific group of farmers. This practice could not possibly be defended as a matter of justice, although it could be defended as a measure that furthered the general good of the community.[45] That is, it supported arguments such as those that prohibited the importation of alien corn in order to preserve the fabric of country estates. Spencer's argument against the general good was not a simple one; his position in "The Proper Sphere of Government" was complex because he believed that farmers did, in fact, possess more rights than anyone else. He credited them with a natural birthright to subsistence derived from the soil.[46] This desert was ordained by God, who had declared that the sweat of the brow gave a person a true ownership to produce.[47] However, Spencer's early theory of the right of ownership did not imply that proprietors could force others to buy the produce of this natural birthright. Since the former possessed rights that were natural rather than political, he concluded that other members of the civil community were not obliged to assist them by supporting high prices for their commodities.

Spencer's argument against the public provision of welfare was phrased in a similar way: that is, he held that there were no natural rights supporting a person's claims on fellow citizens. Spencer could not discover why it was thought that individuals deserved relief from other members of their community when, through either misconduct or improvidence, they had fallen into poverty. The public subsidy of the indigent was not a matter of justice. This hard-fisted stance towards the unfortunate was extreme, because Spencer had added that it did not even matter if indigents' conditions were not caused by their own actions. That is, his rule applied equally to those whose poverty was the result of idleness or vice and to industrious labourers whose distress was not self-inflicted.[48] Neither type of poverty raised an issue of injustice for him. He believed that relief from distress should only be provided if the condition had been caused by a positive act of oppression: that is, if it could be clearly demonstrated that hardship had been caused by a deliberately unjust act.[49] Spencer had two rationales for this restricted definition of justice. The first of these, which was an implication of his general position that artificial legislative actions interfered with the natural workings of society, was that a generous relief for the poor would lead to "unnatural" outcomes; for example, the institution of state charity would cause private benevolence, the more natural form of relief, to wither.[50] Further, Spencer believed that it was unwise to disrupt natural arrangements by providing sustenance to paupers when the very existence of these had been brought about by bad legislation. From his perspective the giving of relief to the poor would be the perpetuation of earlier public error.[51]

Spencer's second rationale for not providing public relief was more novel. It was also strongly indicative of the looseness with which Spencer's early philosophical stance was attached to a natural rights doctrine. He could not clearly specify what rights there were because, while these rested on "that sense of right which God had implemented in man", this was a sense that would only operate in the future when human beings would have evolved out of their darkness.[52] In the present, however, rights could not be delineated with clarity. This meant that Spencer could not appeal to them when addressing an individual's moral sense. In any case, he believed that such an argument would be difficult because it was the community, not a single person, that harboured moral qualities. It was to the community that he gestured when he was warning that the public provision of charity led to a degeneration of the "national character".[53] It was the character of this group identity, rather than the autonomous self, that was at risk of degenerating. This suggests that Spencer should not be recycled by modern philosophers as an individualist because his public rationales were communal, not individualistic. This would not have been confusing in Victorian Britain: to his middle-class contemporaries Spencer's emphasis on the community when protesting against government relief of poverty would have resonated favourably. For many of them it was the central government – not the local institutions that were close to the community – that seemed objectionable after the Poor Law reform of 1834 and after the protests over centralization in 1837–38.[54]

Spencer believed that it was the duty of the state to provide justice "free of cost" to its subjects.[55] Such a benefit would not only eliminate vexatious and expensive court costs – that Dickensian dystopia of legal ruin surrounding the courts of London in the early-nineteenth century – but would necessitate travelling officers-of-the-law, whose circuits would secure remedies for everyone.[56] This recommendation was not simply the advocacy of policing; it was also free legal aid. The judicial officials appointed by Spencer's ideal state would actively search out instances of injustice and remedy them. Rather than justice merely responding to complaints of those rich enough to afford lawyers, Spencer would provide it for everyone. The reformed government would defend the poor from the rich, rather than leaving the former at risk from expensive counter-suits and appeals.[57] Justice would be quick and equal, but would carry a heavy cost to the public, both monetarily and through the undermining of due procedure. For Spencer efficiency in the pursuit of justice could even outweigh the rights of the citizen if these delayed the rectification of injustice. Spencer's enthusiasm for his panacea also caused him to abandon the customary European prejudice against Oriental despotisms. He claimed to admire those Eastern monarchies that *freely* dispensed legal verdicts.[58] This implies that when justice was at stake Spencer condoned coercion, and suspended the checks and balances of the parliamentary system. Even his early democratic ideal of complete suffrage might be put aside.

The primacy of justice in "The Proper Sphere of Government" meant that Spencer's early arguments in favour of democratic reform were somewhat peculiar. His enthusiasm for roving judicial officials and Eastern monarchs sat uneasily with his hopes of empowering the people. Despite being secretary of the Derby branch of the Complete Suffrage Union,[59] he was not sanguine about the wisdom and integrity of ordinary citizens. His defence of the people against constitutional elitists was possibly the most perverse argument he ever made on any subject. He began by noting that the opponents of democracy had said that votes should not be given to those who lacked the qualities of "deep sagacity, and clearness of intellect" that were needed to manage the complicated machinery of society.[60] In reply one could have expected a populist such as Spencer to have argued that the people did, in fact, possess sufficient wisdom to govern themselves or, at least, that they possessed as much as the aristocracy. However, he simply accepted the proposition that the people lacked sagacity and clearness of intellect, and concluded that their lack of capacity did not matter because the complex decision-making of present statecraft reflected the artificial and unnecessary needs of bad government. Once governance was restricted to its proper function then the people would not require wisdom so it would not matter whether or not they voted.[61]

The blunt admission that the masses were incapable of making complex decisions would have been unsettling if it had come from the pen of a modern liberal democrat. However, Spencer's support for the people was not based on the premise that democracy was a forum that reflected a myriad of individual choices. Such a vision could not have emanated from a political environment where voting rights were seldom seen as indefeasible but, instead, reflected contingent methods of choosing representatives who were charged with maintaining civil liberty.[62] Paley and Dymond, the writers on whom Spencer was most reliant in "The Proper Sphere of Government", saw political representation as the maintenance of a balanced constitution rather than the expression of the will of the populace. While Dymond disagreed with Paley on some issues, both philosophers favoured increased civil liberties while not relying on any particular constitutional form or political privilege, such as the type of suffrage that was to be attached to either individuals or communities.[63] That is, to them it did not signify whether parliamentary representatives were selected on the basis of universal suffrage or on voting restricted to ratepayers, corporation members, householders or members of any other group.

The subordination of voting rights to civil liberty can also be seen clearly in the views of Miall, the editor of the newspaper to which Spencer contributed the letters that made up "The Proper Sphere of Government". Miall had seized on comments of another liberal newspaper, which had observed that even the Chartists had wanted to hedge the right to vote with limitations, and had, therefore, admitted that it should be extended only to those who could exercise it competently. In other words, even the wildest speculations contained a kernel of truth, which was that limitations on such rights were neither arbitrary nor unreasonable. Further, Miall thought that such limitations did not undermine the principle that "the suffrage is only a means to good government".[64] For him the basis of politics was not about the representation of rights in an electoral system, but about the supposition that "the thought in the mind of each individual should be able to flow quietly, without discolouration, into the great central reservoir of public opinion".[65] Stating the subordinate nature of voting more plainly, one of the lecturers of the Complete Suffrage Union, Mr Clarke, assured the people of Taunton that the rationale for suffrage was to provide a barrier against the aggressions of the aristocracy. That is, when he demanded the vote for all males above the age of twenty-one, he invoked not their natural right to be represented, but the need to contain the power of the dominant class.[66] Spencer's lack of enthusiasm for democracy not only reflected the micro-context of the Complete Suffrage Union, but, more broadly, the shift of respectable Whig and radical opinion away from democracy by the mid-nineteenth century. In part, this movement had its origins in the disenchantment that British observers had experienced when they reflected on increasing corruption and disputes between political factions in the American republic. American expansionism into parts of Mexico, such as Texas, also offended those radicals who were anti-militarists.[67] Democracy was blamed for these phenomena and, as a consequence, it would have been anomalous to have advocated it in Spencer's circles.

Spencer's early politics reflected the nuances of the defence of civil liberty, not that of consent-based democracy. The niceties of a social contract theory, with its qualified obedience and its insistence on the sanctity of the original compact, did not appeal to a writer steeped in the vulgar discourse of trade and manufacture. Spencer's arguments were underpinned by a simple *quid pro quo*: if men paid their taxes, and submitted themselves to the legislature, then it remained with the legislature to dispense "the benefits of civil order".[68] The provision of order was the chief task of the state, and as soon as this institution returned to its "original" function then it could legitimately command total obedience from the citizenry. To put this proposition in terms of rights, once the citizens had submitted, they ceased to possess rights against the state. Spencer saw submission as an irreversible action: once it had happened the citizens lost their individuality, and could not be permitted to withdraw from their state membership.[69] Since exit from the state was barred, it was a matter of common decency to give justice to citizens. This was, after all, the chief benefit of living under a government. The political theory of submission, and the concomitant idea of involuntary state membership, was a fragment of conventional legal discourse of the first half of the nineteenth century. The only novelties in Spencer's reformulation of this was the entrepreneurial gloss he gave to its expression and the extremism of his handling of Christian utility so as to diminish the role of individual morality in political decision-making.

In "The Proper Sphere of Government", Spencer laid great weight on Paley's maxim "What is expedient is right".[70] Lest his readers take this utilitarian advice to be amoral, Spencer reminded them that expediency should not only refer to immediate effects, but to distant and indirect ones as well. A utilitarian outcome was when the sum total of good results exceeded bad ones in such a way as to include those evils and benefits that would arise in future ages.[71] In adopting this view of utility, Spencer's concern was to stand in-between Paley and Dymond. This stance seemed necessary because the latter believed that Christian utility had erected a principle, or an alternative standard, of right and wrong in opposition to the precepts of the gospel.[72] This was to accuse Paley of heterodoxy. In rebutting this charge Spencer saw his duty as reassuring his readers that they could be Christian and utilitarian at the same time.[73] His position was that moral principles that could be found in orthodox sources, especially the Decalogue, were corollaries of that great fundamental law on which the Christian commandments were founded: namely, "God wills the happiness of man".[74] To Spencer it followed from this that commandments, such as "Thou shalt not steal", were utilitarian because they were extensions of divinely instituted happiness. The injunction against theft was an ordinance signifying that thieves would experience only a temporary gratification from the possession of stolen property and that this would be outweighed by the subsequent increase in the amount of fear concerning similar losses that would be suffered by themselves and by other members of the community.[75] On this point, as with others in "The Proper Sphere of Government", Spencer's communitarianism blunted his individualism. The inclusion of a community's potential suffering in utilitarianism meant that a single individual could not easily employ it in calculating a moral solution. Individuals could not simply ask themselves whether a given political proposition would be likely to bring them pleasure or pain; they would be compelled to consider its impact on the community over an extended period of time. While happiness lay at the basis of Spencer's political speculations, it was not in the individualistic and hedonistic form that would lead to rational maximizing or the utilitarian efficiency employed by modern economists.

In any case, from Spencer's perspective it would have been a waste of effort for individuals to calculate how to maximize their happiness because he always favoured long-term effects over short-term ones, without giving any weight to factors such as the intensity of a pleasure. For example, he could not conceive how a benefit would ensue if the experience of a large amount of immediate happiness outweighed later and consequential pain. This is because Spencer had already seized hold of the idea that guided much of his later moralizing: that all animal creatures, including man, stood in such a relation to the world that their bodily needs had to be satisfied in a moderate rather than an excessive fashion. In addition, he believed that such gratification could not be postponed; bodily needs for food, shelter and sex (Spencer coyly referred to the last as "other natural desires") must be gratified immediately.[76] In pursuing the satisfaction of those needs each animal possessed appropriate organs and instincts (Spencer alternately referred to these respectively as "external apparatus and internal faculties"). Happiness, together with health, was bound up in the perfection and activity of these. If a particular organ or instinct was unnecessary to support an animal's response to its external environment then it would become impaired.[77] From this Spencer concluded that organs and instincts existed only so long as they were required for an animal's needs.[78] This, in turn, led him to believe that man, as an animal, was provided with moral and intellectual faculties that were commensurate with the complexity of his relationship with the external world. Pleasure, therefore, was dependent on whether human faculties were well adjusted to the activity levels.[79] This description of happiness put Spencer in opposition to the hedonism contained in Bentham's definition of utility because any attempt to maximize pleasure through increased intensity, repetition and propinquity could lead to a destabilization of the balance between the exterior world and a human being's organs or instincts.

From the perspective of later utilitarian philosophy Spencer's notion of happiness might seem idiosyncratic, but such a judgement would ignore the historical context in which Christian utilitarians were more influential than Benthamite ones.[80] Spencer's grasp of utility was quite conventional during the 1840s, even though the way in which he applied his theory to his own life often surprised contemporaries. That is, Spencer's beliefs and behaviours were sometimes seen as unusual because they were marked by repeated attempts to adjust his own behaviour and beliefs to the moral precept that happiness involved satisfying all one's organs and instincts in a moderate manner. This led to novelties such as Spencer's rejection of the prevailing Victorian work ethic and his eccentric and outspoken advocacy of recreation and play. His insistence that unhappiness was caused by an excessive focus on the desirability of labour while neglecting leisure could be disturbing to contemporaries. Similarly, Spencer's attempt to provide himself with substitute families and children was a consequence of his desire to exercise his unused parental instincts, a practice that caused some adverse comment in succeeding generations. When Spencer spoke of the need to exercise his "philoprogenitive instinct" on vicarious objects (i.e. he had borrowed little girls to whom he could vicariously give affection because he had no children of his own[81]), he was not merely making a joke: he was referring to his theory of happiness. Since he possessed a particular instinct, he believed that he should act as if his health and happiness required this to be exercised.

Spencer's *Social Statics* was the culminating document of a youthful political theory. Rather than being a model for his later ethical social science, it was the final flowering of 1840s radicalism. Spencer's solution to the problem of aristocratic dominance of parliament was extreme. He advocated the deposing of the elite that unjustly controlled politics, and that hindered social change. The aristocracy had used political power to impose class-biased legislation on the people and the same would happen if any other class acquired influence. Spencer, therefore, would do away with elites and subject society to immutable laws that worked without human direction. These rules could be discovered by the study of the psyche and of the evolution of society. In the future parliament would not instruct us; there would be no need when laws were laid down by nature, and could be enforced without alteration. "[W]e are to search out with a general humility the rules ordained for us – are to do unfalteringly, without speculating as to the consequences, whatsoever these require."[82] Spencer's natural rules projected from a natural theology that was still recognizably Christian. He was yet to break away from his orthodox moorings; the immediate political question that occupied his mind was how to replace the "love of God" with regulations that would be both just and acceptable. These had to be more creditable than the fear of law,[83] which was the secular remedy to the problem of disobedience that had been promoted by Bentham and his disciple John Austin. Spencer believed that proponents of what later came to be called "positive law" were mistaken in regarding civilization as artificial. It was part of nature: "all of a one with the development of the embryo or the unfolding of a flower".[84] The solution of positive law was artificial whereas Spencer wanted governance to grow organically from society. This hope meant that he defined civilization in naturalistic terms: he focused on the "seeds of civilization" that had been long in existence. While these could no longer be regarded as unfolding according to a regular plan, they should still be considered as the beginnings of "a development of man's latent capabilities under the action of favourable circumstances".[85]

Although the notion of development in *Social Statics* was based on an organic metaphor, its source was development psychology and morals rather than biology. Spencer believed that to fail in development was to lack goodness. Conversely, for him, evil resulted from the non-adoption of the constitution to conditions.[86] The metaphorical quality of Spencer's *Social Statics* is emphasized here in order to distinguish it from the detailed analogies that he later drew in his scientific writings between biological and social arguments. His early concept of a constitution was wider than politics. It encompassed the laws of psychology as well as the science of government. Both the mind and the state were entities that had evolved practices that fitted their environments. This idea led Spencer to condone practices that were commonly regarded as immoral. For example, he could argue that the advance of civilization excused the conquest of undeveloped societies because it was a natural process. In distant parts of the globe predatory instincts did not have to be set aside. Instead civilization would function:

> clearing the earth of inferior races of men. The forces which are working out the great scheme of perfect happiness, taking no account of incidental suffering, exterminate such sections of mankind as stand in their way, with the same sternness that they exterminate beasts of prey and herds of useless ruminants.[87]

The harshness of Spencer's sentiments in *Social Statics* was not a result of his nascent scientific interests, but of his belief that the workings of providence were beyond human understanding. Later, when Spencer's social and political discourse had become thoroughly secular and scientific, he reacted against conquest and domination as unnecessary forms of cruelty. However, this ethical sensitivity and humanity did not emerge until Spencer was able to discard the remnants of his religious orthodoxy. His process of secularization was a slow one; there was no conversion, no single moment at which Spencer ceased to be guided by Christian precepts. It was a gradual process, at the end of which providence ceased to excuse evil and cruelty.

While much of *Social Statics* was devoted to castigating the aristocracy and utilitarianism, it also explored Spencer's increasingly ambivalent feelings towards democracy. He was genuinely unable to decide whether or not the popularizing of politics would encourage social progress. In "The Proper Sphere of Government" he had expressed a limited enthusiasm for democracy, and he remained willing to refer to it as "the highest form that a government can assume – indicative, if not of the ultimate phase of civilization, still of the penultimate one".[88] However, he broke away from his initial and partial endorsement of popular government by issuing warnings against the "ultra-democrats" or exponents of pure democracy.[89] The values that he himself had propagated in the early 1840s now seemed dubious. While he could support the initiation of democratic government as a progressive step, he was unwilling to agree that, as a form, it could be justified by philosophical principles. Essentially his objection was to the democratic theorist's contention that a majority of the populace possessed a legitimizing power that they could depute to representatives:[90] according to Spencer this assumption was not based on a political logic, but on a notion that one could transfer sacredness from the spiritual realm to the natural one so that "the voice of the people is the voice of God".[91] To him this was a mere sleight of hand and, as such, jarred with his sense of religiosity. It seemed to him that democratic politicians were masking their prosaic class interests in a mysterious theological language. This offended Spencer, who was already spinning a metaphysical web that would lead Victorian spirituality on a journey into the unknown. Sovereignty – especially in the form of sovereignty of the people – would not be part of this spiritual dimension; it was simply a despotic claim by politicians who were unwilling to think carefully about the likely effects of democracy.

Spencer's preference for distant pleasures over immediate gratification was not solely a product of his personal brand of Christian utilitarianism. It also had roots in his adolescent and untutored reflections on the natural kingdom; he had been shaken by the way in which many aspects of life were accompanied by suffering and death.[92] For Spencer such miserable spectacles raised a problem of theodicy; if one believed in an all-powerful God then one could conclude that He was responsible for what was wrong in the world. While he did not explicitly imply that God was responsible for evil – that would have been a trifle too dramatic for Spencer's cast of mind – he did worry that the presence of so much pain and suffering in the universe could lead to the inference that God was unmerciful.[93] Such a thought was disturbing to the kind of Christian for whom Spencer was writing, so he hypothesized that the evils of nature were only apparent ones. In reality, he argued, they were the collateral result of laws that would eventually generate the greatest amount of health and happiness.[94] This was Spencer's juvenile position: later, in the 1860s, he would discard the ideal that ultimate happiness was obtainable for the organic world as a whole while retaining the belief that the human race was an exception that would ultimately achieve a happiness removed from suffering and death. This later position did not rest on orthodox theology; the philosophy of science developed by Spencer in "A System of Synthetic Philosophy" caused him to treat Life as a force in such a way that no longer echoed Christian conceptions of the Divine. Curiously, this also caused him, in his biology, to elevate the human species above other forms of life. Human beings appeared as alone in the universe – severed from the comforting belief that they were embedded in nature – but, somehow, there was still an optimistic and promissory outcome. Only the human species would move towards a painless future without death. The remainder of the living world would continue to suffer and die.

SEVENTEEN

Sociology as an ethical discipline

Spencer has been ill-served by ethical philosophers. Instead of pondering his serious evolutionary writings in sociology, psychology and biology they have mistakenly restricted themselves to examination of two texts: *Social Statics* and *The Principles of Ethics*. The first of these was published before Spencer developed his detailed evolutionary theories, while the second is partly a product of senescence, written when Spencer had lost sight of his own findings and was desperately attempting to complete "A System of Synthetic Philosophy" before his eightieth birthday. The valuable portion of *The Principles of Ethics* is derived from his sociology;[1] the remainder is a decrepit response to later Victorians, such as Henry Sidgwick, whose philosophies represented the kind of non-scientific scholasticism Spencer detested.[2] Political theorists have done no better than philosophers. They too concentrate on a narrow band of Spencerian literature, ignoring anything complex or systematic about his philosophy or social science. In particular, the excessive attention paid to Spencer's popular essays, *The Man "versus" the State*, has caused the more considered political ideas embedded in his sociology to be overlooked. The result has been the construction of a misleading image of an ageing Spencer, whose youthful radicalism had become buried under a conservative individualism based on an appeal for the preservation of private property. Such an interpretation has marginalized Spencer's political ideas because it allows them to be grouped with a kind of libertarianism that was rare in the late Victorian period. This, in turn, has led careful scholars of ideology to warn that Spencer's stance in the 1880s was untypical. The resulting impression is that by this stage he had stopped incorporating new political ideas into his work, and was content to recycle content from the middle of the century.

This is a debilitating way of interpreting a great liberal: it is more accurate to regard his later political ideas as freshly minted, and not to categorize them as conservative or individualistic. Spencer's later political philosophy is an original discourse, independent of the polemical anti-socialism in *The Man "versus" the State*. His reaction to ideas of economic equality was not the organizing theme of Spencer's political theory, as it would have been if he had been a Hayek or Ludwig von Mises, extending on Adam Smith's *Wealth of Nations*. Spencer's interest lay in sociology, not economic analysis, and he saw political structures as natural growths that could not be overlooked simply because an eighteenth-century economist believed that an "invisible hand" would be the most efficient regulator of the market. Spencer perceived governments, both past and present, as having a variety of valuable non-economic functions. His theory of political evolution was a demonstration that these functions went through a variety of organic developments. Had his perspective been exclusively scientific, Spencer's work would appear as clinical and dispassionate as Karl Marx's studies of the behaviour and ideals of political actors. However, for Spencer, such cold objectivity was implausible; his own attempt to be neutral never protected him from outrage and sorrow. He was unable to dispassionately describe his contemporaries as necessarily filling roles assigned to them by an evolutionary process. For him such a stance was an ethical failure and unobservant. The purpose of his neutrality was to allow him to witness evil without blanching, and to condone it when the eventual outcome was likely to be humane.

Spencer's progressive evolutionary theory has been widely misunderstood as a hard-hearted apology for business. Progressive and anti-capitalist writers perversely credit Spencer with being a prophet of the virtues of competition and industry. Even a normally scrupulous scholar such as Peter Bowler claims – against all evidence – that Spencer believed that *laissez-faire* capitalism was the highest form of society.[3] Similarly, Alain de Botton falsely asserts that, on a visit to America, Spencer was cheered by gatherings of businessmen after flattering them as "the alpha beasts of the human jungle".[4] The prosaic reality is that Spencer's sociology predicted a benevolent and non-competitive future for humanity. In America he addressed capitalists only once and caused much offence with his advice to avoid stressful overwork and seek relaxation. He offered no comfort to either the new capitalist elite or to those who wished to politically expropriate wealth. On the contrary, Spencer detested the prospect of socialist and communist futures: these visions fostered well-meaning, but faulty, doctrines that would allow the politics of industrial societies to fall into the hands of the militarists. The prospect of rulers like Czars and Bismarcks troubled Spencer because, when these captured industrial societies, progress would be arrested. In addition, socialists were not able to defend themselves against militarists. The former's weakness was more disturbing than their visions; without irony Spencer could regret the German Imperial banning of 224 socialist societies, and the suppression of 180 periodicals.[5] To him these repressive actions were similar to the English registration of teachers, inspection of science and art classes, and compulsory insurance.[6] All such regimented measures were to be avoided since social advance would only come if people were left free to choose which utopia, which teacher or which insurance scheme they preferred. Spencer saw the organization of choices by a militarizing society as retrograde; it was the perpetuation of a primitive form of social organization.

Such primitivism was a mistake because it relied on psychological motivations that had already exhausted their functions in the process of the evolution of political institutions. The recent re-emergence of militant societies particularly attracted Spencer's hostility because they served to increase the prevalence of naive notions of personal causation while, at the same time, inhibiting impersonal ones. To him, militant societies were concomitant with the idea that personal will should be determinant. This belief represented both an intellectual and an ethical mistake, because it entailed throwing aside a notion of modern moral causation on behalf of archaic beliefs of how the psyche behaved. Spencer maintained that citizens should by now have realized that functioning political processes were no longer subject to will. It did not matter whether it was seen as emanating from an individual, such as a dictator, or a group such as a democratic majority, the will could no longer effectively dominate others. If totalitarian rule were attempted, progress would be inhibited because there would be a neglect of the impersonal scientific data that ought to inform public decision-making. Totalitarianism would also cause modern people to conceptualize history as if it chiefly consisted of the doings of great men.[7] Spencer's ardent message to his contemporaries was that they should avoid reinflating the primitive belief in will when reflecting on politics. He was so insistent on this point that it is astonishing to note that he has been posthumously conscripted to fight under the libertarian banner in ideological battles. His vision of the end goal for political evolution was not a tableau depicting supreme egos battling for dominance: that scenario belonged to the anarchist world of political romantics, inhabited by the likes of Max Stirner. Spencer's futuristic people were not isolated giants; instead, they were socialized and well-adjusted people locked into ever-specializing developments in social and political structures. It was these structures that would make individual existences bearable.

For the humane Spencer, "primitive man" was a flawed sociological concept drawing on the ersatz militarism and patriotism of his European contemporaries. He preferred to conceptualize "early" cultures as those lacking a sophisticated technology. Earlier, when thinking more exclusively about psychology, he had held harsher views. At that stage he had regarded non-Europeans as savages missing the emotions that had developed as individual minds became better organized: the lowest savages were even without ideas of justice and mercy.[8] When he was writing *The Principles of Psychology* (that is, before 1872), he had not yet crafted the term "pre-social", which he later came to believe was free from the taint of biological realism.[9] Indigenous peoples were still primitives at the lower stages of mental development.[10] The stigma was that their thoughts and imaginations were without originality, while their laws and customs were merely copied from their ancestors.[11] Spencer's early account of the psychology of primitive man was environmentalist, in contrast to his hopes for modern society within which individuals should divest themselves of autochthonous mental habits and of accumulated customary knowledge. To his contemporaries he pleaded that "we" moderns were in many ways still like primitives, and if we wished to progress, it would be necessary to discard much of our mental furniture.[12] There was a biological and scientific facet to this prognosis because the idea of a primitive condition implied that some indigenous peoples had undeveloped faculties due to their limited experience.[13] Spencer did not intend this idea to reinforce the tutorial mission of empires; rather, it was aimed at making Europeans aware that they too were in danger of falling under the sway of ancestral wisdom because of their revival of militarism:

> [I]t cannot be denied that the feelings which impel and restrain men
> are still largely composed of elements like those operative on the savage
> – the dread, partly vague, partly specific, associated with the ideas of
> reprobation, human and divine, and the sense of satisfaction, partly
> vague, partly specific, associated with the idea of approbation, human
> and divine.[14]

It occurred to Spencer that many modern peoples possessed ethics solely because of their prudent fear of divine and human retribution. This seemed so obvious to him that it scarcely required evidence: Spencer could intuit such a truth because he and his family had been of this mind themselves. Spencer was able to remember when his own personal sense of right and wrong had been little more than respective subordination and insubordination to divine and human rulers. This was why he relied on the work of Dymond,[15] a minor Christian apologist who, by the late-nineteenth century, had been forgotten by the public. Great orthodox names, such as Jacques-Benigne Bossuet, Richard Hooker and John Locke, might have impressed his readers, but he was giving a personal testament, which required no evidence from cultural history. Spencer needed only to use his intuition: he could sense no significant moral difference between himself and savages, except when he attempted to exclude the will from his mind.

Spencer's internal scrutiny applied only to his psychological musings; sociology seemed to warrant a more objective examination. Social analysis also reinvigorated his evolutionary theories without undermining his moral stance. Spencer's psychologically driven scepticism over the moral standing of his own generation had been perilously close to denying the moral basis of evolution. This danger was avoided in his sociology, particularly in *Political Institutions*; there, a shift in his ideas allowed him to assert the moral superiority of Europeans, while, at the same time, protecting colonized peoples from harm.[16] This manoeuvre was accomplished by demonstrating that the colonized possessed the same virtues as their more developed masters. That is, by the time he came to craft his sociology, he had found a way of simultaneously defending his idea of progress and equating pre-social societies with social ones. Of course, Spencer still had doubts that progress would continue. He also remained despondent about the frequent lapses in civilization among advanced societies. Often he felt that "social" man had failed to demonstrate any emotional superiority over pre-social precursors.[17] At times, Spencer even believed that the reverse was true: that social members of the species were inferior, especially when the moral standing of modernity was imperilled by imperial aggression. His concern led him to attach a story about the collective punishment of two African towns to his preliminary remarks to *Political Institutions*:

> Anchoring the *Boadicea* and two gunboats off Kribby's Town ("King
> Jack's" residence), Commodore Richards demanded of the king that he
> should come on board and explain: promising him safety, and threatening
> serious consequences in case of refusal. Not trusting the promise, the king

failed to come. Without ascertaining from the natives whether they had any reason for laying hands on Mr. Govier, save the most improbable one alleged by our own people [the king was said to be disappointed at the offer of a sub-factory instead of a full trading factory], Commodore Richards proceeded, after some hours notice, to clear the beach with shells, to burn the town of 300 houses, to cut down the natives' crops, and to destroy their canoes; and then not satisfied with burning "King Jack's" town, went further south and burnt "King Long-Long's" town.[18]

Spencer borrowed his description of this incident from *The Times* (10 September 1880), which he described as the leading journal of those of the ruling classes who believed that there was no distinction between right and wrong except when they were instructed by established theology. The newspaper had irritated him by speculating that, aside from the loss of life, the matter had been humorous.[19] Readers of Joseph Conrad's *Heart of Darkness* will remember a similar gunboat to the *Boadicea*, although the novelist had his vessel carry a French flag rather than a British one. Like Conrad, Spencer was not simply protesting about empire, but against the horror made possible by a mixture of commerce and Christianity. Spencer was an emotional radical who detested the hypocrisy of empires, but his sociology made him observe carefully even while incensed. His ethical sense and his social science were symbiotic; without the latter the former would have been unbearably painful. Objectivity protected him from the pain of being overwhelmed by passionate disgust for his contemporaries.

The ethical dimension of Spencer's symbiosis with science was also vital. It gave substance to scientific observations that would have otherwise been indistinguishable from transitory polemical responses. It is the ethical, not the objective, part of Spencer's sociology that gives it a lasting interest. His values were particularly prominent in *Political Institutions*, the most theoretically inspired portion of his later political writings. Here he attempted a fundamental moral analysis of organizations and behaviour. Whereas *The Man "versus" the State* was a diatribe against the intellectual dominance of socialist thought during the early 1880s, his statements in *Political Institutions* were scientific observations on the past and future of politics. This was why he regarded the book as the most arduous part of his philosophy.[20] Emphatically, this exercise was not value-free; Spencer's sense of objectivity was bound up with his normative values. Both were necessary in order to theorize about social evolution. Only a strong moral sense could provide him with the calmness and neutrality that would shelter him from the suffering he experienced while analysing social development. Moralism gave an explanatory vigour to Spencer's science of society. Unlike Germanic sociologists, who reconciled polar opposites – the communal spirit and the individual spirit – without reference to individual morals, Spencer placed his great sociological antithesis (that between industrial and military societies) in the ethical domain. If his analysis had been accompanied by a clinical coolness this would not have resolved his dilemma because, for many of his contemporaries, science was not politically neutral but a handmaiden of imperial dominion.[21] If a sense of objectivity was compatible with a contemptible activity such as the military conquest of indigenous peoples, then, Spencer believed, it might be regressive: part of a temporary re-emergence of barbaric behaviour. This seemed contradictory to him because he felt that science should be progressive in its outcome. It had, after all, furnished the knowledge that Spencer's *The Principles of Biology* had earlier hailed as the sentient edge of progress itself. That view of knowledge was, however, the long-term hope for the human future. The immediate effect of science on social behaviour could only be interpreted ambiguously, which led Spencer *deliberately* to adopt a normative stance while thinking empirically about human beings. This procedure required some subtlety if it was to be effective; a simple moral plea from Spencer would have been disregarded. His evolutionary sociology was a weapon against empire, but it needed to be tempered so as to conceal its polemical edge. For this reason Spencer stipulated that imperial dominion had indeed brought potential benefits to humanity,[22] while at the same time insisting that such gains could only be relative. As this suggests, Spencer's notion of good was not restricted to the benefits gained by specific nations, but was always measured against the eventual progress of the human species as a whole. His progressive morals made him sceptical about all absolute ethical statements because he came to think that non-relative moral opinions were so ill informed as to be reprehensible.

Spencer's values on this subject were primarily psychological. He evaluated moral outcomes by examining the motivations and indoctrination of the political actors rather than through looking at the social consequences. It was not that his sociology lacked an empirical foundation; rather, it was a science conducted by an investigator whose primary focus was on mental data even when he was handling historical or ethnographical sources. For him, the inherent danger facing the social scientist was the susceptibility of being swayed by popular bias. Thus Spencer suspected that those of his contemporaries who composed the "Jingo" class[23] had been so warped by patriotism and faulty education as to be unable to exclude emotions when contemplating the behaviour of so-called "primitive" peoples. Instead of surveying indigenes in a passionless manner like himself, his jingoistic countrymen had allowed their judgements to be clouded with contempt, disgust and indignation.[24] Their excessively classical education caused them to identify with the Greeks and Romans, seeing their own imperial subjects as analogous to conquered barbarians of the ancient world. To Spencer this was an anachronistic metaphor fostering harmful social attitudes. His objection was not to the possibility that inter-social conflicts had been beneficial in the past,[25] but to their mimicry in modern times. It was too easy to support war on the grounds that a distant country was in the clutches of a defective government that failed to guarantee liberty for its subjects. In any case, there was something to be said on behalf of the relative value of oriental despotisms. The states of the East furnished Spencer with sufficient proof that the form of freedom and its reality were not necessarily commensurate.[26] His admiration of selected non-constitutional governments resembles David Hume's apology for those monarchies that had produced civility without liberty. However, Spencer's intentions were quite different from those of a comfortable enlightenment figure such as Hume.

The latter was politely reminding his readers to shun vulgar prejudices, such as those that deemed the lives of Frenchmen as worthless as those of Turks because lives and property were unprotected by a parliament. Spencer's concern was quite different; he was struggling against a kind of sentimentalism that had not existed during the mid-eighteenth century. During the *ancien régime* philosophers were not disgusted by the customs and laws of indigenous peoples, and they would have been highly sceptical about imperial missions to tutor or protect. However, in Spencer's era doubts about the superiority of European virtues had waned. In their place humanitarian feelings had begun to encourage imperial intervention on the grounds that it elevated the level of culture in backward ethnic groups. While Spencer shared the impulse to spread civilization he insisted that it should not be allowed to extend political dominion. Although, in his sociological theorizing, he felt repugnance when gazing on horrors of oriental despotism, he struggled to subdue this failing. Even memories of torture must not "shut out from our minds the evidence that abject submission of the weak to the strong, however unscrupulously enforced, has in some times and places been necessary".[27]

The humanitarian justification of empire was a serious question for intellectuals of Spencer's generation. If the extension of civilization was justified because it spread progress, then one could ask how much force should be condoned. Spencer's answer to this query was extreme: there was *no* justification for the coercive spread of political power. To him this restraint was specifically modern; progress of civilization had been justified in the past only when it needed sacrifices from pre-social people. However, such an admission was troublesome; Spencer's mind almost overturned while contemplating the miserable lives of the hundreds of thousands of slaves who had been necessary to build the great pyramids or the 300,000 serfs who were said to have perished while building St Petersburg. Still, despite this horror, Spencer felt one should remain passionless: even though the imposition of such servitude seemed to call for condemnation. Spencer believed such a response would be erroneous. Instead, he recommended "an almost passionless consciousness" when contemplating the effects of such tyrannies; this would make them appear as natural phenomena.[28] With such a frame of mind Spencer could maintain his mental balance and conclude that a relative good had flowed from an evil practice. Nonetheless, such relativism always had temporal boundaries: what might have been good about the action of a pharaoh, or a Peter the Great, would not be so if now promoted among a civilized people. Modern civilization should not impose its tyranny, or its institutions, on a "pre-social" people precisely because any putative benefits from such sacrifice had already been achieved by humanity. Further imperial cruelties could not be expected to increase the advance of civilization as a whole.

Spencer's insistence that the relative benefits of imperial dominion applied only to historical conquest goes to the heart of his evolutionary social philosophy. The emphasis he placed on this idea also caused his view of modern society to be disjunctive. He believed that while Europeans had developed a sophisticated code of private ethics, their code of public conduct had lagged behind, remaining comparatively primitive. This uneven progress had been misunderstood by theorists of the state who either subordinated private codes to public ones or else equated them. To Spencer such a conflation was permissible only when considering earlier periods of history during which a citizen's duties had always had a public character because they had been evaluated solely by their impact on tribal warfare.[29] Spencer was particularly critical of German philosophers for construing the Greek idea of freedom as an absolute, instead of a relative, value. This, he believed, had caused them to reverse historical processes and interpolate a modern public–private distinction into ancient Athens. Spencer brushed this aside: instead of proceeding like Hegel and suggesting that private ideals had been subsumed by public ones, he objected that giving public ideas a dominant place in modern society was abandoning the only moral domain – the private one – in which humanity had yet made progress. Spencer's argument was radical for a period in which state theory was usually inclusive of ethics. Novelty was not his only explanatory difficulty; he made his task harder by arguing that evolutionary ethical gain was not the cause of social change, but the consequence. That is, he claimed that modern private morality was the surface, not the substratum, of ethics generally. Civilization, the fruit of the evolutionary process, had been a multi-caused phenomenon and had emerged without an intellectual guide. This achievement could not be used to give direction in the future; nor could private values be a guide for public reform. Civilized and progressive qualities were chiefly personal matters, and could not be forcibly imposed on less developed ethnic groups, especially as individual members of such groups would often have better moral characters than their civilized mentors. Spencer's favourite moral exemplars were the various hill tribes of India, who, he believed, possessed more in the way of simple virtues, such as truthfulness and peacefulness, than either more advanced Indians or the British.[30]

In essence, Spencer's argument was that if, in the process of maintaining an empire, it was proposed to transplant political institutions to a less developed society, one should be aware that such an action would not promote ethical outcomes. Among "primitives" or their pre-social peoples, private and public virtues were the same, and any tampering with public virtues would only erode the private ones. In addition to moral vacuity, modern empires could not truthfully claim a scientific sanction for their dominion because, Spencer believed, the inter-social struggle for existence was no longer the cause of progress that it had been.[31] While it was true that in the past the human species had once progressed because of war, the improved psychological faculties that had been acquired were now available to be employed in less brutal activities.[32] The engine of social change had been nature "red in tooth and claw",[33] but social momentum now continued without this drive. Like a scholar of modern international relations, Spencer observed that frequent or endemic hostilities between small ethnic groups had been replaced by occasional vast battles between modern nations. He believed that these conflicts had become unnecessary; human capacities and technology could now be employed for non-violent ends. This suggestion raised empirical questions about the causes of such changes in behaviour and about the specific nature of new social goals, but Spencer did not provide answers to these as they were not intentionally parts of his enquiry. His claim about the conflict-ridden human past had been a logical implication of his ethical theory. It was dependent on his belief that when we were rationally guided by normative considerations we would never be sufficiently cruel to make progress. This consideration led to a historical gap in morals. Among "pre-social" human beings and then, much later, among advanced ones, moral behaviour was the norm, but between these epochs developing societies would exist that necessarily practised competitive behaviour because without organized violence, "the world would still have been inhabited only by men of feeble types, sheltering in caves and living on wild food".[34] This rustic image was, of course, deliberately borrowed from Socrates. The classics were not allowed to retain their exclusive grasp on metaphor. Spencer constructed *The Principles of Sociology* to demonstrate that his evolutionary approach successfully dealt with the same fundamental questions that traditional philosophers such as Plato and Hobbes had posed, but failed to answer.[35]

Spencer's claim – that violence had caused biological and economic progress, but no longer has this effect – is curious. Why should a universal phenomenon have become inoperative by the early 1880s just when he was thinking about political institutions? His anodyne answer was that the predatory aspect of social organizations had begun to die out as civilization no longer needed it.[36] As a piece of biological reasoning this is unsatisfactory, and resembles the kind of Lamarckianism he had avoided in *The Principles of Biology*. However, Spencer had a more general logic as well, to the effect that such questions need not be answered by empirical evidence, because, properly speaking, enquiries about new goals in human progress did not require answers based on data because they were primarily contemplative. The function of these queries was "to preserve that calmness which is needful for scientific observation",[37] without the observer losing the power of moral feeling. Spencer argued that observers in the modern world needed to retain senses of approbation and disapprobation when they were bombarded by empirical findings that were morally corrosive or counter-intuitive. His purpose here was to insulate the observer, who was the neutral recording instrument of human progress, not to produce an objective scientific method. His psychological strategy would keep ethics in play; evolutionary social philosophy provided a space between scientific observation and ethics. This was a neutral vantage point for the consciousness when contemplating the horrors of both the natural and historical worlds.

Spencer's *The Principles of Sociology* was a piece of political advocacy that carried a normative science. His criticism of empire made evolutionary science less available as a justification for European dominion over others. Spencer's psychology had been useful in providing such justifications, although this had never been his intention; he had been consistently opposed to European expansionism as early as the 1840s, and the occasional racist use of the second edition of *The Principles of Psychology* by others did not signal a temporary sympathy for empire, but was caused by Spencer's carelessness as to how his arguments could be used. While he had frequently praised the characters of aboriginal peoples as better than their civilized overlords, he still casually denigrated them as "rude peoples" who were not equal to the best individuals from civilized nations.[38] Later he excised such bias: in *The Principles of Sociology*, "rude peoples", "primitives" and "savages" were replaced by "pre-social". He also claimed that "pre-social" people possessed the greatest moral advantage when they were not debased by contact with whites. As his most admirable indigenes Spencer favoured hill tribes of India, whom he repeatedly described to be honest, sociable, dutiful, patient, kind and peaceful.[39] Their display of virtues in the face of encroachment by modern Indian peoples and British officials seems to have haunted Spencer. These people retained authenticity because they had escaped the ruin caused by the social stratification and gender divisions of the kind that had corrupted the surrounding militant cultures. The hill tribes were special because their customs prefigured humanity's future: they possessed the right indicators in awarding high status to children and in treating boys and girls equally.[40] Their "pre-social" respect for children and their disregard of gender were identical to the prospects for future humanity that Spencer had propagated in *Education* and in private letters. To him these indigenes demonstrated the qualities that needed to be restored to the human condition.

Desirable moral qualities were not universal. "Pre-social" did not equate to benign morals. Nor was social progress always accompanied by an increase in ethical behaviour. Fiji and Dahomey, in particular, presented Spencer with an analytical challenge. Their inhabitants seemed undeveloped because of their ferocity yet, at the same time, they enjoyed advanced political and social organizations. Since he usually thought of peacefulness as a consequence of life in a highly organized society such examples struck him as anomalous. For instance, Fiji was relatively evolved because it had a complex political system, together with well-organized military forces, elaborate fortifications and a developed agriculture. However, Spencer could not regard Fijians as advanced in their ideas, sentiments and customs because of their practice of regarding treachery, theft and murder as honourable.[41] Spencer's voice here was part of the chorus of Victorian ethnography; before the work of modern anthropologists, such as Ruth Benedict and Franz Boas, it was common to separate aspects of indigenous cultures as if these were located on different levels of sophistication rather than to treat them as parts of an integrated whole. However, a Victorian analysis was not always a deprecating one; Spencer's separation of discrete aspects of a culture did not imply that he despised the fragmented parts. On the contrary, "startling as the truth seems, it is yet a truth to be recognised, that increase of humanity does not go on *pari passu* with civilization".[42] Spencer's contradictory image of Fijians meant that they could be forgiven the more deplorable aspects of their culture because these were features of the aggression common during earlier states of civilization. Since he believed that the possession of humanity was a relative characteristic, its absence in a pre-social group should not be condemned too strongly. However, no similar apology should be permitted for more highly developed English and French civilizations. When these failed to display virtue, while at the same time, indulging in imperial theft and murder, their lack of humanity was culpable. To Spencer any current justification of aggressive empire-building was implausible: Europeans were engaged in mawkish and shabby campaigns that could have been legitimized only in past eras. English predatory behaviour was particularly repellent because of the popular support it had received. This was caused by the dominance – in the parliament and the press – of men whose education was so exclusively classical that they believed Achilles to be one of their contemporaries.[43] Spencer saw an institutional failure here; instead of having the organizations and characteristics proper to an industrial society, Britain had adopted those that had been appropriate for an archaic militant era.

Aping the past was not the only way in which imperialism defended itself. Another justification came in the form of an argument that Europeans were morally obliged to impose their political institutions on backward peoples so as to improve their moral characters. This seemed specious to Spencer because, while he agreed that moral and institutional systems were generally progressive, he doubted that one could specify the nature of the relationship between the systems. Further, he felt that, even if one could clarify the relationship between institutions and morals, this would not justify Europeans placing others in bondage in exchange for tuition. Spencer's objection here was directed against the ideas his friend Lubbock had popularized. Lubbock's *The Origin of Civilization* insisted that savages were utter barbarians who needed education and had to be rescued from their own oppressive customs and laws. It was mistaken, he claimed, to admire savage morals or customs because, like primitive artefacts and machines, their lack of development made them imperfect. Spencer, with an eye on humanity's psychological development rather its technology, was more ambivalent about the uniformity of civilization. If an analysis of progress considered only material culture, then Lubbock would have been right in his objective analysis but, for Spencer, human evolution was partially a matter of moral and psychological growth. He considered development subjectively, believing it to be akin to organic, rather than mechanical, change. Perhaps, Spencer suggested, the modern socialized individual was not emotionally superior to the pre-social one. If that were the case, there was no more reason for believing in European superiority than for accepting the African view that the devil was white.[44]

Spencer's critique of empire and his belief in the moral standing of aboriginal peoples affected how he structured his evolutionary sociology. Assigning a high ethical status to aboriginal people meant that he could not imitate those Victorian anthropologists who redeployed the eighteenth-century notion of stages as a historical template for human progress. If they were not depicted as arranged in stages, indigenous peoples could not be conceptualized as "primitives" who, with tuition, would pass through a barbaric phase before reaching an advanced civilized state. Spencer's refusal to accept commonplace classifications affected his sociology as much as it had his biology. He never seriously defined a group's level of civilization. The level or stage of a culture is not the key to Spencer's sociology any more than palaeontological periodization had been significant to his description of the natural world.[45] This, however, does not mean that Spencer's theory of civilization was wholly prescriptive and without a measured analysis of historical processes. Temporal developments were significant to him, and he was particularly keen to observe that the route towards progress had been through militarized social forms, and that this had repressed virtues. Such experiences should not be replicated because modern society was industrial rather than military even though its individual components possessed characters that resembled those of pre-military people.

Spencer's moralism had come close to undermining his idea of progress: the moral equality between modern and indigenous people had risked derailing his evolutionary theory. These problems emerged in two seemingly contradictory arguments. First, he had repeatedly argued that "pre-social" peoples, some of which were still extant, possessed similar virtues to those found among modern social peoples. Secondly, he also believed that the possession of private morals, such as honesty, kindness and peacefulness, were signs that civilization had been achieved. Taken together, his two positions implied that morals were constant rather than progressive. This, of course, conflicted with one of his evolutionary hypotheses that civilization was possible only for modern people because their ancestors had inherited the effects of being harshly disciplined in the past, and were, subsequently, free from governance by the individual will. He had weighed the freedom in Lubbock's dictum "No savage is free": "All over the world his daily life is regulated by a complicated and apparently most inconvenient set of customs (as forcible as laws), of quaint prohibitions and privileges".[46] Spencer then subtracted it from its amalgam of harsh laws and customs, and concluded that the modern concept of freedom did not imply that a people possessed an increased amount of freedom, or that new virtues had evolved. Instead, freedom was the unhindered enjoyment of simple moral qualities, such as kindness and peacefulness, which had persisted intact from pre-social times. These virtues had only recently resurfaced after being subdued by coercive power during the militant period. The re-emergence of moral qualities would lead to a general improvement in the lives of inhabitants of modern societies. They would be protected by public agencies from aggression by their fellow citizens. In other words, if people displayed virtues, they would no longer be subject to repression, and would require less self-protection. The ruling power and its agents would then administer justice.[47] Spencer's reliance on state administration of justice should alert his readers that his views on the individual were not libertarian. His evolutionary theory posited that the civilized had lost the ability to protect themselves, and were no longer able to administer justice directly.

Spencer's *The Principles of Sociology* and his *The Data of Ethics* are equally enquiries into the moral qualities of social progress, but have different content. The former contains original insights into public and private mores whereas the latter is an academic critique of contemporary morality. Spencer had two targets: absolute ethical principles and the utilitarian calculus. The first was propagated by evangelical Christians, who believed that each person possessed a personal saviour. The second was a secular principle that speculated that a general benefit would result from individuals exercising their faculties in choosing between competing pleasures. To Spencer both alternatives ignored collective lessons that humanity had institutionalized about virtues and laws, and both mistakenly credited individualization as causing ethical progress. Spencer was sceptical of the idea that the culmination of an individual's choices either to adopt an article of faith or to seek pleasure would a have long-lasting effect. In addition, it conflicted with his analysis that public morality had lagged behind private ethics where individual choice was beginning to be important. Both orthodox Christianity and Benthamite utility failed to distinguish between the different kinds of morality and ignored the effect of evolution.

Christianity did not offer much of a challenge to Spencer; he assumed that, by the 1870s, its power had waned and it no longer merited a response. He found the revival of Benthamite utility more discomforting. His reactions to this doctrine were complex. To an extent he was pleased by its reverberations of his own increasingly brusque dismissal of religion, but its re-inflation of the introspective method of classical philosophy offended him because it paid little attention to the new empirical findings of psychology and science. Then, too, he would condone the hedonistic "pig-philosophy" of Bentham if his choice were restricted to either that or the fashionable worship of power and great men.[48] Hedonism alone did not offend Spencer; as he humorously reminded his readers, there were times when our "lower feelings" had to be treated as superior. The examples that occurred to him were standing naked in a sandstorm, spending a week without food or holding one's head under water for ten minutes.[49] In a more serious vein, he agreed with the hedonistic underpinnings of utility because these coincided with his homily that momentary pleasures should not necessarily be sacrificed for future gains.[50] It was one of his complaints about organized religion that it psychologically disfigured its followers by forcing them to live contrary to their desires. Spencer also shared the utilitarian belief that egoism could be valuable. While this was often intended to bolster individualism; he considered that egoistic self-regard more often brought with it a social benefit, because when they displayed a happy self-contentment people often added more to the pleasures of others than they would in providing them with a material benefit.[51]

However, beyond affirming the positive benefits of pleasure and ego Spencer would not seriously endorse any part of utility. On the contrary, he felt the utilitarian attempt to erect a moral philosophy was profoundly mistaken. Selfishness should not be admired:

> To see that those who care nothing about the feelings of other beings are, by implication, shut out from a wide range of aesthetic pleasure, it needs but to ask whether those who delight in dog-fights may be expected to appreciate Beethoven's *Adelaida*, or whether Tennyson's *In Memorium* would greatly move a gang of convicts.[52]

If there was no quality control over the pleasures being experienced, then they were unlikely to have much effect on the direction of human evolution. Spencer had two arguments against utilitarians. First, in response to the idea that satisfying the greatest number of individuals would benefit society as a whole, he simply denied it to be true. He argued that the unintended benefits that resulted from pleasures were so enormous that the measurement of individual satisfaction was not a predictor of social progress. For example, the gratification people received from being married, or from being a parent, was only a small portion of the overall social benefit from such activity.[53] Secondly, since Spencer was struggling to find an index with which to measure civilization, utilitarianism appeared to be yet another of those spurious philosophical lessons gleaned from classical texts that had, so often, irritated him. In his view, only modern science should be seen as propelling the growth of civilization. Philosophy lacked predictive power: further, its basis was merely accumulated observation.[54] If it were accepted as a guide it would encourage the adoption of the conservative belief that welfare would be improved by replication of past results. Philosophy also failed to provide a sense of social change. This seemed nihilistic to Spencer because he placed great faith in the hope that evolutionary knowledge would provide direction in the future.

Utility was inadequate as a social guide because pleasures were not constant. During the course of evolution, some sensations had become greatly "deranged". Spencer believed that, in many cases, pleasures and pains no longer closely corresponded with the social actions that must be performed.[55] This was a sobering reflection that suffering now accompanied civic life, and it was underpinned by his disinclination to accept pleasure as a biological constant. This led him to rebuke one of his followers, Alfred Barratt, for imagining that happiness existed in lower animals, and disregarding the fact that such a sensation was an evolved one.[56] In saying this Spencer distanced himself from the English tradition of natural theology, which had discovered that pleasure was abundant in the universe, especially in its living parts. Where Paley had wondered at what pleasure was felt by an oyster, Spencer simply categorized its sensations as lower than that of a cuttlefish, whose vital activity was greater.[57] Animals such as oysters experienced too little activity for their sensations to be classed as pleasurable. There was an ascent of pleasures in both the evolution of animals and in the development of human societies. As the latter progressed they could no longer be pleased by aggression and crude egoistical display, but, increasingly, would find satisfaction in place and cooperative activity. Eventually, this development would lead to the establishment of morality. "This has, for its subject matter, that form which universal conduct assumes during the last stages of its evolution."[58]

Unfortunately for this hypothesis, humanity was a long distance from its final form. For Spencer, this gap led to an immediate problem of how one was to deal with cultural crimes, particularly those Europeans committed against others who lacked the military capacity to defend themselves. Spencer's quandary was common; he was not unique in musing on the accusations Africans would eventually bring against the white man. Other Europeans also found it obvious that the latter were the true savages. What was peculiar to Spencer was the refusal to take refuge in an historical relativity that would make the crimes of the civilized incommensurable with those of the barbarian. This comfortable acceptance of a non-moral relativism can be found in the sentiments of Max Müller when he speculated on the lack of virtue among European colonists. His audience was not saddened because he had insulated them from moral discomfort. It was Müller's opinion that no notion of morality should equate suffering in one culture with that in another. "The truth is that the morality of the negro and the white man cannot be compared, because their views of life are totally different."[59] Ethics, he proposed, had evolved in such a way that vices and values had no constant value over time: from this one should conclude that savage perceptions were unlikely to contain any normative qualities with which modern peoples could identify. It was this commonplace comfort that Spencer set out to overthrow.

For Victorians the thought that conflicting values existed in different ethnic groups was vexatious. Their minds were packed with ethnographic findings – such as the idea of Turcoman honouring the tombs of dead robbers[60] – which seemed to prove that European moral injunctions, such as that against theft, lacked universality. Then, too, Spencer worried that his contemporaries had been hoodwinked by theories that postulated that the behaviours of both savage and civilized were controlled by atavistic urges and the love of despotic power.[61] Spencer felt that such ideas were morally vacuous and counter-intuitive. It seemed obvious that there were some viable moral statements – he referred to these as truisms rather than principles[62] – that suggest that certain courses of action were evil or examples of wrongdoing. Under these would fall injunctions against fraud, delayed payment of debt, sale of adulterated food to workers, class bias in legislation, bribery of judges, perjury and slander.[63] These values were not universal: Spencer knew that it was possible to find ethnographic examples that countered one or more of such truisms. He had encountered societies that institutionalized fraud, class bias and so on. However, this did not faze him because he could assume that such repellent customs would vanish during the course of human development. Any groups that engaged in such practices would be too dysfunctional to continue long in existence.

Progress was not driven by competitive struggle, but by the planning and mode of behaviour of civilized people increasingly moving into correspondence with each other. Such "coherence" assists people in "making a fortune, founding a family, or gaining a seat in parliament".[64] It also distinguishes between moral and immoral conduct. Moral conduct was self-restrained and regular while immoral behaviour was disorderly in that if the actor could perceive little or no connection between behaviour and outcome, "[t]he sequences of motions are doubtful. He may pay the money or he may not; he may keep his appointment or he may fail; he may tell the truth or he may lie".[65] By analogy with personal behaviour Spencer believed that social obligations developed in proportion to the accuracy with which citizens fulfilled them. Such obligations would appear as citizens acquired duties such as caring for the infirm or participating in political agitation.[66] As the complexity of the action increased and became more coherent with foreseen outcomes there would be moral progress and people would cease to be slaves to desire. This was the opposite of the utilitarian theory, which stressed greatest happiness of the greatest number; Spencerian development would be caused by increased self-restraint, which would severely restrict the kinds of pleasure to be enjoyed. Sensual pleasures, in particular, would diminish with progress. This forecast altered moral judgements; it caused Spencer, for example, to regard theft as primitive: it was an immoral activity directed towards simple and immediate pleasure rather than remote satisfaction.[67] Spencer did not consign such matters to *laissez-faire*: political rules were required because he believed that few dangers were more threatening than the decay of a regulative system before a better one had grown up to replace it.[68]

Spencer's distaste for absolute moral principles[69] still left him with a way of acknowledging the existence of an ethics that crossed cultural boundaries. Essentially his position was that while good behaviour was not unique to a developed society it was certainly more characteristic of it.[70] The moral clashes that his contemporaries saw between cultures were best explained as conflicts within developing industrial-military societies, which resulted in them practising simultaneously two competing modes of conduct. The militant mode would favour external defence and aggressive behaviour, while the industrial mode would sponsor internal cooperation and the maintenance of life. Both fostered a morality that, subsequently, appeared perplexing and inconsistent. However, from Spencer's perspective they were equally justified because they had worked in tandem to further social progress.[71] This optimizing effect would not continue into the future because evolution would cause the militant form to disappear, and the industrial form to be transcended.[72]

Spencer's ethical sociology was fuelled by pleasures that were directed towards improvement of humanity. Emphatically for Spencer, pleasure did not refer to all positive sensations. While Bentham had imagined it to be constant quality of life, pleasure too had evolved. To think otherwise was to assume unscientifically that sensations were intrinsically valuable. From the perspective of the evolving social organism it was not causally plausible to believe that certain types of action were necessarily pleasurable because they provided an immediate pleasing sensation while avoiding painful ones. The reverse was the case. When the human species achieved the ultimate social form, Spencer proclaimed, "we shall see that the re-moulding of human nature into fitness for the requirements of social life must eventually make all needful actions pleasurable, while it makes displeasurable all activity at variance with this requirement".[73] This progressive method of contrasting morals gave him hope that benign social and political institutions such as the family and the state would be reinforced in the future.[74] This prospect did not provide a guiding principle for a given public policy or course of action at present,[75] but it would allow people to formulate a framework for normative conduct in tomorrow's ideal society.[76] Retrodiction from this would permit the designation of certain kinds of current conduct as pathological and to be avoided. While Spencer's reasoning here often failed to answer enquiries about the rightness of particular actions,[77] it did offer reassurance that, in general, civilized conduct was morally justified.

EIGHTEEN

Sociology as political theory

In Spencer's sociology the presence of state-led justice as a goal for human evolution is significant in two respects. First, it reinforces the teleology that came to dominate his evolutionary thought, once he had forgotten the warnings issued against this in *The Principles of Biology*. Secondly, it is a recurrent feature of his political theory that long pre-dates the whole of "A System of Synthetic Philosophy". From the 1840s he had consistently upheld the primacy of institutional justice. Its continued presence in Spencer's political theory is yet another reminder that his individualism was not libertarian. The organizations that would protect individuals in the future would not wither away: the movement towards social justice was irreversible. People had permanently lost the right to protect themselves. Spencer found the notion that modern individuals should resume a right to self-protection as absurd as expecting the state to undertake commercial business as if that were its core activity. Spencer's views are sensible if it is acknowledged that his restrictions on state activity do not develop from a public–private distinction. Such an argument might, as with J. S. Mill, divide human activities between those that intimately invoke self-knowledge and those that imply external or objective knowledge. From Spencer's perspective this bifurcation was based on a faulty description of society. It was not the method of acquiring knowledge that was important, but the study of which human characteristics had already developed and which ones could be retained. He believed that intimate or private choices were more likely to evolve quickly than those formal rights and duties that were embedded in institutional arrangements. This was a sociological, not a philosophical, problem. Changes in one's sense of intimacy or privacy were no more or less objective than those that affected political institutions. The danger in the future was not our ability to control our choices; it was, rather, that the slowness of political evolution was likely to cause reactionary behaviour in archaically educated elites, who would then encourage the very militarism that was endangering the new-found civility of our private lives.

In the pre-social period, before industrialization, cooperation had been facilitated by coercion and military rule.[1] Spencer's views on this were extreme; he saw force as the sole motor power for change.[2] Later, however, when industrial society had emerged, a continued struggle for survival had seemingly become unnecessary. Spencer perceived the shift in human behaviour as having an intrinsic rather than extrinsic cause. His views parallel those of Wallace, who expressed scepticism about the ability of Darwinism to explain the growth of human intelligence by reference to environmental causes.[3] Wallace's doubts appeared in 1869, just before Spencer reissued *The Principles of Psychology* as part of his system. Wallace's views represented a break with Spencer as much as with Darwinism because he began to postulate that this change implied the existence of an "Overruling Intelligence".[4] This was tantamount to reinstituting natural religion, and very distant from Spencer, whose scientific and psychological reasoning had never invoked the image of the Unknown as a divine intelligence. For him the evolution of human psyche had taken place without godly intervention and was simply in response to internal and non-environmental pressures for change. Spencer argued that the "predatory period" had to come to an end because, in the process of evolution, competition had produced individuals who were adapted to the industrial condition, and would, consequently, gain advantages by cooperating and interacting peaceably rather than through violent and competitive behaviour.

The shift from military to industrial society saw a metamorphosis in social conditions. The difference between the two seemed so fundamental to Spencer that a failure to observe it demonstrated an absence of scientific acumen. The fact that it was missing from Comte's sociology signified that the Frenchman was an agent of the French state, rather than a serious scientist.[5] Spencer's social dichotomy, the militant and the industrial, stemmed from his theory of the development of the social psyche at the dawn of the human era. He believed that human beings had been predatory and gregarious. The progress of the species was dependent on both qualities: the predatory or destructive attempts at self-preservation worked in tandem with those cooperative activities that were necessary to maintain domestic relations.[6] Initially, in *The Principles of Psychology*, Spencer posited that these two tendencies had operated together, stabilizing each other and maintaining a mutually beneficial equilibrium, because he suspected that, if either had been pushed too far, society would have shifted and generated a preponderance either of anti-social individuals, or of people who were incapable of resisting aggression.[7] Later, when he wrote his sociology, he jettisoned the idea that in the human condition competition and cooperation were inherently balanced. Instead, he decided that the two qualities were antagonistic tendencies, belonging to different periods of social development.

Spencer's sociology was more consciously driven by political considerations than his psychology had been. There was an intellectual shift that followed the changing international climate. By the late 1870s Britain had been drawn into a race for military supremacy and territorial expansion. In opposing this Spencer attempted to present evolutionary science in a form that could not be co-opted by militarists. Where they had argued that evolutionary theory was prescriptive, he attempted to promote moral neutrality by limiting his commentary to descriptive historical processes. It was a political imperative for him to separate the past from the present. His argument was that while force had led to civilization, this condition could now be peacefully perpetuated. To explain this social transformation Spencer hypothesized that there had been a prehistoric shift in consciousness, which had recognized that the seizure of another's property caused conflict and consequential injuries. Such a development, he believed, would have established the non-predatory custom of assigning whatever is gained through individual work to the labourer.[8]

Spencer rejected social Darwinism as a justification of the need for aggression in *future* social change. However, despite this, his theory of evolution has often been confused with Darwinism, especially by popular scientific writers who wish to protect Darwin's scientific reputation from being stained by racist slurs. They have even suggested that "social Darwinism" be relabelled "social Spencerianism".[9] Such badge-engineering proposes that while, in the modern era, theories of natural selection were used as doctrines of racial competition, this blemish was not Darwin's fault. There is a transfer of blame at work here: Darwin is innocent, therefore the fault must lie elsewhere. At this point Spencer is arbitrarily substituted for Darwin, presumably because he too was well known and, not being a professional scientist, he serves as a more acceptable scapegoat. However, this substitution is unsatisfactory since Spencer was no more racist than Darwin: there is no evidence for Spencer's guilt, and the charge against him is even less well founded. "Social Spencerianism" is even more misleading than "social Darwinism". The latter could plausibly refer to a scientific theory that was misconstrued by Victorian journalists, whereas Spencer's ideas were less available to them.

Since writers as well respected as Stephen Jay Gould and Steven Pinker have misjudged the source of racist social ideology, this is *prima facie* evidence that there is a popular error that is worth correcting. The origin of their mistake lies in the overlap between the scientific and popular understandings of Darwin's natural selection. That theory had specified that species underwent genetic change as a consequence of competition for scarce resources. This process was a natural analogue for the artificial selection of the kind that was undertaken by plant and animal breeders. Although Darwin did not speculate as to whether human races were species-like in his *On the Origin of Species*, he wrote as if his biological theories applied to all organisms, and the breadth of this suggestion could be easily used by popularizers to condone struggle between ethnic groups, classes or nation states. However, in Britain at least, such misapplication of science stayed in the popular realm. Natural selection had little impact on the writings of anthropologists and ethnographers during the late-nineteenth century. Their focus on the history of physical artefacts, and on domestic and political institutions, meant that reductionist biological explanations were not helpful to them. While few other Victorian social scientists were as vehemently anti-racist as Spencer, if one judges them as a group they were possibly less mischievous on racism than modern scientists have been.

Spencer's ideas on social evolution should generally be grouped with those of late-Victorian anthropologists, since most of his data was ethnographic. However, he was writing earlier than they were, and was already speculating about evolution before *On the Origin of Species* appeared. He used both data and opinions from that work to support *The Principles of Biology* without modifying his own theses. In *The Principles of Biology* Spencer did not focus on genetic change in species; nor did he employ natural selection or "struggle for survival" as an organizing concept. Later, writing *The Principles of Sociology*, Spencer did emphasize competition as an important factor in causing social change, but this was not analogous to Darwin's natural selection theory for species change. Spencer's sociological theories largely ignored individual competition, and, instead, focused on the development of social organisms such as tribes and empires. These he viewed as emerging around governing structures that had been initiated in war, and that were capable of corporate action.[10] Early social formations were not caused by competition for scarce resources because they did not occur when food was scarce, or where the population was small and diffusely distributed. According to Spencer, such factors would have hindered cooperation and prevented the establishment of chieftainship, the initial organizing process in a society.[11] Spencer's ethnographic examples included Inuit, Veddahs and Fuegans, none of whom enjoyed the conditions for cooperation and, therefore, could not engage in competitive activity.[12] "Pre-social" groups that could cooperate also constructed administrative groupings, and, subsequently, engaged in war.[13] This militarizing process reached a terminus where modern peoples developed social and industrial structures in addition to the earlier political ones. Since these industrious and life-sustaining structures would be injured by further military competition this would eventually cease to be a feature of human evolution. This hypothesis was a consequence of Spencer's belief that, "With social organisms, as with individual organisms, the structure becomes adapted to the activity".[14] His social organisms, unlike Darwin's species, could be altered by activity. If there was war, or threat of war, the organism could become more militant and revert to behaviour that would have been more appropriate in an archaic past. Organized violence was a remnant of earlier social formations and should be avoided since it no longer served a progressive biological function. Spencer's future was undetermined, and fluid. With a hope that now seems excessively optimistic, Spencer hypothesized that war would cease to be a factor in social change. Whatever the value of this prophecy, it leaves little room for condemning Spencer as a social Darwinist. Spencer did not provide a competitive and individualist account of the origins of social evolution; he did not consistently explain change by reference to the environment; and he constantly expressed his faith that human beings would evolve away from aggressive behaviour. His vision for humanity was a tomorrow without struggle, pain or death.

Viewing the human species as a sociological problem in evolutionary theory, Spencer suggested that the correct procedure would be to begin with a hypothesis that humanity, as a whole, was developing like a social organism.[15] By analogy with other organisms Spencer could then speculate that human organization had begun as a primitive horde with very minimal structure. Spencer thought that since all organisms started with a similar amorphous structure,[16] the proposition must be true about early human societies. These he preferred to call "pre-social groups". As with embryos, the initial forms would be common, even if the adult or mature organisms became dissimilar later in their development.[17] Throughout this process social organisms would display more and more complex structures, the first sign of which would be the formation of a nucleus of leading men.[18] This governing group, in turn, would evolve specialized divisions.[19] Each of these developments would be a spontaneous response to collective needs. The evolutionary process was not the result of planning, rational calculation or individuals struggling for survival. Spencer's insistence that the social organism had evolved spontaneously distinguished his theory from the two most common individualist theories of social change. First, since he rejected rational calculation and planning as causes of social change, he was jettisoning the theory of social contract through which individuals assent to new constitutional frameworks. Secondly, Spencer was throwing aside the individualist paradigm in which it is imagined that progress was the consequence of individual persons engaging in competition with each other. Spencer's social science consigned individualism to the past because, at an early period in human evolution, the social organism had developed to a point where its components could no longer be improved by either internal or external competition. On the contrary, competitive individualistic activity could pose a threat to progress and risk damaging the developing apparatus of governance. Unlike Spencer's data-laden ethnography, his remarks on prehistory were not based on empirical evidence. He did not conduct his usual exhaustive search for supporting material and rarely cited archaeologists, presumably because their writings emphasized technological developments, which would not have furthered his arguments. He had no evidence that social organisms would be damaged if they engaged in competition, and his arguments belong to a domain of speculative organizational theory. In place of empirical evidence he substituted a variety of speculations as to how society might have moved from the primitive governance of leading men towards more articulated and complex administrative structures. His views were the antithesis of those of other nineteenth-century radicals: Spencer's faith in the importance of the developing complexity of government stood in contrast to Friedrich Engels, who dreamt of government becoming minimal, and de Tocqueville, who imagined political progress as a series of tableaux in which each succeeding structure exhibited successively less leadership and social distinction.

While Spencer's chief argument against employing Darwinism in current political debates was anti-individualistic, he offered another objection that was less antipathetic to individualism. He argued that individuals were simply exempt from the competitive forces that had propelled Darwinian evolution.[20] This position was less scientific than the competing one in *Political Institutions*. It was more in the way of being an emotional evocation of political theory in which he hoped that a novel expression of norms would cut across the usual debates concerning the justification of republican liberty and legitimate authority. Such discourses he regarded as reductionist attempts to force political aspirations into an archaic language of the will with its connotations of personal domination. Spencer's "cardinal truth" here was "that while the forms and laws of each society are the consolidated products of the emotions and ideas of those who lived throughout the past, they are made operative by the subordination of existing emotions and ideas to them".[21] This was a personal testament as well as a political one. Spencer had conducted his own life in such a way as to elevate the will so as to repress other emotions. Later, when writing his autobiography, he felt that his conduct had resulted in a cruel as well as false life, and he attempted to ensure that it would not be perpetuated in the political sphere.

Spencer's "cardinal truth" was the kernel of his exposition of a once mighty form of liberalism. This variety has become obscure as memory of Victorian liberal ideas has dwindled to a thin libertarian rendition of J. S. Mill's *On Liberty*. Among the vital strands that have been forgotten is the scientifically focused evolutionary psychology, which provided a humane gloss on the value of tradition and the ancient constitution. For Spencer, "The governing sentiment is, in short, *mainly* the accumulated and organised sentiment of the past".[22] Spencer did not deprecate the importance of governance, but declared its independence from rational calculation. It would be easy at this point to conflate Spencer's liberalism with conservatism. This has been the posthumous fate of Bagehot, whose emphasis on the importance of the "cake of custom" in politics so confused his twentieth-century editor that he eventually suborned the greatest of Victorian liberal journalists into a Tory.[23] If one misperceives Spencer in a similar way, the loss of variety in liberalism would be even greater because his use of tradition was more protean than Bagehot's. Spencer's notion of custom was more than a heavy political constraint on legislation: it had normative features emphasizing the community as the sole source of political power.[24] While the insistence on the community as the ethical source of laws and institutions long pre-dates liberalism, it is unlikely that Spencer would have recognized its medieval and early modern roots. In any case, this past would have been irrelevant to him; his rediscovery of the community as a normative value was not a reiteration of earlier sources; it was a rejection of the modern attempt to replace monarchical domination with the popular will. A democratic theory had arisen that had subordinated communal power to the electoral majority. This was a position that, to Spencer, had no more normative standing than traditional despotism. Spencer's own theory of representative constitutional theory left no gap where popular power could break into the regular functioning of administration. To him, the community's role in government had always been authoritative rather than powerful, and it should remain that way. Unlike J. S. Mill, he did not prophesy that the role of the elected politician would be to occasionally disrupt the smooth running of executive machinery.[25] Disrupting administration was oppositional and antagonistic: a kind of catharsis that was unpalatable to Spencer, who felt that there should be no struggle in governance.[26] Spencer's comments on representative government in *Political Institutions* did not employ the language of Darwinism. Given that he published many of his sociological chapters separately, it is significant that Chapter IX, which dealt with representative institutions, only appeared in this volume. It is likely that Spencer feared that it would offend some of his liberal friends, who were beginning to canvass for working-class votes. For them representative government was increasingly seen as a synonym for democracy.

The practical significance of Spencer's theory of representative government was its requirement that governors receive support from the political elite, and that they should respect and maintain the great mass of institutions and laws produced by past sentiment and ideas.[27] At first sight, such recommendations do not seem progressive or liberal, but the reverse. However, rather than being conservative, they were a projection of his secular and modernizing sociological principles. Spencer's faith in the "Great Architect" and in Romanticism had faded away as his sturdy modernism deprecated his former admiration for Carlyle's reification of departed ideals. The worship of the past was something the author of *Past and Present* had shared with that other fountainhead of modern conservatism, Edmund Burke. These ideologues possessed a common faith in the spirit of departed values that did not distinguish between a private and a public ethos. When Burke considered the workings of the Divine Will, or when Carlyle thundered on about the historical importance of great men, it seemed obvious to them that reflection on previous moral actions had always been, and still was, a guide for statesmen. Spencer found such unqualified respect for the past dangerously naive and, to guard against it, he limited his own advocacy of traditionalism to the public sphere. Morals were recent and personal, so the rules governing human interactions needed to reflect this. On the other hand, public actions had not evolved much in modern times and could be guided by the mass of traditional institutions and laws. In other words, private lives responded to contemporary ethics, while public ones did not.

Spencer's notion of tradition was designed to combat the idea that morals should be affected by theorizing about how rationally to organize the state. From his perspective the different schools of traditional political philosophy from ancient Greece to seventeenth-century England were flawed because of their shared belief that rights and duties were artificial rather than natural injunctions.[28] Political theorists had become confused because they had paid insufficient attention to causation. This was true even of Bentham, the great flail of traditional scholarship, because his utility theory rested on the idea that there was *some* relation between cause and effect, but did not sufficiently specify what this was.[29] Thrusting Benthamite logic aside, Spencer claimed that utility should be considered as an accumulated product that had become independent of conscious experience. The result of social evolution was a bedrock of moral and legal sentiments that had been organized and consolidated through successive generations. As a consequence, human beings now possessed collective emotions that no longer rested on a basis of individual experiences of utility.[30] This in turn meant that people were no longer free to use political organizations to impose whatever laws they conceived as rational without regard to how a society was anchored in its cultural inheritance. Individual liberty was a mirage: political traditions could only be progressively modified when a large number of social restraints had been established, when impulsive conduct was avoided by most of the population, and when social rewards could be clearly identified and achieved. Only then would people be able to predict the effect of their actions and abandon simple pleasures in favour of complex ones. "Only then, too, does there arise a sufficient intellectual power to make an induction from these experiences, followed by a sufficient massing of individual inductions, into a public and traditional induction impressed on each generation as it grows up."[31]

Spencer's political theory is best described as a temporal pluralism that allowed him to advocate a continuing reform of private morals, while insisting that citizenship and international relations belonged to a past that still influenced events and exerted force. This coercion contained a momentum derived from "aggregate feeling, partly embodied in the consolidated system which has come down from the past, and partly excited by immediate circumstances".[32] Since he considered public actions as more primitive than private ones, he was able to distance himself from imperialists who were proclaiming a moral mission to transform savages. While Spencer accepted that the violent formation of larger societies through the absorption of smaller ones had been inevitable, he did not view this process as caused by the superiority of the conquerors, but by their comparative backwardness in public sentiments. This led to a moral quandary. No values were absolute; sometimes even ethical developments in the private sphere were retarded. This stance allowed Spencer a consolation about the evil effects of empire. European despotism over other cultures was wrong, but it would still have a civilizing effect by freeing indigenous peoples from the yoke of personal subjection under which they laboured.[33]

Spencer posited aggregate or group feeling, not law, as the social force framing public actions. This group sensation dictated society's wider activities because it was a vast psychological force compared to which ethical considerations – even when backed by religious sanctions – had no impact. For example, Spencer observed that sacred injunctions and threats of damnation consistently failed to halt iniquitous wars on foreign peoples if these were supported by prevailing passions. In observing this, Spencer was censuring popular British enthusiasm for the Crimean War and German war plans as well as condemning colonial violence.[34] To him conflicts demonstrated the impotence of legal codes and religious creeds in the face of popular sentiment. The latter was regrettable because it led to reaction.[35] Mass opinion was a social force, controlled by an uneducated majority who would always reflect ancient or customary beliefs. Unlike Hegel and many other nineteenth-century political theorists, Spencer did not take refuge in a state theory when explaining how tradition modified and directed collective behaviour. On the contrary, he was contemptuous of his contemporaries who reasoned as if "state instrumentalities" possessed intrinsic power.[36] It was, he believed, a comfortable rational fiction to think that institutions had power rather than the community as a whole.

Spencer, who rejected German idealism in social explanation as he had in biology, had no faith in reason in history. For him it was not the intellect that was "an essential element of political theory" but prevailing social pressure.[37] In proposing this, he was claiming a greater than scientific status for his sociology. He regarded his science of society as more profound than a series of descriptive generalizations based on empirical evidence. Of course, his science could be superficially described as such, but its primary function was the provision of a normative context for his belief that the past provided "consolidated products" that had remained alive, and that had often subordinated our current ideas and emotions. Sociology linked Spencer's personal life with his philosophy. This happened in two ways: first, his devotion to social science clearly marked him out as a secular thinker who, by the time he came to write *The Principles of Sociology*, had abandoned the Christo-centric ideas of society that had underpinned his *Social Statics*. This shift would have been difficult to detect at the time since Spencer had allowed the original text of *Social Statics* to be published as late as the mid-1860s. His nineteenth-century readers were not aware that he had abandoned his conventional religious and ethical beliefs by the time his *Principles of Sociology* began to appear in 1876. However, it is clear that his ethical vision of the past and future of mankind had ceased to be dependent on natural theology. While Spencer still portrayed human development as a natural process, it was now the result of godless development. The second major feature of Spencer's mature political theory was the resistance it offered to materialistic – or non-spiritual – accounts of human development. To him the current shape or structure of human society was not solely the consequence of a series of responses to physical or environmental forces. There would always be a non-environmental extra drawn from the "consolidated product" of humanity. At this point Spencer's non-Darwinian theory separated the idea of the social organism from the concept of organic development that he had used in *The Principles of Biology* where the physical shapes of plants and animals were completely determined by physical forces.

When he divided human beings from the remainder of the living world Spencer was taking a different path than that of the materialists, who insisted that human progress was not exempt from the forces of natural selection. His ideas contrasted sharply with the emphasis Darwinians gave to struggle, and with the fascination Marx and Engels had with class conflict. Their theories were deterministic while Spencer's social evolution gave humanity a limited measure of control over the direction of their future development. His persistent cultural traditions were not functionless echoes of departed utilities; they were cultures that had an independent existence and were able to influence their successors. This depiction of human evolution was not overly optimistic because Spencer also believed that our public actions were currently controlled by archaic mass opinion, and thus only marginally responsive to ethical demands. However, the picture was not as gloomy as materialist accounts of evolution because it was accompanied by prophecies that the future direction of the human species would reflect its generic experiences, and because this overall progress would ultimately encourage private mores at the expense of public values. In the future, Spencer believed, individual life would be less subject to consolidated customs. Tomorrow would be a time of freedom for the individual, although this blessing would not resemble the republican liberties whose disappearance was regretted by many of Spencer's contemporaries. However, Spencer himself did not mourn this loss; to him such liberties had been parts of a defence of a public domain that had become archaic. The republic was a world we should rejoice in having lost, not one that we should wish to recover. Instead of restoring old freedoms, he predicted that, as a species, human beings would inherit a universe in which their lives would be more than a series of automatic or prudent responses to the non-human forces that had dominated their militarized and violent past.[38]

Spencer's sociology not only avoided excessively functional analyses, it was also free from timeless and universal ethical principles. Such norms were only available to some societies – the very primitive and the very advanced – or, in Spencer's terms, the "pre-social" and the modern. In between lay a gap in which organization took a militant form, and where morals were subordinate to the necessities of conflict. While this intermediate period had led to later evolution of morals, Spencer did not assign priority to any particular cause of this process. He simply listed a variety of non-ethical and non-biological causes that he had gleaned from reputable historians. For example, he relied on Bishop Stubbs's prosaic account of the growth of medieval administration as a process motivated by the desire to extend royal power.[39] This historical narrative was a straightforward chronicle of the growth of revenue as caused by a ruler's desire for greater influence. Such an account was not reliant on any descriptions of monarchical virtue or martial valour that one could have discovered in medieval texts, and that might have been adapted for use in a Darwinist interpretation of history.[40] Spencer's practice was to avoid the language of struggle even when he was chronicling tempting examples such as the thirteenth-century unrest that followed the deposition of the Emperor Frederick II. He restricted his comments so they would focus on antagonism between tendencies, not war between "organic" units such as nations, classes or cities. His analysis purported to demonstrate that if the advance of industrialization were checked by a return to militarization, then the growth of popular power would also be arrested.[41] Spencer extended this speculation so that it not only applied to the imperial collapse in Germany and Italy, but to anywhere early industrial societies had been transformed into military ones. Naturally, the histories of France and England struck him as examples of this.[42] If Spencer had a general conception of historical change it consisted of a non-directional account of the varying strengths of industrialism versus militarism, accompanied by an attempt to measure the overall increase of private morals.

There is an occasional use of "survival of the fittest" language in Spencer's political sociology, but this was not intended to advance a Darwinian theory of social evolution. Rather Spencer's purpose was to overthrow the English historiographical tradition of interpreting constitutional development as the fortunate result of parliamentary accidents that occurred in the reigns of King John and Henry III. Similarly, when Spencer refused to accept the conventional analysis of the rise of popular power as a process caused by the alternating succession of wars and alliances that medieval towns had had with neighbouring barons, he was resisting the Whig interpretation of constitutional growth as propounded by Paley, Hallam and Mackintosh.[43] Their progressive histories had relied heavily on the citation of single monumental occasions, such as the summoning of Simon de Montfort's parliament in 1265,[44] which seemed unscientific to Spencer. He himself was in pursuit of grander explanations spanning several countries that possessed parallel social structures. If a historical narrative merely linked together a series of progressive moments, this seemed fallacious to him. It was a fable based on singular data with a substitution of petty anecdotes for a serious causal explanation. For him, true history could only be found in factors such as the relative demographic sizes of medieval populations, the ideas produced by the mode of life and the temporary emotions roused by oppression or distress.[45]

While Spencer's recognition of demography seems to signal his interest in biological explanation, his three-factor social analysis was not scientific in that sense. For example, population density was not given pre-eminence over ideological causes of change. In addition, Spencer advanced alternative rationales for social change that could not be grouped under any of these factors, and such variant explanations were not drawn from the life sciences. For example, Spencer's sociology happily incorporated H. S. Maine's thesis that historical evolution could be described as an institutional shift from a society based on status to one based on contract. This suggests that he was susceptible to any meta-historical explanation that did not overemphasize the roles of reason and individual choice in social evolution.

For Spencer, the direction of social change was indicated by the needs of political organizations, not by deliberation. This was the kind of teleological or goal-focused progress that Spencer had avoided when speculating about animal development in *The Principles of Biology*. In his sociological writings he favoured the idea that growth towards a developed structure was the purpose of evolution, and that this took place at the expense of the "pre-social" mass of humanity. When he focused on institutions he foresaw "finished forms" as outcomes for both the militant society and the industrial one. He found this classification process difficult: these forms resembled species and seemed to be more than the artificial classifications he had favoured in his biology. Each of the types had acquired their form slowly and many societies did not display all the organizational characteristics of their type.[46] Casting aside his earlier scientific hostility towards idealism, Spencer searched for evidence that a particular type of society was "ideally" organized for war before attempting to identify the characteristics of real social entities. Similarly, when he contemplated the industrial type of society he mentioned its habitual form, but also suggested that, in actuality, it was hard to disentangle a type from its opposite.[47] He could only distinguish between his types by examining their goals. However, since he had already dismissed the notion that intellectual propositions guided change he could not, therefore, emphasize any intrinsic thought within a social movement, and he had, as a result, no evidence that his types were more than synthetic constructs. This difficulty left him as the sole arbitrator of where one should classify types that possessed no unique characteristics. Spencer's problem here is not merely taxonomic; he had stumbled into the realm of biological idealism, which had so repelled him when he had earlier knocked against it in the writings of Richard Owen.

His only guide in identifying the direction of change was his belief that political and administrative structures had developed at the expense of equity and popular control. This put Spencer at odds with the conventional scholarly opinion of his period because what he regarded as progress, many of his contemporaries lamented as the unfortunate consolidation of dynastic power at the expense of free peoples and parliaments. In the Victorian period it was common for English historians to regret the loss of demotic power in early European societies. This was characterized as a decline in France and in various states in Italy, Germany and Spain; nations were depicted as experiencing an interruption in their growth towards freedom. Spencer was critical of such accounts because they were based on what he regarded as a non-scientific idea: that a desire for an outcome could be the cause of the changes. He also suspected that since freedom, like popular power, was always associated with an increase in trading activities, it was a co-relative phenomenon rather than a causal one.[48]

In Spencer's evolutionary sociology, it was unimportant for him to establish an original cause of organic change. He simply posited the existence of a social organism by analogy with biology to the effect that while adult organisms have little in common, in their juvenile stages they were very similar to each other, and, in fact, had started with a common structure.[49] Spencer's analogy organized the external relations between social organisms as well as internal arrangements.[50] These social organic structures did not necessarily work for, or against, the individualization. For example, citizens did not inevitably become more independent from, or subordinate to, their political institutions during the course of evolution. Rather, Spencer's thesis was restricted to emphasizing that both civic political structures and subordinate internal organizations always developed towards complexity, and away from simple structures in which the mass of the population was able to exert control. This theory has little in common with later political science because it is not driven by a practical enquiry. Twenty-first-century analysis is often applied scholarship – for example, a social scientist might analyse how the Chinese state was able to perpetuate itself or how the United States was able to control capital markets by using international organizations – but such analyses might have little bearing on theoretical knowledge of the long-term growth of institutions in general. In such studies the historical evolution of particular organizations would be irrelevant, so long as their functioning produced outcomes that were efficient from the perspective of a member of the Chinese elite or a World Bank economist. Spencerian analysis also presumes efficiency – otherwise, he believed, institutions would not have evolved – but would be additionally concerned to demonstrate that because institutions were like organisms, the primary scientific effort should concentrate on demonstrating how a collection of historically determined characteristics had preserved themselves. The explanation would revolve around the question of whether additional political goals would impact on existing institutional structures in such a way as to either destroy them or make them more complex and less demotic. For Spencer no other outcomes were possible. This was a dramatic argument during the late-nineteenth century because it ran contrary to the growing belief in the benefit of popular governance. Spencer's insistence that institutions could only adapt by becoming more complex in structure suggested that progress would be inhibited if power were placed in the hands of the people.[51]

In Spencer's eyes, democratic control was always primitive. It referred back to human origins of unstable hordes and the emergence of a controlling nucleus.[52] This structure would be spontaneously produced; it would not be the consequence of agreement. Spencer drew examples from regions as dissimilar as India,

Madagascar and New Zealand to demonstrate that early organizations always produced three classes consisting of a leader, chiefs and a general assembly,[53] a division that resembled that of medieval Europe without the clergy.[54] If Spencer's simple progressive division is compared to the more complex binary analysis used by twentieth-century commentators, such as Emmanuel Le Roy Ladurie and Barrington Moore Jr., it suggests that, unlike theirs, Spencer's primary object is not the explication of conflict. If historical evidence is deployed to demonstrate that two classes – lords and peasants – have some permanence, then the analysis will focus on long-lasting antagonisms. Contrary to this, Spencer was uninterested in the long-term consequences of class war. Instead, he was concerned with elucidating the natural phenomenon of the exclusion of low-status groups in society from deliberation and power. Each early society that he had examined possessed a passive non-political class who were excluded from government. At first, this group was just composed of women, but afterwards it would come to incorporate slaves and other dependents.[55] Political evolution would eventually ensure that most people were subordinated.

The impetus driving social evolution was the need to ensure that members of a ruling group would become obedient to a head.[56] The process had not come about because of a simple biological determinant such as the exercise of greater strength. There is little in Spencer's writing to support a theory of the psycho-pathology of dependency; he did not favour Hegel-like examples showing that slaves would become subject to masters to whom they lost in combat. Instead, he offered a psychological rationale that, in the past, disobedience to a ruler would have been punished; loyalty had therefore become habitual. Spencer's suggestion about the origin of obedience was directed against the thesis that political power is naturally inherent in an institution. The chief purpose of such an idea had been to arrest the development of popular government. However, since reactionaries had been unsuccessful in preventing the outcome they feared, there was no reason to reiterate their beliefs. In any case, Spencer thought that it was unreasonable to link discussions about how people responded to power to an analysis of pre-modern or modern institutions. Spencer believed that one should not assign power to institutions because it was a quality that refers to a sensation of awe that pre-dated the existence of government of an organized kind.

When Spencer quoted Lubbock's maxim "No savage is free", he was not just echoing the latter's barb that indigenous peoples were more severely constrained by their customs than Europeans were by laws. Lubbock had divided his "savages" from Europeans in time or space, but Spencer's sense of modernity would not allow for this separation. The distinctive high-points in European civilization were undermined in Spencer's sociology: what was true of savages also applied to developed and militant societies, such as Italian city-states during the Renaissance. In these societies individuals only valued freedom in order to pursue their private ends once the civic unit had no further need for their services. If duties were required then the actions of citizens were forced to conform to the public will at the expense of their private desires.[57] In early modern societies the individual was owned by the state; it would be mistaken, therefore, to emulate their liberties.

Freedom could only exist in recent industrial societies in which individuals were not required to risk their lives while destroying others in war.[58] Although freedom was a recent phenomenon, the same could not be said about politics. This had existed as long as institutions had in the form of a governing sentiment that was "the accumulated and organised sentiment of the past".[59] Such sentiment was neither intellectual nor instinctive; rather, it was evolved sediment of social beliefs that allowed the feelings of the dead to control the actions of the living.[60] Since the process of evolution produced new aspirations in private lives, individuals could exercise spontaneity in that domain. In their public activities, however, people were still likely to be controlled by the past.

Curiously, for an evolutionary theory, Spencer's sociology was neither progressive nor regressive in a political sense that would have been familiar to his contemporaries.[61] For him social change was not directional in a temporal sense. This was in contrast to Lecky, Hegel and other historically bounded intellects who believed that societies had always possessed a kernel of freedom, but were moving towards its fuller realization. Nor did Spencer's views parallel to those of a reactionary, such as Joseph de Maistre, who believed that humanity had degenerated. Spencer had no arrow of time; he was unable even to equivocate on the issue after the fashion of Tylor by imagining social change as an elevator that both raised and lowered culture. Spencer's temporal neutrality was a late emendation; it only appeared with *The Principles of Sociology*. Before then his political and scientific writings had been progressive, regarding social change as an upward movement. However, by the time he came to write the political section of his sociology, he was no longer convinced that evolutionary processes were uniform. From his mature perspective the valuable traits of humanity were virtues together with the individuality that had largely been vitiated during most of the time period for which there were written records. During this long fallow period human beings had lost their individuality through being compelled to cooperate.[62]

Social coercion had been necessary in order to ensure military subordination. Spencer believed that the bearing of arms had severely restricted liberty; giving individuals only freedoms that were compatible with military obligations.[63] To his contemporaries – who perversely admired the ancient Greek ideas of freedom – this was shocking. Spencer had reversed their values. Instead of arms-bearing being the criterion of citizens' freedom, it was a sign that they were socially underdeveloped. In Spencer's view people were only free to pursue private goals when the social organization to which they belonged, such as a tribe or nation, had no need of them. When, however, the organization required individual servitude, actions were required to conform to the public will.[64] Spencer was extreme on this subject, insisting that, during the lengthy militant period of human development, the individual was owned by the state.[65] This proprietorial relationship was not softened by a social contract as it had been for Thomas Hobbes, whose recommendation of a Leviathan erected by, and on, an equal citizenry was an example of mistaken evolutionary theorizing. For Spencer, a militant form of society was one that functioned in association with status, not contract. When this form tended towards an egalitarian despotism, as it would with a Leviathan, this would be a sign of decay rather than of a progressive development towards an industrial society.[66] Spencer found Hobbesian logic flawed because its psychology was based on a defective analogy between the operation of the reason and the law, both appealing to the will. Contact theory functioned by the pretence that a rational understanding of a personal interest could be made permanent by a promise. In opposing this Spencer appealed directly to emotions. For him politics was primarily a matter of reconciling differing emotional demands, not of satisfying each individual's reason. An adequate explanation of political change should refer only to non-rational and emotional causes.

Reasoned appeals for change, such as those offered by advocates for democracy or responsible government, were insignificant. Political development would only be caused by an emotional movement in the mass of the population.[67] In the course of modern history there had been a weakening of sentiments supporting governance; these were feelings that would be offended by breaches of authority. This change had been accompanied by a growth of those feelings that reacted to "injuries to the individual *and to the community*".[68] Spencer bracketed the community with the individual in order to demonstrate that his philosophic stance was not libertarian, but corporate. His beliefs about the weakening of authority specifically referred to the diminishing of the standing of gods and monarchs who had symbolized a larger than superior political will; there is no support here for the ideal of a lone person who draws strength from being isolated from society.[69] Individuals, in the Spencerian future, would still be subject to law. This would not be a legal code or jurisprudence, but a revival of the consensually based rules that had existed before political and/or military structures had forced the pace of social evolution.[70] In this modern age there would be a return to the kind of law that had been prominent before personal authority had arisen.[71]

Spencer's theoretical politics was in stark contrast to modern democratic theory, which rests on majority will. It was also quite distinct from J.-J. Rousseau's dislike of majority will because it did not exclude passion from political calculation. The Genevan placed his entire trust in the will only once it was unencumbered by feelings. Spencer, however, wished to minimize the will itself because it repressed emotions. While not relying on explicit decisions by the greatest number, Spencer's argument was based on the general will.[72] As with Rousseau, the rationale for invoking the general will was the desirability of emphasizing the public good in decision-making. However, unlike the eighteenth-century author of *The Social Contract*, Spencer did not identify the general will with the sovereign because that would have anchored it in a conceptual tradition that modern societies should be shedding. A sovereign would be unnecessary in a future where legal sanctions no longer required support from an embodied authority. It was imperative that Spencer make this point because his contemporaries, such as T. H. Green, were in the process of reviving "classical" political theory. This political discourse had traced the source of law to the will of someone specific, a monarch, and democrats were beginning to imagine that a numerical majority of the people would fulfil that role. For Spencer the revival of classical political thought and the new popular sovereign were equally anachronistic and would hinder progress.

For Spencer, democratic theories were crude transitional beliefs. He saw them as similar to the notion of transitional species, which still intrigues evolutionary biologists. They hope that such missing links will display the characteristics of future species, although, disappointingly, these often show more affinity with even earlier ones than with the future. In a similar vein, Spencer perceived democratic theory as displaying some characteristics of dated political formulas, but not yet demonstrating the qualities needed for tomorrow's governance. It was not the people's will he hoped to promote, but their collective interest: "The ultimate theory, which it foreshadows, is that the source of legal obligations is the *consensus* of individual interests itself, and not the will of the majority determined by their opinion concerning it; which may or may not be right".[73] The notion that a political theory was transitional, and that its successor might be the ultimate form, could be taken as suggesting that Spencer possessed an explanatory framework similar to those that palaeontologists have used when they were speculating about the Devonian period, or prehistorians when musing about the Iron Age. However, these hypotheses refer to specific successive eras while Spencer's temporal periods were variable and reversible in sequence. "Pre-modern" could mean archaic, undeveloped or prior to the modern state. It could also refer to the prehistoric origins of humanity, to still extant indigenous peoples or to relatively recent European history. As an example of recent pre-modernism, Spencer endorsed French natural law theory of the sixteenth and seventeenth centuries in order to underscore his point that individual claims, together with any social welfare supported by such claims, preceded political authority. Spencer also noted that when such sources discussed religious dissidence they provided individuals with rights against the law even when that expressed the will of the majority.[74] To him this was evidence that, even in pre-revolutionary France, the will of the majority was merely an antiquated reference to monarchical prerogative.

Spencer completely lacked sympathy for those writers who were attempting to provide either a royal or popular pedigree for current practices in governance. He was also impatient with any attempt to distinguish between elected and appointed officials. Both were necessary, and the distinction was a functional one rather than fundamental.[75] Such issues were marginal to Spencer; modernity could not depend on the popularity of governments because to him the expression of will was no more legitimate in chorus than it was in solo voice. Instead of sourcing legitimacy in expressions of will, Spencer suggested that, in a developed industrial state, law-making was justified only when it enforced the conditions of harmonious social cooperation.[76] Such legislation should be impersonal in the sense that it would not derive from the expression of a sovereign's will, and would – in its final form – be merely an applied system of ethics.[77]

It is in their ethical, rather than their descriptive, functions that Spencer's forms assume an idealistic quality of the kind he had repudiated in *The Principles of Biology*. When it was a question of the evolution of social life he assigned to human intelligence an influence on future outcomes. Even though he believed that the individual consciousness was incapable of rationally planning civil futures, he still believed that the increasing "sentiency" of humanity would have a positive effect. This process would be a matter of psychological growth as, in the course of human evolution, immediate goals, which were frequently oppressive, would be subordinated to future ones.[78] Increasingly, distant ends would be fostered by humanity adopting ideal motives. This was a development over which people could assert a measure of control by avoiding aspirations that produced only short-term benefits in favour of higher ones.

At this point in his argument, Spencer erected an ideal form to be the goal for humanity. This form would never actually be achieved because of the continued aggressions that political communities would commit against each other,[79] but it would exist in the collective consciousness. Spencer's future social state would remain unachieved even if human beings ceased to have conflicts. He even feared that a purely industrial society that had eliminated aggressive activity would be insufficiently peaceful. While he argued that industrialism would ultimately require that citizens live without violence, there was no insistence that they should act immediately to improve the lives of other citizens.[80] Spencer judged that if people lived inoffensive lives while scrupulously fulfilling their contracts, but gave nothing to others beyond the terms agreed on, then they would fall short of the highest degree of life, which was the gratuitous rendering of those services that made corporate life possible.[81] In a pure industrial form of society everyone would suffer from the prevailing selfishness and the public interest would be harmed. "The limit of evolution of conduct is consequently not reached, until, beyond avoidance of direct and indirect injuries to others, there are spontaneous efforts to further the welfare of others."[82] The final social form would be one in which beneficence is added to justice. This future would be reached as human faculties evolve to a point that an action that brought positive benefits to others would satisfy them as most selfish actions had in the past.

Spencer's idea that a legal system would have a final form further suggests that he had reverted to the teleological language he had set aside when thinking about biology during the 1860s. In politics he was not content with merely providing scientific taxonomy of progressive trends: that might encourage the majority to consider will as the engine for social change, and Spencer had become increasingly convinced that the will – in any formulation – was a vacuous philosophical idea. It seemed merely an imaginary picture of inner essences that purportedly drove social progress. Its proponents, who included Goethe and Carlyle, worshipped a spiritual humanity, which they had extracted from dead historical texts. Spencer's teleology was designed to avoid any historically determined spirits: he looked to the future for his direction, not the past. His visionary final form of politics focused on impersonally derived laws that would give expression to the consensus of individual interests.

Spencer's philosophy on political evolution not only eradicated the language of will, it constructed a social distinction between compulsory and voluntary cooperation. This is well known because his popular text *The Man "versus" the State* expressed a vehement opposition to compulsion. As an implication of this, he also rejected socialism and communism because these ideologies shared with Bismarck's Germany the belief that permanent improvements could be achieved by force.[83] When he was in his polemical mode, labels did not matter to Spencer: whatever compulsory cooperation was called it would still regulate status, maintain inequality and enforce authority. His adamant stance on this subject alienated many potential allies who shared Spencer's progressive ideals on other subjects. Like him they were hostile to aristocratic privilege, but his arguments went further, insisting that the liberal campaign against status be conducted without surrendering power to the people. Whatever support he gave to egalitarianism was limited to his agreement that equity was an admirable ideal, rather than it being acceptable within current political processes. No olive branch should be extended to the masses; democratic control of the political representation would encourage the same sort of compulsory cooperation that had existed in the aristocratic past.

Opposition to the ideals of compulsory cooperation should not be taken to imply that Spencer was an anarchist and wished for the abolition of government. He did not share Peter Kropotkin's belief that natural or spontaneous cooperation would be a sufficient regulator of human affairs: the Spencerian world required a type of cooperation focused on public ends.[84] When speculating about voluntary social action he did, of course, draw on ideas of unplanned economic efficiencies, but he regarded Adam Smith's "invisible hand" as a ghostly distraction from the real business of thinking about political organization. There would always be a social organization consciously exercising a directive and restraining function in public activity.[85] His mature political theory did not see voluntaryism as supplanting the state's positive and negative functions; Spencer should not be confused with those mid-Victorian Manchester liberals whose advocacy of *laissez-faire* hinted at a minimal theory of government. Spencer's admonishments against excess governmental activity were usually practical, rather than theoretical, strictures. For example, he criticized the British government in India for imposing heavy taxation and excessive restraints because these caused their subjects to move to native states in order to avoid them.[86] More commonly, his objections to state activity were caused by his fear that this placed an industrial government at risk of being transformed into a militant one. He desired to lessen the inspection of science and arts classes, the registration of teachers, the inspection of public libraries, and he rejected compulsory insurance.[87] To him state regulations always suggested the militarization of social life. There was even militarization in parliamentary practices; Spencer was concerned about the use of the caucus system to control MPs, a practice that even the Liberals had adopted.[88] Since he had always supported the Liberal Party as an organization promoting peace, such regimented discipline struck him as contradictory and repellent. However, not all examples of control gave him pain; he found considerable amusement in a French proposal that a government should provide good music in order to supplant bad melodies such as those by Offenbach.[89]

Spencer's fear of compulsory cooperation referred more to the future than the present. In his own time the greatest threat was democracy. The concern that he continuously expressed in his ethical sociology echoed the intense feelings in *An Autobiography*, where Spencer claimed that it was cruel and retrograde for persons to govern their emotions by erecting a presidential chair over their personality so as to force a consensus on it by the exercise of will. Any such domination was contrary to his belief that a true and healthy psychological consensus should not rely on reason imposing itself on the emotions. An internal consensus was similar to a communal one; the parallel suggested to Spencer that a psyche was like a political governor, having no mandate to repress internal or minority voices in the name of the majority's conception of the general good. Only informed agreement by all the interested parties could legitimize a consensus.

Like consensual agreement, rights would have a privileged place in the future of Spencer's industrial societies. In this harmonious tomorrow, when life, liberty and property were secure, and when all interests were justly regarded, one would possess perfect rights.[90] Rights had not existed for individuals in the past, either in a state of nature or a result of a grant from a sovereign. However, they hovered distantly in front of humanity as a goal for further evolution: as advances were made so rights would become more perfect. Actual rights, and the respect accorded to them, were indicators of how far a society had developed. Significantly, Spencer did not define rights, nor specify them in a way that social contractarians had done; that would have assigned them functions that antedated industrial societies. It would have also mistakenly identified individuals as rights-bearing entities who, somehow, existed outside evolutionary processes.

Spencer's ethical sociology did not rest on individuals, but on the consensus among individuals; it aimed at developing corporate rather than individualist values. This was why Spencer could not place the welfare of citizens on the summit of state policy: "The life of the social organism must, as an end, rank above the lives of its units."[91] While the end result of social evolution would be an increase in individualization, this transformation was not rooted in rights theory. Spencer's position as a corporate theorist who was also a utopian individualist would have been easy to understand if A. V. Dicey had not brought confusion to social theory in the Edwardian era by treating collectivism and individualism as if they were polar opposites. Such an ideological dichotomy is misleading when one is considering Victorians; while it is true that most of Spencer's contemporaries by the late-nineteenth century had defined themselves as collectivists, the opposing category was an empty one. It would not have included important liberal figures such as de Tocqueville, J. S. Mill and John Morley. Nor should it include Herbert Spencer, unless he is viewed solely as the author of one tract: *The Man "versus" the State*. If one puts Dicey's misleading dichotomy to one side, then Spencer is revealed to be a progressive liberal whose core values of peace and altruism are the chief legacy of nineteenth-century liberalism. His discourse is still significant because, after the lingering deaths of most progressive doctrines, liberalism is the sole survivor of ideologies that promoted a secular humanitarian vision for the future. Modern liberalism, however, has become moribund, and it too will expire unless it recovers some of the variety it once possessed. This process could be assisted by the revival of Spencer's ethical sociology and his theories about the primacy of communal decision-making. Otherwise, liberalism will continue to wither into a rendition of negative liberty based on a thin reading of J. S. Mill.

While ethical considerations about peace and altruism occupied a prominent place in Spencerian political sociology, the other insistent message carried by his liberalism was the warning that freedom should be rescued from democracy. This was not unusual; Spencer's contemporaries had similar reflections on the irresistible tide of popular dominance. Some, such as J. F. Stephens, railed against the current; others, such as J. S. Mill and Ruskin, attempted to tame the movement with education. Spencer alone tried to stem the flood. He believed that democracy, with its borrowed and antique republican finery, was an avoidable mistake. It became his mission to breathe life into modern liberty while preventing electoral masses from destroying the evolutionary gains that civilization had so far achieved. In this way the public domain would be protected and laws would not be framed by a majority that forced its will on a minority. Instead, governance would emanate from the consensus of individual interests. Such a consensus would free people from domination even if they were unable to defend themselves or to express their will. In Spencer's future, government would protect individuals even if their interests were only felt, rather than rationally articulated.

NINETEEN

Progress *versus* democracy

Spencer's scientific views on politics emerged effortlessly from his philosophical labours. His ideas were disconnected from the great political events of his era: he felt nothing over the failure of the 1848 revolutions; nor did he desire or fear any future ones. It was the growth of conventional politics that intrigued him, but his curiosity was not aroused by the detail of parliamentary affairs and the intrigues of politicians. Rather, he was interested in the theoretical possibilities of parliamentary evolution: this was not a practical stance. When reflecting on the prospects of representative government Spencer was not timely; he completed his diagnosis several years before the confused and passionate debates over electoral reform that unsettled the other Englishmen in 1866 and 1867. The same was true of his well-known essays of 1884, *The Man "versus" the State*. These were not topical reactions to the revival of socialism, but reiterations of ideas he had first elucidated over quarter of a century before. In contrast to the protean qualities of Spencer's psychology and biology, his political logic remained constant: once he had determined his views on this subject they were subjected only to minor modifications and recycled either as essays or as parts of *The Principles of Sociology*. Indeed, Spencer often made little distinction between these formats, and many chapters of *The Principles of Sociology* first appeared as essays in popular journals.

Spencer's uniform values were minted in the late 1850s, and were quite distinct from the ideas he had earlier promulgated in "The Proper Sphere of Government" and *Social Statics*. As is clear from *An Autobiography*, Spencer still held to his ideal for the future, but its realization seemed more remote. He now regretted that his youthful indignation against social wrongs had made him too eager to force the reorganization of government.[1] His juvenile radicalism on this subject had been softened into a more comfortable liberalism.[2] In his early work he had espoused the popular cause in politics, and adhered to natural rights. His mature politics saw a departure from this stance. Spencer came to reject the idea that democracy was a progressive force; instead, he interpreted it as a reactionary one that was barely contained by the constitutional forms of representative government. Spencer's later rights theory also bears little resemblance to his early one. Changes in his philosophical and religious views meant that he was no longer able to support rights as necessary for the well-being of either individual citizens or the people as a whole. The "essential" aspect of natural rights had become objectionable: to Spencer it was reminiscent of the Platonism that his philosophy of science had attempted to overturn. Rather than perceiving rights as natural qualities that had always been seated within individuals, Spencer had begun to see them as present or future entitlements that a constitution should protect.[3] Rights were not the substance of politics, nor would the desire for them generate more liberty. "[T]he degree of liberty a people is capable of in any given age, is a fixed quantity."[4] Not only was freedom dependent on the prior development of political forms, but these would be vacuous in the absence of an advanced national character.[5] A constitutional form, such as representative government, was no more independent from social development than individual characters were. If this was not understood, and Spencer's contemporaries continued to insist that the progress in politics could be achieved by the exercise of will, then there would be social retardation. The old notion of unqualified duty had been replaced by the idea that "the will of the citizen is supreme and the governing agent exists merely to carry out their will". As a result it had become necessary to resist a responsible government as strongly as irresponsible ones.[6] The notion of "responsible government" irritated Spencer in its advocacy of legislation on a host of measures including government purchase of railways, restriction of alcohol consumption, regulation of safety in mines, factories, ships and lodging houses, institution of public health measures such as prescribing water closets in private homes, and the edification of the populace by free libraries and museums.[7] These demands were merely the corporate personality exercising its will, and Spencer viewed this as likely to result in an ill-informed and class-based despotism.

Spencer was dissatisfied with how his contemporaries conceptualized political theory. For most people politics was still illuminated by a residual Lockeanism, which had captivated his countrymen during the previous century. Scarcely anyone read Locke closely, yet he was often cited, and a vestige of his ideas lingered on. There remained a ghost of the social construct based on individual rights whose quasi-legal language offended Spencer's sense of modernity. He believed that discourse on justice should be restricted to principles of equity and fairness without reference to archaic norms and psychology. To cleanse language of error was his contribution to reform: he assumed the burden of correcting the myths of contractarians; their outmoded ideas would be replaced with a social science based on the developmental realities of social structures. Spencer believed that he would uncover the optimum path on which national life had organically emerged in the past. Such knowledge would prevent both practical politicians and utopian speculators from recommending artificial solutions to political problems. He disliked the way in which his contemporaries parroted artificial theories of politics sanctified by classical texts. A science of social structures should reveal natural structure, not dignify unjust traditional wisdom. As a liberal, Spencer continually railed against mindless perpetuation of aged political structures based on aristocratic status and military values. For him true reforms could not be accomplished by legislation that equated monarchs or lords with popular sovereignty. Sovereignty of any kind was a claim to privilege, and, as such, was as archaic as nobility, valorous leadership or the honour evoked in the keeping of promises. It was these ancient values that sustained the social contract and he believed that they should be seen for the nonsense they were.

For Spencer contractarianism was anachronistic. Moreover, it embodied an intellectual error, addressing individuals' reason as if this action could alter the direction of social change. Spencer held that modern knowledge had clearly established that national life could not be initiated by agreement between individuals; nor could it be understood by references to their intentions.[8] Spencer's political essays consistently hammered out an anti-contractarian doctrine. This refrain was taken up in his sociological works, and ultimately worked into a coherent political discourse. His task was to defeat classical political theory by replacing it with an account of the scientific functioning of modern institutions and mores. Spencer was not a pioneer in this endeavour; there had been eighteenth-century attempts to overthrow contract theory, but these had been based on moral philosophy, not sociology. From Spencer's perspective moralists had made the situation worse because, especially in the hands of Bentham's followers, they had initiated a revival of Hobbes's philosophy and an instance that a rational psychology should underpin political will. If, it was argued, each individual possessed a sovereign will that could govern their actions, then each could forge binding agreements with others. Spencer saw this as a baseless analogy between how individuals' reasons ruled their emotions or feelings, and how sovereigns ruled fellow citizens. The first part of the analogy was an unlikely and cruel account of the psyche that relied on the reason forcing other emotions to do its bidding. The second part was an attempt to convince modern citizenry to abandon the sophisticated national life that protected them by replacing it with an artificial account of how they should be ruled by the exercise of will. Not only was the analogy founded on bad psychology, its sociological basis was flawed. It overemphasized the importance of individuals in the administration of justice, the only area of politics Spencer thought was useful in furthering the development of national life. Those who conceptualized justice from the perspective of individuals overlooked the fact that justice too had evolved. Viewing equity and fairness from the point of view of the individual was a kind of recapitulation of departed societies. In the modern world individuals were primitive remnants; their desire for recognition and power should be set aside in favour of the principles encapsulated in social structures. Today interference with such structures by either a single despot or a democratic mass of individuals would be retrograde. It might even return humanity to a past in which social arrangements were dictated by personal will rather than principles.

Spencer's hostility to the collective doctrines of socialism and communism is so well known that it has led to the false impression that he was an ideological warrior belonging to the opposite camp: individualism. However, this is so far from the case in his mature political and sociological writings that he could be more accurately described as an anti-individualist. There is no good rationale to perpetuate the division of social thought as either collectivist or individualist as if this always leads to a useful analysis when, on the contrary, this approach generally lacks utility when one is classifying Victorian social theory. Spencer's political discourse, in particular, is ill served by this distinction. While it is true that he was consistent in his distaste for socialism and communism, this feeling was a subset of his hostility to democratic politics; he believed that the working classes would attempt to legislate advantages for themselves just as the upper classes had done.[9]

Spencer always avoided seeing the individual as an ideal, or seeing individual participation in politics as progressive; instead he regarded it to be primitive.[10] Individualism was a feature of an undeveloped society in which everyone was a warrior, hunter, angler or tool-maker. In such a society the only base differentiation in function was sexual. In evolutionary terms individuals were less important than the governing bodies, which showed "a continuity of life exceeding that of the persons constituting them".[11] Of these social entities the lesser ones such as private associations and local public bodies were more likely to suffer decay. Therefore, Spencer assigned more permanent progressive qualities to a national government that was likely to maintain its integrity as it developed in mass and structure.[12] He supported his thesis with an analogy between a biological organism such as a crustacean and a social organism such as a medieval state. Neither could develop without a central vascular system and this would be at the expense of autonomous component parts.[13] Spencer was a corporate theorist who saw improvement in individual lives as resulting from more sophisticated institutional arrangements. It was not a social gain if individuals directly or indirectly participated in such arrangements. On the contrary, Spencer regarded the theory of democracy as yet another revival of classical political theory that would hinder the development of justice and peace.

Spencer's pessimism on the subject of popular governance was directly proportional to the scope of its rule; the more institutions were controlled by democratic practices the more likely it would be that they were reactionary. Since universal suffrage had not yet become a matter for serious debate, he focused on the property qualifications an elector should possess. Although Spencer worried that a low qualification would encourage the poor to promote laws benefiting their class, he was unwilling to discuss the minutiae of suffrage, such as the relative disadvantages of "five-pound" householders voting compared to "six-pound" ones.[14] All such qualifications seemed to have the same faults as those they were to replace; none ensured that electors would possess the specialized knowledge needed when regulating complex public affairs. In Spencer's Britain parliament was inhabited by ill-informed members of the upper class, but the addition of ignorant representatives from other classes would not improve matters. In any case, changing the class bias of representation would have little effect since Spencer believed that the different classes were uniform in morals.[15]

Any change in voter qualifications would only have the minor effect of making politics even more faction-ridden, which he, in common with many of his contemporaries, regarded as a deficit. The notion that politics would be improved by functioning on party lines was still in the future. Thus for Spencer the problem of the popular selection of parliamentary representatives was a special case of the general danger that loomed over all organizations with elected governors. Municipal government, especially in New York, coloured his thinking here,[16] but he did not restrict his comments to civic institutions. Spencer believed that the same flaws would emerge in any organization that had an element of popular control. Even the harmless and voluntaristic Mechanics' Institutes crumpled under his gaze: they had begun with benevolent people combining their efforts to increase general literacy, and to gain other advantages such as the popular circulation of scientific journals. Such aims seemed sensible since these social goods were beyond the reach of individuals. However, Spencer observed that shortly after their initiation the organizations were blighted by indifference, stupidity and party disputes.[17] The more decent people among the membership – those who had the ability to administer – were driven from committees that left control in the hands of a clique whose humdrum decisions would fail to satisfy the original aims of the membership. Instead of an organization effectively accomplishing its goals, the result was the unwanted creation of a middle-class lounge.[18] Spencer did not confine his criticism to one class; he was equally scathing about the affairs of working-class trade unions and of upper-class organizations such as athenaeums and philosophical societies. He was not suggesting that cultural goals would be more effectively accomplished if one left them to private enterprise in the form of Mudie's Lending Library.[19] Spencer's Mechanics' Institutes were exemplars of why elective governance should be restricted to intervention in areas where even very stupid voters could reach agreement on collective needs.

Spencer's contempt for the average intellect of the people remained constant. He nurtured the prejudice from the early 1840s when he wrote in "The Proper Sphere of Government", where he ambiguously favoured democracy. His scorn had not diminished in the 1870s when he issued the early parts of *The Principles of Sociology*. It even ranked ahead of his constant bias against ceremony, which he usually saw as a prop decorating the primitive aspirations of the militarized upper classes. He was kinder to the bottom rungs of the social ladder: when thinking of the limited capacity of the working classes Spencer condoned political ritual as a necessary evil: "It needs but to name certain of our lower classes, such as colliers and brick makers, whose relations to masters and others are not such as to leave them at all restrained, to see that considerable evils arise from a premature decay of ceremonial rule".[20] Even by Victorian standards those views were trenchant. Except for Robert Lowe, the fierce albino opponent of 1860s electoral reform, few liberals had more pejorative opinions about the political acumen of the labouring classes than Spencer.

This bias did not usually interfere with Spencer's hopes for a progressive future. He was not constantly pessimistic about the advent of popular governance because he did not believe the problem to be worrisome. He did not consider that a group's lack of ability in following an argument or in assessing evidence should exclude them as citizens. While their mediocrity of intellect would be likely to result in the election of representatives who would be unable to oversee the administration of national life, these representatives could still be useful. For example, they could promote legislation against obvious social ills such as the loss of property through theft.[21] Spencer thought that even incapacitated voters and their representatives could perceive the need for protective legislation. The legislators could, if they followed the lead of wiser colleagues, be usefully employed by extending regulations laid down over many generations in the areas of constitutional law and administrative practice.[22] While national administration and policy were too complex for elected representatives to initiate new measures, they would be sufficiently skilled to reinforce well-understood ones such as those protecting citizens from internal and external aggression.[23]

The masses were not only unintelligent but, in Spencer's view, they lacked the capacity to express political will. This argument was directed at a common reform theory to the effect that political corruption and class bias would lessen as the interests of representatives became similar to those of the electorate. This, of course, implied that both individuals and classes were self-interested. The best-known version of this theory – that developed by James Mill – speculated that if the electorate were sufficiently large there would be a natural identity between the will of legislators and that of voters. Spencer rejected such political logic because it conflicted with his belief that several classes of electors possessed little or no will.[24] There were reasons for this: some voters prided themselves on being indifferent to matters that did not concern them; some, such as shopkeepers, simply cast their votes to please patrons, while others allowed their votes to be bought.[25] It seemed obvious to Spencer that many voters would favour representatives whose interests were at odds with their own. This was why the House of Commons was stuffed with members of the nobility, military officers and lawyers. Spencer calculated that MPs included 98 sons of peers or Irish peers and a further 138 were closely related to the nobility. Then, if one counted members of the armed services, one discovered 106 active or retired army and navy officers, with an additional 64 who were in the militia or yeomanry. Ninety-eight lawyers should be added to these tallies leaving only 250 parliamentarians (from a House of Commons of 654) whose interests Spencer believed to be a tolerable representation of the general public.[26] Spencer's analysis of officers and lawyers suggests that he considered them as groups that were similar to social classes in having strong collective interests, separate from the general public. Furthermore, he believed that officers would agitate for war while lawyers would militate against reforms of the legal system that threatened to decrease their income. It was clear to him that the composition of the House of Commons signified that most electors failed to recognize the "self-evident" truths that we may most safely trust people whose interest is identical to our own, and that it is dangerous to place confidence in those whose interests are antagonistic to ours.[27]

Since voters did not behave rationally one might have expected Spencer, as a liberal educationalist, to promote the teaching of civics to electors and their representatives in modern social sciences so that they could clearly perceive and promote national interests. Such a strategy would have synchronized with his objection that the archaic training parliamentarians had received in classics had crippled them. "None have that competent knowledge of science in general, culminating in the Science of Life, which can alone form a basis for the Science of Society."[28] Since, for Spencer, all political phenomena were natural ones and subject to the laws of life, it would have been plausible for him to have proposed that prospective legislators be trained in biology and organizational theory. He believed this would be an improvement on the spectacle of modern statesmen parroting the wisdom of Aristotle and Thucydides.[29]

Spencer was silent on the training of legislators. While he criticized the harm they had suffered from their archaic education, he offered no remedy. However, on the subject of educating the masses he was very vocal, declaring it a mistake. Their political stupidity could not be alleviated by increasing their literacy and numeracy. Instead of regarding working-class education as progressive, Spencer saw only a danger that would increase as learning spread downwards. Unlike J. S. Mill and Ruskin, Spencer found no positive correlation between giving the lower classes better grammar or mathematics and the discipline that was necessary for the people to perform their civic duties in a satisfactory manner. Instead of reading useful literature, many citizens might only read works that appealed to their class prejudices, and that offered fallacious arguments.[30] Workers would thus remain as ignorant as members of the upper classes, whose political information hailed from ancient Greece.[31] Spencer saw this behaviour as indicative of other classes who would be equally lacking in ability to foresee their own peril. Spencer feared that if the working classes gained power while holding their current views they would disastrously extend the scope of state administration.[32] When he elaborated on these views of the working classes in *The Principles of Sociology* it was an extension of his belief that education could not produce equality. "Human beings are unalike in the amounts of sensation and emotion producible in them by like causes: here great callousness, here great susceptibility, is characteristic."[33] Then, too, people who laboured mechanically were less sensitive than those who worked with their brains.[34] Education would not be able to overcome such psychological and occupational defects. While Spencer believed that society existed for the benefit of all its numbers,[35] it was obvious to him that many were unfit for any civic role.

Extended state administration meant despotism to Spencer. Not that despotic regimes, such as that of Napoleon III, were without advantages; they were able efficiently to accomplish goals that elected regimes could not. Despots were "business-like" in their conduct of public affairs because of their use of expert advice and technical knowledge.[36] To be "business-like" meant to demonstrate leadership, and to avoid the use of committees, which accompanied any kind of representative institution. In deprecating the organizational advantages of top-down management Spencer was parting company with those nineteenth-century liberals who perceived leadership as a panacea. It was Spencer's friend Morley who formulated the idea that liberals should be a vanguard who spearheaded national reform. In this image leaders would forge ahead of the less-enlightened populace, who could keep to the path to progress once it was clear of impediments. Spencer felt that such metaphors were not progressive, but dangerously archaic. They stemmed from liberals, such as J. S. Mill, reading too much Carlyle. The great Scottish sage had deluded people into subordinating themselves to single individuals with his talk of "hero-worship". Spencer's world contained no heroes, and he believed that the feelings that had created them were merely blind awe and a fear of power, which belonged to the dawn of humanity.[37] Such primitive emotions presupposed inferior natures in both ruler and subject. The ruler was imagined to be a cold and unsympathetic figure demanding that others sacrifice their wills to him, while the subjects were mean cowardly creatures.[38] This division of humanity into two repellent types did not belong to the present.

Behind Spencer's concern to prevent history from dominating modern society was the political theory he had begun to develop during the late-1850s and which later determined the direction of his sociological writings. This was based on an insistence that politics should not be regarded as a series of responses to individual character as if ruling and subordinate groups were permanent features of the human species that could not be transformed by social development. As an alternative, Spencer's notion of change relied on corporate transformation in which the psychological desires and fears of individuals were insignificant. At most, an individual's character could interpose a momentary gain or hiatus in a process that was as gradual as a geological one.[39]

Classical accounts of citizenship, such as Plato's, had erred in asserting that the state had grown from human character, and had mistakenly asserted that aspects of social organization, such as the division of labour, could be established by reason. Reliance on such flawed accounts was the explanation why there was still a prevailing conception that societies could be artificially arranged.[40] This presupposed that human characters were constant whereas for Spencer they were plastic features of humanity that had evolved as social functions changed. There was no reason to suppose that civic qualities that had been admirable in ancient Greece or republican Rome would also be constructive in modern society. On the contrary, Spencer objected to perpetuating an ideal of citizenship that would promote independent cunning creatures who loved conquest. The heroic qualities of Odysseus had been valuable in Homeric times, but now society required a citizenry that possessed communal values. If Spencer had a literary muse here, it was not Carlyle, but the Poet Laureate: "It is true, not only in the sense which Tennyson writes it, but also in a higher sense that – 'The individual withers, and the world is more and more'."[41] Present-day rulers who attempted to copy the individualistic virtues of the past would unwittingly promote personal characteristics that had become unpleasant as well as unnecessary.

Emulation of ancient virtues would bring to prominence those artful and crafty types who admired conquest and who had an implacable hatred of enemies. The result would be a misfortune for the modern world, which needed administrators who possessed a philosophic culture together with a desire for the general happiness of others.[42] Of course, Spencer knew that such requirements were utopian and could not be satisfied. It was, he feared, absurd to hope that cultivated and kindly administrators would either be born heirs to thrones or be elected by millions of voters.[43] He assumed that no human being would be found who was fit to reign over the workings of a modern complex society.[44] The training of princes, like the education of the masses, was unlikely to be successful so the optimum solution was to keep amateurs away from governance.

Individualism was encumbered with antediluvian notions about the importance of personal character in politics. Therefore, for Spencer, it was a doctrine to be excluded from constitutional discussion. As a psychologist he could not condone naive descriptions of the human mind to masquerade as political norms. Instead, he attempted to inform his contemporaries about the true nature of constitutional forms, which were hinted at in the biological world. They had the advantage of providing a practical account of how a social organism had acquired its current structure and functions. The modern organism was representative government because it was more capable than other forms of maintaining equitable relations between citizens,[45] although to Spencer, these individuals lacked the ability to foresee and support most national objectives. Curiously, from the progressive point of view of later liberal democrats, his preferred form was not the most efficient or intelligent option. It was simply the only one that would not be dominated by simple individuals who, as a rule, were untrustworthy.[46] Spencer believed that the mediocrity of intellect that accompanied popular rule would be grossly incompetent in administering the processes of national life, but this was a cost that should be borne as efficient government would be more disruptive to long-term social progress.

Spencer's scepticism about the effectiveness of popular government was labelled administrative nihilism by his friend Huxley. The latter was absolutely certain that popular education, together with energetic and strong leadership could overcome any impediments. Since Huxley was the scientist who had once guided Spencer through the complexities of biology, he had no patience with the claim that political processes were evolutionary developments that could not be hastened. He accused Spencer of diminishing the functions of government to a police force. Spencer was equated with Humboldt or J. S. Mill as a philosopher who feared that the state would grind freedom and variety out of human life.[47] Huxley dignified Spencer as "one of the profoundest of living English philosophers" and then denigrated him as a *laissez-faire* liberal who believed that self-interest would provide all the administration needed by modern society by preventing individuals from harming one another. The polemical crudeness of Huxley's characterization was typical of the way in which he confused his own bias with the national interest. He would have known that referring to his friend as a simplistic Manchester exponent of *The Wealth of Nations* was rude and inaccurate, but ridicule was his intention. In any case, caution was brushed aside since the stakes were high; the bellicose Huxley was on a mission to promote the value of scientists in a new arms race with Prussia. Accuracy was unimportant given that the purpose was to ensure that the target would suffer collateral damage. This was also the rationale of Huxley's comparison of Spencer to the French anarchist Proudhon, whose anti-militarism was as well known as Spencer's. Huxley was giving his readers the opportunity of recoiling from a non-patriotic and socially irresponsible radical. Spencer's continuous insistence that a strong military was antithetical to industrial progress was seen as anarchistic by Huxley, whose position as a scientific adviser in the new industrial–military complex gave him a theory of strong governance.[48]

Spencer was taken aback by this assault from a man who had been one of his closest supporters only a decade earlier. The shock made him provide a clear exposition of how his liberalism differed from that of his contemporaries. He was so surprised by Huxley's yoking him together with Mill and Gladstone that, for once, he expressed himself in a way that was unencumbered by ethnographic and psychological data. Spencer made it clear which liberal values he did not share with Huxley's other targets. Hostility to state intervention was not, Spencer protested, caused by his adoption of Humboldt's theory of "police government".[49] He had not read the Prussian, and, in any case, he did not support a *laissez-faire* theory of government. On the contrary, he had always claimed that he was in favour of the state's engaging in more active regulation.[50] Spencer actively insisted on the need for positive, as well as a negative, state activity; both rested mainly on his conception of justice. The law should be proactive in searching out and remedying abuses. Instead of the state being content with punishing petty crimes, such as the theft of firewood, it should investigate and correct more important misdemeanours, such as great frauds.[51] Another positive function for the state was the need to defend the community against external aggression. This extended the need for state regulation; national preservation was an indispensable state function because humanity had not yet evolved beyond the predatory period in its development.[52] Spencer's arguments against extensive state activity were largely practical. For example, his stricture against banking regulations was based on a belief that the state produced inadequate policies in this area. Similarly, his general position that the state should chiefly focus on justice was pragmatic. If, he argued, the state avoided entangling itself in most areas of regulations then it would become more efficient in promoting justice.[53] There is little similarity between J. S. Mill's principled reasons for the need for a sphere to protect individuals and Spencer's statement: "Not only do I contend that the restraining power of the State over individuals, and bodies or classes of individuals, is requisite, but I have contended that it should be exercised much more effectively, and carried out much further than at present".[54] To Spencer the protection of individuals was not the purpose of a representative assembly. Instead, the business of a parliament was to regulate class interests: its task was "to average the interests of the various classes in a community".[55] Spencer refers to *averaging* as "balancing" interests, which make it clear that his electoral politics is pluralistic. He had avoided the extremes of anarchism and Huxley's directorial governance in favour of an ideal of politics as the reconciliation of conflicting group interests. This was, and possibly still is, a conventional liberal perception of the function of elected assemblies.

Huxley had missed the point of his friend's evolutionary politics; Spencer was not opposed to the practical intrusions in governance by the new professional administrators, such as Huxley, but he was against ill-informed democracy, which would, however well educated, have a regressive effect on progress. Populist aspirations would push the state into adopting generous welfare policies instead of concentrating on its core business of improving justice. At a superficial level, Spencer's views resemble those held by twentieth-century rights theorists who wished to protect the individual against state intrusion. However, Spencer would have regarded such aspirations as unfavourably as he did socialism. Individualism was yet another antique dogma that had nothing to offer the modern world. To Spencer it was equally anachronistic whether individuals were reinflated as Carlyle's heroes or transformed into democratic citizens exercising their sovereign wills.

There is a rift between *The Man "versus" the State* and Spencer's politics in general. This was caused by his fear that democracy, which had suddenly emerged around him, would cause society to revert to a condition in which slavery rather than liberty would be the norm. The tensions within Spencer are more visible in this late work than in his other writings. In it he only occasionally felt the compassion for the dispossessed that so marked his other writings. In its place was an increasing loathing of the masses. It was no longer their lack of intelligence that affronted him; it was their lack of moral fibre. He began to rail against the "good-for nothing", the parasites living from the labour of others.[56] This intemperate moral language did not penetrate deeply into *The Principles of Sociology* – the momentum and scale of his scientific writings were too vast to be altered – but, henceforward, it directed polemics. His followers began to warn that the cumulative effect of the many small legislative changes fostered by the working classes meant "the gradual and insidious advance of a dull and enervating pauperism".[57] This had happened in England before the old Poor Law had been changed, when it had caused a "weakening of the fibre of character".[58]

The new Spencerian harshness began to compete for space with the progressive sentiments that he had borrowed from his sociological writings, where his "passionless neutrality" and his sense of justice had informed his benevolence. *The Man "versus" the State* is not a coherent inspired outpouring like J. S. Mill's *On Liberty*; it is a synthetic product in which increasing irritations with the masses jostle for space with mature fragments from his ethical sociology and humane remnants from his earlier writings. For example, he could still support the idea of women escaping from tutelage, and could claim, enthusiastically, that he had always been a radical on this subject. Similarly, there was little novelty in Spencer's objection to colonial tyranny under which a single murder would provoke a wholesale massacre of indigenes. Then, too, when *The Man "versus" the State* objected to rights language and to the social contract theory of Hobbes and Rousseau, his contemporaries would have been struck by a sense of *déjà vu*. Nonetheless, this work possessed a stridency that was missing from both his earlier political essays and his sociology. Where he has been content to wait for freedom to emerge in the course of future evolution, he was now anxious to protect it in the present. Now his fellow liberals had relinquished their grasp on the reform agenda when pandering to the mass voter.[59] Reformers had noticed that, in the past, legislation had indirectly produced benefits for the populace; now they mistakenly thought that these side benefits were their goal, and that liberals should aim for popular good.[60] Spencer believed that reform politicians had become confused by the need to seek electoral endorsement, and, these, strayed from the true path of their beliefs. It became his task to remind them that it was unimportant if officials were elected; the only significant matter in politics was whether or not there was interference with the lives of citizens.

One aspect of Spencer's argument here was consistent with the idea that had guided his political speculations for two decades. That is, there was no novelty in his statement that those who use their liberty by surrendering it are no less dominated than any other slaves.[61] Spencer had frequently objected to the primitive practice of submitting political decisions to the exercise of will. It seemed to him that an expression of will through popular sovereignty was as anachronistic as the exercise of prerogative by a monarch; neither added legitimacy to a decision. When Spencer refused to accept that a citizen's liberty was enhanced by being derived from an elected government,[62] he was treading a familiar path. When, however, *The Man "versus" the State* declared its support for negative liberty, it was venturing on new territory. Before, Spencer had insisted that he was indifferent as to the extent of state regulations so long as they furthered justice. This permissive stance towards government was now replaced by a refusal to condone any restraints on individuals other than those necessary to prevent them from committing aggression against others.[63]

The intrusion of an element of negative liberalism undermined the coherency of Spencer's politics. Outside *The Man "versus" the State* his political philosophy more or less harmonized with his sociology. Now, when he offered a scientific aside, such as that all societies have structures and laws that override individuals, it was disjunctive. Spencer could no longer reconcile the ethical qualities of individuals with their scientific ones. While he still scorned the facile way in which other philosophers had ascribed either natural or artificial rights to persons, it was difficult to see how Spencer would protect them without such useful devices. In addition, Spencer's inability to see legitimacy in a particular form of government made it difficult for him to identify despotism as anything other than a state that gave preference to a group, which, in modern times, would be the working class. Without his scientific ballast, Spencer's philosophy was not very useful. Negative liberals could not have found much guidance in his cautionary statement that legislators should refrain from meddling with individuals because these were the origin of societies, and it should be assumed that "there must be some kind of order".[64] Since this order projected from life itself, someone seeking further instruction from Spencer would have had to turn to his scientific works on biology and sociology, where the vital impulse was surveyed. This, of course, meant that prospective Spencerians would have to take their bearings from older texts that did not support negative liberalism.

A reader would be well advised not to interrogate *The Man "versus" the State* too closely. Its flaws were not caused by Spencer's difficulties in writing a political philosophy without help from convenient fictions such as rights and utility; rather, the contradictions in the work went deeper and stem from Spencer's emotional retrieval of his own family's involvement in welfare relief. He had been provoked into dropping his scientific coolness because he was enraged by political demands from "good-for-nothings". It was not the rational plans of socialists that had irritated him; he had been calmly objecting to these for years. It was the prospect of a selfish group successfully demanding welfare relief. This struck Spencer as a coercive attack on the natural balance of politics, and superheated his response.

Rather than granting claimants some immediate relief, Spencer wanted them to be faced with such unhappiness that they would be reformed.[65] This recommendation was a return to the harsh Christian doctrine that the Reverend Thomas Spencer had enforced with the administration of the Poor Law when Herbert was a child. Although he had earlier rejected his uncle Thomas, as well as Christian industry as cruel and uncivilized, it was now as if they had never been cast aside. The ambivalence Spencer had felt towards his uncle in *An Autobiography* vanished; Herbert had forgotten how inhumane Thomas had been.[66] *The Man "versus" the State* saw welfare in terms of early-nineteenth-century charity. Another early message also intruded on Spencer's consciousness. Parliament had not, in 1833, foreseen the negative consequences that public education would have half a century later.[67] An unintended consequence of their policy was the election of working-class parliamentarians, who were making economic and social decisions that were likely to cause the nation to return to a condition of domination similar to that produced by military societies in the past. Such a danger could only be avoided if politicians refrained from pandering to their popular electorates.[68] The novel demands of popular politics annoyed Spencer because, unlike his own early scheme of the nationalization of land, they were not designed to increase social justice. It irked him to be lumped together with H. M. Hyndman, who was pleading for the interests of a single class.[69] If socialism was not about the just apportioning of property, then it must be concerned with control. Spencer did not need to produce sociological evidence for this proposition because, as a psychologist, he could see directly into the modern psyche, where lurked a desire to force some of the population to labour under coercion.[70] If he had possessed his customary phlegmatic calmness he would have noticed the contradiction that undermined *The Man "versus" the State*. That is, he would not have been able to reconcile his own wish to reform people by forcing them to work with his condemnation of this desire in others. If he had been consistent with the views he had expressed in his scientific work and *An Autobiography* he could have never punished his fellow citizens by condemning them to industry. Spencer's new punitive attitude occupied the space left vacant by the disappearance of his enthusiasm for nationalism. His political thought had been guided by his dismissal of electoral systems and of other artificial solutions in favour of a natural social harmony that informed both national society at large and its bureaucratic apparatus, as long as both kept to their allotted function of enforcing justice. While he had always been suspicious that popular participation had the potential of disturbing this balance, in *The Man "versus" the State* this was transformed into a fear that the people were on the brink of seizing electoral power. To Spencer this hinted at a catastrophe that would, at least temporarily, arrest natural progress. In his sociological writings – whether these were issued before or after *The Man "versus" the State* – nationalism had been the corporate expression of life. With this faith in evolutionary force he could overcome his fear of the people; they would be guided towards a social condition where they would be exempt from the selfish and destructive attitudes that had marked upper-class politics. In his maturity, he had two competing political discourses: an optimistic one that continued to inform his sociological writings, and another, mildly pessimistic, which was a polemic against the working classes. There was a limit to his fear of the populace because he was grudgingly still able to envisage a future when humanity would evolve away from its predatory and individualistic origins, but this goal now seemed further away, and would have to be achieved without bureaucratic help. He had come to the conclusion that the progress of social justice could not be forced.

Conclusion

Spencer took Life in general, and his own life in particular, too seriously. Even his attempts to enjoy leisure were forced and slightly desperate. However, excessive earnestness is not necessarily repellent, and, as his biographer, I have adopted a neutral or non-critical approach when assessing his gestures at humour or his awkwardly playful relations with others. I have also tried to give him due credit for the successful construction of a personality open to feelings and sociability, qualities that were often missing in members of his family and in many of his acquaintances. While most people can be said to naturally possess a personality, Spencer consciously contrived one. This was a defensive move on his part; he was excessively sensitive and needed to fortify himself with the only capacity he possessed in abundance: the ability to scrutinize his own psyche and those of others. His personality, like his writings, was a creative achievement that gave him the standing to interact successfully with strangers. Since Spencer was aware of his limitations when relating to others, this also provided him with an analytical tool to compare the desires and sentiments of other individuals and societies with his own feelings. This was never cold; his psychological analysis did not reduce people to objective and impersonal data with no connection to himself.

In assessing Spencer's artificial personality I have avoided the temptations of either sympathy or hostility. Instead, I attempt to adjudicate between his own explanations and the probable truth. When, for instance, I explain what parts of his reputation for eccentricity were deserved, I focus alternately on his sometimes improbable attempts to reform or educate and his genuine lapses into quirky behaviour, which occurred when he lost sight of the fact that his actions invited ridicule. Rather than offering a consistent interpretation that would overlook the nuances and contradictions of parts of his character, I have simply offered the best-case scenario for each fragment; the result is a curate's egg whose synthesis owes a small debt to biographical art and a great deal to Spencer's intentions. The latter, of course, are difficult to decipher. He spent much of his adult life defending his shifting psychological attitudes; it is, therefore, not surprising that these appear in his philosophical works as well as in his autobiography. It is no easy matter to decide at what points Spencer's behaviour lacked a deliberate rationale. At times, when his rationalizations for actions seem improbable, the poised neutrality I adopt is hard to maintain. For example, when I contemplate the unpleasant qualities associated with Spencer's hypochondria, or his justifications for kissing young girls when this made them feel uncomfortable, I have been tempted to brush these aside as personal foibles. Occasionally I would have wished to chronicle only his more benign and acceptable beliefs, such as his desire to save both human beings and animals from pain and cruelty. However, such a partial strategy would have resulted in a division between Spencer's individual psycho-pathology and the adumbration of a universal ethics of which he would have disapproved. It also would fly in the face of Spencer's habit of fashioning his life and philosophy so they were parts of an integrated whole. I have tried to respect Spencer's wishes here, and, as a consequence, take his self-justifications and apologies quite seriously. As much as possible, his statements have been tested for veracity and plausibility, but never in such a way that falls outside his own mental universe.

Spencer's own painstaking autobiographical analysis should be accorded a measure of regard even when it appears as failing in sensibility or decency from the perspective of the twenty-first century. His courage when analysing the faults and weaknesses of himself or his society deserves a better reward than being made the subject of casual censure, or furnishing an illustration of the coyness of Victorian sexuality. There should be no moral to Spencer's story; only a truthful and historically nuanced account of his life and ideas. The need to provide a reckoning of a human life is not an ethical imperative, and even if it were Spencer would be an unlikely choice for a moral exemplar. Yet, even if it lacks normative value, Spencer's life should not lack a conclusion. That is, even if there is no ethical lesson to be learned from a biography, it should not only be a series of amusing and quaint tableaux about a man's personal struggles and achievements.

The need for a non-moralistic ending to this book raises a query as to what could possibly serve as a conclusion to an intellectual biography. The separate sections of this work are largely historical, and, as Michel Foucault has observed about method, discovering the past involves an archaeological exploration of horizontal layers of time.[1] Applied to Spencer's writings this suggests that the layers of his knowledge about psychology, biology and so on are properly explicable in terms of their temporal contexts, each of which lasted less than a decade. There is no single pattern in the "linear successions" of Spencer's ideas over his whole life. While his personality demonstrated some repetitive features, his philosophical work was protean and displayed discontinuous features in successive decades, even when he believed in its constancy. As a biological subject he lived through and beyond the slices of time during which his ideas had a context. So my conclusion of his biography must reach outside historical method in order to provide meaningful patterns. That is, while I have my feet firmly planted in the clay of historical fact and fancy I feel obliged to provide something extra: a biographical explanation of how Spencer existed as a person living outside, and through, his contexts. If for no other reason, this is desirable to prevent future readers from innocently accepting Spencer's own account of his significance.

Spencer twice wrote the conclusion to his own intellectual biography. He did this briefly in 1890 in "The Filiation of Ideas" which was included in a posthumous book that he controlled by issuing precise instructions to his executors.[2] Spencer also reflected on himself at length in his enigmatically titled *An Autobiography*, which he rearranged and polished until 1889. Both Spencer's conclusions stressed his originality as the primary explorer of human and biological evolution. He also gave himself credit for being a self-created novelty, whose scientific and philosophical ideas had been acquired without the assistance of a formal education and outside the framework of the academy. In addition to these claims Spencer listed his staunch adherence to a political radicalism under whose banner he had warned against the evils of state, militarism, colonialism and cruelty to children or animals. Finally, he applauded himself for being consistently hostile to the aristocracy, and for his depiction of women as the current repository of non-aggression and civility, which in the future would spread to humanity in general.

Spencer's self-images were designed to protect his future reputation. No other interpretations were to be countenanced, which was why he went to such extraordinary lengths to control his posthumous reputation by repossessing and destroying the vast bulk of his personal and family correspondence. However, his precautions were futile, because, by the time Spencer employed them, he no longer attracted sophisticated readers. At the end of the nineteenth century most of his devotees were self-educated folk who could not easily penetrate Spencer's clouds of empirical detail and ironical self-deprecation. Nor were they capable of the analytical scrutiny that he expected his followers to apply to philosophical and literary texts. All they could perceive was the prime claim: that Spencer invented evolution, and was therefore responsible for Darwinism and the harsh social philosophies that were built on it. Spencer's other pretensions were disregarded: *An Autobiography* and his most impressive writings, such as *The Principles of Biology*, *The Principles of Psychology*, and the first two volumes of *The Principles of Sociology*, were seldom read. If these works had been carefully analysed by late Victorians, and succeeding generations, then they would have found voluminous textual evidence of Spencer's intentions to demonstrate that life in general, and humanity in particular, was evolving away a cruel and predatory past. It would have been clear that rather than regarding competition and natural selection as principles guiding current human development, Spencer had consigned these to the pre-institutional past of humanity.

Spencer himself is partly responsible for the way in which he has been misrepresented. His memory was poor, and, when he wrote the conclusion to his intellectual life, he could no longer accurately reflect on the philosophical biases he had adopted in the 1850s, nor his anti-teleological and anti-Darwinism stance of the 1860s, nor that the racist message carried by his psychology had had to be corrected by his sociological writings during the 1870s. His habit of reissuing unaltered texts, and of appending corrections to these without removing discarded ideas, allowed him the false boast of being consistent as well as constant in his values. Instead of this claim embellishing Spencer's posthumous reputation, it has encouraged the mistaken belief that any of Spencer's statements stood for the whole, which, in turn, made him deceptively easy to criticize. The resulting scholarly confession has obscured the study of Spencer, but it has been less of a hindrance than his mental condition in the years 1889 and 1890 when he completed his memoirs. His summations are the feeble products of a man in his late sixties who felt ill and vulnerable to attack. He was troubled that he had been consigned to the past by a new generation of critics, who could scarcely distinguish between his work and that of contemporaries such as Comte, J. S. Mill and Darwin. Spencer was poorly placed to respond to such misunderstandings: he could scarcely remember how, in fact, his ideas had related to those of other eminent Victorians. He was also likely to take any critical comment as a personal attack on his claims to have priority in scientific discovery, when his defence should have been to argue that his originality had been in his philosophical acuity. He had forgotten what it was he had attempted to do; his priority claims about the discovery of knowledge was a myth that would have made sense only if he had been a working scientist.

In reality Spencer was a philosopher, not a scientist, and his productivity had come to an end in 1886 with his severe depression. While he continued to issue parts of "A System of Synthetic Philosophy" for another decade, these last effusions lacked verve. Spencer had lost the capacity to process and respond to the masses of empirical information produced by the newly formed professional cadres of biologists, social scientists and philosophers of ethics. Those parts of *The Principles of Sociology* and *The Principles of Ethics* that first appeared in the 1890s were motivated by a sense of duty, and divorced from the concerns of his contemporaries. Spencer's last works were usually ignored by contemporaries, although some praise still came his way because Spencer had lived long enough to become so famous that it scarcely mattered what he said. His decrepitude was a pity because his belated comments on sociology and ethics included the subjects of industrial institutions, justice and altruism; if he had still been capable of breathing life into these ideas then he would not have been so disregarded by intellectuals of the *fin de siècle*, who were, in general, enthralled by any new work that discoursed on industrialism and social justice.

Spencer's responsibility for the misapprehension of his life and works leaves his biographer with an intriguing task. Obviously, his intellectual life has to be rescued from his autobiographical ruminations and provided with a conclusion that makes sense of the precise intellectual contexts in which he flourished. It is crucial to disentangle Spencer's arguments and attitudes so they variously represent the periods during which they were contributions to, or reactions against, political and scholarly debate. Most of these contexts were mid-Victorian ones that had lost their vitality by 1889–90. That is, Spencer is historically meaningful when, as a philosopher, he was able to respond to, and amplify, the concerns of metaphysicians, psychologists, biologists and sociologists who were thinking about their disciplines between 1855 and 1880. However, by itself, historically salvaging Spencer can do little more than correct error while reimagining why many Victorians placed their spiritual trust in him and how it was that his sense of modernity captivated so many of them. At his apex in 1872 Spencer was painted by John Bagnold Burgess.[3] This portrait of the philosopher has him gazing calmly and beatifically at his observers. By his side are some sea shells, typical of those reassuringly humble works of nature whose symmetries had so pleased him in *The Principles of Biology*. At the time of sitting for this artist Spencer had just completed *The Principles of Psychology*. He was also about to publish *The Study of Sociology*, and was anticipating making further pronouncements on the future direction of humanity. It was this contemplative Spencer who was both feared and admired as the epitome of progress. Such sentiments did not stir philosophical admirers only, such as Fiske; the most sensitized Spencerians were poets and novelists who reacted to their intellectual leader as if he were a seer:

> I talk'd with Bastian, who affirms
> Spontaneous generation proven,
> And, prone with Darwin, watch'd the worms
> Wriggling like live eels in an oven
> Then finally, in sheer despair,
> Burn'd deep with Scepticism's caustic,
> Found Spencer staring at the air,
> Crying God knows if God is there!
> Robert Buchanan, "The First Christmas Eve"
> I thank them, sage and seer, that thou has brought
> The widely wandering intellect of man
> Back to where its childhood's walk began;
> Where the reality we long have sought
> Through clouds and mists and vapours phantom-fraught
> Is found upon the firm familiar earth,
> While flowers are bright and children make their mirth,
> And mines of knowledge ask but to be wrought.
> Joseph John Murphy, "On Reading Herbert Spencer's
> Psychology", in *Sonnets and Other Poems* (1890)

Mary Augusta Ward could blame Spencer for a temporary fad at Oxford, "an overblown rationalism", which had caused a reaction to a religious romanticism that flourished in trivial souls.[4] Olive Schreiner, who was more sympathetic to novel spiritual yearnings than Mrs Ward, gave Spencer's philosophy pride of place as the transition between God-inspired traditional wisdom and scientific knowledge. Schreiner's Spencerian spokesman was Waldo's stranger who was given to lecturing those still mired in religion: "In the end experience will inevitably teach us that the laws for a wise and noble life have a foundation infinitely deeper than the fiat of any being, God or man, even in the ground work of human nature".[5]

Looking back from the twenty-first century, it is hard to hear these antique rhapsodies on the tensions between knowledge based on faith and knowledge gained through scientifically grounded experience. Like agnosticism and Carlyle's "nescience", Spencer's spiritualism only had significance when Christianity and science were both laying claim to certainty. Spencer's "Unknown" took its meaning from a contemporary agony of indecision over how to balance divine inspiration and human experience. However, such halfway houses no longer offer any shelter: religious faith is no longer bedevilled with doubts, while the life sciences, which Spencer searched for ethical meaning, are now more likely to raise moral quandaries than scientifically guaranteed solutions. Where Spencer and his early sympathizers often felt a sense of ethical satisfaction while looking at the far edge of human knowledge, the twenty-first century is likely to feel spiritually puzzled at advances in areas such as genetic modification and DNA mapping. Scientific knowledge is increasing, but, somehow, it is more unlikely to offer a meaningful substitute for religious feelings. Although it is still possible to feel uncertainty when choosing between God's wisdom and scientific knowledge, this sensation is not accompanied by great anxiety. The dynamic between religious faith and science has changed; much of what now passes for religion no longer establishes serious knowledge claims, while science does not fill its devotees with hope for the future. In brief, Spencer's evolutionary philosophy can no longer substitute for a theology. Further, his scholarly comments on botany, sociology and human prehistory have also dated. While they were often perceptive, they are not easy to locate within the current literatures on these subjects without doing serious damage to Spencer's intentions. For example, while it would be possible to reconstruct Spencer's evolutionary altruism so as to challenge a reductionist sociobiology, this strategy would never have occurred to most of Spencer's generation, for whom Darwinism was a theory restricted to evolution in the animal kingdom.

My conclusion – that Spencer's prophetic role has come to an end – contains greater truth if one is thinking as a historian. That is, it has more validity if one is focusing on subjects that seem locked in the past. It would be difficult, for instance, to reinflate Spencer's aesthetic sensibilities by insisting that art should be non-functional and display harmony and variation. While he believed these qualities were the essences of modernism, subsequent forms of artistic expression have thrown them aside. It would also be difficult to imagine any form of twentieth-century aesthetic modernism that would countenance Spencer's separation between instinctive drives and beauty. Similarly, it would be unlikely that Spencer's "prevision" (his attempt to predict which line of empirical investigation will be successful) or his exciting and lucid theories of biological classification will be revived. It would be even more improbable to suggest that Spencer's variety of association psychology will live again when one considers that William James was the last well-known exponent of this kind of discourse. Finally, a detailed account of why Spencer's sociology made an impact on Durkheim and American ethnographers, but not on Weber, should not be part of a conclusion to a biographical work on Spencer. Sociology developed as a discipline after Spencer ceased to write, and it did so in a way that was disconnected from his metaphysical and ethical philosophy. It is, of course, possible to write a history of sociology that would account for the fact that its professional development reflected late-Victorian philosophical ideas rather than those of Spencer, but this task belongs exclusively to the realm of intellectual history. In any case, statements about Spencer's influence on the development of aesthetics, empirical philosophy, psychology and sociology cannot form the summation of an intellectual biography, because while this is a genre that uses archival tools, it has to escape from the discipline of history.

It is not that conclusions of historical studies lack value. On the contrary, they illuminate the past when they deal with a great statesman, a decisive event or the mobilization of a large population. The goal of a historical analysis might very well be to say something imperative or typical about social or political change: this cannot, however, descend to the level of a single individual even if, like Spencer, he were widely read and discussed. There is an additional complexity in that Spencer was, in many ways, so untypical as to be considered divorced from contemporary attitudes and mores. While his personality was within nineteenth-century norms, his ideas were not. Rather, they were expressions of alienation from common aspirations. If he had been able to conform to the sentiments and behaviour of his contemporaries then he would not have been alternately greeted with praise or ridicule.

Historical considerations can inform an intellectual biography, but for conclusions one has to look elsewhere. What is required is an assessment of which of his ideas could possibly be a useful legacy for the twenty-first century. The candidates are few since Spencer was an empirically minded philosopher who based his ideas on collections of data that are now divorced from research. Then, too, while one can admire Spencer's honesty and humanity, it is difficult to infer that he therefore possessed a personality that should be emulated. His personal values were so inextricably linked with qualities of hypochondria, aesthetic absolutism and idealization of gender as to make them hard to admire. As a result, it seems that neither Spencer's philosophy nor his persona can be transfigured into universal lessons to be studied and absorbed by successive generations. In any case, to treat him as a "classic" would offend against his evolutionary philosophy, which was resolutely opposed to the elevation of universal and traditional values at the expense of transitory ones. What is required for a conclusion to this book is a survey of Spencer's reflections on humanity and politics in order to extract any that may still impact on the modern reader, even if they do not excite admiration.

Spencer saw the future of humanity as one of sentient harmony with the rest of the organic universe. While the grandeur of this vision fascinated Victorians the modern reader is likely to be unresponsive. A claim that the human species is the pinnacle of an organic universe is now unlikely to attract the scientifically minded, while the religiously inspired have little patience with philosophical exposition or empirical data on natural processes. It is unlikely, therefore, that Spencerian cosmologists will ever again become as important as they were at the close of the nineteenth century. However, this major shift in intellectual desires does not completely dispose of Spencer as his ideas remain valuable in two other discourses that still possess relevance. That is, Spencer has left us with two non-cosmic legacies that may be of value for humanity at large. The first of these emerged from the clash between his personal life and his technical writings. Since he never became inured to emotion or pain he eventually modified his evolutionary philosophy to cope with these. That is, he decided his early progressive aspirations for humanity had been at the expense of individuals' feelings, and that he should encourage human emotions so as to promote the future development of humanity towards a condition where pain was to be banished. Spencer's second legacy was a reconceptualization of liberalism so that it could not be absorbed by democratic politics. "Classical" liberalism was usually constructed on the importance of choice-making by individuals, which meant that it has often been conscripted as a defence of democracy. In opposition to this Spencer's ethical liberalism was couched in a way that avoided both individualism and choice-making. Liberalism was to be a corporate doctrine stressing the need to protect private lives even if, as was likely, these served no observable public function. This was a critical matter to Spencer because he could not conceptualize a way of giving privacy a functional role in maintaining a political society. Privacy was simply a consequence of a gain in civilization, and would be threatened if it were pressed into any public use. In any case, Spencer did not idealize political activity: politics, whether conventional or rebellious, was a sphere of activity that harboured primitive values. The evolutionary gains of private lives were unlikely to find support in a political arena that was evocative of the coercive and predatory behaviour of the past. Further, Spencer was largely uninterested in offering a principled defence of this particular evolutionary gain; such an argument would be too easily confused with individualism with its justifications based on egoism or selfishness and its cult of great men. In Spencer's altruistic and non-aggressive future such ideas would seem as detestable as any other reactionary political values. Egoism and the ideal of the great man focused excessively on the importance of single persons; as a prophet of social evolution Spencer could hardly uphold the primacy of the individual when he viewed that as a condition of pre-social humanity.

When it is realized that Spencer was a corporate thinker rather than an individualist, then his argument for the need to give a paramount place for the emotions becomes more easily explicable. It is the feelings that must be protected from the rational will even though philosophers had often argued for the rationale of subjecting desire to reason. To Spencer emotions, such as love and grief, did not exist in a vacuum, but were socialized responses possessing the same moral standing as reason. When philosophers used terms such as "will" they had done so in order to associate it constantly with one feeling, reason. Spencer believed that this was an intellectual conceit that elevated reason above other feelings, in contradiction to evolution, which had placed it at the same level in the human psyche. Spencer was adamant that reason was no more, or less, than an emotion, and was no more reliable when it came to the kind of decision-making that would deliver long-term benefits to either the individual or the group. It was obvious to Spencer that some emotional responses were antiquated and should be avoided, but this did not indicate to him that these were naturally subordinate to reason: that too had undergone evolutionary development. Spencer's target here was the habit, which some philosophers had popularized, of importing an archaic political theory into the philosophy of mind so that the reason was treated as the ruler, a supreme or dominant feeling, who could command the obedience of the other emotions even if the consequences of this were painful. He could see no rationale for adopting such a psychological description other than adherence to the outmoded belief that the mind worked like a monarchy under which private desires should be sacrificed to the public good. According to this model the ruler could cruelly instruct the citizens to endure pain and suffering in order to achieve his or her long-term goals. Since the analogy was between the mind and a primitive model of a polity, Spencer felt that it should be avoided when speculating about the most significant kind of evolution: the future adaptation leading to human emotions becoming more harmonious with the social environment. Spencer steadfastly refused to accept ancient political metaphors as part of his vision for the future. While other psychologists saw the expression of emotion as a struggle for freedom, and speculated that individuals reached maturity when they exercised control over their emotions, these ideas were not shared by Spencer. Instead, he portrayed emotional contention as a striving for development of potential and for perfection. Human progress could not, he felt, be achieved through a renunciation of emotions of the kind that happened when duty and privation overcame desire and grief.

In resisting the despotism of the rational will, Spencer was not advocating cerebral anarchy; instead, he was pursuing the evolutionary ideal of a futuristic world populated by non-impulsive and restrained people who were not coerced into this condition, but who were internally governed by a collective decision of their feelings. This is a sort of psychological democracy with each person having a mental council before which every individual action shall have been fully and calmly advocated. In this deliberative process the voice of reason would not be imperative: if it occupied the "presidential chair" of the mind it would occasionally have to give way to emotions rather than acting as a tyrant. Spencer had replaced the rational self-control of his youth with a representative government that would avoid the imposition of pain.

Spencer never felt that the imposition of pain was justified in either psychological or political governance. Pain, whether it was caused by deferred pleasure or by cruelty, was something he believed would diminish in the course of evolution. Its lingering presence in modern times merely showed that social institutions and individual psyches were still immature in their development. Since Spencer recommended an increasing pursuit of pleasure it would be tempting to classify him as a hedonist, but to do so would be to misunderstand his intentions. In his future, the psyche's decision to act, or to refrain from action, would not be based solely on rational calculation on how to achieve happiness, but would be partly governed by emotions that would be modified through socialization and the evolution of instinctive drives. This would be only partly voluntaristic; Spencer's human being would continue to exercise only partial control over his or her actions. His stance here was extremely optimistic as the only barrier against human reversion to violent and pathological behaviour was his speculation that social evolution would have largely eliminated these over the course of time, and that institutional systems would develop in order to control any remnants of recidivist behaviour.

Spencer's conception of a parallel evolution of psyche and society went to the heart of how he perceived the progress of civilization. In the past human development had often been a violent process in which the will of single individuals had been impressed on the populace. However, such personal tyranny, and the way in which domination was exercised, represented a transitory phase in human development. Spencer did not perceive the existence of domination in the past as suggesting that there was a general social law that made the imposition of a governor's will on a modern people or a nation a progressive development. On the contrary, such an action was now likely to undermine civilization rather than to propagate it.

The "will", whose awakening romantic nationalists had taken to be the beginning of modernity, was merely a pre-institutional and outmoded psychological concept, whose recapitulation would lead to unnecessary conflict. If Spencer had been better schooled in the sources of German nationalism he might have blamed Fichte and Herder for dressing the notion of the will in modern garb but, in common with other Britons, he was better acquainted with Goethe and Carlyle. Accordingly, he focused on them as the authors of the notion of the will as the engine of progress. They had claimed to map the path of civilization by plotting the historical movement of the will as it appeared in a series of great men, enabling them to subject their contemporaries to spiritual or political discipline. Spencer saw the reinscription of tales of greatness and heroic images as giving undue credit to archaic cultural practices that, in reality, were repellent. To him the resurrection of this past was dangerous as well as offensive; it encouraged a psychological passivity, a mental state in which people imagined themselves as civilized to the degree that their emotions obeyed the commands of the sovereign will.

Domination was not the only problem Spencer discovered in the work of Goethe and Carlyle. He also condemned them for their attempt to promote civilization by recycling one of the key doctrines of Christianity in a secular dress. That is, even though God was missing from their visions they had assigned the doctrine of renunciation an important role in improving humanity. This difference was that instead of sacrifice being desired by a higher being it was now demanded as the price of progress. Such a substitution offended against both Spencer's academic views on psychology and his heartfelt personal discovery of the effects of pain and loss. Spencer felt it imperative to warn of the misery caused by renunciation. Whether it was desired by the individual or the human species, pleasure should be the immediate as well as the distant goal. To Spencer such enjoyment was not the equivalent to the satisfaction that followed the exercise of instinctual drives. The latter included sexual reproduction and the appropriation of nourishment, but they were mere responses to environmental pressures that did not deserve to be called pleasures. More importantly, while instinctual gratifications guided most human actions, they did not propel human evolution. To Spencer this was a matter of measuring progress; this had to be true or there has been no improvement in the human condition. If the only pleasures were instructional drives then the quality of pleasure would have been the same whether or not social evolution had occurred. Pleasure, as distinct from satisfaction, was an extra benefit that existed just beyond biological imperatives; it derived from the enjoyment of beauty and of the products of a developed civilization. The experience of pleasure was a consequence, rather than a cause, of human development; it was also contingent on being placed outside the control of political institutions. Spencer felt that since formal political activity often involved archaic attitudes and behaviour, it was unlikely to possess the capacity to protect evolutionary gains. Since benefits such as privacy and pleasure had no obvious political value they could not be plausibly defended by arguments for public utility. This means that Spencer's defence of private pleasure was not an echo of republicanism like so much of classical liberalism: unlike other liberals he did not argue that the pleasure experienced by individuals was accompanied by the need to participate as a citizen and support the state. Nor did Spencer agree that what makes life pleasurable was the capacity of making choices, which is a belief that, as I have already remarked, is easy to combine with democratic theory. Spencer's defence of pleasure would work even in a society in which individual participation and political choices were rare. However, Spencer's lack of enthusiasm did not make him less of a liberal; on the contrary it makes his liberalism all the more valuable because it is unencumbered with the individualist and democratic qualities that have often overwhelmed more common varieties of that doctrine.

Its uniqueness makes Spencer's defence of liberalism an especially valuable legacy: it is also worthwhile because it is difficult to confuse with either libertarianism or socialism. His doctrine was an ethical and humane approach to future social development, which prohibited dominance and aggression towards dependent persons or groups, even if it could be demonstrated that the long-term result would be beneficial. This restriction on the methods that could be used to encourage social change made Spencer oppose militarism and socialism within Europe, but he was even more concerned to attack imperial rationales that justified the forceful extension of modern political freedom to Africa and Asia. Essentially, these were meaningless to him because the core value of his liberalism was the defence of private lives as something external to politics. This is in distinction to the more common variety of liberalism, which has defended the desirability of a private sphere of life because this encourages an essential freedom in individuals' way of making choices. From this it can be inferred that a liberal political system is one in which a primary task is to protect each individual's freedom. The reverse argument also seems compelling to many liberals: that privacy should effectively encourage choices that improve the quality of lives and, simultaneously, maintain the polity that enables such freedom. At this level of abstraction conventional liberalism harnesses the perfectibility of each individual's existence to freedom as a public good. This, in turn, implies that liberalism, through education and participation, should conceptualize individuals as citizens whose potential is fulfilled as they develop into effective political actors. In extreme cases such arguments could legitimize coercive attempts to transform the politics of ethnic or national groups so that they too will behave in such a way as will enhance their political freedom.

Spencer's variety of liberalism could not be suborned in this way. Since he did not promote the value of individuals' private lives as a political freedom, he could not encourage a sacrifice of one for the other. For him the enhancement of private lives was not valuable because it supported communal values that he saw as outmoded; nor would he have wished to promote individualism, which he believed to be a repellent form of selfishness fated to disappear in the course of social evolution. It was simply that private lives were intrinsically valuable without a public function. They were the unintended benefits of earlier social evolution; the only political consideration they raised was the desirability of preventing the nation-state from claiming the rights of medieval sovereigns, and thus reducing their peoples into abject dependence. Spencer noted with alarm that the new democratic breed of politicians was as eager as aristocratic ones to exercise absolute power without recourse to either the cumulative wisdom of law or to developed institutional traditions. Law and institutions showed evidence of social evolution whereas popular political discourse perpetuated the archaic values of monarchs and republics. It was important, he believed, to value the former over the latter if modern societies were to retain their advantages over pre-social ones.

The chief value that modern peoples found in their private lives was the contingent one of happiness. Rather than seeking this in their role as citizen, it could only be gained through individuals exercising their faculties, avoiding stressful work and engaging in play. Since these benign activities occurred outside structured social life, Spencer was adamant that it was only on rare occasions that pleasures could be produced by either governments or other large-scale organizations. Institutions imposed an organizational rationality, and the imitation of this within individual decision-making would not be pleasurable. The result of this was that each of us subjected ourselves to a rational governor who would overrule emotional needs, and place these under governance: the result would lead to psychological tyranny rather than to happiness. The control of emotions through duty and abnegation might encourage the improvement of a person's character, but Spencer preferred a more gradual progress through socialization.

While socialization was less damaging to happiness than duty and self-restraint, it too imposed barriers to pleasure. In addition, Spencer was keenly aware that individual development was often a process in conflict with social evolution, and that the former had often been harmed in the course of human development. This thought was crucial to Spencer's philosophy; if his liberalism could be said to have a prime principle it would be that the happiness of an individual life should never be easily sacrificed to further general progress. Spencer was constant in his bias in favour of the individual, because, except before he began "A System of Synthetic Philosophy", he had never been enthusiastic in promoting progress *per se*. To him that would have meant accepting that an increase in economic propriety and technological expertise was equivalent in value to the moral improvement of a society. Yet, as far as Spencer could see, this had not happened: his "pre-social" peoples were ethically on all fours with European societies. The only important advantage of modern society he could find was that, on average, the normative quality of life had increased even if individual characters of modern peoples were the same or worse. This suggested that a modern person's greater probability of avoiding brutality and cruelty was not owing to an increase in virtue, but was an accidental matter, or good fortune. Moral improvement had not occurred because the individuals had become superior in making moral judgements, but because a greater portion of their activity took place outside the public domain, which allowed them the luxury of scruples. Avoiding the traditions of Plato and Rousseau, Spencer refused to conflate private morals with the public good. He regarded these as so dissimilar that he suspected that they belonged to different time periods. This, in turn, implied to him that the public good would always be archaic, and could only be promoted at the expense of more recent private ethics.

Spencer's insistence on ranking private values above public ones would be surprising if he were a political theorist accustomed to regard each person chiefly as a citizen, but it is less startling if one considers that evolutionary psychology was always his primary area of investigation. For him the chief burden of any ethical analysis was how changes to a society affected the mental well-being of the individuals who composed it. It was this stance that drove him to condemn emergent European democracies. He could not see beyond their fashionable rhetoric that the people's will was sovereign. To Spencer such a claim could only be a stratagem to hoodwink his contemporaries into granting sovereignty to elected officials, giving them an authority that would only be appropriate to a primitive monarch. Modern politicians should not have this power, but only preside over government in accordance with tradition, which Spencer defined as "the accumulated and organised sentiment of the past".[6] He viewed present-day governance as the province of skilled professional administrators who knew the regulations and probable consequences of their actions. Any attempt by delegates of the people to seize power over government machinery would be a *coup de main* likely to reduce the effectiveness and value of the institutions, while, at the same time, cultivating a dangerous yearning for arbitrary power among both the rulers and the ruled.

This was a repeated warning given in all Spencer's sociological and political writings. After his democratic youth, he was a consistent critic of popular control of politics. Paradoxically, however, he was an advocate of democracy in his psychology. In his technical writings and in *An Autobiography* he refused to countenance the way in which the varied human emotions had been subjected to the dominance of the will. This personal exercise of sovereignty would threaten the civilized meaning that personal lives had acquired during the slow and unplanned accretion of manners and mores that had accompanied social development. Instead of interrogating citizens as to their decision-making on laws and choices of governors, Spencer's democracy focused on each individual's emotions having a voice in deliberation. This would replace the rational will, with its sense of duty and accompanying emotional passivity and deferred pleasure, with an open debate between equals. The different emotions would all have voices and, as a result, truly experience the freedom that modern societies offered. This kind of democracy would be successful in increasing the ethical quality of life because it would be supported by the masses of the people, who, as Spencer noted, quite properly sought pleasure while resisting submission to duty. In this, if not in his deprecation of democratic politics, Spencer spoke for the common man's hope for the future.

Notes

Introduction

1. This was J. D. Y. Peel, *Herbert Spencer: The Evolution of a Sociologist* (London: Heinemann, 1971).
2. Greta Jones and Robert Peel (eds), *Herbert Spencer: The Intellectual Legacy* (London: The Galton Institute, 2004), which contains the proceedings of the centennial conference on Spencer, has noted (p. xiv) that a new life of Spencer is overdue.
3. For example, Stanislav Andreski, a sociologist who was influential in drawing attention to the history of his discipline in the 1970s, depicted Spencer as fulminating against bureaucracy in contrast to Marxism, defined as an ideology of the bureaucratic intelligentsia (in "Introductory Essay", in *Herbert Spencer: Structure, Function and Evolution*, S. Andreski (ed.), 7–32 [London: Michael Joseph, 1971], 29.) This distinction might have some value as part of a critique of Marxism, but is merely confusing on the subject of Spencer, who did not fulminate against bureaucracy, but against the aristocracy, the middle class and the workers. Spencer's general stance was hostile to any popular interference in governance, so long as that governance was restricted to administering justice. Andreski's thesis could only be sustained if Spencer's writings were very selectively edited.
4. See Lawrence Goldman on the British neglect of Spencer compared with Weber and Durkheim, in *Science, Reform and Politics in Victorian Britain: The Social Science Association* (Cambridge: Cambridge University Press, 2002), 322.
5. Alan Macfarlane, *The Origins of English Individualism* (Oxford: Blackwell, 1978). Macfarlane does cite English figures such as R. H. Tawney, but only as an echo of Weber. The resulting bias is that only Weber's views on ideology are taken seriously, which results in individualism being an antithesis of nineteenth-century corporate theory.
6. Graham Wallas, *Men and Ideas* (London: Allen & Unwin, 1940), 90.
7. Charles Darwin, *The Illustrated Origin of Species*, abridged and introduced by Richard E. Leakey (London, Faber & Faber, 1979), 66.
8. Later in this work I offer further examples, but these are not to be taken as a substitute for a study of varieties of Spencerism, a topic that is in great need of thorough investigation.
9. Letter from George Eliot to Sara Hennell, 10 July 1854, in *The George Eliot Letters*, 9 vols, G. S. Haight (ed.) (New Haven, CT: Yale University Press, 1954–78), vol. II, 165.
10. Hector Macpherson, *Herbert Spencer: The Man and His Work* (London: Chapman & Hall, 1901), 226.
11. *Ibid.*, 236–7.
12. Leonard Courtney, Funeral address, 14 December 1903, *Ethical Review* 4, Herbert Spencer Memorial Number (December 1906), 45.
13. "Men of the Day", no. 198, *Vanity Fair* (26 April 1879).
14. Spencer's complaints about Gladstone were meant for his posthumous readers. When it was useful politically to Spencer he was capable of being very pleasant to the Liberal leader. Spencer breakfasted with the Gladstones during the 1870s and 1880s, and they exchanged copies of their books (BM

Gladstone Papers, #44784,25; 44785,205; 44475,265; 44475,323; and 44523,187). The high point in their acquaintance came in 1882, when Spencer was eliciting Gladstone's sympathies for the Anti-Aggression League (see letter from Herbert Spencer to Henry Richard, 25 June 1882, National Library of Wales).

15. Herbert Spencer, *An Autobiography*, 2 vols (London: Williams & Norgate, 1904), vol. I, 193.

16. *Ibid.*, vol. I, 457.

17. *Ibid.*, vol. I, 87–8.

18. *Ibid.*, vol. I, 71, 74.

19. *Ibid.*, vol. I, 89. Between intervals of working as an engineer and a journalist, Spencer was still collecting botanical specimens at the age of twenty-one. He complained later that the knowledge he gained from his classifications was limited by Linnaean taxonomy, and that the work of the Jussieus had been unknown to him. *Ibid.*, vol I, 191, 157–8.

20. In Spencer's early writings the word "life" often carries a heavy freight making it equivalent to nature or even God. It referred not only to biological organisms, but to the human psyche and even to knowledge itself. I have capitalized Life when Spencer uses it in this sense, but not in other senses such as when he speaks of the "definition of life". The word "unknown" has also been capitalized when it seems to function as a synonym for God. However, the Unknown did not have the significance of Life for either Spencer or his Victorian readers. The capital letter also tended to be abandoned earlier.

21. Most of Spencer's philosophical works appeared as "A System of Synthetic Philosophy" (short title, "A System of Philosophy") in ten volumes between 1862 and 1896. These were *First Principles* (1862), *The Principles of Biology* (2 vols, 1864 and 1867), *The Principles of Pscyhology* (2 vols, 1870 and 1872), *The Principles of Sociology* (3 vols, 1876, 1879 and 1896) and *The Principles of Ethics* (2 vols, 1892 and 1893). Initially "A System of Synthetic Philosophy" was funded by subscribers, who received each work in separate parts, which were also published as separate volumes for general sale. Thus *The Data of Ethics* (1879), for example, was Part I of Volume 1 of *The Principles of Ethics*, and *Political Institutions* (1882) was Part V of Volume 2 of *The Principles of Sociology*. The prospectus for subscribers was issued in 1860, and is Appendix A of Spencer's *An Autobiography*, vol. I, 479–84.

22. John Cartwright, *Evolution and Human Behaviour: Darwinian Perspectives on Human Nature* (Cambridge, MA: MIT Press, 2000), 18, 192. Whewell's stance was not so much a Kantian one, as Cartwright would have it, as an explanation of the space between Platonism and modern science.

23. For reasons of vanity Spencer rather enjoyed his later intellectual isolation. It flattered his sense of originality, and it meant that he could be derivative without being condemned for it. When late-nineteenth-century critics assailed Spencer for the purloining of ideas from Kant or Comte, their accusations were so misplaced that he could always successfully protest his innocence.

24. There is useful literature on Ruskin's modernism. See Michael Wheeler, *Ruskin's God* (Cambridge: Cambridge University Press, 1999) and Dinah Birch (ed.), *Ruskin and The Dawn of the Modern* (Oxford: Clarendon Press, 1999). It is a reminder that modernism is very much a Victorian phenomenon.

25. That most cultivated historian of sociology, Wolf Lepenies, carelessly puts Spencer into the twin Procrustean beds of collectivism and individualism by suggesting that he was a utopian and a fellow traveller with collectivists. (*Between Literature and Science: The Rise of Sociology* [Cambridge University Press, 1992; originally published 1985], 58). Lepenies's Weberian definition of collectivism is so wide as to encompass the ideas of any social scientists who studied corporate structures and activities. Similarly, his concept of utopianism is too broad, as it implies that futuristic and rationally planned socialist communes should be equated with a theory of social evolution that was hostile to these.

26. In 1884, in *The Man "versus" The State* (D. Macrae (ed.) [Harmondsworth: Penguin, 1969]), Spencer did write as if the state and the individual were antinomies, but this momentary lapse on his part did not extend over eighteenth-century state theory, nor to the romantic revival of Wilhelm von Humboldt's individualism. Spencer rarely subscribed to negative liberty; that badge was usually displayed by followers of J. S. Mill, who wished to give permanence to a type of liberalism based on individual choice. Negative liberty and the individualism that accompanied it were the antitheses of the socialist doctrines that had a popular renaissance in the 1880s. Spencer also reacted strongly against collectivism at this time, but not as an individualist. By then he had forged his theory of a corporate society on a scientific basis that did not rest on the satisfaction of individual happiness.

27. For example, Jock Gunn, in the course of debating with John Brewer, suggests that there were eighteenth-century English ideas of the state and concludes with a reference to the individualism of

J. S. Mill, a very Victorian philosopher (J. A. W. Gunn, "Eighteenth-Century Britain", in *Rethinking Leviathan: The Eighteenth-Century State in Britain and Germany*, J. Brewer and E. Hellmuth (eds), 99–125 [Oxford: Oxford University Press, 1999]). The search for the concept of the state has been pushed back to before the eighteenth century. Quentin Skinner finds it between the thirteenth and sixteenth centuries (*The Foundation of Modern Political Thought*, 2 vols [Cambridge: Cambridge University Press, 1978]). However, this broader ranging historical enquiry does not suggest that state theory has always been accompanied by individualism. It is too early to say what Skinner's final word will be on the subject of the origin of state discourse because he is still writing on this subject.

28. Istvan Hont (*Jealousy of Trade: International Competition and the Nation-State in Historical Perspective* [Cambridge, MA: Harvard University Press, 2005], 464–8) discusses the development of state theory in a way that minimizes the antithesis between the individual and the state so that it is only one aspect of the conceptual growth of sovereignty. This is a more promising line of investigation than those in accounts that explained the conceptual development of the state exclusively in terms of the antithesis. However, Hont's explanation requires that one believe that successful states intended to make their populations homogeneous, that Hobbes was influential in the process, and that individuals accepted political obligations in return for safety. From the historical perspective of the nineteenth century – a period of rapid state growth – these conditions are hard to sustain.

Chapter 1. A portrait of a private man

1. Havelock Ellis, *The Nineteenth Century: A Dialogue in Utopia* (London: Grant Richards, 1900), 166.
2. *Ethical Review*, no. 4, Herbert Spencer Memorial Number (December 1906).
3. Modern literary scholarship is suspicious of the homespun quality of Spencer's scientific prose. It has been suggested that his use of passive and flat construction was an example of the successful illusion of letting facts speak for themselves. See James G. Kennedy, *Herbert Spencer* (Boston, MA: Twayne Publishers, 1978), 30.
4. David Duncan, *The Life and Letters of Herbert Spencer* (London: Methuen, 1908), 43, 51.
5. *Ibid.*
6. When Spencer had no family letters on which to rely he could remember only scattered incidents (Spencer, *An Autobiography*, vol. I, 118.) The poorness of his memory relates to the adult period of his life; Duncan had evidence that his ability to memorize material as a child was not below average (Duncan, *The Life and Letters*, 9).
7. Spencer's account of his father's beliefs and attitudes relied on the published work of Thomas Mozley, an Anglican cleric, who annoyed him enormously (Spencer, *An Autobiography*, vol. I, 44–53, 82–3). Also see Beatrice Webb, *My Apprenticeship* (London: Longmans, 1926), 138.
8. Spencer, *An Autobiography*, vol. I, 7.
9. Spencer, *An Autobiography*, vol. I, 12.
10. *Ibid.*, 56. When speculating about sociology Spencer used the terms "conformity" and "nonconformity" as opposites. See, for example, *The Principles of Sociology*, 2nd edn (London: Williams & Norgate, 1877), vol. 1, 587. In *The Study of Sociology*, 6th edn (London: Henry S. King, 1877), 238, Spencer also criticized Matthew Arnold for using the word "nonconformity" to refer to a phase of Protestantism when it should refer to any dissent, whether Jewish, Greek, Pagan or Christian. This should have made scholars suspicious when reflecting on Spencer's italics in his statement that "Our family was essentially a *dissenting* family" ("The Filiation of Ideas", in Duncan, *The Life and Letters*, 537.)
11. There was an entire tradition of Spencer scholarship that accepted the theory that since Spencer boasted of his nonconformist ancestry it must be the basis both of his personal life and of his ideas. See for example, David Wiltshire, *The Social and Political Thought of Herbert Spencer* (Oxford: Oxford University Press, 1978), 12–14. This tradition is still alive; Tim Gray has recently combined the Protestant reformation and nonconformity with an interpolation of Spencer as a liberator and rebel (*The Political Philosophy of Herbert Spencer* [Aldershot: Avebury, 1996], 74–5). This argument is difficult to sustain since Spencer was personally and politically passive and unrebellious.
12. Spencer, *Political Institutions*, 722.
13. John Sturrock, *The Language of Autobiography: Studies in the First Person Singular* (Cambridge: Cambridge University Press, 1993), 52.
14. James Collier, "Reminiscence of Herbert Spencer", in *Herbert Spencer: An Estimate and Review*, Josiah

Royce (New York: Fox Duffield, 1904), 212. Collier had worked for Spencer for nineteen years, first as his secretary then as his assistant. Spencer, when writing a reference for Collier in 1882, just before the latter emigrated to New Zealand, claimed that he had worked for him for ten years (Herbert Spencer to Walter Mantell, 25 July 1882 [Alexander Turnbull Library, Wellington, New Zealand, ms: 83-382:3]).

15. Duncan, *The Life and Letters*, 6.

16. Spencer, *An Autobiography*, vol. I, 24. To be a Methodist did not imply adherence to radical causes. A large group of Methodists who followed Jabez Bunting were Tories who hated democracy.

17. *Ibid.*, 26.

18. According to his son, George Spencer resigned from the committee after arguments with the other members, who were all ministers and who objected to his desire to purchase further scientific works. This library's journals included the *Athenaeum* and *Chambers's Journal* (Spencer, *An Autobiography*, vol. I, 87–8.) Disagreements over the choice of books and periodicals was one of the examples Herbert Spencer gave for doubting the wisdom of controlling institutions by committees or representative governments ("Representative Government" [1857], in *Essays: Scientific, Political and Speculative* [London: Williams & Norgate, 1868], vol. I, 167–8.)

19. The original membership of this society had been largely medical, with additions of gentry, clergy, manufacturers and lawyers. It was the Anglican, rather than dissenting, clergymen who joined the society, although it was a multi-denominational body. See R. P. Sturges, "The Membership of the Derby Philosophical Society, 1787–1802", *Midland History* IV(3&4) (1978), 212–29.

20. Thomas Mozley, *Reminiscences, Chiefly of Towns, Villages, and Schools*, 2 vols (London: Longmans, 1885), vol. II, 172–3. Erasmus Darwin's *Zoonomia, or The Laws of Organic Life* was published between 1794 and 1796, shortly before his death in 1802.

21. Paul Elliott, in his interesting study of the Derby Philosophical Society ("Erasmus Darwin, Herbert Spencer, and the Origins of the Evolutionary Worldview in British Provincial Scientific Culture, 1770–1850", *Isis* 94 [2003], 1–29), emphasizes the deistical tone of the membership prior to Spencer's birth in 1820. While I agree with Elliott that Spencer's slight mention of this society was part of a general downplaying of the significance of his Derby background, I think that it is imperative to avoid giving too secular an account of this. While scholars such as G. Kitson Clark can be faulted for over-emphasizing nonconformity as a general explanation of Victorian intellectual developments, one should also avoid focusing too earnestly on any sign of disbelief as a harbinger of a monolithic secular and scientific future.

22. Duncan, *The Life and Letters*, 9.

23. Spencer, *An Autobiography*, vol. I, 151.

24. Isaac Watts, *Divine and Moral Songs, Attempted in Easy Language for The Use of Children* (Derby: Thomas Richardson, 1829), ix.

25. Anne Taylor, *Hymns for Infant Minds by the Authors of "Original Poems": Rhymes for the Nursery*, 13th edn (London: Holdsworth, 1821), v.

26. Isaac Watts, *Divine and Moral Songs* (Devizes: J. Harrison, *c.* 1830), 14.

27. *Ibid.*, 25.

28. *Ibid.*, 26.

29. Taylor, *Hymns for Infant Minds*, 66.

30. Victorian readers of this book saw it as reinforcing the need for discipline. See John R. Reed, "Learning to Punish: Victorian Children's Literature", in *Culture and Education in Victorian England*, P. Scott and P. Fletcher (eds) (Lewisburg, PA: Bucknell University Press, 1990), 99–101. However, Spencer's father, George, who attempted to censor his son's early reading, was possibly giving him a measure of secular freedom in directing him towards Thomas Day. Day had been acquainted with Erasmus Darwin and other free spirits around Lichfield and Derby. Day was an emphatically secular author.

31. Spencer, *An Autobiography*, vol. I, 77.

32. Thomas Day, *The History of Sandford and Merton: A Work Intended for the Use of Children*, 9th edn (London: John Stockdale, 1812), vol. I, 28–9.

33. Spencer, *An Autobiography*, vol. I, 102.

34. On the subject of class, historical reality was more complex than literary conventions adopted by novelists such as Trollope and Gaskell would suggest. Class divisions between Wesleyans and Primitive Methodists might be more extreme than those between Wesleyans and Anglicans. Further, class divisions in Derby seemed less important than those in neighbouring cities such as Lichfield. Finally,

Spencer's youth might pre-date strong class divisions between church and chapel. As Alan Everitt remarks (*The Pattern of Rural Dissent: The Nineteenth Century* [Leicester: Leicester University Press, 1972], 42), the rigid class distinction between Anglicans and Dissenters that was a commonplace of nineteenth-century fiction seemed less important at the time of the evangelical awakening.

35. Forster married the sister of Matthew Arnold.

36. Thomas Mozley and his father, who was an Anglican clergyman and one of the few conventionally religious members of the Derby Philosophical Society, thought that George Spencer seldom went to a place of worship, and possibly had no belief at all (Thomas Mozley, *Reminiscences Chiefly of Oriel College and the Oxford Movement*, 2 vols [London: Longmans, 1882], vol. I, 149 and Mozley, *Reminiscences, Chiefly of Towns*, vol. I, 147–8, 151). Although Mozley's views about George Spencer's lack of religious beliefs were too extreme, they do indicate that a large chasm existed between him and orthodox Wesleyanism. However, George Spencer believed in some sort of revealed religion as late as 1860 (Spencer, *An Autobiography*, vol. I, 549–56.)

37. Shelley himself, in the preface to *Prometheus Unbound*, observes that the only imaginary figure at all like Prometheus was the fallen archangel in Milton's *Paradise Lost*, and that both had courage, majesty and firm but patient opposition. The difference between the two was that the former was exempt from ambition, envy, revenge and the desire for personal aggrandisement. It was the possession of these faults, according to Shelley, that make Satan uninteresting. (*The Poems of Percy Bysshe Shelley*, Thomas Hutchinson (ed.) [Oxford University Press, 1919], 201.) On Spencer's imitation of Shelley, see Chapter 2.

38. *The Diary of Beatrice Webb*, N. MacKenzie and J. MacKenzie (eds) (London: Virago in association with the London School of Economics, 1982), vol. I, 169.

39. Spencer, *An Autobiography*, vol. I, 295. Spencer's lack of enthusiasm for Carlyle's *Oliver Cromwell's Letters and Speeches* was not a sign that he was generally hostile to Carlyle during the 1840s. He read writings such as *Sartor Resartus* with interest (*ibid.*, vol. I, 230, 242.) He also sent Carlyle a copy of one of his early pamphlets. Spencer was briefly acquainted with Carlyle in 1851 (MS Memorandum to Herbert Spencer "About Carlyle", 1882–83 [Athenaeum Club, Spencer Papers]). It was only late in his life that he perceived Carlyle as a symbol of everything to which he himself was opposed. For instance, it was then that he commented adversely on Carlyle's admiration for a schoolmaster who treated the stupid harshly (*An Autobiography*, vol. I, 121).

40. *Ibid.*, vol. I, 243.

41. See R. K. Webb, "The Faith of Nineteenth-Century Unitarians: A Curious Incident", in *Victorian Faith in Crisis: Essays and Continuity and Change in Nineteenth-Century Religious Belief*, R. J. Helmstadter and B. Lightman (eds), 126–49 (London: Macmillan, 1990) and Richard J. Helmstadter, "W. R. Greg: A Manchester Creed", in *Victorian Faith in Crisis*, Helmstadter and Lightman (eds), 187–222.

42. Spencer, *An Autobiography*, vol. I, 253.

43. *Ibid.*

44. Macpherson, *Herbert Spencer*, 8–9. Spencer took a "kindly" interest in Macpherson's book and allowed questions on topics such as whether the impact of Wesleyanism was the same on him as it had been on George Eliot.

45. Spencer, *An Autobiography*, vol. I, 331. On occasion he even regarded his multi-volumed philosophical system, "A System of Synthetic Philosophy", as an unfortunate drain on his finances (*ibid.*, vol. II, 64–5, 135–6, 455–6).

46. This was *Descriptive Sociology; or Groups of Sociological Facts*, a series of elephant folios containing tables of data concerning national and ethnic groups. Spencer organized and paid for the series. It was published from 1873 to 1881.

47. Duncan, *The Life and Letters*, 44–5.

48. Spencer, *An Autobiography*, vol. I, 52.

49. Duncan, *The Life and Letters*, 29, 51.

50. *Ibid.*, 47.

51. Spencer, *An Autobiography*, vol. I, 28–9. Thomas Spencer was an enthusiastic convert to the ideals of the New Poor Law. At an earlier period, he had practised a less hard-fisted kind of charity. Herbert Spencer, as a boy, remembered Thomas giving many gifts from his own pocket and organizing a clothing club to distribute clothing to the needy. On the connection between religiosity and welfare economics of this period see Boyd Hilton's *The Age of Atonement: The Influence of Evangelicalism on Social and Economic Thought, 1785–1865* (Oxford: Clarendon Press, 1988).

52. Spencer, *An Autobiography*, vol. I, 34. The harshness of Herbert's condemnation of his uncle might have been prompted by the fact that in the early 1840s he had shared his views. In the *Nonconformist* II(63) (22 June 1842), 47, Herbert had taken a stand against the Poor Law on the grounds that no individual was entitled to publicly subsidized relief. Even if an individual's distress was not brought on by improvidence or misconduct but was solely the result of misfortune, relief was not a matter of justice. Also, in the *Nonconformist* II(68) (27 July 1842), 506, Spencer argued that it was morally wrong to make arrangements that prevented a great part of the people from earning their bread by the sweat of their brow. This was the Christian argument that industry was a virtue in itself. In later repudiating his uncle's views as inhuman, Herbert was also discarding his own juvenile beliefs.

53. It was important for Herbert to reprint his revengeful obituary in *An Autobiography* because his aunt, or one of her friends, had softened the original when it had appeared in a newspaper. He recorded this episode in the phrase, "And thus I was defeated" (Spencer, *An Autobiography*, vol. I, 38). By including the unexpurgated version in his autobiography, Herbert ensured that his posthumous revenge upon his uncle could not be thwarted twice. Herbert Spencer felt little in the way of ordinary gratitude towards either of the uncles who left him substantial sums. Thomas left him money in 1859 and William in 1860. In both cases he used the money to speed up his publishing plans.

54. Spencer, *Political Institutions*, 700.

55. George Spencer became conscious of his "abnormal lack of control over his temper" in 1821 and 1822. Other comments suggest that he began to recover his equipoise during the period he spent away from teaching, that is, from 1824 to 1827 (Spencer, *An Autobiography*, vol. I, 55, 69).

56. *Ibid.*, vol. I, 39–40.

57. *Ibid.*, vol. I, 38.

58. *Ibid.*, vol. I, 150.

59. *Ibid.*, vol. I, 143.

60. Duncan, *The Life and Letters*, 58.

61. Richard Le Gallienne, *The Romantic '90s* (London: Robin Clark, 1993). This volume was first published in 1926, and the incident referred to took place when Le Gallienne was a child. He presumably had the story from Harrison, as it is given in the gossipy *ingénu* style that Harrison used when he was young. However, when the old Harrison published his rather ponderous memoirs the story did not appear. Instead, Harrison only mentioned that he was on the most friendly terms with Spencer until the latter's death and that too much had been made of Spencer's egoism and eccentricities (Frederic Harrison, *Autobiographic Memoirs*, 2 vols [London: Macmillan, 1911], vol. II, 113).

62. Henrietta Octavia W. Barnett, *Canon Barnett, His Life, Work and Friends, By His Wife*, 2 vols (London: John Murray, 1918), vol. I, 231.

63. W. H. Hudson, *Herbert Spencer* (London: Archibald Constable, 1908), 13.

64. Spencer, *An Autobiography*, vol. I, 39.

65. *Ibid.*, 40.

66. His mother lost approximately six infants, and a daughter at the age of two. Only one cousin is mentioned by Herbert in *An Autobiography*.

67. Duncan, *The Life and Letters*, 16.

68. Spencer, *An Autobiography*, vol. I, 54–5. Harriet seems to have been a conventional Wesleyan who had little interest in her husband's reform ideas. Someone remembered that, before her marriage, her most prominent characteristic had been sweetness (*ibid.*, vol. I, 57).

69. Herbert Spencer, *Social Statics: Or the Conditions Essential to Human Happiness Specified, and the First of them Developed* (London: John Chapman, 1851), 65, and *Social Statics, Abridged and Revised* (London: Williams & Norgate, 1892), 74.

70. Spencer, *An Autobiography*, vol. I, 67. Thomas Mozley, who knew George Spencer well between 1818 and 1830, said that the latter always protested against all charity, whether public or private. This suggests that his son was chiefly complaining about George giving away time rather than worldly goods. This suggestion finds confirmation in Herbert's complaint about George giving free tuition. See Mozley, *Reminiscences, Chiefly of Towns*, vol. I, 165 and Spencer, *An Autobiography*, vol. I, 51.

71. *Ibid.*, vol. I, 93.

72. *Ibid.*, vol. I, 94.

73. *Ibid.*, vol. I, 96–7.

74. *Ibid.*, vol. I, 98.

75. *Ibid.*

76. *Ibid.*, vol. I, 98–9.

77. Duncan, *The Life and Letters*, 13. This control did not lead to Herbert making progress in his schooling; he failed to learn any Greek, Latin or French.
78. *Ibid.*, 16.
79. Spencer, *An Autobiography*, vol. I, 105.
80. *Ibid.*
81. Duncan, *The Life and Letters*, 17.
82. Duncan was also following the lead of Spencer's old friend, T. H. Huxley, who had remarked, when reading an early draft of *An Autobiography*, that men with the force of character of the father and uncles would have played the deuce with the young Herbert had they been less restrained (*ibid.*, 12.) Huxley's typically brazen remark might have been made with less confidence if he had seen the extra material that Duncan later printed.
83. *Ibid.*, 17.
84. *Ibid.*, 18.
85. Spencer, *An Autobiography*, vol. I, 116.
86. *Ibid.*, vol. I, 77.
87. Herbert Spencer, "The Rights of Children", in *Social Statics*, 185.

Chapter 2. The longing for passion

1. Spencer, *An Autobiography*, vol. I, 41.
2. *Ibid.*, vol. I, 4.
3. *Ibid.*, vol. I, 41; also 85–6.
4. *Ibid.*, vol. I, 41.
5. *Ibid.*, vol. I, 44, 52.
6. *Ibid.*, vol. I, 47, 52. J. W. Burrow, in *Evolution and Society* (Cambridge: Cambridge University Press, 1968), 184–5, misquotes a printed source to the effect that George Spencer *refused* to address people as "Mr" when it stated that George would *only* address them as "Mr". Burrow then uses this error to help him construct a theory concerning the importance of Herbert Spencer's background in religious nonconformity. This mistaken thesis has helped mislead scholars such as J. D. Y. Peel and David Wiltshire. On a trivial political level, George Spencer's views were those of a mild provincial follower of Rousseau rather than of a religious man. It is notable, for example, that he resisted canvassing at elections. Also, his habit of addressing people as "Mr" never seemed to cause alienation, and possibly was considered polite enough in Derby. On a more important level, Burrow, and those who follow him, have given too much importance to general religious phenomena when explaining small localized groups of people. While it is true that many late Georgians and Victorians were periodically in a state of religious crisis or enthusiasm, this did not apply to everyone.
7. Spencer, *An Autobiography*, vol. I, 52.
8. *Ibid.*, vol. I, 106.
9. *Ibid.*, vol. I, 47–8.
10. *Ibid.*, vol. I, 52–3.
11. *Ibid.*, vol. I, 48.
12. *Ibid.*, vol. I, 23. Also see Herbert Spencer, "Manners and Fashion", in *Essays: Scientific, Political and Speculative* (London: Longman, 1858), 110.
13. Duncan, *The Life and Letters*, 35.
14. Beatrice Webb, *My Apprenticeship*, 25. Spencer's care over the neatness of his costume can be seen in his refusal to be seen walking to his parents' house on a Sunday in the summer of 1876. His costume was "dilapidated" from a fishing holiday (Duncan, *The Life and Letters*, 84).
15. Collier, "Reminiscence of Herbert Spencer", 190.
16. Spencer, *Political Institutions*, 329.
17. Duncan, *The Life and Letters*, 185–6. Spencer was capable of abandoning his repugnance to *haute tone* if the invitation came from one of his liberal friends. For example, he attended Monckton Milnes's select literary soirée for the King of the Belgians. From a later perspective this would make no sense. That is, eventually there would be no perceptible difference between the Emperor of Russia and the King of the Belgians. The latter's treatment of the inhabitants of the Congo would make him a repressive symbol as fearsome as the emperor. However in the mid-1870s it was still possible to see the King of the Belgians as liberal.
18. *Ibid.*

19. Spencer, *An Autobiography*, vol. I, 129.
20. Herbert Spencer, *The Principles of Sociology*, 2 vols (London: Williams & Norgate, 1893), vol. II, 232. This reference is to Part V of *The Principles of Sociology*, which originally appeared in book form as *Political Institutions* in 1882.
21. Elizabeth Barrett Browning, "Grief", in *The New Penguin Book of English Verse*, P. Keegan (ed.) (London: Allen Lane, 2000), 698. I have cited Barratt Browning's poem "Grief" here because she was a friend of Spencer. The same association between being "passionless" and moral depth can be found in her "Hiram Powers' Greek Slave" and "An Essay on Mind". Other poets, such as Tennyson, Charles Mackay and Robert Montgomery, also used the word "passionless" to signify deep and pure emotions.
22. The last chapter of the second edition of Spencer's *The Principles of Psychology* was titled "Aesthetic Sentiment". In the plan for *The Principles of Sociology*, the work was to end with a section on "Moral Progress", but the penultimate section was to have been on "Aesthetic Progress".
23. Duncan, *The Life and Letters*, 46.
24. The first edition of Spencer's *The Principles of Psychology* appeared in 1855, seven years before the first volume of "A System of Synthetic Philosophy".
25. Herbert Spencer, *The Principles of Psychology* (London: Longman, 1855), 564.
26. Spencer, *An Autobiography*, vol. I, 33.
27. *Ibid.*, vol. I, 54–5.
28. Herbert claimed to have six siblings in *An Autobiography* (vol. I, 61). Duncan corrected this to eight. Spencer seemed unconcerned with the number; he was chiefly concerned with his misfortune in having no sisters.
29. Duncan, *The Life and Letters*, 30.
30. *Ibid.*, 31.
31. Spencer, *An Autobiography*, vol. I, 184.
32. *Ibid.*, vol. I, 185.
33. Webb, *My Apprenticeship*, 25.
34. Kitty Muggeridge and Ruth Adam, *Beatrice Webb: A Life, 1853–1942* (London: Secker & Warburg, 1967), 41.
35. *Ibid.*
36. *Ibid.*, 30. On reviewing her diaries Beatrice was surprised that, in considering how intimate they were, she had not read any of Spencer's books until she was eighteen years old.
37. *Ibid.*, 14; and Hyppolite Taine, *Notes on England*, W. F. Rae (trans.) (London: Strahan, 1873), 93.
38. Muggeridge & Adam, *Beatrice Webb*, 11. Rosalind, who was the mother of Kitty Muggeridge, was presumably responsible for the poisonous character sketch of Laurencina in Kitty's biography of Beatrice Webb. Unlike Beatrice, Rosalind never seems to have forgiven her mother.
39. Spencer, *An Autobiography*, vol. I, 260.
40. Barbara Caine did not agree with Spencer and the Potter daughters, all of whom saw Potter as amiable. Instead, she preferred to rely on the views of T. B. Macaulay. Macaulay, who was Potter's brother-in-law, thought of him as a harsh and even violent man who was capable of being a bully. The basis for Macaulay's views was his belief that Potter had treated carelessly a nephew who was financially dependant on him. Since there was no confirmation of Macaulay's information in the daughters' correspondence its source is possibly Richard's poor relationship with his sister, Mary, when they were children. For Caine's views, see *Destined to be Wives: The Sisters of Beatrice Webb* (Oxford, Clarendon Press, 1986), 19–20.
41. Spencer, *An Autobiography*, vol. I, 311.
42. Webb, *My Apprenticeship*, 24, 56.
43. Caine, *Destined to be Wives*, 17, 23.
44. Webb, *My Apprenticeship*, 10.
45. *Ibid.*, 24.
46. Duncan, *The Life and Letters*, 493.
47. G. P. Gooch, *Life of Lord Courtney* (London: Macmillan, 1920), 179.
48. Georgina Meinertzhagen, *From Ploughshire to Parliament: A Short Memoir of the Potters of Tadcaster* (London: John Murray, 1908), 261. Georgina was another of Richard Potter's daughters. It seems significant that while she reports on the politics of her grandfather, uncles and cousins, the only mention of her father's political views comes in quotations from letters to his own father when he was a young man of radical reforming views.

49. *Ibid.*
50. Webb, *The Diary*, vol. I, 59.
51. Barnett, *Canon Barnett*, 231, 242.
52. Muggeridge and Adam, *Beatrice Webb*, 72. Spencer was, of course, being literal and not sarcastic when he used the word awe. He was later to say that the only two things that excited awe in him were "a great mountain, and fine music in a cathedral" (*An Autobiography*, vol. I, 431). His comment about "monotony" was also meant to be serious rather than jocular. His own aesthetic theories insisted that art and music should display variety.
53. Caine, *Destined to be Wives*, 16.
54. Gooch, *Life of Lord Courtney*, 102.
55. *Ibid.*, 183.
56. *Ibid.*, 184. Curiously, head coverings often appeared in Spencer's life. When he had encouraged too much rebellion among the Potter children they repaid him by petting him with beech leaves, and stealing his hat. (Kate Courtney's account in Duncan, *The Life and Letters*, 510.)
57. Muggeridge and Adam, *Beatrice Webb*, 39.
58. Spencer, *An Autobiography*, vol. I, 204.
59. *Ibid.*, vol. I, 267.
60. *Ibid.*, vol. I, 366. Even in his club he deprecated any show of enthusiasm. For example, he became particularly annoyed if anyone exclaimed amazement to see a philosopher playing billiards.
61. *Ibid.*, vol. I, 55–6.
62. *Ibid.*, vol. I, 267–8.
63. If Spencer had been familiar with Hegel's theory of lordship and bondage he could have spoken of the loss one suffered when one's identity was only achieved by the reflection glimpsed in an inferior. However, he was not schooled in the phenomenological tradition and his notion of identity was, in any case, strongly resistant to the central place that the will served in that tradition.
64. Ibid., vol. I, 279. Spencer's objection to Kant had, to some extent, a similar basis. That is, he objected to Kant's faith in the *supremacy* of the reason (*ibid.*, vol. I, 252–3.) In *An Autobiography* Spencer resisted anything that interfered with his idea that various emotions were on equal political footing with reason, and with the implication that any emotion may hold sway with as much legitimacy as reason.
65. Herbert Spencer, "A Theory of Population, Deduced from the General Law of Animal Fertility", *Westminster Review* 57(112) (April 1852) (new series I[II]), 469); also *ibid.*, 497–8.
66. *Ibid.*, 469.
67. Spencer, *The Principles of Psychology* (1855), 420.
68. *Ibid.*, 461.
69. In the preface to the second edition of *Erewhon* (June 1872) Samuel Butler seized hold of the fact that some of his critics had chastised him for reducing Darwin's theory to absurdity in his chapters on machines. He accomplished his goal by dint of mentioning Darwin's name several times while saying he gleaned his "specious misuse of analogy" from a book written by another person whom the reader will recognize. This "person" was the Joseph Butler who wrote the *Analogy of Religion*, a book much loved by the orthodox, whom Samuel Butler enjoyed annoying. Henry Festing Jones, *Samuel Butler*, 2 vols (London: Macmillan, 1919), vol. I, 156–7.
70. One of the few places it persisted was in the last section of his *The Principles of Biology* (2 vols [London: Williams & Norgate, 1864 and 1867]), where, as he noted, he had copied some passages verbatim from his 1852 article on population. See Spencer, *The Principles of Biology*, vol. II, 499, 500n. This inclusion meant that he was forced to correct himself, and reassure the readers of this work that their future would not be a "a mentally laborious life", but one of spontaneity and pleasure (*ibid.*, vol. II, 503).
71. Spencer, *An Autobiography*, vol. I, 279.
72. *Ibid.*, vol. I, 280.
73. *Ibid.* The "renunciation" upon which Spencer focused was given to the English-speaking world by Carlyle. (One of its other sources was Goethe's *The Renunciants*.) Spencer knew that Carlyle had combined this doctrine with his strong hostility to utility (*ibid.*, vol. I, 281), but although Spencer was an anti-utilitarian himself, he was outraged by Carlyle's extreme and dogmatic insistence that happiness was of no importance.
74. John Morrow, *Thomas Carlyle* (London: Hambledon Continuum, 2006), 56, suggests that Carlyle was able to incorporate aspects of historical culture into current civilization by claiming that human

beings faced recurrent problems. To Spencer – when he had developed his system – this would have been merely an expression of ignorance in the face of evolutionary change, which would, of course, modify the problems.

75. Thomas Carlyle, *Translations from the German: Wilhelm Meister's Apprenticeship and Travels*, 2 vols (London: Chapman & Hall, 1868), vol. II, 208, 321. Carlyle began to translate *Wilhelm Meister* in 1823 (his translation appeared in 1824), and the work immediately became a formative influence on him. A dramatic version appeared in English in 1879, and it seems to have been this that crystallized Spencer's antipathy towards Carlyle's views. The reaction was a highly personal one because, in his youth, Spencer had accepted many of Carlyle's views, together with those of Carlyle's friend Ralph Waldo Emerson.

76. On Carlyle's religious beliefs see J. A. Froude, *Thomas Carlyle, A History of the First Forty Years of His Life, 1795–1835*, 2 vols (London: Longman, 1882), vol. II, 2–3 and Fred Kaplan, *Thomas Carlyle, A Biography* (Ithaca, NY: Cornell University Press, 1983), 114, 143, 208.

77. In 1845 Spencer said that Shelley's *Prometheus Unbound* was the only poem about which he ever became enthusiastic (*An Autobiography*, vol. I, 261, 295 and Duncan, *The Life and Letters*, 53). While he later lost some of his enthusiasm for Shelley, it was never completely abandoned. As Duncan noted, the fundamental idea behind Shelley's "Revolt of Islam" and Spencer's *Ecclesiastical Institutions* (London: Williams & Norgate, 1885), was the same (Duncan, *The Life and Letters*, 52, n.1). Also, when Spencer was compelled to read poetry aloud in a tense social situation in 1880 it was to Shelley he turned. He read the poet with much gentleness and feeling (Barnett, *Canon Barnett*, vol. I, 299).

78. Spencer, *The Principles of Sociology* (1893), vol. II, 232.

79. This wisdom was to be rejected by his protégée, Beatrice Webb. Her choice to be a socialist was as much a reaction to Spencer's individualism as it was to her perception that her father's capitalist activities were amoral. Significantly, Beatrice expressed her conversion to the secular faith of socialism in terms of her need for "renunciation". This was probably a reaction to Spencer's *An Autobiography* (she had read the draft carefully since she had initially been his literary executor), and to George Eliot's *The Spanish Gipsy* (Webb, *The Diary*, vol. I, 130–31). Since Beatrice had often been compared to Eliot by Spencer, her hatred of Eliot's idea of inherited fate had an intense personal quality. Beatrice's rejection of individualism and egoism, and her desire for renunciation, were a revolt against fatalism.

80. An important part of Spencer's reputation as a progressive figure for late-nineteenth-century radicals was due to his demand that children be freed from oppression. At the end of the nineteenth century his ideas were incorporated into the education of working-class children by Margaret McMillan. See Carolyn Steedman, *Childhood, Culture and Class in Britain: Margaret McMillan, 1860–1931* (London: Virago, 1990), 44, 74, 236, 277 n.54. Spencer's *Education: Intellectual, Moral and Physical* (London: G. Manwaring, 1861) was chosen as the textbook for the teacher-training colleges of England. The selection of his book came after a row in which his supporters were opposed by an ecclesiastic who wanted the study of Locke rather than Spencer. The issue was put to a vote at the twenty-six colleges, and decided when twenty-five of them supported Spencer over Locke ("Two", *Home Life with Herbert Spencer* [Bristol: J. W. Arrowsmith, 1906], 105–6).

Chapter 3. The problem with women

1. Spencer, *The Principles of Biology*, vol. II, 486.
2. Spencer, *An Autobiography*, vol. II, 246.
3. *Ibid.*, vol. I, 365.
4. Duncan, *The Life and Letters*, 30.
5. Spencer, *An Autobiography*, vol. I, 168.
6. "Two", *Home Life with Herbert Spencer*, 126.
7. Spencer, *An Autobiography*, vol. I, 168.
8. *Ibid.*, vol. I, 57–8.
9. *Ibid.*, vol. I, 52.
10. *Ibid.*, vol. I, 57.
11. See Mary Poovey, *Uneven Developments: The Ideological Work of Gender in Mid-Victorian England* (London: Virago, 1989), 25–8, on the 1840s.

12. Spencer, *An Autobiography*, vol. I, 55. The "individuation" he saw in his father was, like his mother's altruism, a sign of evolutionary progress for Spencer. Since this general improvement might be consistent with grief in his personal life, he was unable to take much solace in the operation of his system.

13. In his late work, *The Principles of Ethics*, 2 vols (London: Williams & Norgate, 1892 and 1893), vol. II, 194–8, Spencer queried whether the giving of voting rights to women would be expedient. It was then in his opinion that women lacked freedom from passion and were excited by temporary causes as particular objects. In particular, he believed that women were more attracted to socialism than men, and that they were more likely to worship power. This interpretation was more motivated by Spencer's increasingly anti-socialist political stance than it was by his biological views on gender. From the mid-1880s his views on political subjects were increasingly filtered through an imperative need to combat socialism. On the subject of the women's movement and the socialist parties of the 1890s see Sally Ledger, "The New Women and the Crisis of Victorianism", in *Cultural Politics at the Fin-De-Siècle*, S. Ledger and S. McCracken (eds), 22–44 (Cambridge: Cambridge University Press, 1995), 38–9.

14. On Octavia Hill and the charitable labours of middle-class ladies among the distressed of London's East End, see Judith R. Walkowitz, *City of Dreadful Delight: Narratives of Sexual Danger in Late-Victorian London* (London: Virago, 1994), 54–5.

15. Barnett, *Canon Barnett*, vol. I, 230. Barnett's unfavourable opinion of the ghost theory as an explanation for the origins of religion was common. It was one of the weakest aspects of Spencer's sociology. The hypothesis attracted unfavourable reaction from most serious late-nineteenth-century scholars of religion, including Durkheim, who was generally well disposed towards Spencer's theories.

16. *Ibid.*, vol. I, 231.

17. *Ibid.*, vol. I, 241. In *An Autobiography* (vol. II, 340) Spencer's only comments on the turmoil were that he had had morbid fancies that had led him to have "ill-balanced estimates and consequent unwise judgements".

18. Walkowitz, *City of Dreadful Delight*, 59–60.

19. Barnett, *Canon Barnett*, vol. I, 243. Spencer was obviously completely nonplussed by the presence of a confident woman who was not conversant with science. Henrietta Barnett was a practical intellect who wrote books on baby-minding and home-making.

20. Malcolm Muggeridge, *Chronicles of Wasted Time: The Green Stick* (London: Collins, 1972), vol. I, 144.

21. Muggeridge & Adam, *Beatrice Webb*, 41.

22. Duncan, *The Life and Letters*, 31.

23. Muggeridge, *Chronicles of Wasted Time*, 144. Lewis Carroll scholars do not accept this identification. Presumably, there were alternative, and more plausible, models for the mad hatter among Carroll's Oxford colleagues.

24. Muggeridge & Adam, *Beatrice Webb*, 40.

25. "Two", *Home Life with Herbert Spencer*, 145.

26. Spencer, *An Autobiography*, vol. II, 498.

27. *Ibid.*, vol. II, 411.

28. *Ibid.*, vol. II, 412.

29. "Two", *Home Life with Herbert Spencer*, 145–8.

30. Spencer, *The Principles of Biology*, vol. I, 250–52.

31. Herbert Spencer, *First Principles*, 3rd edn (London: Williams & Norgate, 1875), 123.

32. Spencer, *An Autobiography*, vol. II, 357. The particular pathos that had moved Spencer at this point was Francis Bacon's warning that to have children was to give hostages to fortune. This phrase has attained proverbial status, but began as a sentiment in Bacon's essay "Of Marriage and Single Life".

33. Duncan, *The Life and Letters*, 191. Beginning in 1856 Spencer stayed with Mr and Mrs Octavius Smith on periodic visits over twenty years. Mrs Smith, like Laurencina Potter, was a woman whose married status made her a safe recipient for his affections.

34. On the use of the "birth metaphor" by Eliot see Christopher Ricks, "George Eliot: 'She Was Still Young'", in his *Essays in Appreciation*, 206–34 (Oxford: Oxford University Press, 1996), 218–22.

35. Eliza Lynn Linton, *My Literary Life* (London: Hodder & Stoughton, 1899), 18–32, portrays the young (*c.* 1850) Lewes as a bold and audacious sensualist: a free-thinker and not only on theological

questions. Rosemary Ashton has noted that Lewes and his friend Thornton Hunt were rumoured to be devotees of Shelley's doctrine of "free love" and that Lewes generally enjoyed the reputation of being a libertine (*G. H. Lewes: A Life* [Oxford: Clarendon Press, 1991], 5, 56–60, 66, 153).

36. Herbert Spencer, draft of part of *An Autobiography* (British Library [BL] Add. ms 65530 f. 62). This was a subdued form of the charge made by another of George Eliot's early acquaintances, Eliza Lynn Linton, who vociferously protested the famous novelist's adoption of respectability when she was living in common law with a worthless man whom she had snatched from his wife. Lynn Linton's complaint had more emotional justification than Spencer's because she, like the young George Eliot, had found the courage to live a Bohemian life, while Spencer had frequently bowed to convention.

37. *Ibid.*

38. *Ibid.*

39. *Ibid.*, f. 63.

40. *Ibid.*

41. Spencer, *An Autobiography*, vol. I, 95.

42. *Ibid.*

43. *Ibid.*, vol. I, 397.

44. One of George Eliot's modern biographers, Frederick Karl, also notes that Spencer's comments on George Eliot's supposed masculinity appealed to gender stereotypes, but also concealed features of Spencer's concern about his own gender (*George Eliot: A Biography* [London: HarperCollins, 1995], 148–9).

45. Letter from Herbert Spencer to George Eliot, 30 September 1859 (BL Add. ms. 59789, f. 2).

46. Sandra M. Gilbert and Susan Gubar, *The Madwoman in the Attic: The Woman Writer and the Nineteenth-Century Literary Imagination* (New Haven, CT: Yale University Press, 1984), 466.

47. Webb, *My Apprenticeship*, 29, and Muggeridge & Adam, *Beatrice Webb*, 42.

48. Spencer, *An Autobiography*, vol. I, 492.

49. G. H. Lewes was already married, and George Eliot's use of the name "Mrs Lewes" in her private life was part of a successful attempt to prevent her alliance with him from being regarded as notorious by her contemporaries. Spencer was very defensive about the Leweses' relationship, and, as a consequence, forced himself to attend Lewes's funeral in 1878 even though he disapproved of burial services on principle. He suspected that his non-attendance would have lent weight to the unfavourable comment that had circulated about Lewes's relationship with George Eliot (*ibid.*, vol. II, 318).

50. *Ibid.*, vol. I, 399. Spencer also used his early "official" biographer, Hector Macpherson, to deny these rumours. See Macpherson, *Herbert Spencer*, 54.

51. Gordon S. Haight, *George Eliot: A Biography* (Oxford: Oxford University Press, 1978), 115. In 1884 Spencer wrote to J. W. Cross that he had dropped no word – not even to his father – of his relations with Eliot.

52. See Eliot, *The George Eliot Letters*, vol. II, 1852–1858, 37.

53. Rosemary Ashton's account of Eliot's feelings differs from mine in crediting her with calmness and more conventional responses. In Ashton's account Eliot always keeps her pride and eventually and "resignedly withdrew her affections" from Herbert (*George Eliot: A Life* [Harmondsworth: Penguin, 1997], 99).

54. Three other friends, T. H. Huxley, John Tyndall and Richard Potter, had also been sent copies of this letter (see Duncan, *The Life and Letters*, 266.)

55. Letter from Herbert Spencer to E. L. Youmans, 3 February 1881 (BL Add. ms. 65530, f. 18). A shortened and sanitized version of this letter is in Duncan, *The Life and Letters*, 267–7.

56. Spencer to Youmans, 3 February 1881 (BL Add. ms. 65530, f. 17), and letter from Edward Lott to Herbert Spencer (BL Add. ms. 65530, f. 26). Kathryn Hughes attributes the introduction of Lewes and Eliot to John Chapman (*George Eliot: The Last Victorian* [London: Fourth Estate, 1998], 184). There remains some confusion as to when the two met. My account ignores any previous casual social encounter and follows Spencer's view that he introduced them. This is supported by Lewes's own journal, which credits Spencer with initiating his acquaintance with Eliot (see Karl, *George Eliot*, 280). Haight dates this journal entry as 28 January 1859 (Haight, *George Eliot*, 272).

57 Eliot married the young J. W. Cross after the death of her common-law husband, G. H. Lewes.

58. Duncan, *The Life and Letters*, 267.

59. Within the key passage of Cross's biography of Eliot, Spencer figures as "my 'excellent friend, Herbert

Spencer', as Lewes calls him" (George Eliot, *George Eliot's Life, As Related In Her Letters and Journals*, arranged and edited by her husband, J. W. Cross, 3 vols [Edinburgh: William Blackwood & Son, 1885], vol. I, 278). Cross did not correct the misimpression that Spencer was only Eliot's friend with the later remark that it was Spencer who had first made Lewes acquainted with George Eliot (*ibid.*, vol. III, 22); this statement was about priority, not intimacy.

60. *Ibid.*, vol. I, 278, 281.
61. Spencer to Youmans, 3 February 1881 (BL Add. ms. 65530, f. 18).
62. *Ibid.*
63. Kathryn Hughes refers to Eliot's affair with Brabant as an intoxication (*George Eliot: The Last Victorian*, 92). The affair with Chapman has been recorded by all Eliot scholars since Gordon Haight. Eliot may have also loved Charles Bray, the husband of her friend Cara, but little substantive evidence for this has been discovered. The keenest supporter of this particular speculation is Kathryn Hughes (*ibid*, 79, 85–6).
64. Spencer to Youmans, 3 February 1881 (BL Add. ms. 65530, f. 18).
65. *Ibid.*
66. *Ibid.*
67. *Ibid.* Given how innocent Spencer was in sexual matters it seems unlikely that he knew that Lewes was characterized by other friends as a lascivious man who coarsely boasted about his sexual conquests. Two modern scholars have dismissed such rumours as typifying Lewes's early (pre-George Eliot) character or have challenged their veracity. See Hock Guan Tjoa, *George Henry Lewes: A Victorian Mind* (Cambridge, MA: Harvard University Press, 1977), 28, and Ashton, *G. H. Lewes*, 57–9. However, they are not able to produce any evidence to support their belief in respectability. For Spencer the true nature of Lewes was not an issue; since he did not gossip he tended not to know about the private lives of contemporaries. If Spencer had known of such rumours he would have been shocked and his emotional betrayal of Eliot would have been even darker.
68. See Hughes, *George Eliot: The Last Victorian*, 165.
69. Letter from Marian Evans to Herbert Spencer (undated, written at 142 Strand) (BL Add. ms. 65530, f. 2).
70. Letter from Marian Evans to Herbert Spencer (undated, written at Chandos Cottage) (BL Add. ms. 65530, f. 3).
71. *Ibid.*
72. Letter from Marian Evans to Herbert Spencer (undated, written at Broadstairs) (BL Add. ms. 65530, f. 8).
73. *Ibid.*, f. 10.
74. *Ibid.*
75. *Ibid.*
76. Letter from Herbert Spencer to George Eliot, 30 September 1859 (BL Add. ms. 59789, f. 2). Eliot was so pleased to receive this letter that she put aside her thoughts that its author had been cool towards her and Lewes over the preceding months (Eliot, *The George Eliot Letters*, vol. III, 169).
77. George Eliot, *Adam Bede* (Edinburgh: William Blackwood [1859], 1901), 34.
78. *Ibid.*, 157.
79. *Ibid.*, 155.
80. *Ibid.*, 180.
81. This useful distinction between anti-sensualism and pro-sensualism is the property of Michael Mason. See his *The Making of Victorian Sexual Attitudes* (Oxford: Oxford University Press, 1994).
82. George Somes Layard, *Mrs Lynn Linton* (London: Methuen, 1901), 251. Also see Nancy Fix Anderson, *Women against Women in Victorian England: A Life of Eliza Lynn Linton* (Bloomington, IN: Indiana University Press, 1987), 115–16, 181.
83. Harrison, *Autobiographic Memoirs*, vol. II, 108.
84. See Kate Flint, *The Woman Reader, 1837–1914* (Oxford: Clarendon Press, 1993), 179–80.
85. Webb, *My Apprenticeship*, 30.
86. Webb, *The Diary*, vol. I, 100 (27 December 1883).
87. *Ibid.*
88. "Two", *Home Life with Herbert Spencer*, 49.
89. *Ibid.*, 119–20.
90. Spencer, *Education*, 187–8.
91. *Ibid.*

92. *Ibid.*, 106.
93. Spencer's formal work of psychology was relatively free from this habit of generalization from his personal experience to that of everyone. In his *The Principles of Psychology* (London: Williams & Norgate, 1970 and 1872), vol. II, 576, he saw sociability, permanent sexual relations and "double parental relations" (a married couple with children), as coexisting with high intelligence. In such pairings Spencer thought that the sympathetic and frequent operation of sociability in the presence of high intelligence would strengthen bonds. Intelligence is here seen as a positive feature and one that would coexist with sex.
94. Webb, *The Diary*, vol. I, 131–2. Elaine Showalter, *Sexual Anarchy: Gender and Culture and the Fin-de-Siècle* (London: Virago, 1992), 62–3, has stressed the ways in which Beatrice positively identified with Eliot. I am suggesting the opposite; Beatrice adopted an ideological stance as socialist in opposition to the idealized model of Eliot that Spencer constantly placed in front of her.
95. Eliot scholar Rosemary Ashton agrees with Spencer's favourable assessment here. However, others go to great pains stressing Eliot's passionate refusal to believe that love can be contained by sympathy and rationality. I agree with the latter scholars and suspect that Spencer and Ashton have underemphasized Eliot's dramatic side.
96. Coventry Patmore, *Memoirs and Correspondence of Coventry Patmore*, Basil Champneys (ed.) (London: George Bell, 1900), vol. I, 14.
97. R. W. Emerson, "Love", in *Essays, First Series* (1841), 159–79 (Boston, MA: Houghton, Mifflin, 1884), 172. When Emerson toured England popularizing his views he was particularly appealing to young men in the Midlands. Spencer read Emerson's *Essays* in 1844 and was then enthusiastic about Emerson's belief that the true sentiment of love between a man and a woman arose from each serving as the representative of the other's ideal (Spencer, *An Autobiography*, vol. I, 242–3, 267).
98. Emerson, "Love", in *Essays*, 178, and "Friendship", in *Essays*, 190.
99. Emerson, "Art", in *Essays*, 341.

Chapter 4. Spencer's feminist politics

1. Mill wrote "The Subjection of Women" in the winter of 1860–61, but delayed its publication until 1869, when, he believed, public opinion would be more favourable to its ideas. See Stefan Collini, "Introduction", in *On Liberty and the Subjection of Women*, J. S. Mill (Cambridge: Cambridge University Press, 1989), xviii. Spencer, not being a practical politician like Mill, seldom displayed this sort of caution. Harriet Taylor's article on the "Enfranchisement of Women" was in circulation from 1851. See Jo Ellen Jacobs, *The Voice of Harriet Taylor Mill* (Bloomington, IN: Indiana University Press, 2002), 218–19.
2. George Barnett Smith, *The Life and Speeches of the Right Hon. John Bright, MP* (London: Hodder & Stoughton, 1881), vol. I, 142.
3. Spencer, *Social Statics*, 155. In his later revised edition of *Social Statics* (1892), Spencer retained in an abridged form some of the arguments supporting women's rights. However, in the 1864 American edition, which was also circulated in England from 1868, he added a new preface stating that he now had qualifications about the logical aspects of his chapter on "The Rights of Women". He also noted in 1892 that his reader should refer to his *The Principles of Ethics* for his current views.
4. Spencer, *Social Statics*, 169.
5. *Ibid.*
6. *Ibid.*, 158.
7. *Ibid.*, 156.
8. See Susan Kingsley Kent, *Sex and Suffrage in Britain, 1860–1914* (Princeton, NJ: Princeton University Press, 1987), 193.
9. The National Democratic Union is discussed in Chapter 16.
10. Spencer, *Social Statics*, 167.
11. *Ibid.*, 168.
12. Philip Pettit (*Republicanism: A Theory of Freedom and Government* [Oxford: Oxford University Press, 1997]) has advanced a theory of anti-domination that draws on Spencer's views about despotism.
13. Spencer, *The Principles of Ethics*, vol. II, 196–7. (This reference is to Part IV of *The Principles of Ethics*, which was first published in 1891.)
14. David Weinstein, *Equal Freedom and Utility, Herbert Spencer's Liberal Utilitarianism* (Cambridge: Cambridge University Press, 1998), 125, 127, supports David Wiltshire's general thesis that Spencer

was drifting towards political conservatism (*The Social and Political Thought*). For Weinstein, Spencer's late hostility to women's suffrage is part of a larger ideological shift. T. S. Gray, "Herbert Spencer on Women", *International Journal of Women's Studies* 7 (1984), 221, partially supports the argument that the late Spencer drifted towards conservatism.

15. J. S. Mill, *Collected Works of John Stuart Mill*, 33 vols, J. M. Robson (ed.) (Toronto: University of Toronto Press, 1963–90), vol. 16, 1299–1300.

16. Duncan, *The Life and Letters*, 138–9. In his scholarly rather than political works Spencer was much less likely to distinguish between male and female natures than he was when writing about politics. For example, in *The Principles of Sociology* (1877), vol. I, 747, Spencer assigns primitive women a nature equally brutal to that of primitive men. He suggests that if women do not torture enemies it is because of a lack of power, not lack of will. *The Principles of Ethics* should be classed with Spencer's political writings when it deals with subjects such as the status of women.

17. Duncan, *The Life and Letters*, 139.

18. Mill, *The Collected Works*, vol. 17, 1615, emphasis added.

19. Collini, "Introduction", xx–xxi.

20. Weinstein, *Equal Freedom and Utility*, 214.

21. Spencer, *The Principles of Ethics*, vol. II, 160.

22. See Christine Bolt, "The Ideas of British Suffragism", in *Votes for Women*, June Purvis and Sandra Stanley Holton (eds), 34–56 (London: Routledge, 2000), 39, where she suggests that moderate suffragettes thought that the key political goal was to secure property and wages.

23. Spencer, *The Principles of Ethics*, vol. II, 160.

24. *Ibid.*, vol. II, 166.

25. It is intriguing that Spencer was advocating citizenship based on military service at approximately the same time as G. G. Coulton – the scourge of pacifists – began to do so. Coulton began to worry about the small size and lack of preparedness of the British army in 1888. He also began to see the Swiss citizen militia as an object lesson for a democratic country. See G. G. Coulton, *Fourscore Years: An Autobiography* (Cambridge: Cambridge University Press, 1944), 198–9, 265–6, 278.

26. Spencer, *The Principles of Ethics*, vol. II, 166.

27. See Philippa Levine, *Victorian Feminism, 1850–1900* (London: Hutchinson, 1987), 72.

28. Gray, "Herbert Spencer on Women", 221, notes these inconsistencies as the same sort as those that appeared between Spencer's *Social Statics* and *The Principles of Ethics*. However, Gray is mistaken. While it is true that Spencer's *The Principles of Ethics* contains so many inconsistencies about the rights of women citizens that it demonstrates that he was not making serious claims about rights, the same cannot be said about *Social Statics*.

29. Spencer, *The Principles of Ethics*, vol. II, 196.

30. *Ibid.*, vol. II, 196–7.

31. *Ibid.*, vol. II, 197.

32. *Ibid.* Since Spencer believed that pre-social societies treated women better than modern ones (*The Principles of Sociology* (1877), vol. I, 756) he could not rely on social pressure as an argument to explain women's conservatism in this case. Therefore, he needed to employ a psychological language of instincts.

33. Spencer, *The Principles of Ethics*, vol. II, 196.

34. Spencer, *The Principles of Sociology* (1877), vol. I, 766 n. Spencer thought that this information was so valuable that he had to cancel several stereotyped plates in order to include it in this work.

Chapter 5. Culture and beauty

1. Herbert Spencer, "Progress: Its Law and Cause" (1857), in *Essays* (1868), vol. I, 1–60.

2. Spencer, *The Principles of Psychology* (1872), vol. II, 647.

3. *Ibid.*, vol. II, 645.

4. *Ibid.*

5. *Ibid.*, vol. II, 647.

6. The novelty of Spencer's aesthetics stands out sharply against the conventional Romanticism of his friend G. H. Lewes who contented himself with condemning realism and reiterating Goethe's idealism. See Peter Allan Dale, "George Lewes' Scientific Aesthetic: Restructuring the Ideology of the Symbol", in *One Culture, Essays in Science and Literature*, George Levine (ed.), 92–116 (Madison, WI: University of Wisconsin Press, 1987), 100–101.

7. Spencer, *The Principles of Psychology* (1872), vol. II, 636.
8. Herbert Spencer, "The Origin and Function of Music" (1857), in *Essays* (1868), vol. I, 235.
9. Herbert Spencer, "Use and Beauty" (1852–54), in *Essays* (1868), vol. I, 432.
10. William Paley wrote about this. See Mark Francis, "Naturalism and William Paley", *History of European Ideas* 10(2) (1989), 203–20.
11. Spencer, "The Origins and Function of Music", 232, his italics. It should be remarked that he seldom used this form of emphasis.
12. *Ibid.*, 237.
13. Spencer, "Use and Beauty", 430.
14. *Ibid.*
15. *Ibid.*, 432.
16. *Ibid.*
17. Spencer, *Education*, 102.
18. Susan P. Casteras, "Pre-Raphaelite Challenges to Victorian Canons of Beauty", *Huntington Library Quarterly* 55 (February 1992) 13–35, esp. 13.
19. George Eliot appears to have parted company with Spencer on this point. While she shared his concerns with the problematic nature of biological beauty, she wanted art to portray common coarse people who had no place in picturesque wretchedness. That is, she believed that art should have a function: that of evoking sympathy (*Adam Bede*, 181).
20. Spencer, "Use and Beauty", 433. On David Masson's views see Casteras, "Pre-Raphaelite Challenges", 13.
21. Spencer, *Education*, 38.
22. *Ibid.*, 37.
23. Eliot, *Adam Bede*, 155.
24. On the distinction between necessary and sufficient see Spencer, *The Principles of Psychology* (1872), vol. II, 653.
25. *Ibid.*, vol. II, 627.
26. *Ibid.*, vol. II, 628.
27. *Ibid.*
28. *Ibid.*, vol. II, 633. Spencer did not always remember his distinction between function and beauty. It was one of his misfortunes at the end of his life that he could no longer revisit the central discovery of his own autobiography. When he published a late and brief essay on "The Pursuit of Prettiness" he simplistically claimed that feelings concerned with visible beauty originated with sexual admiration (in *Facts and Comments* [London: Williams & Norgate, 1902], 84).
29. Spencer, *The Principles of Psychology* (1872), vol. II, 634.
30. *Ibid.*
31. Spencer, *Education*, 88.
32. *Ibid.*
33. Spencer, *The Principles of Psychology* (1872), vol. II, 647.
34. Trevor H. Levere, "Coleridge and the Sciences", in *Romanticism and the Sciences*, Andrew Cunningham and Nicholas Jardine (eds) (Cambridge: Cambridge University Press, 1990), 298–300.
35. Herbert Spencer, "The Sources of Architectural Types" (1852–54), *Essays* (1868), vol. I, 434.
36. *Ibid.*
37. *Ibid.*, 435.
38. *Ibid.*
39. *Ibid.*, 436.
40. *Ibid.*
41. *Ibid.*
42. *Ibid.*
43. *Ibid.*, 439.
44. *Ibid.*
45. J. Arthur Thomson, *Herbert Spencer* (London: Dent, 1906), 42, 47.
46. Spencer mentioned *The Stones of Venice* when he visited the former maritime republic in 1880 (*An Autobiography*, vol. II, 343–6). However, his comments should be taken to refer to all of Ruskin's aesthetic views subsequent to the appearance of *Modern Painters* in 1843.
47. Herbert Spencer, "Gracefulness" (1852–54), in *Essays* (1868), vol. II, 315–17.
48. Spencer, *The Principles of Psychology* (1872), vol. II, 638.

49. J. T. Boulton's introduction to Edmund Burke's *A Philosophical Enquiry into the Origin of our Ideas of the Sublime and Beautiful* (London: Routledge & Kegan Paul, 1958), xxvi.
50. Spencer, *The Principles of Psychology* (1872), vol. II, 639. Like Spencer, Alexander Bain took the opportunity of enunciating his views of why curves are beautiful (*Mental and Moral Science: A Compendium of Psychology and Ethics* [London: Longmans, 1868–72], 314). However, unlike Spencer, he equates function and beauty. Bain's rationale for this seems to be his annoyance with Ruskin's unscientific notion of the purity of visual pleasures.
51. Spencer, *The Principles of Psychology* (1872), vol. II, 641.
52. Spencer, *An Autobiography*, vol. I, 204.
53. Casteras, "Pre-Raphaelite Challenges", 29.
54. Spencer, *Education*, 41.
55. Casteras, "Pre-Raphaelite Challenges", 29.
56. Spencer, *Education*, 187.
57. Spencer, *The Principles of Psychology* (1872), vol. II, 632.
58. *Ibid.*, vol. II, 633.
59. Hostility to incongruity was not unique to Spencer. This sentiment was also the basis of Macaulay's criticism of Walter Scott's novels, where he condemned them for importing grotesque impressions from actual historical documents when, as works of fiction, they should be harmonious. On this aspect of Macaulay see Mark Phillips, "Macaulay, Scott and the Literary Challenge to Historiography", *Journal of the History of Ideas* 50 (1989), 117–33.
60. Spencer, *Education*, 41.
61. *Ibid.*
62. Spencer, *An Autobiography*, vol. I, 207.
63. *Ibid.*, vol. I, 233–5. Spencer's views on the appropriateness of the location of light and darkness are reminiscent of those of Owen Jones. See the comments on Jones in *The Art of All Nations, 1850–73: The Emerging Role of Exhibitions and Critics*, E. G. Holt (Princeton, NJ: Princeton University Press, 1982), 55–6.
64. Spencer, *An Autobiography*, vol. I, 261–3.
65. Spencer, "Progress: Its Law and Cause", 24.
66. *Ibid.* On "variety" in eighteenth-century art theory, see Boulton's introduction to Burke, *A Philosophical Enquiry into the Origin*, xvi.
67. Herbert Spencer, "The Genesis of Science" (1854), in *Essays* (1858), 189–90.
68. Spencer, *An Autobiography*, vol. I, 274.
69. *Ibid.*, vol. I, 309. This fancy was a product of 1846.
70. Spencer's earlier writings had stressed the need for genius in artistic production, but this was when he was still under the influence of Carlyle's ideas.

Chapter 6. Eccentricities: health and the perils of recreation

1. Edith Sitwell, *The English Eccentrics* (London: Faber & Faber, 1933), 217–18. Republication of Sitwell's book has ensured that stories about Spencer's eccentricity have stayed in circulation in popular accounts of the subject. See, for example, David Weeks and Jamie James, *Eccentrics* (London: Phoenix, 1996), 74, on Spencer's unusual dress and velvet earplugs.
2. Herbert Spencer, "The Proper Sphere of Government, Letter VII" (19 October 1842), in *Spencer: Political Writings*, John Offer (ed.) (Cambridge: Cambridge University Press, 1994), 34.
3. *Ibid.*, 35.
4. Spencer, *An Autobiography*, vol. I, 208.
5. On his habit of constantly feeling his pulse in old age see Webb, *The Diary*, vol. II, 90.
6. Duncan, *The Life and Letters*, 414.
7. "Two", *Home Life with Herbert Spencer*, 140–41, 185.
8. *Ibid.*, 158–9.
9. *Ibid.*, 170. Harold Nicholson (*Small Talk* [London: Constable, 1937], 34–5) was highly amused by Spencer's one-piece suit, and noted that the part in the front, which was laced together, would now be fastened by "what I believe is classed the zip method".
10. Herbert Spencer, "Manners and Fashion", in *Essays* (1858), 109.
11. *Ibid.*, 111.
12. *Ibid.*, 133.

13. *Ibid.*, 142.
14. *Ibid.*, 143.
15. *Ibid.*, 145.
16. Muggeridge & Adam, *Beatrice Webb*, 39.
17. The "two" were sisters named Sickle. They were a lively pair who enjoyed teasing the lonely old philosopher, and arguing with him over household arrangements. Spencer's only record of them was, "I am asking that not very wise but very good creature, Miss Charlotte Sickle, to play housekeeper as she did last winter" (Letter from Herbert Spencer to Mary Cross, 28 October 1894 [Knox College, Galesburg, Illinois]).
18. Muggeridge & Adam, *Beatrice Webb*, 40, Sitwell, *The English Eccentrics*, 219–21. Spencer's own account of using a hammock on a train is prosaic. He simply says that from 1886 ill health forced him to conduct his railway travel in a hammock slung diagonally across an invalid carriage (*An Autobiography*, vol. II, 411).
19. A typescript of Troughton's memoir was held by the Athenaeum Club, London, when I consulted it.
20. Duncan, *The Life and Letters*, 505.
21. Spencer, *The Principles of Biology*, vol. II, 323.
22. Duncan, *The Life and Letters*, 510. These "technical" subjects of design were much discussed in public during the 1850s, when Spencer was still impressionable, in printed debates and correspondence relating to the Society of Arts and the South Kensington institutions. On public discussions about carpets see Winslow Ames, *Prince Albert and Victorian Taste* (London: Chapman & Hall, 1967), 95-6.
23. Duncan, *The Life and Letters*, 286, 317–18. Spencer's theory about children's clothing was quite complex, and relied on a belief that "uneven circulation" caused blood to be pushed into body cavities, which became congested. This problem could be overcome by ensuring that clothing did not unevenly cover the body.
24. *Ibid.*, 495.
25. *Ibid.*, 510.
26. There is a vivid description of how "water" was used in hydropathic establishments from the 1840s in England in Robert Bernard Martin; *Tennyson: The Unquiet Heart* (London: Faber, 1983), 276–81.
27. The *Lancet* stopped attacking hydropathy in 1852, and the practice had been supported in 1846 by Sir John Forbes in articles in his journal, *The British and Foreign Medical Review*. See Robin Price, "Hydropathy in England 1840–1870", *Medical History* 25 (1981), 269–80, esp. 274, 278.
28. Spencer, *An Autobiography*, vol. I, 449.
29. W. H. McMenemy ("The Water Doctors of Malvern, with Special Reference to the Years 1842 to 1872", *Proceedings of the Royal Society of Medicine*, Section on the History of Medicine, 46 [1952], 1–7, esp. 4) notes that the Malvern establishments not only banned alcohol and tobacco, but also spices, pickles and fruit. This strictness was obviously an attempt to emulate the simplicity and harshness of Vincent Priessnitz's original hydropathic establishment in Austria.
30. Price, "Hydropathy in England", 278.
31. Spencer, "The Proper Sphere of Government", in *Spencer: Political Writings*, 53–4.
32. See E. S. Turner, *Taking the Cure* (London: Michael Joseph, 1967), 200.
33. Duncan, *The Life and Letters*, 79–80.
34. *Ibid.*, 80 and Spencer, *An Autobiography*, vol. II, 319.
35. Spencer, *Social Statics*, 70.
36. Duncan, *The Life and Letters*, 505.
37. *Ibid.*
38. Spencer, *Political Institutions*, 271 n.
39. Barnett, *Canon Barnett*, vol. I, 242.
40. Duncan, *The Life and Letters*, 502.
41. *Ibid.*, 414.
42. Barnett, *Canon Barnett*, vol. I, 231 and Duncan, *The Life and Letters*, 498.
43. Duncan, *The Life and Letters*, 498.
44. Jo Manton, *Mary Carpenter and the Children of the Streets* (London: Heinemann Educational, 1976), 62.
45. Ralph Colp, *To Be an Invalid: The Illness of Charles Darwin* (Chicago, IL: University of Chicago Press, 1977), 15–16, 18, 24, 31, 32.
46. Sturrock, *The Language of Autobiography*, 220–21.

47. *Autobiography of Charles Darwin*, Nora Barlow (ed.) (London: Collins, 1958), 144, 243.
48. As an example of the ease with which the Darwin family dealt with the valetudinarian, see the account by Charles's grand-daughter, Gwen Raverat, of Aunt Etty in *Period Piece: A Cambridge Childhood* (London: Faber, 1955), 121–4.
49. Duncan, who had access to manuscripts that were later destroyed, notes that Spencer worried about his health even early in his life when there was no need (*The Life and Letters*, 78). This probably referred to the period immediately after 1855, as Spencer himself was surprised not to have suffered any consequences to his health from his three-day walk in 1833. The fact that he could not remember this subject and had to reconstruct it from letters suggests that his later obsession about well-being was not present in his youth.
50. The use of opium as an orally taken medication was common in the nineteenth century although not everyone approved of it. Alfred Tennyson thought it was the worst kind of millstone a man could hang around his own neck (Martin, *Tennyson*, 42).
51. Spencer, *An Autobiography*, vol. II, 340, 356.
52. T. Wemyss Reid, *The Life, Letters and Friendships of Richard Monckton Milnes, First Lord Houghton*, 2 vols (London: Cassell, 1890), vol. II, 338.
53. *Ibid.*, vol. I, 446.
54. Spencer, *Ecclesiastical Institutions*, v.
55. *Herbert Spencer on the Americans, and The Americans on Herbert Spencer, Being a Full Report of His Interview, and of the Proceedings at the Farewell Banquet of Nov. 9, 1882* (New York: D. Appleton, 1882), 18. Spencer's attempt to clarify his views on the limits of government have, unfortunately, seldom been listened to; the false portrayal of him as a thoroughgoing advocate of *laissez-faire* has lasted well into the twentieth century and has been used to justify extreme libertarian positions that neither he nor any other prominent nineteenth-century English liberal would have adopted. For an account of some of this literature see M. Francis, "Herbert Spencer and the Myth of Laissez-Faire", *Journal of the History of Ideas* 39(3) (1978), 317–28.
56. Spencer, *Herbert Spencer on the Americans*, 22–4.
57. *Ibid.*, 32.
58. *Ibid.*, 34.
59. *Ibid.*, 34–5.
60. *Ibid.*
61. *Aphorisms from the Writings of Herbert Spencer*, selected and arranged by Julia Raymond Gingell (London: Chapman & Hall, 1894), 156.
62. Spencer, *An Autobiography*, vol. I, 36, 413–14.
63. *Ibid.*, vol. I, 66–7. This objection to overwork was not unique to Spencer. A fear similar to his can be found in Leslie Stephen's entries in the original *Dictionary of National Biography* for contemporaries such as W. K. Clifford and Alfred Barratt.
64. Spencer, "A Theory of Population", 499. This article was so distant from his later views that it was one of the few major articles never to be reprinted by him in his collections of essays.
65. Spencer, *An Autobiography*, vol. I, 413.
66. *Ibid.*
67. *Ibid.* There is a strong parallel here with J. S. Mill's early desire to find his "poetic" faculty, and rescue something of himself from work. See John Stuart Mill, *Autobiography*, 3rd edn (London: Longmans, Green, Reader & Dyer, 1874), 36, 136–41.
68. Spencer, *An Autobiography*, vol. I, 331.
69. *Ibid.*, vol. I, 448.
70. Spencer's own account of the wet flannel nightcap appears in *An Autobiography*, vol. I, 474. Kitty Muggeridge's recitation of it as a hitherto buried secret is an interesting example of the twentieth-century intolerance towards any deviant behaviour that does not involve sexuality (see Muggeridge & Adam, *Beatrice Webb*, 39).
71. George Eliot (*The George Eliot Letters*, vol. II, 233) recorded that Spencer's condition at this time prohibited him from reading for more than a quarter of an hour at a time.
72. Fraser obtained the chair without this reference. The fact that Spencer was asked to write a testimonial is somewhat curious as his religious views would have been extremely radical for Edinburgh. Spencer's attempts to ingratiate himself with Fraser's mentor, Sir William Hamilton, had brought a rebuff (Duncan, *The Life and Letters*, 74), and it is unlikely that Spencer himself would have been successful in the "Athens of the north".

73. Spencer, *An Autobiography*, vol. I, 483.
74. *Ibid.*, vol. I, 494.
75. Webb, *My Apprenticeship*, 37. As was mentioned above, Spencer avoided medicines during bouts of ill health when he was young. His use of an opiate when he was old represents a change in practice, but not one that would have caused much comment or disapproval among contemporaries. Even at the end of the nineteenth century, doctors prescribed opiates as freely as they had done earlier, and the problem of addiction was not the subject of much discussion. See Virginia Berridge, "Victorian Opium Eating: Responses to Opiate Use in Nineteenth-Century England", *Victorian Studies* 21(4) (Summer 1978), 439–61, esp. 460–61.
76. Beatrice Webb, *Diary*, vol. II, 90. The ladies who kept house for him between 1889 and 1897 said that the well-known photographs of Spencer did not show "The fresh ruddy colour in his cheeks which made him look so well and belied his bad health" ("Two", *Home Life with Herbert Spencer*, 16). Walter Troughton, who was Spencer's secretary from the beginning of 1888, noted that Spencer, unsatisfied with his local medical man, resorted several times to consulting Sir Andrew Clark about an imaginary heart condition (Walter Troughton's Memoirs of Herbert Spencer, typescript, Athenaeum Club). Despite Sir Andrew's assurance that he had "the heart of a bullock", Spencer refused to believe in his own good health. Other doctors who were consulted confirmed Sir Andrew's diagnosis, and rejected Spencer's lengthy memoranda that questioned it.
77. Spencer, *An Autobiography*, vol. I, 480.
78. *Ibid.*, vol. I, 481.
79. "Two", *Home Life with Herbert Spencer*, 217.
80. Herbert Spencer, *The Principles of Sociology*, 3 vols (London: Williams & Norgate, 1896), vol. III, 157. The third volume of Spencer's *The Principles of Sociology* appeared almost two decades after the first. It was the final original part of "A System of Synthetic Philosophy", although part VI, *Ecclesiastical Institutions*, was published separately in 1885.

Chapter 7. The New Reformation

1. Some twentieth-century scholarship on working-class radicalism has emphasized Spencer's role as a scientific source for agnosticism. See, for example, Bernard Lightman's "Ideology, Evolution and Late Victorian Agnostic Popularizers", in *History, Humanity and Evolution*, James R. Moore (ed.), 285–309 (Cambridge: Cambridge University Press, 1989), 294. However, this seems to refer to the 1880s and 1890s, and has little to do with the meaning and context of Spencer's metaphysics, in which science and religion were symbiotic.
2. Hebert Spencer, letter rejecting the charge of atheism in *The Principles of Psychology*, *Nonconformist* (23 January 1856), 62.
3. Letter from Herbert Spencer to G. J. Holyoake, 11 April 1861 (Manchester Co-operative Union, Holyoake Papers, #1306).
4. Spencer, "Progress: Its Law and Cause", 59.
5. *Ibid.*
6. Spencer, *An Autobiography*, vol. I, 386. This comment was repeated without critical scrutiny by Duncan (*The Life and Letters*, 65) and Gordon S. Haight (*George Eliot and John Chapman, with Chapman's Diaries* [New Haven, CT: Yale University Press, 1940], 65).
7. *An Autobiography* was begun in 1870 and continuously modified until 1894.
8. The Haythorne papers were published in Spencer, *Essays* (1858), which was actually published late in 1857. The *Leader* reviewed it on 19 December 1857.
9. Ashton, *G. H. Lewes*, 86–8. It should not be thought that the *Leader* took up religious liberalisation merely because the editors had found a backer who cared about the subject. In the early 1850s unorthodox religious speculation was more fashionable and unsettling than debates on suffrage and the tenure of parliament.
10. *Leader* II(61) (24 May 1851), 488; *Leader* II(72) (9 August 1851), 765–7; and *Leader* II(76) (6 September 1851), 852–3. Spencer became friendly with Lewes during the summer of 1851. See Ashton, *G. H. Lewes*, 119.
11. *Leader* II(78) (20 September 1851), 897.
12. G. H. Lewes to Herbert Spencer, 9 July 1853 (Ms. letter in the Athenaeum Club).
13. *Leader* IV(185) (8 October 1853), 976–7.
14. *Leader* IV(192) (28 November 1853), 1147–8.

15. *Leader* IV(170) (25 June 1853), 619–20.

16. *Leader* I(1) (30 March 1850), 1. This quotation headed all issues of the *Leader* until July 1858.

17. *Ibid.*, 18.

18. *The Apprenticeship of Life* was a badly bred hybrid of J.-J. Rousseau's *Confessions* and Froude's *Nemesis of Faith*, and amply confirms Lewes's reputation as a poor novelist.

19. The first two essays of Mill's posthumous three essays on religion (*Nature, The Utility of Religion and Theism being Three Essays on Religion*, 2nd edn [London: Longmans, 1874]) were written during February and March 1854. See John Stuart Mill, *Essays on Ethics, Religion and Society*, J. M. Robson (ed.) (Toronto: University of Toronto Press, 1969), cxxii, cxxviii.

20. The Christian Socialist periodical *Politics for the People* refrained from discussing religion.

21. The *Leader* began to refer to itself as the organ of "The New Reformation" in May 1850; "An Attack from an Atheist", *Leader* I(7) (11 May 1850), 155.

22. *Leader* I(4) (20 April 1850), 87. Lewes greeted an anonymous tract with the comment that the author must be one of the restless clergy, daily becoming an important class and swelling the ranks of "Catholicism" or spiritualism (a word better characterizing the body than infidelity, seeing as the old secessionists were merely negative) out of profound dissatisfaction with the Church. The word "spiritualism" had not yet acquired its other-worldly connotations. This development came in the 1860s with the controversial advocacy of the spirit-world by men such as Alfred Russel Wallace and Robert Chambers. Wallace's claims were hotly disputed by T. H. Huxley. See the letter from Robert Chambers to A. R. Wallace, 10 February 1867 (British Museum [BM] Add. 46439, 15).

23. *Leader* I(2) (6 April 1850), 39.

24. Letter from F. W. Newman to Thornton Hunt, March 1850 (BM Add. 38110, f.360). Francis William Newman (1805–97) was the academically brilliant younger brother of J. H. Newman. After taking his degree, Francis Newman adopted an extreme evangelicalism, and went to Persia as a missionary. On his return to England he was unable to conform to orthodox Christianity, and began to explore radical religious doctrines. His pronounced radicalism was expressed in a great variety of ways, from vegetarianism to sympathy with Hungarian revolutionaries. Newman's strenuous literary efforts resulted in books in the fields of logic, political economy, classics, biblical criticism and theology, although it was only in the last of these that he became prominent.

25. *Leader* I(2) (6 April 1850), 39.

26. *Ibid.*

27. *Leader* I(8) (18 May 1850), 181.

28. *Leader* I(11) (8 June 1850), 256–8. On 15 June 1850, Lewes wrote that it was a hopeful "sign of the times" that Newman's *Phases of Faith* was available in the circulating libraries. "The book which of all others most penetratingly and securely saps the foundation of reigning dogmas, which unequivocally says that the doctrine taught in churches and high places is no longer the doctrine to animate men and society...." Lewes concluded in thrilled tones, "to think of such a book being read openly twenty years ago".

29. It is somewhat cumbersome always to mention both religion and philosophy, but it is important to stress that the *Leader* usually regarded them as the same phenomenon. For example, Lewes, while reviewing R. W. Mackay's *The Progress of the Intellect*, remarked that philosophy and religion have one common aim; they are but different forms of answer to the same great question, that of man and his destination; *Leader* I(20) (10 August 1850), 472.

30. *Leader* III(129) (11 September 1852), 878. On F. W. Newman's ideas on first principles, see pp. 139, 141.

31. *Leader* I(19) (3 August 1850), 445–6.

32. The choice of Masson (later known as a Milton scholar) would, at first sight, appear curious as he was not a contributor to the *Leader*, nor, although a liberal, did he hold any advanced religious or political opinions. Masson was simply a friend of Lewes's (see the joy with which Lewes greeted Masson's appointment as professor of English literature at University College London; *Leader* III[139] [20 November 1852], 1117). It was typical of the cosy and personal way in which the *Leader* was conducted that open letters could be addressed to obscure friends. Probably the choice of Masson was dictated by a conversation he and Lewes had had at one of the innumerable literary clubs to which they both belonged (see David Masson, *Memories of London in the Forties* [Edinburgh: William Blackwood & Sons, 1908], 211).

33. *Leader* I(20) (10 August 1850), 469.

34. *Ibid*, 470.

35. *Ibid.* This quotation has been given at such length for two reasons. First, it is the clearest expression of the New Reformation's ideals during the early days of the *Leader*. Secondly, Lewes's use of the term "sentiment" is parallel to the use Spencer made of "sentiment" in *Social Statics* (pp. 96–8), which was receiving its final revisions in 1850. Both Lewes and Spencer claim absolute liberty without any sensible restrictions, because at another level they believed society to be naturally harmonious. Both men rely on Adam Smith's *Theory of Moral Sentiments*. During 1850 Spencer and Lewes spent much time together on Sunday rambles, which they both enjoyed for the "mental results" that were produced (Spencer, *An Autobiography*, vol. I, 376).

36. *Leader* I(21) (17 August 1850), 493–4. Hunt identified himself as a member of an anti-utilitarian group that included Thomas Hodgskin and Spencer. What was new in Thornton's exposition was the addition of religious radicalism. Hunt came out of a Unitarian tradition that regarded utilitarianism as an unfortunate extension of the work of Joseph Priestley. This form of Unitarianism can best be seen in the early writings of James Martineau and in Leigh Hunt's *Religion of the Heart*.

37. There has been some controversy on the subject of whether one utilitarian, J. S. Mill, was religiously indifferent. It has been suggested that he had some positive religious feelings: Robert Carr, "The Religious Thought of John Stuart Mill: A Study In Reluctant Thought Scepticism", *Journal of The History of Ideas* 23(4) (October–December 1962), 475–95. However, Carr had no evidence for his suggestion, other than Mill's posthumous *Three Essays on Religion*, which do not seem to support this suggestion. Mill was notoriously secretive about his religious feelings. Even acknowledged friends, such as Bain and Masson, were uncertain of Mill's position. In view of the lack of private or public statements by Mill, it was reasonable for contemporaries to class him with his fellow utilitarians, such as George Grote, who were avowedly hostile to religion.

38. *Leader* I(21) (17 August 1850), 493–4.

39. *Ibid.*

40. *Leader* I(22) (24 August 1850), 517.

41. *Ibid.*, 517–18.

42. Francis E. Mineka, *The Dissidence of Dissent: The Monthly Repository 1806–1839* (Chapel Hill, NC: University of North Carolina Press, 1944), 385–92, condemns both Horne and Leigh Hunt for having little interest in political matters, and remarks that Hunt's radicalism mellowed into a kind of sentimental idealism. One suspects, however, that Horne and Leigh Hunt were attempting to initiate the kind of cultural and religious reform attempted by the *Leader*. Leigh Hunt's views were remarkably consistent over the last quarter century of his life, and his *Christianism* (1832) contains the same views as his *Religion of The Heart* (1853). Leigh Hunt had carefully inculcated these views in his son Thornton and in the young Lewes.

43. *Leader* I(25) (14 September 1850), 590.

44. Lewes first noticed the work unfavourably in his "Literature" column on 22 February 1851. On 1 March he published his review. Usually the waiting time for a book was much longer. For instance, Spencer's *Social Statics* was mentioned in the issue of 18 January, but was not reviewed until 15 March. The *Leader*'s debate with Martineau was not accompanied by personal hostility; the same issue that contained this adverse review of *Letters on Man's Nature and Development* also carried no. VII of her "Sketches from Life".

45. *Leader* II(49) (1 March 1851), 201.

46. *Ibid.* 201–2.

47. In this period, before Harrison's generation adopted Comte, Lewes was one of his important English sympathizers. The other two were Martineau and J. S. Mill. In the early 1850s it was Lewes who organized the collection of funds for the importunate Comte. Mill's relationship with Comte was complicated by the latter's egocentric behaviour. Comte was unable to tolerate disagreement; and if an adherent's money was not totally at his disposal he regarded this as betrayal.

48. *Leader* III(106) (3 April 1852), 327.

49. *Ibid.*

50. *Ibid.*, 327–8. The *Leader* never tired of propounding this idea. When a reader asked plaintively whether the soul existed after life, he was answered, "As a reader of this journal, our correspondent must be aware than on questions which transcend human Logic, we neither offer nor accept the arguments of Logic; but as we often insist the Soul of man is larger than Logic, and that Soul, God, Love, Life, Immortality, which Logic can neither shake nor support" (*Leader* III[141] [4 December 1852], 1166–7). This quotation is noteworthy because if reveals that the *Leader* had yet to realize that

a good defence of the New Reformation could be constructed from the "necessary idea", even though Whewell and "necessary ideas" had been canvassed in the *Leader* the month before, the newspaper was still using F. W. Newman's "soul" on which to found its metaphysics. Lewes and Spencer did not adopt "necessary ideas" until late 1853, although this development derived as much from Scottish philosophy as from Whewell.

51. I do not regard Spencer as a continuator of the traditions of the Enlightenment because of his distrust of reason. However, if one agreed with John Robertson (*The Case for the Enlightenment: Scotland and Naples, 1680–1760* [Cambridge: Cambridge University Press, 2005], 30) that what characterized the Enlightenment was simply the primacy accorded to human betterment, then Spencer could be included.

52. *Leader* III(113) (22 May 1852), 496–7.

53. *Leader* VI(261) (24 March 1855), 281.

54. Moral earnestness, the subordination of science to morality and an enlarged conception of man's destiny, were the hallmarks of the New Reformation.

55. This concept was borrowed from Leigh Hunt's *Religion of The Heart*, but it also could be found in other works. For example, advocacy of a genuine and undefined feeling that was not distorted by excessive rationality can be found in Isaac Taylor, *Natural History of Enthusiasm*, 6th edn (London: Holdsworth & Ball, 1832), esp. 9, 21–4.

56. Tyndall followed Spencer on this matter, and advocated moral evolution *vis-à-vis* intellectual development as advocated by Buckle. See John Tyndall, *Address Delivered Before the British Association Assembled at Belfast* (London: Longmans, Green, 1874), 61. One of the few neutrals in this dispute was Sara Hennell, who thought that there was a basic agreement between Spencer and Buckle. See Sara S. Hennell, *Thoughts in Aid of Faith, Gathered Chiefly from Recent Works in Theology and Philosophy* (London: George Manwaring, 1860), 318–24.

57. *Leader* III(113) (22 May 1852), 497.

58. Herbert Spencer, "The Use of Anthropomorphism", *Leader* IV(189) (5 November 1853), 1076–7.

59. *Leader* III(113) (22 May 1852), 496. Comte himself used the word "metaphysical" to describe outdated and pre-scientific thinking.

60. F. W. Newman together with many other spiritualists were published in Chapman's "Catholic Series". Of course, this use of the world "catholic" was an implausible attempt to claim ownership of a conventional religious term.

61. *Leader* IV(149) (25 January 1853), 111.

62. *Ibid.*, 111–12.

63. *Leader* III(113) (22 May 1852), 497.

64. *Leader* III(137) (6 November 1852), 1070.

65. Leigh Hunt, *The Autobiography of Leigh Hunt*, J. E. Morpurgo (ed.) (London: Cresset Press, 1949), 445; *The Correspondence of Leigh Hunt, edited by his eldest son* [Thornton Hunt], 2 vols (London: Smith, Elder, 1862), vol. II, 213.

66. Leigh Hunt, *The Religion Of The Heart: A Manual of Faith And Duty* (London: John Chapman, 1853), 18.

67. Edmund Blunden, *Leigh Hunt* (London: Cobden-Sanderson, 1930), 321–2.

68. Anon. [J. H. Leigh Hunt], *Christianism: Or Belief and Unbelief Reconciled: Being Exercises and Mediations* (Not for Sale; Only Seventy-Five Copies Printed, London, 1832), xi.

69. *Ibid.*

70. *Ibid.*, 57.

71. *Ibid.*, 7, 11.

72. Hunt, *The Autobiography*, 378.

73. In *The Religion of the Heart* Hunt used the terms "moral sense" and "heart" interchangeably (e.g. *The Religion of the Heart*, x).

74. Hunt, *The Religion of the Heart*, 1–2.

75. *Ibid.*

76. *Ibid.*, 254.

77. Hunt uses both these terms: *The Religion of the Heart*, xii, and *The Autobiography*, 446. These two works were largely written during the late 1840s and early 1850s. Although the latter term, the "New Reformation", was obviously one of his coining, his friends and followers during the 1850s were apt to use it with additional meanings to his.

78. Hunt, *The Religion of the Heart*, xix–xx.

79. *Ibid.*, 89 n.
80. *Ibid.*, 221–2.
81. *Ibid.*, 111.
82. *Ibid.*, xx.
83. *Leader* IV(192) (28 November 1853), 1147.
84. *Ibid.*, 1148. Lewes, who wrote this review, was in a transitional state. He had lost his belief in Baconian or Comtean empiricism, but had not fully worked out a new philosophical position. Lewes was later to follow Spencer on the subject of innate ideas, but in 1852 and 1853 he still thought of himself as an opponent of them. In his review of Whewell's philosophy, he had classed himself with J. S. Mill as arguing that ideas are given in experience, and that they were not truths necessarily commanding this assent of the mind (*Leader* III(138) (13 November 1852), 1095–7).
85. This essay had been written by the editor of the *North British Review*, Alexander Campbell Fraser (1819–1914). Fraser had been an early student of Hamilton's, and a follower of Chalmers. He was a minister of the Free Church.
86. *Leader* V(241) (4 November 1854), 1045.
87. *Leader* V(242) (11 November 1854), 1070–71.
88. When Spencer's *The Principles of Psychology* first appeared, Eliot reported that Lewes was "nailed" to it (letter from George Eliot to Sara Hennell, in *The George Eliot Letters*, vol. II, 213).
89. *Leader* VI(292) (27 October 1855), 1036–7.
90. The ideological exclusivity of the group can be seen in the treatment of W. R. Greg, whose status as a radical of Unitarian origins might have been thought to have assured his inclusion. However, the *Leader* intellectuals demanded a precise identification. E. F. S. Pigott – a key spiritualist and an editor – wrote that he had received two articles from Greg that were good in manner and matter, but that had a radical defect for the *Leader*: "They are not abreast of the *Leader* in their treatment of the subjects. The *"Leader"* has long ago said all they say and much more beyond: in fact they write *up to* a point *from which* the *Leader* started long ago". Pigott concluded that Greg had not read the *Leader* carefully enough, and that his services would not be required unless he did (letter from E. F. S. Pigott to G. J. Holyoake, 7 March 1856 [Manchester Co-operative Union, Holyoake Papers, #839]).
91. Edward Frederick Smyth-Pigott (1824–95) owned the *Daily News*, a London newspaper, with his friend Thomas Spencer Baynes. Pigott was appointed Examiner of Plays in 1874 and died on 23 February 1895. See John Skelton, *The Table-Talk of Shirley: Reminiscences of and letters from Froude, Thackeray, Disraeli, Browning, Rossetti, Kingsley, Baynes, Huxley, Tyndall and others* (Edinburgh: William Blackwood & Sons, 1895), 44 and n. Pigott is not mentioned in the old *Dictionary of National Biography* but does figure in Frederic Boase's *Modern English Biography*.
92. Letter from E. F. S. Pigott to G. J. Holyoake, 27 July 1859 (Manchester Co-operative Union, Holyoake Papers, #1121).
93. Thomas Archer Hirst (1830–92) was an enthusiastic member of a literary club in the north of England during the late 1840s. In the early 1850s he followed his friend John Tyndall to Germany, and became a student at the University of Marburg. Hirst succeeded Augustua De Morgan in the chair of mathematics at University College London. With Tyndall, Huxley and Spencer, Hirst formed the backbone of the "X-Club", which was the self-appointed dictator of English science during the late-nineteenth century. See J. Vernon Jensen, "The X Club: Fraternity of Victoria Scientists", *British Journal for the History of Science* 5(17), part I (June 1970), 63–72 and Roy M. MacLeod, "The X-Club, A Social Network of Science in Late-Victorian England", *Notes and Records of the Royal Society of London* 24(2) (April 1970), 305–22. More recently the X-Club has received much attention from Ruth Barton.
94. John Tyndall (1820–93). As a great expositor and popularizer of science in the late-nineteenth century, Tyndall was only rivalled by Huxley. Tyndall was also a fitting successor to Michael Faraday as superintendent of the Royal Institution of Great Britain. Due to his unfortunate death by poisoning, Tyndall did not receive the customary full-dress Victorian life and letters. It was not until 1945 that a biography was written on him, and that, because of the premature death of its first author, was badly executed. See A. S. Eve and C. H. Creasy, *Life and Work of John Tyndall* (London: Macmillan, 1945). There was a revival of Tyndall scholarship in the last quarter of the twentieth century.
95. Tyndall had heard Emerson lecture at Halifax, and bought his books (Eve & Creasy, *Life and Work of John Tyndall*, 19). Hirst mentions that he was reading Emerson in February 1849, and that, as always, Emerson's writings harmonized with "one's inner nature, the inner *me*". He also praised what he guessed was Emerson's Unitarianism (journal entry 18 February 1849, Journals of T. A. Hirst,

vol. I, 360 [Tyndall Collection, Royal Institution of Great Britain]). Reading Emerson was an ideal preparation for understanding the *Leader*'s religious writings.

96. Journal entry 11 April 1850, Journals of T. A. Hirst, vol. I, 594 (Tyndall Collection, Royal Institution of Great Britain).

97. *Ibid.*, vol. I, 597 (journal entry, 20 April 1850).

98. He was so absorbed with the newspaper that he always read it the day it arrived. If the *Leader* was a day late in arriving he would note this fact in his journal (*Ibid.*, vol. II, 687 [journal entry 29 December 1850]).

99. Hirst had been a regular reader of *Howitt's Journal* and the *People's Journal*. Occasionally he had also read the *Eclectic*, *Truthseeker*, *Tait's Magazine*, *The Economist* and the quarterlies.

100. *Ibid.*, vol. I, 614 (journal entry 18 June 1850).

101. Hirst, "Educational Societies Considered in their Relation to Individual Culture", speech delivered to the Halifax Franklin Society, October 1850, bound in with Journals of T. A. Hirst, vol. II, 690–97, esp. 692, 693.

102. *Ibid.*, vol. II, 694. By "knowledge" Hirst meant our "theory of valuables", the end to which all things point.

103. *Ibid.*, vol. II, 717 (journal entry 14 April 1851).

104. *Ibid.*, vol. II, 1044 (journal entry 26 December 1852).

105. Tyndall mentioned that he purchased the *Leader* every week (letter from John Tyndall to T. A. Hirst, 26 May 1852 [Tyndall Collection, Royal Institution of Great Britain]). Tyndall aspired to be more than a reader. In October 1850 he wrote an article for the *Leader* (Journal entries 11–17 October 1850, Journals of John Tyndall, vol. II, 513 [Tyndall Collection, Royal Institution of Great Britain]).

106. Journals of John Tyndall, vol. II, 533 (journal entry 9 April 1851).

107. *Ibid.* This quotation was taken from "Letters on Man's Nature and Development", *Leader* II(49) (1 March 1851), 201–3. Although he only quoted it briefly, many of Tyndall's ideas were borrowed from this issue of the *Leader*. In particular he borrowed the antithesis between transcendental and mechanic, and the refusal to dogmatize on what we cannot know. It was significant that the second statement was adopted because it represented the most elaborate statement the *Leader* offered during 1851 on the value of spiritualism as against varieties of atheism.

108. Journals of John Tyndall, vol. II, 533 (journal entry 9 April 1851).

109. John Hall Gladstone (1827–1902) was a chemist who had studied, as Tyndall and Hirst had, in Germany. Tyndall later wrote that Gladstone's fundamentalist Christian beliefs had been upset in Germany, and that at times he had a fervent belief in God, and at other times none at all. Both Gladstone and Tyndall were professors at the Royal Institution of Great Britain. See "A Chat with Professor J. H. Gladstone", *Westminster Gazette*, 8 December 1893.

110. Letter from J. H. Gladstone to John Tyndall, 15 August 1851 (Tyndall Collection, Royal Institution of Great Britain).

111. Letter from John Tyndall to J. H. Gladstone, 17 August 1851 (Tyndall Collection, Royal Institution of Great Britain).

112. Leigh Hunt's *Religion of the Heart* was not published until 1853, but its ideas and terminology received almost weekly adumbration in the *Leader* from 1850.

113. Letter from John Tyndall to J. H. Gladstone, October 1851 (Tyndall Collection, Royal Institution of Great Britain). Tyndall's comments on religious symbols were drawn from the discussion of anthropomorphism in *The Leader*.

114. Letter from John Tyndall to Hector Tyndall, 27 July 1855 (Tyndall Collection, Royal Institution of Great Britain).

115. The only philosophical comment Tyndall published during this period was "Physics and Metaphysics" (1860).

116. The prospectus for Spencer's *First Principles* and "A System of Synthetic Philosophy" was issued on 27 March 1860 by John Chapman's successor, Manwaring. It was accompanied by a list of subscribers, which included the old editors of the *Leader*: Lewes, Thornton Hunt and Pigott. This was remarkable because these three had become estranged over financial and personal difficulties. Their names contrasted oddly with the names of the other subscribers, many of whom were eminent Fellows of the Royal Society whom Huxley had bludgeoned into supporting Spencer (Prospectus, Printed for Private Circulation [Huxley Collection, Imperial College, vol. 7, nos. 106–7]; also see Spencer, *An Autobiography*, vol. II, 479–84).

Chapter 8. Intellectuals in the Strand

1. Spencer, *An Autobiography*, vol. I, 347. Spencer wrote of "weekly soirées" at Chapman's house on the Strand, but there is some evidence to suggest that they were more frequent.
2. Ralph Waldo Emerson, *English Traits* (Boston, MA: Houghton, Mifflin, 1884), 239.
3. Letter from Thornton Hunt, 3 February 1851 (Manchester Central Library). Their newspaper also gave frequent puffs to Chapman's publishing endeavours: *Leader* II(66) (28 June 1851), 609; *Leader* II(82) (18 October 1851), 995; *Leader* II(92) (27 December 1851), 1236 *Leader* III(94) (10 January 1852), 37–8; *Leader* III(132) (2 October 1852), 949–50; *Leader* IV(182) (17 September 1853), 904–5; *Leader* IV(223) (1 July 1854), 617.
4. William Rathbone Greg (1809–1881) was a Unitarian industrialist. When his business failed, he, like his father-in-law James Wilson, turned to journalism. In the 1840s and 1850s Greg was well known as a writer for the quarterlies. Greg was regarded as a typical early Victorian liberal by Lord Morley, Lord Acton, and Sir Mountstuart Elphinstone Grant Duff.
5. Haight, *George Eliot and John Chapman*, 52, 177, 205. This work by Haight was an immensely valuable piece of scholarship when it appeared. However, it must be used with care as Haight's animus against Chapman was as strong as his love for George Eliot. The book is full of unsupported comments on Chapman's lack of moral fibre. For example, Haight refers to Chapman as a notorious philanderer, apparently because he kept a mistress. Haight's sense of outrage cannot be supported by contemporary evidence. Indeed, this points in the other direction. William Hale White, who worked and lived with Chapman at the same time Eliot did, wrote that Chapman had "liberal" notions about the relationships between the sexes, but was not a libertine (*The Autobiography of Mark Rutherford, Dissenting Minister* [London: Trübner, 1881], 158). White's negative evidence is made stronger by his dislike of Chapman, and his puritan code of morality. If Chapman had been a philanderer, White would have been the first to comment on it. Haight seems to have allowed his contempt for Chapman to dominate his research and writing. Haight's view that Chapman was a sexual predator has been accepted by Adrian Desmond (*Huxley: The Devil's Disciple* [London: Michael Joseph, 1994], 185, 216) and Hughes (*George Eliot, The Last Victorian*, 136), who refers to him as a "Don Juan". Other recent Eliot scholars, such as Ashton and Karl, have more nuanced and less theatrical views of Chapman's character. This, of course, is a necessary part of rescuing Eliot as a woman in charge of her own emotional destiny. If Chapman was a Don Juan then she would be a seduced victim.
6. Spencer, *An Autobiography*, vol. II, 38. According to Barbara Bodichon, Spencer had told Pigott (the editor of the *Leader*) that he knew the identity of the author and had seen her recently. Since the Pigotts had deduced that George Eliot had been writing a novel, the secret was guessed (letter from Barbara Bodichon to George Eliot, 28 June 1859, *The George Eliot Letters*, vol. III, 103).
7. Harrison, *Autobiographic Memoirs*, vol. I, 205–7; MS Memorandum as to Herbert Spencer by Frederic Harrison (deposited in the Athenaeum Club). *Essays and Reviews*, 5th edn (London: Longman, Green, Longman & Roberts, 1861) was a famous controversial book, first published in 1860. Its editor is unknown, but was probably Mark Pattison.
8. MS Memorandum as to Herbert Spencer by Frederic Harrison (deposited in the Athenaeum Club).
9. Letter from John Tyndall to John Chapman, October 1860 (London School of Economics and Political Science [LSE], Harrison Papers, c. 84–90); letter from T. H. Huxley to John Chapman, October 1860 (LSE, Harrison Papers, c. 84–90).
10. Hunt, *The Correspondence of Leigh Hunt*, vol. II, 120.
11. Francis Espinasse, *Literary Recollections and Sketches* (London: Hodder and Stoughton, 1893), 231; also 250.
12. *A List of Works Published by John Chapman (late John Green)*, appended to *Human Nature* (London: John Chapman, 1844).
13. Preface and catalogue prefixed to *The Philosophical and Aesthetic Letters and Essays of Schiller* (London: John Chapman, 1845).
14. "A List of Mr. Chapman's Publications", appended to F. W. Newman's *Lectures on Political Economy* (London: John Chapman, 1851).
15. Letters from John Chapman to Mark Pattison (Bodleian Library, m.s. Pattison, vol. 50, ff. 301–2, 305, 402–3, 427–8, 430, 431, and vol. 51, ff. 386–90).
16. Letter from T. H. Huxley to John Tyndall, 17 October 1854 (Huxley Collection, Imperial College, 8–16). During most of 1854 Huxley had written all the science articles in the review. In this letter

he persuaded Tyndall to do half the work, that is, the half relating to physics and chemistry. Unlike Pattison, the theological editor, Huxley was subject to little control.

17. Chapman was twenty-two in 1843 when he purchased John Green's business, and he had had little formal education or business experience. Business experience did not improve him because after nine years as a publisher he was forced to ask his wife to write apologies to his creditors. Susanna Chapman admitted that her husband had no business education, but said in his favour that he had *now* mastered the science of book-keeping (letter from Susanna Chapman to George Combe, 14 July 1852 [National Library of Scotland]).

18. Linton, *My Literary Life*, 92.

19. *The Correspondence of Thomas Carlyle and Ralph Waldo Emerson, 1834–1872*, C. E. Norton (ed.) (London: Chatto & Windus, 1883), vol. II, 141; also see vol. II, 73, where Chapman is described as "a tall lank youth of five-and-twenty; full of good will".

20. Hunt, *The Correspondence of Leigh Hunt*, vol. II, 120. Hunt did not mention Chapman by name; he simply wrote that he was going to meet Emerson at "the most tremendous of booksellers, in the Strand". Emerson was staying with Chapman.

21. James A. Secord's account of Chapman's circle (*Victorian Sensation: The Extraordinary Publication, Reception and Secret Authorship of "Vestiges of the Natural History of Creation"* [Chicago, IL: University of Chicago Press, 2000], 486–9) does not distinguish, as I have done, between it and Lewes's writing in the *Leader*. However, Secord is primarily interested in the popular perspective on scientific ideas. It is unlikely that such a distinction would be useful to him.

22. Carlyle, *The Correspondence of Thomas Carlyle*, vol. II, 73. Carlyle was referring to the only book that Chapman ever wrote: *Human Nature*.

23. J. S. Mill to John Chapman, 9 June 1851 (LSE, Mill–Taylor Collection, vol. 9, no. 2).

24. *Ibid.*

25. James Martineau (1805–1900) was the best known Unitarian theologian of the nineteenth century. He first sprang to prominence as a preacher in Liverpool during the 1830s. In the 1840s and 1850s he was known in London literary circles for his unorthodox faith. His theology went through several phases, and was expressed in a vast number of books and articles that span the whole reign of Queen Victoria. Today, he is known only as the brother of Harriet Martineau.

26. Neither J. Estlin Carpenter, *James Martineau, Theologian and Teacher: A Study of His Life and Thought* (London: Philip Green, 1905) nor R. V. Holt, *The Unitarian Contribution to Social Progress in England* (London: Lindsey Press, 1952) discussed Martineau's *Endeavours After the Christian Life*, although the work could have been used to further their theses. There is some mention of it in James Drummond and C. B. Upton, *The Life and Letters of James Martineau* (London: James Nisbet, 1902), vol. I, 136.

27. James Martineau, *Endeavours After the Christian Life: A Volume of Discourses* (London: J. Green, 1843), 233.

28. *Ibid.*, 233–5.

29. *Ibid.*, 236.

30. The copy of *Human Nature* in the Cambridge University Library was, except for the title page, completely uncut when the author first read it on 30 October 1970.

31. Anon. [John Chapman], *Human Nature* (London: John Chapman, 1844), 1.

32. *Ibid.*, 2–3.

33. *Ibid.*, 5. For a short period subsequent to the publication of *Endeavours*, Martineau became more sympathetic to rational enquiries into the nature of religion. This was when he published the third edition of *The Rationale of Religious Enquiry* in 1845. However, Martineau's temporary change of mind had nothing to do with Chapman, but was a result of a correspondence with his Unitarian friend, J. Blanco White. See the letter from White at the end of the third edition of James Martineau, *The Rationale of Religious Enquiry: On the Questions Stated of Reason, the Bible, and the Church; In Six Lectures*, 3rd edn (London: John Chapman, 1845) and J. B. White, *The Life of the Rev. Joseph Blanco White written by himself with portions of his correspondence*, John Hamilton Thom (ed.) (London: John Chapman, 1845), vol. II, 9. Later, Martineau disavowed *The Rationale*.

34. *Ibid.*, 8.

35. *Ibid.*, 10.

36. *Ibid.*, 7.

37. *Ibid.*

38. *Ibid.*, 11.

39. White, *The Autobiography of Mark Rutherford*, 161. In this, the figure of Chapman is thinly concealed by the pseudonym of Wollaston. The aspect of psychology that interested Chapman was the general evolution of intelligence and mind. First, he believed, there is a development of life, "then of *sensation*, then of *intelligence*, and finally of *mind*. The laws of mutation of species and progressive development of vegetable and animal forms, as taught by M. Geoffrey Saint-Hilaire and Lamarck, and which, with some modification, characterize the 'Vestiges of the Natural History of Creation', are emphasized as true, and are expounded in language and style intimating a most minute and profound knowledge of the facts upon which the theory is founded" (John Chapman, preface to Andrew Jackson Davis, *The Principles of Nature, Her Divine Revelations, And a Voice to Mankind*, 2 vols [London: John Chapman, 1847], vol. I, 20).

40. Haight, *George Eliot and John Chapman*, 5.

41. Eliot, *The George Eliot Letters*, vol. I, 348 n.5, 350 n.1.

42. [George Eliot], *An Analytical Catalogue of Mr. Chapman's Publications* (London: John Chapman, 1852), 16.

43. *Ibid.* Eliot made a similar statement about most of the authors she reviewed in this catalogue. Her review of *Social Statics* emphasized that Spencer, in using reason to trace the causal connections between God, morality and human progress, fulfilled contemporary demands for a rational religion.

44. *Ibid.*, 17.

45. Drummond and Upton, *The Life and Letters of James Martineau*, vol. I, 93, 123–5.

46. White, *The Life of the Rev. Joseph Blanco White*, vol. II, 92. Also see the letter from White to Martineau appended to *The Rationale*.

47. In the 1840s Unitarians still referred to themselves as Christians.

48. Martineau, *The Rationale*, viii–ix.

49. *Ibid.*, 12.

50. *Ibid.*, 18.

51. *Ibid.*, 174–5.

52. *Ibid.*, 72 n.

53. *Ibid.*, 72.

54. Francis Newman described his own religious position in the following terms: "if I am prominent in anything, it is not a Rationalism (a word which does not imply anything devotional or even emotional, but simply the explaining away of religious miracle by intellectual considerations) but it is by my *spiritual* writings; and my distinction that I am among the few writers, who, renouncing the miraculous pretensions of Christianity, yet retain and revere that part of it which is purely spiritual, and in great measure (compatible) with Judaism or with the best of Mohammedanism. My doctrine is Spiritual Theism, as opposed to Atheism, Pantheism or Old Deism or German Rationalism" (letter from F. W. Newman to G. Griffen, 16 January 1860 [BM Add. MS 28, 511 f. 215]). Newman's correct interpretation that he was only important as a religious thinker was ignored by his early biographer, who did not deal with his spiritualism, except to refer to it briefly as an agnostic aberration (I. Giberne Sieveking, *Memoirs and Letters of Francis W. Newman* [London: Kegan, Paul, Trench, Trübner, 1909], 339). That account is wrong because agnosticism was developed after the 1850s, that is, after the period in which Newman's religious writings were influential. Further, agnosticism was a movement that was out of sympathy with Newman's views.

55. Newman was powerless to help Chapman financially, because the latter continually borrowed money against the *whole* value of his stock, without calculating overheads. This meant that he always lost money, and had to borrow more to pay earlier loans. See letter from F. W. Newman to G. J. Holyoake, 23 August 1854 (Manchester Co-operative Union, Holyoake Papers, #865).

56. William Robbins, *The Newman Brothers: An Essay In Comparative Intellectual Biography* (London: Heinemann, 1966), xi.

57. F. W. Newman, *A History of the Hebrew Monarchy*, 2nd edn (London: John Chapman, 1852), xv.

58. F. W. Newman, *The Soul, Her Sorrows and Her Aspirations: An Essay towards the natural history of the Soul as the true basis of theology*, 2nd edn (London: John Chapman, 1849), 242.

59. *Ibid.*, 249–50, 252, 257. Newman's first work was a logic concerned with the science of evidence. This made him critical of the nature of historical proofs.

60. *Ibid.*, 249–50.

61. *Ibid.*, 254, 257.

62. F. W. Newman, *Phases of Faith; Or, Passages From the History of My Creed* (London: John Chapman, 1850), 106–13.

63. Letter from F. W. Newman to G. J. Holyoake, 1851 (Manchester Co-operative Union, Holyoake Papers, #452).

64. *Ibid.*

65. Letter from F. W. Newman to Mrs James Ambrose Morris, 10 October 1855 (UCL, Add. MS).

66. F. W. Newman, *Theism, Doctrinal and Practical; or Didactic Religious Utterances* (London: John Chapman, 1858), 27–30.

67. Newman's admiration for J. D. Morell was shared by Spencer. Morell was one of the few who received complimentary copies of two of Spencer's early works, *The Principles of Psychology* (1855) and *Essays* (1858) (Longmans Ledger #C9, p. 307 and #C10, p. 549 [Longmans Archives, Harlow, Essex]).

68. F. W. Newman, *Theism, Doctrinal and Practical*, 101 n.

69. In a letter written during 1855, Newman listed the known principles of morality in the order of preference for "Spiritual" morality: "The Right (which is supreme) the Expedient (or profitable) which is second in value; and the Pleasant, which is the lowest of the Three" (F. W. Newman to G. J. Holyoake, 23 August 1855 [Manchester Co-operative Union, Holyoake Papers, #784]).

70. F. W. Newman, *Theism, Doctrinal and Practical*, 101 n. Newman's endeavour to develop a philosophy of intuition was misunderstood by William Robbins. Robbins regards *Theism, Doctrinal and Practical* as "evidence of Frank's earnest, life-long effort to preserve the passionate intuition of God derived from Evangelical experience" (Robbins, *The Newman Brothers*, 139). This interpretation is forced, and ignores the religious changes described in Newman's own *Phases of Faith*. The description of *Theism, Doctrinal and Practical* as evangelical owes more to Élie Halévy's emphasis on the cardinal importance of evangelization in Britain as a whole than to any evidence in Newman's thought in the 1840s and 1850s. Further, his intuition was not passionate, but enervated and philosophical.

71. [George Eliot], "W. R. Greg: The Creed of Christendom", in Chapman, *An Analytical Catalogue*, 5.

72. W. R. Greg (*The Creed of Christendom: Its Foundations and Superstructure* [London: John Chapman, 1851], v, vii–viii, 300 n.) refers to Martineau and Newman in his introduction, and uses Newman's *The Soul* to buttress his own conclusions.

73. W. R. Greg to Samuel Greg, 25 February 1827, Greg Correspondence (lent by Andrew Greg of Trinity College, Cambridge).

74. W. R. Greg, *Enigmas of Life* (with a Prefatory Memoir Edited by his Wife) (London: Kegan, Paul, Trench, Trübner, 1891), xxxix.

75. Greg criticized Charles Babbage's *The Ninth Bridgewater Treatise* (London: John Murray, 1837), which attempted to show that miracles may be the exceptional expressions of a natural law expressly provided for beforehand (Greg, *The Creed of Christendom*, 192). However Greg's own conclusions are expressed in the same form as Babbage's: "We hold that God has so arranged matters in this beautiful and well-ordered, but mysteriously governed universe, that one great mind after another will arise from time to time, as such are needed, to discover and flash forth before the eyes of men the truths that are wanted" (*ibid.*, 233). There is no reason why one could not support miracles by Greg's argument, which relies on the exceptional expressions of natural law to produce a genius such as Jesus of Nazareth.

76. Greg, *The Creed of Christendom*, x.

77. *Ibid.*

78. Greg, *Enigmas of Life*, xvi–xvii.

79. *Ibid.*

80. Greg, *The Creed of Christendom*, xi.

81. *Ibid.*, xii n.

82. Greg repeated these statements when discussing the future life, which he regarded merely as a hope, a faith or an earnest desire. Christianity, he thought, added no knowledge to this hope (*The Creed of Christendom*, 279). Whenever Greg attempted to make out logical grounds for his hope, he destroyed it (*ibid.*, 301). Instead, he took refuge in Newman's *The Soul*, which allowed him to abandon inference and induction, and which permitted the soul to take counsel with itself (*ibid.*, 300 n.).

83. The influence and reputation of Robert William Mackay (1803–82) was limited to Chapman's circle during the 1850s. Otherwise, he was unknown. Although he was an original member of the

Athenaeum Club, by the time of his death he was so obscure as to receive only one column for his obituary in the *Athenaeum* (no. 2836) (4 March 1882). The writer of this obituary dwelt on Mackay's extremely difficult and condensed prose style, and suggested that any popularization of his work would have to be several times longer than the original. In comparison to the brilliant talkers of Chapman's salon, such as Froude and Greg, Mackay was shy and retiring.

84. Letters from John Chapman to Mark Pattison, January 3 and November 17, 1854. The Bodleian Library, MSS Pattison 50, #301-2 and #427-8.

85. Letter from John Chapman to Mark Pattison, 1 December 1854 (Bodleian Library, MSS Pattison 50, f. 431).

86. R. W. Mackay, *The Progress of the Intellect as exemplified in the religious development of the Greeks and Hebrews* (London: John Chapman, 1850), vol. I, 18–20.

87. *Ibid.*, vol. I, 22.

88. *Ibid.*, vol. I, 35.

89. *Ibid.*, vol. I, 36.

90. *Ibid.*, vol. I, 49.

91. *Ibid.*, vol. I, 38. See also R. W. Mackay, *A Sketch of the Rise and Progress of Christianity* (London: John Chapman, 1854), v.

92. Letter from John Chapman to Mark Pattison, 9 October 1857 (Bodleian Library, MSS Pattison 51, #386–390).

93. *Ibid.* Chapman did have a destructive power that was recognized by his contemporaries. Charles Kingsley wrote that there was "a great and growing movement" that was destroying orthodoxy: "Onward to Strauss, Transcendentalism – and Mr John Chapman's *Catholic Series* is the appointed path, and God help them!" (W. Hale White, *The Early Life of Mark Rutherford, by Himself* [Oxford: Oxford University Press, 1913], 74–5).

94. W. H. Dunn (*James Anthony Froude: A Biography*, 2 vols [Oxford: Clarendon Press, 1961–63], vol. I, 278) states that Chapman sold the *Westminster Review* to G. Manwaring in 1860. This is wrong; it was only the publishing business he sold. He continued to control the review.

95. The 1860 prospectus for Spencer's *First Principles* and "A System of Philosophy" was published by G. Manwaring, who had purchased Chapman's firm. Manwaring soon succumbed to Chapman's debts.

Chapter 9. The genesis of a system

1. Longmans Ledger #C9, 307 (Longmans Archives, Harlow, Essex). "C" stands for commission publication, which meant that Spencer had to pay for paper and printing, and give ten per cent of the gross return to the publisher.

2. Longmans Ledger #C10, 549 (Longmans Archives, Harlow, Essex).

3. Letter from Herbert Spencer to G. J. Holyoake, 22 April 1860 (Manchester Co-operative Union, Holyoake Papers, #1210).

4. Spencer, *An Autobiography*, vol. II, 52–3. Eliot had been informed that Spencer was in good spirits because of the numbers of the American subscribers: 140 out of what she believed was a total of 600 (letter of 6 August 1860, Eliot, *The George Eliot Letters*, vol. III, 329; see also vol. III, 437), but these numbers seem inflated. Possibly because Eliot always criticized Spencer for being pessimistic – or excessively critical – he was inclined to put a positive gloss on his communications to her.

5. "Parker – P & P" Ledger, 135, 140, 198, 199, 240 (Longmans Archives, Harlow, Essex).

6. Letter from Herbert Spencer to Professor Wilson, 12 August 1851 (National Library of Scotland).

7. Letter from Herbert Spencer to the editor of *Blackwood's Magazine*, 11 May 1857 (National Library of Scotland).

8. Letter from Herbert Spencer to the editor of *Blackwood's Magazine*, 2 February 1858 (National Library of Scotland).

9. Through most of the 1850s Spencer entertained the notion of acquiring a lucrative official position (*An Autobiography*, vol. II, 51).

10. This was the impression Spencer had left on Charles Kingsley although they had only met once (letter from Charles Kingsley to T. H. Huxley, 31 October 1860 [Imperial College, Huxley Collection, vol. 19, #195]). Kingsley subscribed to Spencer's "A System of Synthetic Philosophy" because he felt *the* question "must be faced sooner or later" (letter from Charles Kingsley to Herbert Spencer, 15 March 1860 [Athenaeum Club, Spencer Papers]).

11. Huxley asked Tyndall to proofread the preface to his *Critiques and Addresses*, mentioning that he was particularly anxious not to annoy Spencer in any way (letter from T. H. Huxley to John Tyndall, 1 April 1873 [Imperial College, Huxley Collection, vol. 8, #138]).

12. Letter from Herbert Spencer to T. H. Huxley, 25 September 1852 (Imperial College, Huxley Collection, vol. 7, #94).

13. Letter from Herbert Spencer to T. H. Huxley, 31 December 1858 (Imperial College, Huxley Collection, vol. 7, #104). This letter was not published in Leonard Huxley's *Life and Letters of T. H. Huxley*, although it was obviously of great importance, and typified the early relationship between Huxley and Spencer. (Spencer had also humbly expressed his sense of reliance on Huxley in two other letters, one dated 6 October 1860 and the other dated 1 January 1873, which referred to the previous twenty years [Imperial College, Huxley Collection, vol. 7, #112, and Athenaeum Club, Spencer Collection]). One can only assume that Leonard Huxley was afraid of the embarrassment that would have been caused by the publication of a letter in which Spencer had been shown to rely on someone. Even as it stood, the *Life and Letters of T. H. Huxley* had to be modified in order to soothe Spencer, who had been angered by the harmless and basically correct comment that Huxley had read Spencer's proofs. Spencer perversely chose to interpret this as meaning Huxley had read *all* of his proofs. See the preface to the second edition of *Life and Letters of T. H. Huxley*, 3 vols (London: Macmillan, 1903).

14. Letter from T. H. Huxley to Herbert Spencer, 3 September 1860 (Athenaeum Club, Spencer Collection). This letter was published by Leonard Huxley, *Life and Letters of T. H. Huxley*, vol. I, 212–13.

15. For example, see William Irvine, *Apes, Angels and Victorians: A Joint Biography of Darwin and Huxley* (London: Weidenfeld & Nicolson, 1955), 30, and S. R. Letwin, *The Pursuit of Certainty* (Cambridge: Cambridge University Press, 1968), 328.

16. Letter from T. H. Huxley to Henrietta Heathorn, 18 October 1847 (Imperial College, Huxley Collection, 1847, #3).

17. Huxley's autobiographical statement prefixed to Volume I of his *Collected Essays* was only a seventeen-page outline. It does, however, contain the interesting comment that he could not count his various honours as marks of success "if I could not hope that I had somewhat helped the movement of opinion which has been called the New Reformation" (T. H. Huxley, *Collected Essays*, 9 vols [London: Macmillan, 1904], vol. I, 17). Huxley was conscious of the origins of the term "New Reformation" because he also refers to Strauss as one of the protagonists of the New Reformation (*Collected Essays*, vol. V, vi).

18. Imperial College, Huxley Collection, vol. 47, #73–4. This quotation is from the manuscript Part II of Huxley's "Mr Balfour's Attack on Agnosticism". Part I was published in the *Nineteenth Century* (March 1895), but Part II remained unpublished because Huxley died before he was able to revise it.

19. T. H. Huxley, "On the Advisableness of Improving Natural Knowledge", in *Lay Sermons, Addresses and Reviews*, 3–22 (London: Macmillan, 1870), 20. Before the essay on natural knowledge, most of Huxley's publications were technical ones.

20. *Ibid.*, 22.

21. Huxley's book on Hume was published in John Morley's English Men of Letters series in 1878. Whatever its scholarly merits as a study of Hume, it was the result of more effort than most of the books in this series. He also wrote "The Evolution of Theology: An Anthropological Study", which was more hostile to religion than his earlier writings. Huxley's late position was that theological writings were suitable only for study by anthropologists (Huxley, *Collected Essays*, vol. IV, 287). He also expressed great admiration for the Alexandrian Philo, because he had suggested that God had no predicate beyond existence. Huxley made this into a criticism of Spencer: "That is to say, the Alexandrian Jew of the first century had anticipated the reasonings of Hamilton and Mansell in the nineteenth, and, for him, God is the Unknowable in the sense in which that term is used by Mr Herbert Spencer" (Huxley, *Collected Essays*, vol. IV, 365). Huxley later had to apologise for misspelling Mansel's name, because some unkind critic remarked that it indicated that Huxley had not read Mansel (T. H. Huxley, "Mr Balfour's Attack on Agnosticism", *Nineteenth Century* 37 [March 1895], 534)).

22. Huxley, *Collected Essays*, vol. V, 311, 311 n.

23. Mario A. di Gregorio, "The Dinosaur Connection: A Re-interpretation of T. H. Huxley's Evolutionary View", *Journal of the History of Biology* 15 (1978), 397–418, esp. 397–8.

24. Recent commentators on Huxley seem to overemphasize his scientific standing at the expense of his cultural interests. See, for example, Paul White, *Thomas Huxley: Making the "Man of Science"* (Cambridge: Cambridge University Press, 2003), 74. Given Huxley's importance in forcing Spencer to think systematically about both metaphysics and biology this seems curiously narrow. Huxley's role as a midwife to Victorian philosophical culture needs further exploration.

25. Letter from Frances Hooker to John Tyndall, 1859 (Royal Institution of Great Britain, Tyndall Collection, vol. 8, #2556).

26. This is the last paragraph from a letter from Herbert Spencer to John Tyndall (1858 or 1859), written at Loudoun Road, St John's Wood, where Spencer had moved during 1858 to be near Huxley (Royal Institution of Great Britain, Tyndall Collection, vol. 3, #1175). This quotation has been given at length because it was omitted by Duncan when he published the letter's first three paragraphs (Duncan, *The Life and Letters*, 104).

27. Letter from John Tyndall to Herbert Spencer, 7 April 1863 (?) (Royal Institution of Great Britain, Tyndall Collection, vol. 3, #1028). This letter was omitted from Spencer, *An Autobiography*, and Duncan, *The Life and Letters*, even though Spencer had read and annotated the letter at Mrs Tyndall's request after her husband's death in 1893. One can only suppose that Spencer omitted the letter because he was offended by the mild hint of criticism at the beginning; he was notoriously sensitive to criticism in his later years. The criticism, as can be seen from a subsequent letter, relates to Spencer's chapters on the theory of physics. Later Tyndall wrote that, "I have often wished to say to you that your chapters on the Persistence of Force etc. were never satisfactory to me" (letter from John Tyndall to Herbert Spencer [Royal Institution of Great Britain, Tyndall Collection, vol. 3, #1031; Mrs Tyndall dated this letter 1864, but Duncan, *The Life and Letters*, 428, correctly dated it 1873). Tyndall thought Spencer's area of scientific competence was biology, and that he had cleared away the blemishes of Darwin's and Huxley's view of evolution (letter from John Tyndall to Herbert Spencer, 1864 [Royal Institution of Great Britain, Tyndall Collection, vol. 3, #1030]).

28. John Tyndall, "Physics and Metaphysics", *Saturday Review* 10 (4 August 1860), 141.

29. Paul L. Sawyer, "Ruskin and Tyndall: The Poetry of Matter and the Poetry of Spirit", in *Victorian Science and Victorian Values*, James Paradis and Thomas Postlewait (eds), 217–46 (New York: New York Academy of Science, 1981), 221.

30. For a correction to the caricature see Frank M. Turner, "John Tyndall and Victorian Scientific Naturalism", in *John Tyndall: Essays on a Natural Philosopher*, W. H. Brock, N. D. McMillan and R. C. Mollan (eds), 169–80 (Dublin: Royal Dublin Society, 1981), 171.

31. Sawyer, "Ruskin and Tyndall", 234–5.

32. Anna Therese Cosslett, "Science and Value: The Writings of John Tyndall", in Brock *et al.* (eds), *John Tyndall, Essays on a Natural Philosopher*, 181–91, esp. 185.

33. *Ibid.*, 186–8 and Sawyer, "Ruskin and Tyndall", 236.

34. Tyndall used to be regarded as a thoroughgoing exponent of materialism. But more recently scholarship has critically drawn attention to aspects of his work that were congenial to religion. See Turner, "John Tyndall and Victorian Scientific Naturalism"; Cosslett, "Science and Value"; Sawyer, "Ruskin and Tyndall"; and James R. Moore and Colin Chant, "The Metaphysics of Evolution", in *Science and Metaphysics in Victorian Britain*, 3–38 (Milton Keynes: Open University Press, 1981), 26–30.

35. This address was given before the mathematical section of the British Association at Norwich on 19 August 1868. It was added to a similar discourse, which Tyndall delivered to the British Association in Liverpool in 1870, and published under the title *Essays on the Use and Limit of the Imagination in Science* (London: Longmans, Green, 1870).

36. Tyndall, *Essays on the Use*, 49.

37. *Ibid.*, 30.

38. Letter from John Tyndall to Professor Blackie, 15 October 1868 (National Library of Scotland).

39. *Ibid.*

40. Tyndall, *Address*, vii, 56.

41. *Ibid.*, 47, 49, 57.

42. *Ibid.*, 57. The connection between the Belfast Address and Tyndall's earlier philosophical views has usually been ignored, but Tyndall's friend, Sir Frederick Pollock (lawyer, Spinoza scholar and literary gossip) commented on it: "The much misunderstood Belfast Address (British Association) was only the climax of a series of papers on the philosophy of science which went back to the early sixties". Pollock also remarked that far from being a materialist, Tyndall was "against mechanical views of

spiritual things" (Sir Frederick Pollock, "Tyndall as Worker and Teacher", lecture given at St George's Hall, Langham Place to the Sunday Lecture Society, 20 October 1895 (given earlier at Toynbee Hall and the South Place Chapel) (Royal Institution of Great Britain, Tyndall Collection, manuscript).

43. Tyndall's training in physics and chemistry indicated to him that science was a matter of experiments, and it was his opinion that "the true man of science" could not prove evolution by experiment. Evolution, therefore, was an intellectual matter rather than a scientific one (Tyndall, *Address*, 56).

44. John Tyndall, *Fragments of Science*, 5th edn (London: Longmans, Green, 1876), 328.

45. Sara S. Hennell (1812–99) lived in Coventry, and was the sister-in-law of Charles Bray, the philosophical editor of the *Coventry Herald*. Haight gave a short biography of her in the introduction to *The Letters of George Eliot* and Basil Willey mentions her in *Nineteenth Century Studies* (London: Chatto & Windus, 1949). However, both these scholars focused on Hennell as a friend of Eliot, and did not dwell extensively on her philosophical affiliations, nor on her devotion to Spencer.

46. Hennell had previously written two minor works for the Baillie Essay Prize: *Christianity and Infidelity: An Exposition of the Arguments on Both Sides. Arranged According to a Plan Proposed by George Baillie, Esq.* (London: Arthur, Hall, Virtue, 1857) and *The Early Christian Anticipation of an Approaching End of the World, and its Bearing Upon the Character of Christianity as a Divine Revelation* (London: George Manwaring, 1860), written during 1850 and 1857, respectively.

47. Letter from Sara Hennell to Harriet Martineau, 13 April 1860 (Birmingham University Library, Harriet Martineau Correspondence, H.M.428). This unpublished letter has been quoted at such length because it is one of the few long statements about the early Spencer by someone outside his immediate London circle.

48. *Ibid.* Hennell enclosed her first chapter of an unpublished work with the letter.

49. Hennell, *Thoughts in Aid of Faith*, 137.

50. *Ibid.*, 138.

51. *Ibid.*, 143.

52. *Ibid.*, 148–9.

53. *Ibid.*

54. *Ibid.*, 148–9; also see pp. 174, 272–3, 338, 342–3.

55. *Ibid.*, 270. In the 1880s Hennell lost her faith in Spencer. His sociological work had generated views – such as the notion that all religion is ancestor worship – that she found wholly antipathetic to her views and to his own early ones. Smarting from a sense of betrayal, she formally declared her opposition to his philosophy (Sara S. Hennell, *Present Religion*, 3 vols [London: Trübner, 1887], vol. III, 233–50). This opposition was difficult to express as Spencer never properly revised his early works. These were frequently republished in an unchanged form, and were the cause of much confusion and contradiction.

56. Charles Bray, *The Philosophy of Necessity: or, The Law of Consequences: As Applicable to Mental, Moral and Social Science*, 2 vols (London: Longman, Orme, Brown, Green & Longmans, 1841), vol. I, iv, 2.

57. Charles Bray, *The Philosophy of Necessity: or, Natural Law as Applicable to Moral, Mental and Social Science*, 2 vols (London: Longman, Green, Longman & Roberts, 1863), vol. II, 219 n.

58. *Ibid.* This is the same quotation Sara Hennell had used in *Thoughts in Aid of Faith*, 272–3, which is not surprising since Bray and Hennell worked out their views together. What is surprising is that he does not acknowledge his debt to Hennell. He was ready to express his agreement with Spencer (Charles Bray, *Phases of Opinion and Experience during a Long Life: An Autobiography* [London: Longmans, 1884], 205) but he ignored his sister-in-law.

59. Alfred Russel Wallace (1823–1913) is usually known as the co-discoverer with Darwin of natural selection. He was also extremely active in the Land Nationalization movement, and in the investigation of other-worldly phenomena.

60. This was daring because Thomas Sims, like most of Wallace's other relations, was extremely orthodox.

61. British Museum Add. MSS 39, 168 f. 27v. This passage was printed without some of the punctuation and emphasis by James Marchant, *Alfred Russel Wallace, Letters and Reminiscences*, 2 vols (London: Cassell, 1916), vol. I, 82–3.

62. Marchant, *Alfred Russel Wallace*, vol. I, 147. Wallace's relationship with Spencer is complex: his recent biographer claims that Wallace admired Spencer's philosophy, but not his specific biological comments (Michael Shermer, *In Darwin's Shadow: The Life and Science of Alfred Russel Wallace – A Biographical Study on the Psychology of History* [Oxford: Oxford University Press, 2002], 285–6).

Since this comment seems to refer to Wallace before 1859, and therefore long before Spencer's own biology was published, it is puzzling. It would be more appropriate to say that Wallace took his early metaphysical and religious bearings from Spencer, but as the latter became increasingly secular he would have had less relevance to Wallace, whose soul searching eventually inclined him towards a kind of natural religion.

63. Marchant, *Alfred Russel Wallace*, vol. I, 188.

64. Alfred Russel Wallace, *My Life: A Record of Events and Opinions*, 2 vols (London: Chapman & Hall, 1905), vol. II, 23.

65. Marchant, *Alfred Russel Wallace*, vol. I, 150–51. Neither Hennell nor Wallace, who were outside the London literary circles in which Spencer moved, would have realized that his treatment of the nebular hypothesis, which was based on the work of William Herschel and Pierre-Simon Laplace, was a commonplace one. For example, Chapman had previously developed this theme using Herschel, Laplace and *Vestiges of the Natural History of Creation* (see his preface in Davis, *The Principles of Nature*, vol. I, 15, 15 n.).

66. Marchant, *Alfred Russel Wallace*, vol. I, 151; punctuation as original BM Add MSS 46434 f. 34.

67. Wallace used Spencer's *First Principles* and *The Principles of Biology* to reinforce his claim that evolution was guided by general laws of the equilibrium or harmony of nature as distinct from special intervention by the creator. Also, to Wallace human beings were exempt from the process of natural selection both in the growth of their intelligence and in the development of their moral sense. Alfred Russel Wallace, *Contributions to the Theory of Natural Selection* (London: Macmillan, 1871), 267–371.

68. Jim Endersby, "Escaping Darwin's Shadow", *Journal of the History of Biology* 36(2) (2003), 385–403, esp. 391–2.

69. This comment about the extent of Darwinism in Wallace's writing is truer about his ontological beliefs than it is about his technical work in biology. However, since it was the former that chiefly excited Wallace it should not be overlooked, as it is by those recent historians of science who disregarded theories that seem to have had little impact in the development of genetics.

70. Alfred Russel Wallace, *Man's Place in the Universe: A Study of the Results of Scientific Research in Relation to the Unity or Plurality of Worlds* (London: Chapman & Hall, 1903), 320–24.

71. Webb, *My Apprenticeship*, 20.

72. *Ibid*, 96.

73. Lord and Lady Amberley, *The Amberley Papers: The Letters and Diaries of Lord and Lady Amberley*, 2 vols, Bertrand Russell and Patricia Russell (eds) (London: Leonard and Virginia Woolf at the Hogarth Press, 1937), vol. I, 35.

74. The Metaphysical Society was founded by Tennyson, and dominated by Catholics such as Archbishop Manning and W. G. Ward, and journalists such as Walter Bagehot and R. H. Hutton. The society was remarkable for the number of eminent Victorians who refused to become members. This list included J. S. Mill, Spencer, Benjamin Jowett, Arnold, Mansel and Bain. The society was first documented by Alan Willard Brown in *The Metaphysical Society: Victorian Minds in Crisis 1869–1880* (New York: Columbia University Press, 1947).

75. Hutton's talk was published under the title "A Questionable Parentage for Morals" in *Macmillan's Magazine* 20 (July 1869), 266–73. Hutton was challenged by Spencer, and the result was a controversy that lasted several years. See R. H. Hutton, "Mr Herbert Spencer on Moral Intuitions and Moral Sentiments", *Contemporary Review* 17 (July 1869), 463–72, and Herbert Spencer, "Morals and Moral Sentiments", *Fortnightly Review*, new series 9 (1 April 1871), 419–32.

76. Sir Mountstuart E. Grant Duff, *Notes from a Diary, 1851–1872*, 2 vols (London: John Murray, 1897), vol. I, 231–2.

77. Frederick Pollock, *Personal Remembrances of Sir Frederick Pollock, Second Baronet, Sometime Queen's Remembrancer*, 2 vols (London: Macmillan, 1887), vol. II, 123.

Chapter 10. Common sense in the mid-nineteenth century

1. It would be unfair to credit Hamilton with all his effects. Mansel's great work, *The Limits of Religious Thought Examined in Eight Lectures, Preached Before the University of Oxford, In the Year MDCCCLVIII, on the Foundation of the Late Rev. John Bampton, M.A.* (Oxford: J. H. & Jas. Parker, 1858) (the Bampton Lectures), was published in 1858, two years after Hamilton's death, so one does not know how Hamilton would have received it. Also he was capable of rejecting his own students;

J. F. Ferrier's *Institutes of Metaphysic* was dismissed by Hamilton as made up of baseless paradoxes. See J. F. Ferrier, *Scottish Philosophy: The Old and the New* (Edinburgh: Sutherland & Knox, 1856), 16.

2. Here prominent means in widespread public recognition or a professorial chair.

3. Thomas Spencer Baynes, *An Essay on the New Analytic of Logical Forms, Being that which Gained the Prize Proposed by Sir William Hamilton, in the Year 1846, for the Best Exposition of the New Doctrine Propounded in his Lectures: With an Historical Appendix* (Edinburgh: Sutherland & Knox, 1850), viii.

4. The old *Dictionary of National Biography* noted that Baynes translated Antoine Arnauld and Pierre Nicole's *Port-Royal Logic* in 1851. This date is that of the second edition, which, like the first, was published by Sutherland and Knox of Edinburgh. The catalogue of the Edinburgh University Library lists the apparently rare first edition as "Logic; or the art of Thinking: being the Post-Royal Logic. Trans … with an introd. by T.S. Baines [*sic*]. Edinburgh, 1850". The 1851 second edition, which spelt Baynes's name correctly, is much more common. The entry for Baynes in *The New Cambridge Bibliography of English Literature* contains no mention of any edition of *The Port-Royal Logic*.

5. Also, the busy Hamilton had done Baynes the unparalleled favour of proofreading for him (Thomas Spencer Baynes, *The Port-Royal Logic*, 2nd edn [Edinburgh: Sutherland & Knox, 1851], iii, viii).

6. Skelton, *The Table-Talk of Shirley*, 41.

7. Baynes, *The Port-Royal Logic*, viii.

8. Thomas Spencer Baynes, "Sir William Hamilton", in *Edinburgh Essays by Members of the University, 1856*, 241–300 (Edinburgh: Adam & Charles Black, 1857), 244.

9. Veitch was only sixteen, but, according to G. E. Davie, *The Democratic Intellect: Scotland and Her Universities in the Nineteenth Century* (Edinburgh: Edinburgh University Press, 1961), most Scottish university students at this time began their studies when they were in their mid-teens.

10. Mary R. L. Bryce, *Memoir of John Veitch, L.L.D.* (Edinburgh: William Blackwood, 1896), 49.

11. *Ibid.*, 59.

12. *Ibid.*, 111.

13. Veitch also produced the standard life of Hamilton.

14. John Veitch, *Speculative Philosophy, An Introductory Lecture Delivered at the Opening of the Class of Logic and Rhetoric* (at Glasgow), 1 November 1864 (Edinburgh: William Blackwood, 1864), 13.

15. Bryce, *Memoir of John Veitch*, 68. This point has to be made because G. E. Davie (*The Democratic Intellect*) has overemphasized the connection between Presbyterianism and mid-nineteenth-century Scottish metaphysics.

16. Veitch, *Speculative Philosophy*, 19.

17. *Ibid.*, 15.

18. *Ibid.*, 20.

19. *Ibid.*, 21–2.

20. A. C. Fraser, *Biographia Philosophica: A Retrospect* (Edinburgh: William Blackwood, 1904), 57.

21. *Ibid.*, 73–4.

22. Hamilton had been successful in weaning Veitch and Fraser away from Chalmers's Free Church doctrine, but in the case of another student, Henry Calderwood, his efforts backfired and produced a Presbyterian monster in the shape of Calderwood's *The Philosophy of the Infinite*.

23. Fraser, *Biographica Philosophica*, 61–2.

24. *Ibid.*, 62.

25. A. C. Fraser, *Rational Philosophy in History and in System: An Introduction to a Logical and Metaphysical Course* (Edinburgh: Thomas Constable, 1858), 3.

26. *Ibid.*, 98–9.

27. *Ibid.*, 99 n.

28. *Ibid.*, 104.

29. *Ibid.*

30. Fraser, *Rational Philosophy*, 106.

31. A.C. Fraser, *Thomas Reid* (Edinburgh: Oliphant, Anderson & Ferrier, 1898), 144.

32. David Masson, *Recent British Philosophy: A Review, with Criticisms; Including Some Comments on Mr. Mill's Answer to Sir William Hamilton* (London: Macmillan, 1865), 258. Masson was an alumnus of the University of Aberdeen, but he studied under Hamilton at Edinburgh after his graduation.

33. John Brown, *Horae Subsecivae, Locke and Sydenham with Other Occasional Papers* (Edinburgh: Thomas Constable, 1858), 189.

34. *Ibid.*, 191.
35. This is how Bernard Lightman describes Mansel in "Henry Longueville Mansel and the Origins of Agnosticism", *History of European Ideas* 5(1) (1984), 45–64. Lightman, who oddly describes Spencer as an agnostic, uses such a wide definition that he is also able to include Huxley's Humean scepticism, and the ideas of Kant and J. S. Mill.
36. J. W. Burgon, *Lives of Twelve Good Men*, 2 vols (London: John Murray, 1888), vol. II, 185.
37. Burgon, *Lives of Twelve Good Men*, vol. II, 189.
38. William Whewell, *William Whewell, DD: An Account of His Writings with Selections from his Literary and Scientific Correspondence*, 2 vols, I. Todhunter (ed.) (London: Macmillan, 1876), vol. I, 341. Whewell claimed that fears here were for religion, not for philosophy, which means that unlike the New Reformation he did not closely link these two endeavours.
39. Fraser, *Biographica Philosophica*, 174.
40. Mansel, *The Limits of Religious Thought*, v.
41. *Ibid.*, ix.
42. *Ibid.*, 67–8.
43. *Ibid.*, 72.
44. Mansel wished to defeat the new spirituality as represented by Strauss, Baden Powell, and F. W. Newman for excessive reliance on reason (Mansel, *The Limits of Religious Thought*, 182). He also feared godless philosophy, which dreamed of the subordination of the individual to the universal. The latter wanted "to be swallowed up in the formless and boundless universe". The godless philosophers in question were Schelling, Hegel, Schleiermacher, Comte and H. G. Atkinson (*ibid.*, 88, 88 n.).
45. *Ibid.*, 89.
46. Mansel used the image of a painted curtain that the godless philosophers had mistaken for the real object.
47. Mansel used the word "belief" here in the same way as Hamilton and Spencer did.
48. Mansel also stated his central position that "we are bound to believe *that* God exists; and to acknowledge Him as our Sustainer and our Moral Governor; though we are wholly unable to declare *that* He is in His own Absolute Essence" (*ibid.*, 170–71).
49. *Ibid.*, 145–6.
50. *Ibid.*, 146.
51. *Ibid.*, 235.
52. *Ibid.*, 189–90.
53. *Ibid.*, 390 n.26, quoting Baden Powell, *Christianity without Judaism*.
54. Mansel, *The Limits of Religious Thought*, 388 n.21.
55. Pietro Corsi is one of the few scholars who has commented perceptively on the theological and philosophical prospects debated by Mansel and Powell. However, as Corsi notes, these ideas need much more exploration (*Science and Religion: Baden Powell and the Anglican Debate, 1800–1860* [Cambridge: Cambridge University Press, 1988], 208).
56. Mansel, *The Limits of Religious Thought*, 189.
57. Although a large section of *The Limits of Religious Thought* was devoted to destroying Newman, it does not inspire conviction. One suspects that the two, especially on the matter of eternal evidence, were closer in outlook than Mansel would have liked.
58. John Stuart Mill, *An Examination of Sir William Hamilton's Philosophy and of the Principal Philosophical Questions Discussed in his Writings*, 3rd edn (London: Longmans, Green, Reader & Dyer, 1867), 625.
59. Mill, *Autobiography* (1874), 271–3; Alexander Bain, *John Stuart Mill: A Criticism, with Personal Recollections* (London: Longmans, Green, 1882), 118.
60. Mill, *Autobiography* (1874), 272.
61. Mill, *An Examination of Sir Williams Hamilton's Philosophy* (1867), vii.
62. For an account of this incident see Fraser, *Biographia Philosophica*, 175.
63. Ernest Nagel, *John Stuart Mill's Philosophy of Scientific Method* (New York: Hafner, 1950), xvi.
64. Peter Heath, "Introduction", in *On The Syllogism and Other Logical Writings*, Augustus De Morgan, vii–xxx (London: Routledge & Kegan Paul, 1966), xvi.
65. Fraser, *Biographia Philosophica*, 136.
66. *Ibid.*
67. James McCosh, *The Method of Divine Government, Physical and Moral*, 2nd edn (Edinburgh: Sutherland & Knox, 1850), 296.

68. *Ibid.*, 87 n., 526.

69. *Ibid.*, 526.

70. James McCosh, *The Intuitions Of The Mind Inductively Investigated* (London: Macmillian, 1865), 341.

71. *Ibid.*, 343.

72. Mill, *An Examination of Sir William Hamilton's Philosophy*, 3rd edn (1867), 613, 623. Also see J. S. Mill's notes to James Mill, *Analysis of The Phenomena of The Human Mind*, a new edition with notes by Alexander Bain, Andrew Findlater and George Grote, 2 vols (London: Longmans, Green, Reader & Dyer, 1869), vol. I, 352 n., and Mill, *Autobiography* (1874), 273.

73. There had been earlier French and American editions.

74. See, for example, Nagel, *John Stuart Mill's Philosophy*, xxvii–xxviii.

75. See comments on the followers of Hamilton in Masson, *Recent British Philosophy*.

76. McCosh, *The Intuitions of The Mind*, 345.

77. *Ibid.*, 346.

78. *Ibid.*, 347.

79. Mill, *An Examination of Sir William Hamilton's Philosophy*, 2nd edn (London: Longmans, Green, 1865), 150, 561. This seems to have been aimed more at Mansel, whom Mill thought responsible for "the most morally pernicious doctrine now current" (*ibid.*, 90), than at the dead Hamilton. This matter is, however, a difficult one to decide because Mill holds Hamilton directly responsible for Mansel's opinions, *and* doubts that Mansel reasons correctly from Hamilton's premises (*ibid.*, 88–91).

80. H.L. Mansel, *The Philosophy of The Conditioned, Comprising Some Remarks on Sir William Hamilton's Philosophy and on Mr J. S. Mill's Examination of that Philosophy* (London: Alexander Strahan, 1866), 39 n.

81. Duncan, *The Life and Letters*, 73–4. This letter was mild by Hamilton's usual combative standard. He once caused great distress to Ferrier, a personal friend and the most rigorous metaphysician of the period, by remarking that he had no philosophical talent.

82. Herbert Spencer, "Theological Criticism", *Athenaeum* 1830 (22 November 1862), 663.

83. Letter from J. A. Froude to Herbert Spencer, 16 March 1860 (Athenaeum Club, Spencer Papers). Froude no longer had sympathy with religious doubt or questioning. The Socinianism that distressed Markham Sutherland, the hero of Froude's novel *The Nemesis of Faith*, was buried in Froude's past. From his letter it would appear that religious and metaphysical enquiry was no longer of interest to Froude, and that he was subscribing to Spencer's philosophy from politeness.

84. Letter from Sir John Herschel to Herbert Spencer, 17 March 1860 (Royal Society of London, Herschel Papers, vol. 16).

85. Letter from J. S. Mill to Herbert Spencer, 3 April 1864 (Athenaeum Club, Spencer Papers). The bulk of this letter is given in Duncan, *The Life and Letters*, 114, but it was peculiarly edited, either by Spencer or Duncan, so that the reference to Hamilton was deleted. *The Later Letters of John Stuart Mill* contains the correct version of this letter (Mill, *The Collected Works*, vol. 15, 934), the editors having found a copy and a draft of the letter at Northwestern University.

86. Mill, *An Examination of Sir William Hamilton's Philosophy* (1865), 150.

87. See Geoffrey Scarre, *Logic and Reality in the Philosophy of John Stuart Mill* (Dordrecht: Kluwer, 1989), 165.

88. Mill, *An Examination of Sir William Hamilton's Philosophy* (1867), 175 n.

Chapter 11. From philosophy to psychology

1. Thomas Reid, *The Works of Thomas Reid DD now Fully Collected with Selections from his Unpublished Letters*, 3 vols, Sir William Hamilton Bart. (ed.) (Edinburgh: Maclachlan, Stewart, 1846), 18. Dugald Stewart's desire to rescue Reid from obscurity set the stage for later scholars such as Hamilton, who subjected both Stewart and Reid to a more technical examination than they had received before.

2. *Ibid.*, 28.

3. This development was overlooked by J. S. Mill, who accused Hamilton of disregarding the science of psychology. This was presumably because Mansel and Veitch did not publish the third part of Hamilton's lectures, which dealt with this subject. However, Mill should have known better as his friend Bain praised Hamilton's psychology, and admitted borrowing heavily from it.

4. Also see James McCosh, *The Scottish Philosophy* (London: Macmillan, 1875), 458. McCosh believed

that the metaphysician should enter the physiological field. "He must, if he can, conduct researches; he must at least master the ascertained facts" (*ibid.*).

5. Most of Hamilton's students followed him in stressing the importance of psychology. See Fraser, *Rational Philosophy*, 122; Veitch, *Speculative Philosophy*, 13; and Baynes, "Sir William Hamilton", 245.

6. H. L. Mansel, *Psychology, The Test of Moral and Metaphysical Philosophy; An Inaugural Lecture Delivered in Magdalen College*, October 1855 (Oxford: William Graham, 1855), 9–10.

7. This was John Earle, professor of Anglo-Saxon, in a letter written in 1874 quoted by Burgon, *Lives of Twelve Good Men*, vol. II, 183.

8. Thomas Reid, "An Inquiry into the Human Mind, On the Principles of Common Sense", in *The Works of Thomas Reid*, esp. 110.

9. *Ibid.*, 117. Spencer, like Reid, explicitly included all species of life in this claim.

10. *Ibid.*, 130.

11. Thomas Reid, *Essays on the Intellectual Powers of Man*, in *The Works of Thomas Reid*, 231. Or see Thomas Reid, *Essays on the Intellectual Powers of Man: A Critical Edition*, R. Derek Brookes & K. Haakonssen (eds) (University Park, PA: Pennsylvania State University Press, 2002), 40.

12. Reid, *The Works of Thomas Reid*, 249.

13. Reid used the words "faculty", "power" and "sense" as synonymous (see *ibid.*, 221), and he was followed in this practice by nineteenth-century philosophers such as Mansel and Spencer. The term "*active* faculty" had the effects of introducing the *will* and the freedom of the will into psychology.

14. Hamilton generated an idea that the mind could actively "touch" objects.

15. *Ibid.*, 294. Reid distinguished between himself and Hume by noting that the latter did not leave room for active faculties. "This author leaves no power to the mind in framing its ideas and impressions; and no wonder, since he holds that we have no idea of power; and that the mind is nothing but that succession of impressions and ideas of which we are intimately conscious" (*ibid.*).

16. James Mill had only minor reservations to make about Hume's categories of association, and J. S. Mill refrained from adding a note to this section of his heavily annotated edition of his father's psychology (James Mill, *Analysis of the Phenomena*, vol. I, 106–13).

17. Fraser, *Rational Philosophy*, 124 n.

18. Mill, although disagreeing with Spencer's *First Principles* and his universal postulate, approved of his handling of association doctrine in his psychology. See Letter from J. S. Mill to Alexander Bain, 1864 (LSE, Mill-Taylor Collection, vol. 48, no. 48) and letter from J. S. Mill to Herbert Spencer, 3 April 1864 (Athenaeum Club, Spencer Papers) (printed in Duncan, *The Life and Letters*, 114–15).

19. John Skorupski's detailed modern discussion of the Mill–Hamilton debate refers to Reid and his followers arguing against Mill's "psychological method" (John Skorupski, *John Stuart Mill* [London: Routledge, 1989], 227). This comment is historically confusing if one remembers that Spencer, McCosh, Mansel and so on all thought of themselves as followers of Hamilton and psychologists at the same time. Skorupski also credits Mill here with psychologically reducing the data of consciousness to the minimum number of elements, but this was true of all association psychologists.

20. Thomas Reid, "Of First Principles in General", in *The Works of Thomas Reid*, 441.

21. *Ibid.*

22. Reid's distinction between necessary and contingent appears to be confused at this point because he has described the nature and being of the Supreme Being as necessary. However, this question is beyond the scope of this book as Spencer ignored Reid's statements on the Supreme Being.

23. Reid, "Of First Principles in General", 442, emphasis added.

24. *Ibid.*, 445.

25. *Ibid.*, 447.

26. *Ibid.*, 452–3. Reid admitted that he was going against the majority by including morals and taste among necessary truths. He gave as his defence the statement that if there is judgement in our determinations on these matters then they must be necessary (*ibid.*, 454).

27. Reid, "An Inquiry into the Human Mind", 130.

28. Reid, *Essays on the Active Powers of Man*, in *The Works of Thomas Reid*, 521.

29. *Ibid.*

30. "For my own part, I am decidedly of opinion, that, as the great end – the governing principle of Reid's doctrine was to reconcile philosophy with the necessary convictions of mankind …" (Hamilton, "Note B", in Reid, *The Works of Thomas Reid*, 820).

31. Hamilton, "Note A", in Reid, *The Works of Thomas Reid*, 742.

32. *Ibid.*, 743 n.
33. *Ibid.* Hamilton notes here that this is his objection to the account of Reid given by Dugald Stewart.
34. Hamilton could have easily forgotten the purport of his earlier statement since his dissertations were written as annotations to his edited *The Works of Thomas Reid* over a number of years, and eventually published in an incomplete and unrevised form.
35. *Ibid.*, 754.
36. Undoubtedly, Hamilton and, following him, Spencer saw necessity in the truths of mathematics, and this would have troubled Mill. Whether or not he was fair to his opponents in ascribing to them the fallacy of imagining that a proposition was necessary if it is "not only false but inconceivable", he was certain that mathematics was non-empirical. See Philip Kitcher, "Mill, Mathematics, and the Naturalist Tradition", in *The Cambridge Companion to Mill*, J. Skorupski (ed.), 56–111 (Cambridge: Cambridge University Press, 1998), 98–9.
37. Herbert Spencer, "The Genesis of Science", *British Quarterly Review* 20 (July 1854), 108–62, esp. 108; reprinted in *Essays* (1858), 158–227.
38. Spencer, *The Principles of Psychology* (1855), 486; also see *ibid.*, 564.
39. Spencer, *An Autobiography*, vol. II, 463.
40. Letter from Herbert Spencer to Thomas Hodgskin, 10 April 1855 (Knox College, Galesburg, Illinois).
41. Spencer, *The Principles of Psychology* (1855), 578–9.
42. Spencer, *Social Statics*, 27–8.
43. *Ibid.*, 27.
44. *Ibid.*, 28.
45. *Ibid.*, 29.
46. *Ibid.*, 30.
47. *Ibid.*, 96.
48. *Ibid.*, 97.
49. He mentioned reading Smith during the 1840s (*An Autobiography*, vol. I, 229), but there is little evidence that Spencer was well acquainted with the work of other eighteenth-century philosophers at that time.
50. Spencer, *Social Statics*, 215.
51. Spencer later replaced this theological conception of a moral governor with a biological conception of the organism. Both allowed for the organization of society without resort to crude force or violence. Spencer's fear of force and violence was one of the few constants that permeated his whole life.
52. Spencer, *An Autobiography*, vol. I, 378–9.
53. Herbert Spencer, "The Universal Postulate", *Westminster Review* (October 1853), 513–50.
54. Hegal and Fichte were also rejected because, by their analyses, metaphysics could be nothing but an analysis of knowledge by means of knowledge, a procedure that took for granted what was at question. This argument was a common mid-nineteenth-century one for a man attempting to establish common sense partly on experience (*ibid.*, 514).
55. *Ibid.*, 516.
56. *Ibid*, 520.
57. Reid, "An Inquiry into the Human Mind", 206, 206 n.
58. It was only as a moralist that Spencer ignored Whewell. In the philosophy of science he was much closer to the Cambridge don.
59. Spencer added that he thought little of Reid's refutations of scepticism, but much of his contributions to psychology (Spencer, "The Universal Postulate", 515). However, in *The Principles of Psychology* Spencer borrowed heavily from Reid when supporting his own refutation of scepticism.
60. Spencer thought of his task as replacing Descartes's first principles (Spencer, "The Universal Postulate", 518; *The Principles of Psychology* (1855), 11). His friend and rival, Alexander Bain, agreed that this was Spencer's goal, but that it was a mistake. "[Descartes's] attempt to get a fundamental datum, and minimum of assumption, is illustrated by the parallel attempt of Spencer in the Universal Postulate. Both have failed but no one else has succeeded" (letter from Alexander Bain to G. C. Robertson, 14 May 1865 [UCL, Robertson Papers]).
61. Spencer, "The Universal Postulate", 518.
62. *Ibid.*, 519.
63. *Ibid.*

64. Hamilton, "On the Philosophy of Common Sense; or Our Primary Beliefs as the Ultimate Criterion of Truth", in *The Works of Thomas Reid*, 743.
65. *Ibid.*, 743 n. This passage has been quoted at length because it contains both the question that Spencer's metaphysics attempted to answer and the emphasis on classification that lay at the core of Spencer's philosophy of science.
66. Spencer, "The Universal Postulate", 528–9.
67. *Ibid.*, 530, original emphasis.
68. *Ibid.*, 532; also see *ibid.*, 549. Mansel, *The Limits of Religious Thought*, 182–3, later developed a similar point. He avoided excessive reliance on reason because of Hume. Instead he believed that philosophers must make a direct and immediate appeal to the mind of the common man. Mansel allowed no inferences leading from established principles because casual links depend on reason. All philosophical structures were replaced with a simple reference to *belief*, which had only a practical and regulative value. This last differed from Spencer's postulate, because for Spencer, during the 1850s, such *belief* did have theoretical or philosophical value.
69. Spencer, "The Universal Postulate", 548.
70. *Ibid*, 540. Spencer demonstrated that the subject can be known only as an object. In this he was following Reid, who also refused to give any content to the ego. Spencer's satisfaction with his philosophy can be seen by the fact that he made no changes in it for the rest of the decade. He developed these ideas in his argument against Rudolf Lotze, Hegel and Comte in 1854. In 1855 the "The Universal Postulate" was repeated with one minor addition as the core argument of *The Principles of Psychology*.
71. Spencer, "The Universal Postulate", 527.
72. These statements are some half-developed elements borrowed from Hamilton's philosophy of the relative, and were inconsistent with the rest of Spencer's argument.
73. *Ibid.*, 530.
74. *Ibid.*, 550.
75. Herbert Spencer, *First Principles* (London: Williams & Norgate, 1862), 85.
76. *Ibid.*, 86.
77. *Ibid.*, 4, emphasis added. Significantly, the whole of this quotation, and the paragraph of which it was a part, were dropped from the sixth edition of *First Principles* (London: Williams & Norgate, 1900). Spencer's continuing drift away from common-sense philosophy made it increasingly inconvenient to have as part of his opening statement.
78. Spencer, *First Principles* (1862), 36–7.
79. *Ibid.*, 37. Here, Spencer was following Hamilton's maxim that one should search for an ultimate law.
80. *Ibid.*, 37–8.
81. *Ibid.*, 39–43.
82. *Ibid.*, 44.
83. *Ibid.*, 87.
84. *Ibid.*
85. *Ibid.*, 88.
86. The "indefinite consciousness" stayed in all editions of *First Principles*, even the sixth edition, which showed signs of alteration on other matters (Spencer, *First Principles* [1900], 65).
87. Herbert Spencer, "Mill *versus* Hamilton – the Test of Truth", in *Essays* (1868), vol. II, 390. Spencer was referring to J. S. Mill, *An Examination of Sir William Hamilton's Philosophy* (1865), 150.
88. Spencer, "Mill *versus* Hamilton", 391.
89. In *The Principles of Psychology* Spencer argued that experience was a register of objective facts and, as mentioned above, objective facts in the universal postulate were intuitions (Spencer, *The Principles of Psychology* [1855], 22).
90. *Ibid.*, 22–3.
91. *Ibid.*, 23 n.
92. Any empiricism that did not accept Spencer's formulation of the test of truth was referred to as "pure empiricism", and faulted because it must presuppose more than Spencer's version did (Spencer, "Mill *versus* Hamilton", 398). This criticism of Mill was phrased obscurely, presumably because Spencer also wished to emphasize that he and Mill agreed on many points (*ibid.*, 413). This critique of Mill was a common one. It can also be seen in the works of McCosh, and in the introduction to the later editions of Lewes's *The History of Philosophy*.

93. Spencer, "Mill *versus* Hamilton", 395.
94. Spencer, *The Principles of Psychology*, (1870, 1872). "The Universal Postulate" was absent from all editions of Spencer's *Essays*. This is notable because he usually reprinted essays even when their material also appeared in his philosophical works. One must presume that in this case he felt he had made some very important improvements to his argument.
95. Spencer, *The Principles of Psychology* (1855), 65.
96. Spencer, *The Principles of Psychology* (1872), vol. II, 406–7.
97. *Ibid.*, vol. II, 408–9.
98. *Ibid.*, vol. II, 494.
99. Herbert Spencer, "Replies to Criticisms", in *Essays* (1874), vol. III, 284.
100. T. H. (Paul) Ribot, *English Psychology* (London: Henry S. King, 1873), 125.
101. *Ibid.*, 126, 192. Ribot thought that Spencer was an even richer source of inspiration than Comte because the latter had rejected research into final and first causes as nonsense.
102. A recent historian of nineteenth-century British psychology (Rick Rylance, *Victorian Psychology and British Culture* [Oxford: Oxford University Press, 2000], 161) has described Spencer as a maverick in psychological theory. This is a curious judgement given that much of Spencer's work in this area was straightforward and orthodox association psychology combined with a fashionable evolutionary theory. The basis for Rylance's judgement seems to be J. S Mill's choice of Bain as his standard-bearer in psychology instead of Spencer. However, this is a retrospective comment that works better for 1880 (the end of Rylance's survey) than it would before that date. Contemporaries, such as Ribot, cited Mill's writings, which praised Spencer as one of the small band who was a peer of Comte. H. Taine (*On Intelligence* [London: L. Reeve, 1871], part II, 273) cited Bain, Mill and Spencer as if they were interchangeable. The distance between Spencer and Bain has little to do with Mill's 1865 criticism of Spencer's links with Hamilton and Mansel. It began in 1860 when Spencer published a discourteous review of Bain focusing on his philosophical naivety. When Bain eventually criticized Spencer, the latter responded with an appendix to the third edition of his pamphlet *The Classification of the Sciences* (London: Williams & Norgate, 1871), which made his rival seem to be merely a pupil of Comte.
103. Alexander Bain, "On the Correlation of Force in its Bearing on Mind", *Macmillan's Magazine* 16 (May–October 1867), 372–83.
104. Bain was usually sympathetic to phrenology, particularly that of Dr Alexander Combe, in his earliest writings. See Alexander Bain, *On the Applications of Science to Human Health and Well-Being* (London: John J. Griffith, 1848), 12, and *On the Study of Character, Including An Estimate of Phrenology* (London: Parker, 1861). In the second work he criticized phrenology, which was already very much in decline, while expressing gratitude for the good service it had done. In *The Senses and the Intellect* (London: J. W. Parker, 1855) Bain took his lead from physiologists such as Johannes Müller and William Pulteney Alison.
105. Samuel Bailey, *Letters on the Philosophy of the Human Mind*, 3rd series (London: Longman, 1863), 1.
106. Herbert Spencer, *The Classification of the Sciences* (London: Williams & Norgate, 1864), 40.
107. Herbert Spencer, "Progress: its Law and Cause" (1857), in *Essays* (1868), vol. I, 59.

Chapter 12. On goodness, perfection and the shape of living things

1. John Lubbock, *Pre-Historic Times, As Illustrated by Ancient Remains, and the Manners and Customs of Modern Savages*, 5th edn (London: Williams & Norgate, 1890), 601.
2. R. Jackson Wilson, *Darwinism and the American Intellectual: An Anthology* (Chicago, IL: Dorsey Press, 1989), 80. Richard Hofstadter thought that Sumner could be excused his confusion since the combination of Christian ethics and democratic heritage in America made it easier to focus on the present conflict and warfare rather than upon the distant future. On this see Richard Hofstadter, *Social Darwinism in American Thought*, 2nd rev. edn (Boston, MA: Beacon Press, 1955), 85–7. However Stanley Elkins and Eric McKitrick have observed that Sumner was the *only* American intellectual of the first rank to have expounded a truly social Darwinist system so he cannot contextually be pardoned for confusing altruism with struggle in the way Hofstadter had hoped. See S. Elkins and E. McKitrick, *The Hofstadter Aegis: A Memorial* (New York: Knopf, 1974), 305.
3. Herbert Spencer, "The Development Hypothesis" [1852], in *Essays* (1858), 391, original emphasis. Spencer referred to this early writing in *The Principles of Biology*, vol. I, 333 n., where he frequently

drew on Darwin's ideas of natural selection and "struggle for survival" as mechanisms to explain the structures and shapes of animals and plants. This would have helped to create the conditions that led some readers to confuse him with Darwin. Some of these would have missed Spencer's comments about the limits of natural selection. They may have also misunderstood the significance of his reliance on his friends, Huxley and Joseph Dalton Hooker, who stressed that any theory of progressive modification must be compatible with persistence without progression through indefinite periods of time. In 1864 Spencer was taking a sceptical stance on "natural selection" as a general explanation of evolution. See *The Principles of Biology*, vol. I, 324, 326.

4. Herbert Spencer, "The Filiation of Ideas", Appendix B, in Duncan, *The Life and Letters*, 540. While this comment refers to *Social Statics*, which was published in 1851, Spencer also claimed that in late 1850 and early 1851 he rejected the views set forth in *Vestiges of the Natural History of Creation* because he then thought that functional adaptation was the sole cause of development (*ibid.*, 541). If this statement is considered together with his belief that his views in "The Development Hypothesis" (in Essays [1868], vol. I, 377–83; originally published in the *Leader* [January 1852–May 1854]) were derived from the general aspects of organic nature ("The Filiation of Ideas", 543), then it is clear that he was not engaged in a belated effort to claim priority of discovery of "natural selection". On the contrary, he was emphasizing that his theories were of generalizations about organic life in general with no particular attention paid to historical adaptation of species.

When Spencer used various ideas of Darwin's freely during the 1860s he mostly borrowed the theory of natural selection. However, he also relied on Darwin to reinforce theories that competed with natural selection, such as the belief that some evolution was caused by the effect of the use or disuse of certain organs. For example, Spencer could draw on Darwin to bolster the idea that there are sightless moles because their ancestors did not use their eyes (Spencer, *The Principles of Biology*, vol. I, 246–7).

5. This is different from claiming that either Spencer or Darwin was more important. While Darwin has had a greater posthumous reputation, there have always been some who claim that Spencer was the greater of the two. This minority view could be found among Spencer contemporaries, such as W. H. Hudson (*Men, Books, and Birds* [London: Jonathan Cape, 1928], 219).

6. Letter from George Eliot to Sara Hennell, 22 November 1857, in *The George Eliot Letters*, vol. 2, 405.

7. Spencer, *The Principles of Biology*, vol. I, 253.

8. *Ibid.*, vol. I, 97.

9. *Ibid.*, vol. I, 160. Henri Milne-Edwards suggested the utility of geological study for determining the natural affinity of animals, but unlike scientists who believed in recapitulation he did not believe that the embryo moved through the "species" forms of inferior creatures, but that it rose from the more general to the more particular forms of a given species. See Robert J. Richards, *The Meaning of Evolution: The Morphological Constrictions and Ideological Reconstruction of Darwin's Theory* (Chicago, IL: University of Chicago Press, 1992), 134–6. As Richards observes, Darwin adopted Milne-Edwards's views in a progressive way. What I am observing is that Milne-Edwards was useful for Spencer because his argument made species classification subordinate to a theory of differentiation rather than supporting an argument of historical evolution.

10. W. B. Carpenter used Milne-Edwards heavily. He cited him in *Animal Physiology* (London: Willian S. Orr, 1843), v–vi. It was noticed by a contemporary that he also copied many of the illustrations from Milne-Edwards's *Elementary Causes of Natural History* (see T. Lindley Kemp, *Agricultural Physiology* [Edinburgh: Blackwood, 1850], xi–xii). Sherri Lyons ("The Origins of T. H. Huxley's Saltationism: History in Darwin's Shadow", *Journal of the History of Biology* 28 (1995), 463–94, esp. 480) has noticed that William Carpenter, in his *Principles of General and Comparative Physiology, Intended as an Introduction to the Study of Human Physiology and as a Guide to the Philosophical Pursuit of Natural History* (3rd edn [London: John Churchill, 1851]), followed Karl Ernst von Baer's idea that the undifferentiated germ became specialized as development proceeded and applied it to the fossil record. He suggested that the fossil record showed a progression from a generalized archetype to a more specialized form. In this book Carpenter was grudging about progressive development, but in other works he was one of its advocates.

11. Thomas Rymer Jones, *The Natural History of Animals, Being the Substance of Three Courses of Lectures Delivered Before The Royal Institution of Great Britain*, 2 vols (London: John Van Voorst, 1845–52), vol. II, 1.

12. Spencer, *The Principles of Biology*, vol. I, 451–6.

13. *Ibid.*, vol. I, 428.
14. *Ibid.*, vol. I, 449.
15. By remarking that Bagehot and Butler were *non-scientific* commentators on evolution I do not wish to add my voice to the debate about the professionalization of science during the mid-Victorian period. As Endersby has lucidly observed, the term "professional scientist" is inappropriate even as late as the 1860s ("Escaping Darwin's Shadow", 396–7). It is also misleading when thinking about Spencer and his circle. Even leaving the adjective "professional" and the noun "scientist" to one side, it is an unhelpful discussion. Spencer was not seen by contemporaries as a botanist or a physiologist, but as a philosopher.
16. A comparison between Lamarck's view of biological change and Darwin's should not focus excessively on either the transmission of acquired characteristics or on natural selection. Perhaps the most important difference, as has been noted by Sandra Herbert ("Introduction", in Charles Darwin, *The Red Notebook of Charles Darwin, Bulletin of the British Museum [Natural History] Historical Series* 7 [24 April 1980], 4–29, esp. 7–8), is that Darwin chose to avoid a Lamarckian understanding of the boundary of a species. While Lamarck saw the boundaries of species as blurred, Darwin perceived important differences between closely related species, and stressed a notion of representative species. Although Darwin rejected Lamarck's theory that one species gradually changed into another in response to "degenerative circumstances" such as a gradual change of climate, he shared Lamarck's belief that present-day species were descended from earlier related forms.
17. Herbert Spencer, "A Theory of Population", 478–9, and *The Principles of Biology*, vol. I, 217.
18. Spencer, *The Principles of Biology*, vol. I, 216. Spencer defined "genesis" here as a process of negative or positive disintegration as opposed to the process of integration that occurred in individual evolution.
19. Spencer's connection with Gideon Mantell has been overlooked because he lost touch with the Mantell family when they moved to New Zealand. In a late letter to Walter Mantell, Spencer refers to the long-past days when they used to meet at the Trelawney's and elsewhere (letter of Herbert Spencer to Walter Mantell, 23 February 1883 [Alexander Turnbull Library, Wellington, ms papers: 83-382:4]). John Trelawney was an early subscriber to Spencer's "A System of Synthetic Philosophy", as was one of the Mantells (Alexander Turnbull Library, Wellington, M.S. papers: 83-382:2). Gideon Mantell has been credited with being the midwife to the birth of historical palaeontology in England. He was also the scientist who gave Lyell a copy of Lamarck's *Philosophic Zoology* to read. See Pietro Corsi, "The Importance of French Transformist Ideas for the Second Volume of Lyell's Principles of Geology", *British Journal for the History of Science* 11(39) (1978), 221–44, esp. 224.
20. Spencer, *The Principles of Psychology* (1855), 394.
21. Many "typical" Victorians, such as John Herman Merivale, E. B. Tylor and Charles Dilke, had doubts about progress during the late-nineteenth century. That is, it is not true that progress was doubted only by exceptional and overly sensitive figures such as "B.V.", the author of *City of Dreadful Night*.
22. Spencer, "A Theory of Population", 501.
23. As he did with the "Unknown", in the 1850s Spencer usually capitalized "Life". In doing this he was emphasizing both as successors to the Christian God.
24. Herbert Spencer, "The Genesis of Science" (1854), in *Essays* (1858), 220, and Spencer, *The Principles of Psychology* (1855), 409.
25. On Spencer's personal reaction against excessive optimism see Chapter 1. His essay on population was the only major essay he never reprinted despite the favourable reception it received at the time, and the fact that it led to his friendship with Huxley. See Spencer's "Filiation of Ideas" in Duncan, *The Life and Letters*, 543. In his essay "A Theory of Population" he was largely indebted to the physiological ideas of Carpenter, although in "Filiation of Ideas" the only credit Spencer gives to physiological work is to Milne-Edward's phrase "the physiological division of labour" (*ibid.*, 546). Spencer's claim to have reused some of his essay on population at the end of volume II of *The Principles of Biology* (500 n.; also see 391 n.) is true only if one does not notice that he has omitted innate tendencies, perfection and teleology. It would also be necessary to qualify the claim by Spencer's newfound distrust of the benefits of industry and of self-denial (*ibid.*, vol. II, 503).
26. J. F. W. Herschel, *A Preliminary Discourse on the Study of Natural Philosophy* (London: Longman, 1831), 8.
27. Spencer, "A Theory of Population", 469.
28. *Ibid.*

29. *Ibid.*
30. *Ibid.*, 496–7.
31. *Ibid.*, 471.
32. Cuvier's law is quoted by Nicolaas Rupke, *The Great Chain of History: William Buckland and the English School of Geology, 1814–1849* (Oxford: Clarendon Press, 1983), 131; for Rupke's comments on Mantell, see *ibid.*, 19, 20, 92, 124, 244. Mantell's early association with Spencer has been overlooked.
33. Georges Cuvier, *The Animal Kingdom, Arranged after its Organization, Forming a Natural History of Animals*, W. B. Carpenter and J. O. Westwood (eds) (London: William S. Orr, 1849), 43; these editors believed that Cuvier had sanctioned this proposition by his remarks on linear arrangement.
34. Spencer, *The Principles of Psychology* (1855), 376.
35. Goldwin Smith, *Reminiscences of Goldwin Smith*, Arnold Haultain (ed.) (New York: Macmillan, 1910), 146, n.1.
36. Spencer, *The Principles of Psychology* (1855), 381.
37. *Ibid.*
38. Spencer, *Social Statics*, 69.
39. Spencer, *The Principles of Biology*, vol. I, 87.
40. Herbert Spencer, "A Criticism on Professor Owen's Theory of the Vertebrate Skeleton", *British and Foreign Medico-Chirurgical Review* (October 1858). This article was reprinted as "Appendix B" in Spencer, *The Principles of Biology*, 2 vols (New York: D. Appleton & Co., 1898), vol. II, 517–35.
41. Spencer, *The Principles of Psychology* (1855), 383.
42. *Ibid.*, 421.
43. *Ibid.*, 464–5.
44. *Ibid.*, 482. The term "man of science" was used by mid-Victorians instead of "scientist".
45. *Ibid.*, 341.
46. J. S. Mill, *A System of Logic*, 4th edn, 2 vols (London: John W. Parker, 1856), vol. I, 93.
47. Spencer's theory here could be seen as close to Whewell's since the latter also held that he had transformed induction so as to give the mind an active rather than a passive role. (The mind would work dynamically to connect facts.) Since Whewell also repeated Bacon's warning about the dangers of a "method of anticipation" he would have been hostile to Spencer's "pre-vision" theory. On Whewell's theory of induction see Richard Yeo, *Refining Science: William Whewell, Natural Knowledge and Public Debate in Early Victorian Britain* (Cambridge: Cambridge University Press, 1993), 12–13, 97. On Whewell's fundamental "ideas" that were not derived from experience see Richard Yeo, "William Whewell, Natural Theory and the Philosophy of Science in Mid-Nineteenth-Century Britain", *Annals of Science* 36 (1979), 494–516, esp. 500–503.
48. Spencer, *The Principles of Psychology* (1855), 348.
49. Spencer, "The Universal Postulate", 519. This article was later reused as the introductory section of *The Principles of Psychology* (1855).
50. Spencer, "The Universal Postulate", 522. J. S. Mill had broadened the notion of induction by emphasizing that the universality of an idea of cause and effect depends on our experience; see Mill, *A System of Logic*, vol. I, 97. Ultimately this position is derived from Hume's philosophy but Spencer, like many of his contemporaries, was more familiar with such arguments as they were found in the writings of critics of Hume such as Thomas Reid.
51. Whewell, *The Philosophy of the Inductive Sciences*, 2 vols (London: J. W. Parker, 1847), vol. I, 247; also see *ibid.*, 64.
52. A modern account of the Whewell–Mill debate (Gerd Buchdahl, "Deductivist Versus Inductivist Approaches", in *William Whewell: A Composite Portrait*, Menachem Fisch and Simon Schaffer [eds] [Oxford: Clarendon Press, 1991], 331–44) suggests, rather unkindly, that each philosopher possessed a metaphysical basis unrecognized by the other, and that neither had a defensible and clear account of induction. This suggestion is not a useful comment on an historical controversy because it could be made about any past philosophical dispute when the antagonists did not share a modern commentator's interests and language about induction.
53. Whewell, *The Philosophy of the Inductive Sciences*, vol. I, 61.
54. Spencer, *The Principles of Psychology* (1855), 340.
55. Spencer, "The Genesis of Science", in *Essays* (1858), 167–8, 171–83.
56. *Ibid.*, 179.
57. *Ibid.*, 165.

58. Spencer, *The Principles of Psychology* (1855), 347.
59. *Ibid.*
60. Spencer's views on scientific method were later relegated to a miscellany. They only reappeared as "An Element in Method" in *Various Fragments* (London: Williams & Norgate, 1897), 3–13.
61. As Nicolaas Rupke notes with reference to British geology – the premier natural science of the 1820s and 1830s – the major controversies were not part of pan-European movements, but insular arguments such as that between the Oxford and Edinburgh schools of earth sciences (*The Great Chain of History*, 82).
62. Whewell, *The Philosophy of the Inductive Sciences*, vol. I, 74.
63. Spencer, "A Theory of Population", 472.
64. *Ibid.*, 473.
65. Carpenter, whose physiological work was an important source for Spencer, speculated about the correlation of living and physical forces and about the mechanistic functioning of nervous systems, in such a way as to indicate that his views were not mechanistic. See Roger Smith, "The Human Significance of Biology: Carpenter, Darwin, and the *Vera Causa*", in *Nature and the Victorian Imagination*, U. C. Knoepflmacher and G. B. Tennyson (eds), 220–28 (Berkeley, CA: University of California Press, 1977).
66. Spencer, "A Theory of Population", 473.
67. Spencer, *The Principles of Psychology* (1855), 347.
68. *Ibid.*, 349–50. Spencer's inability to distinguish between animal instinct and human intelligence is also a feature of French psychologists, such as Théodule Ribot, who used Spencer's views to reinforce their racial doctrines. On this latter subject they seemed to have drawn on the harsh views of the first edition of *The Principles of Psychology*, and to have ignored the addition of relatively more humane and tentative statements about human inequalities that appeared in the second edition (1870, 1872) and in sociological works that began appearing in 1873. (Spencer's *The Principles of Psychology* did not appear in a French translation until the mid-1870s.) If one was debating Spencer's impact on evolutionary psychology one could find reasons for claiming both that he fostered racial categories and that his work militated against them. However, in neither case should one agree with Hilary Rose ("Colonizing the Social Sciences", in *Alas, Poor Darwin: Arguments Against Evolutionary Psychology*, H. Rose and S. Rose (eds), 127–54 [New York, Harmony Books, 2000], 132), and claim that what attracted Spencer was the importance Darwin gave to competition as the mechanism for natural selection. As a social scientist Spencer's grand theory was not centred on competition.
69. Spencer, *The Principles of Psychology* (1855), 350.
70. *Ibid.*, 353.
71. This "musketeers" phrase can be found in Nicolaas Rupke, *Richard Owen: Victorian Naturalist* (New Haven, CT: Yale University Press, 1994), 204; the other musketeers are Huxley and Lewes. In the last case, as with Spencer, Rupke's view is too simplistic. As was hinted by Tjoa (*George Henry Lewes*, 100), positivism does not coexist easily with a form of vitalism that rejects mechanistic explanations for complex and organic phenomena, and that refused to reduce psychology to physiology. The last point was the basis of Spencer's objection to Comte. It put him at odds with Huxley, who criticized Comte for saying that biology could not be an experimental science. Huxley also produced a single scientific method applicable to all science (Thomas H. Huxley, *On The Education Value of the Natural History Sciences* [London: John Van Voorst, 1854], 18, 32 n.B).
72. Spencer, *The Principles of Psychology* (1855), 441.
73. Spencer, *The Principles of Psychology* (1855), 353. Spencer comments here that Coleridge plagiarized his view of "Life" from Schelling. In this prejudice he was probably drawing on Fraser's comments on Coleridge. The definition of life as the tendency to individuation was part of Coleridge's 1816 comments on John Hunter's views of the idea of life. Coleridge's comments were published in 1848 as *Hints Towards the Formation of a More Comprehensive Theory of Life* (in S.T. Coleridge, *Shorter Works and Fragments*, 2 vols, H. J. Jackson and J. R. de J. Jackson (eds) [Princeton, NJ: Princeton University Press, 1995], vol. I, 481, 510–11). While Coleridge's view on individuation appeared in "A Theory of Life" in *The Friend* (3rd edn, 3 vols [London: Williams Pickering, 1837], vol. III, 204), he had already observed that "the whole purport and function" of the understanding consisted of "individualization".
74. Herbert Spencer, "The Social Organism" (1860), in *Essays: Scientific, Political, and Speculative*, second series, 143–84 (London: Williams & Norgate, 1863), 150.

75. Spencer, *The Principles of Psychology* (1855), 354. This was also the definition of life offered by Milne-Edwards in *A Manual of Zoology*, 2nd edn, R. Knox (trans.) (London: Henry Renshaw, 1863), 3; the first edition was published in 1856.

76. See Pietro Corsi, *The Age of Larmarck: Evolutionary Theory in France, 1790–1830* (Berkeley, CA: University of California Press, 1988), 264.

77. For a well-informed discussion of this debate see Franck Bourdier, "Geoffrey Saint-Hilaire versus Cuvier: The Campaign for Paleontological Evolution (1825–1838)", in *Toward a History of Geology: Proceedings of the New Hampshire Inter-Disciplinary Conference on the History of Geology*, C. J. Schneer (ed.), 36–61 (Cambridge, MA: MIT Press, 1969).

78. See G. H. Lewes, *The Life and Works of Goethe*, 2 vols (London: David Nutt, 1855), vol. II, 148. Lewes had been Spencer's companion on his botanizing rambles but, by the time the former's study of Goethe appeared, Spencer had become increasingly hostile towards transcendentalist science.

79. As William Coleman observes (*Biology in the Nineteenth Century: Problems of Form, Function and Transformation* [Cambridge: Cambridge University Press, 1977], 145–6), vitalism had many forms, not the least of which played a role in generating the subject of organic chemistry. This was a long way from speculative botany.

80. Spencer, *The Principles of Psychology* (1855), 354.

81. *Ibid.*, 355. In this definition Spencer was partly shifting away from his former reliance on the language of mechanics.

82. Lewes seems to have been nettled by this comment (see G. H. Lewes, *The Physiology of Common Life*, 2 vols [Edinburgh: William Blackwood, 1866], vol. II, 426) but had no plausible response. He continued to define life vaguely as a dynamical condition that induced decay, even though his main purpose was to distinguish living from non-living things.

83. Spencer, *The Principles of Psychology* (1855), 357.

84. *Ibid.*, 352. If one takes William Paley as typical of orthodox natural theology, then Spencer, in dividing vital entities from non-vital ones such as watches, was taking a radical approach here. In his *Natural Theology, Or Evidence of the Existence and Attributes of the Deity* (London: C. J. G. & F. Rivington, 1830), Paley referred to natural contrivances in animal physiology in the same way as he did to a watch. That is, he saw both as equally mechanical, even though the former were often badly designed.

85. *Ibid.*, 358.

86. Spencer, "The Social Organism", 155–7.

87. Spencer, *The Principles of Psychology* (1855), 359.

88. Spencer, "The Social Organism", 151–4. The first of these differences – that which relates to "form" – is to be explained by Spencer's rejection of Platonic types and archetypes in physiology. From the later 1850s he found it increasingly difficult to maintain his notion of form in his general philosophy.

89. *The Principles of Psychology* (1855), 359.

90. *Ibid.*, 388.

91. It is possible that this idea was borrowed by E. B. Tylor and used in his definition of culture. If the idea was not original to him then it may explain the mystery of why Tylor's definition of culture did not structure any of his specific theories about indigenous cultures. It may also be significant to Tylor's "culture" that Spencer emphasized non-essential attributes or "survivals" when analysing change (Spencer, *The Principles of Biology*, vol. I, 309). On Tylor's concepts of culture see George W. Stocking, Jr, *Race, Culture and Evolution: Essays in the History of Anthropology*, 2nd edn (Chicago, IL: University of Chicago Press, 1982), ch. 4, and *Victorian Anthropology* (New York: Free Press, 1987), 304–5.

92. The problem of machines displaying life-like behaviour has continued to exercise scholars long after Spencer's death. See R. L. Gregory, *Mind in Science: A History of Explanations in Psychology and Physics* (London: Weidenfeld & Nicolson, 1981), 82.

93. Spencer, *The Principles of Psychology* (1855), 364, emphasis added.

94. *Ibid.*

95. See William Kirby, *On the Power, Wisdom and Goodness of God, As Manifested in the Creation of Animals, and In their History, Habits and Instincts*, 2 vols (London: Henry G. Bohn, 1854), vol. II, 381. This new edition of a Bridgewater Treatise was edited by Thomas Rymer Jones and defended Gideon Mantell's palaeontological work against biblical fundamentalists. Spencer's acquaintance with these scientists is pertinent. It was their orthodox combination of religion and science that he and the New Reformation were determined to replace.

96. Spencer, *The Principles of Psychology* (1855), 482.

97. *Ibid.*, 486.

98. *Ibid.*, 364.

99. Spencer's rejection of Owen's Platonism in 1858 has been commented on by Rupke (*Richard Owen*, 206). Curiously, Rupke has ignored the effect of Platonism on Spencer's own early evolutionary theory and treats Spencer's rejection of Platonism as an amateurish imitation of Huxley's comments. Rupke is mistaken to see Spencer as an amateur philosopher in comparison to Huxley. Of course, Rupke could have meant to imply that Spencer was a physiologist, but that would be a greater error.

100. Spencer's psychology was ultimately derived from the common-sense school of philosophy that began with the work of Reid.

101. Whewell, *The Philosophy of the Inductive Sciences*, vol. I, 59.

102. *Ibid.*, vol. I, 61.

103. Whewell's own idea of life was rather tentative. He suspected that science was gradually moving towards a clear *conception* of life, but had not arrived at this yet. This meant that he thought it premature to speak of an *ideal* of life (*ibid.*, vol. I, 545).

104. Spencer, "The Universal Postulate", 524, and *The Principles of Psychology* (1855), 7.

105. In "The Universal Postulate" Spencer conceded that this procedure was only a psychological procedure (p. 530), but in *The Principles of Psychology* he did not make this qualification.

106. Spencer, *The Principles of Psychology* (1855), 367.

107. *Ibid.*

108. *Ibid.*, 368.

109. *Ibid.* In the 1864 *Biology* this interrogative phrase was changed to the neutral "how it was."

110. A heavy emphasis on Spencer's supposed Lamarckianism has featured in the writings of a number of historians of science. Most recently, it has captivated Peter J. Bowler (*The Invention of Progress: The Victorians and the Past* [Oxford: Blackwell, 1989], 137, 143, 153–4). Primarily this identification is misleading in that it relies on a dated view of the French scientist. Recent work on Lamarck is less hostile, as well as more sophisticated, than that during the mid-twentieth century. It is now suggested that Lamarck's ideas were not teleological. If one views Lamarck through this improved perspective then one might say that in his early writing he was attempting to arrange the plant kingdom on the criterion of *decreasing* structural complexity. His later work on vertebrates was considerably more complex and covered many aspects of palaeontology, geology, embryology and related subjects. Since Lamarck was at the centre of a complex debate about the extinction of species, and Spencer seemed uninterested in this debate, there is no reason to refer to him as Lamarckian in either the old or new uses of this term. On Lamarck see Richard Burkhardt, *The Spirit of System: Lamarck and Evolutionary Biology* (Cambridge, MA: Harvard University Press, 1995), and Corsi, *The Age of Lamarck*. Paul Weindling's unpublished remarks on this subject have been helpful to me.

111. Spencer, *The Principles of Psychology* (1855), 369–70. This comment about the medusa was included in *The Principles of Biology*, vol. I, 76, where the name of the animal was changed to "polype".

112. Spencer, *The Principles of Psychology* (1855), 369–70.

113. *Ibid.*, 370.

114. *Ibid.*, 371.

115. *Ibid.*, 375.

116. *Ibid.*, 370.

117. *Ibid.*, 375.

118. On this subject see Adrian Desmond, *The Politics of Evolution: Morphology, Medicine, and Reform in Radical London* (Chicago, IL: University of Chicago Press, 1989).

119. This change can be seen if one compares Spencer's *The Principles of Biology* (vol. I, 75) with *The Principles of Psychology* (1855), 369. In the later work he no longer insisted that an organism must be in correspondence with its environment in order for it to survive. By 1864 his views on biology were more conventional and less dependent on analogies with psychology.

120. M. J. S. Hodge, "Darwin and the Laws of the Animate Part of the Terrestrial System (1835–1837): On the Lyellian Origins of His Zoonomical Explanatory Program", in *Studies in History of Biology*, vol. 6, W. Coleman & C. Limoges (eds) (Baltimore, MD: Johns Hopkins University Press, 1983), 2.

Chapter 13: The meaning of life

1. Spencer was too modern in his outlook to be inspired by the work of Comte, whose materialism recapitulated the views of the Marquis de Condorcet, William Robertson and Hume. Spencer's ideas

were closer to those of Henri de Saint-Simon, for whom an acquaintance with biomedical science undermined the classic Enlightenment view. That is, Saint-Simon thought that it was old-fashioned and excessively analytical to divide the science of man from natural science. See Barbara Haines, "The Inter-Relations Between Social, Biological and Medical Thought, 1780–1850: Saint-Simon and Comte", *British Journal for the History of Science* 11i(37) (1978), 19–35, esp. 32.

2. Spencer, *The Principles of Biology*, vol. I, 59.

3. Spencer, *The Principles of Biology*. Spencer had uttered the same warnings about the dangers of relying on contemporary definitions of organic phenomena in the first edition of his *Principles of Psychology* (1855), 353, but it was not then central to the main thrust of his ideas.

4. Spencer, *The Principles of Biology*, vol. I, 332.

5. *Ibid.*, vol. I, 286, 326. Although Spencer was keen to make some use of the theory of natural selection to support his own evolutionary ideas, he was aware that Darwin's idea of "blending" inheritance implied that spontaneous variation must have simultaneously affected a large number of animals (*ibid.*, vol. I, 450–51). This implication was implausible, and was repeatedly remarked on by Victorian critics of Darwin. For the views of some of these critics see D. L. Hull, *Darwin and his Critics: The Reception of Darwin's Theory of Evolution by the Scientific Community* (Cambridge, MA: Harvard University Press, 1973).

6. Spencer, *The Principles of Biology*, vol. I, 286, 289. In this work Spencer tended not to use the phrase "vital force"; he simply contrasted internal forces or energies, which off-set, or were overcome by, external ones.

7. *Ibid.*, vol. I, 206.

8. *Ibid.*, vol. I, 216–17.

9. *Ibid.*, vol. I, 220–21.

10. *Ibid.*, vol. I, 101, 234–5. Spencer presented Darwin's natural selection theory in terms of structure and function and absorbed it into his own explanatory framework as an anti-teleological argument. Since he also feared that Darwin's use of the notion of "species" concealed teleological remnants, his view of Darwin's *On the Origin of Species* was fundamentally ambivalent.

11. *Ibid.*, vol. I, 341–3.

12. *Ibid.*, vol. I, 340.

13. *Ibid.*

14. *Ibid.*, vol. I, 298–300.

15. *Ibid.*, vol. I, 408.

16. *Ibid.*, vol. I, 189 n., 449.

17. *Ibid.*, vol. I, 300–302. In the early 1850s Spencer had interpreted von Baer's ideas on the increasing heterogeneity of structure as progressive because he had been dependent on Carpenter; see Carpenter, *Principles of General and Comparative Physiology*, ix, 575–6. In the 1864 edition of this work Carpenter too began to show "non-progressive" tendencies. He also moved away from his partial 1851 endorsement essence within organisms. Carpenter was periodically hesitant in his affirmation of vitalism. Sometimes, it was merely a verbal expression (see *A Manual of Physiology* [London: John Churchill, 1846], 17). At other times he attempted to establish a basis in physics for vitalism (See *Memorial of William Benjamin Carpenter*, private circulation, printed by H. K. Lewis [Bodleian ref 1891 e.339], 39).

18. The exact nature and extent of Huxley's early reservations about natural selection is the subject of a specialized literature in the history of science. For an able summary of the literature see Lyons, "The Origins of T. H. Huxley's Saltationism", 463–94. If Adrian Desmond is to be believed it would have been impossible to be a Darwinian until about 1870 because it did not harden into a doctrinal creed until then (*Huxley: Evolution's High Priest* [London: Michael Joseph, 1997], 11; *also* ibid., 71).

19. Spencer, *The Principles of Biology*, vol. I, 323.

20. *Ibid.*, vol. I, 234.

21. *Ibid.*, vol. I, 323.

22. *Ibid.*

23. Herbert Spencer, "Illogical Geology" (1859), in *Essays*, second series (1863), 57–104, esp. 64, 71. On Huxley's scepticism about geological classification see Desmond, *Huxley: The Devil's Disciple*, 204.

24. Spencer, "Illogical Geology", 69, 75.

25. *Ibid.*, 70. The Neptunist theory was that successive worldwide floods had deposited regular layers of sediment, producing successive strata that resembled the skins of an onion.

26. *Ibid.*, 71. Spencer's harshness on the subject of unsupported speculation about the meaning of science was not limited to his condemnation of geology. He was similarly dismissive in articles dealing with other scientific subjects. For example, he thought that rather than cosmology becoming easier, the "nebular hypothesis" made the universe harder to think about ("The Nebular Hypothesis" [1858], in *Essays*, second series [1863], 1–56, esp. 55). Like some modern philosophers he thought that his task was to ensure that science was well grounded in logic.

27. Spencer, "Illogical Geology", 82–4.

28. *Ibid.*, 90.

29. William Whewell, *History of the Inductive Sciences from the Earliest to the Present Time*, 3 vols (London: J. W. Parker, 1857), vol. III, 518.

30. Spencer would not have needed Whewell to remind him that geology should not be pursued on a grand scale. As a young man he had been a habitué of the house of Gideon Mantell, a geologist who was famous for his studies of specific locales in England.

31. Spencer, *The Principles of Biology*, vol. I, 324, emphasis added.

32. *Ibid.*, vol. I, 336. Also see Spencer, *First Principles*, §9.

33. See M. Francis, "Herbert Spencer and the mid-Victorian Scientists", *Metascience* 4 (1986), 2–21.

34. Spencer, *The Principles of Biology*, vol. I, 324.

35. *Ibid.*

36. *Ibid.*, vol. I, 326.

37. *Ibid.*, vol. I, 403–4.

38. Spencer, *The Principles of Biology*, vol. II, 73.

39. This was Spencer's revolt against orthodox systematic botany. John Lindley (*The Vegetable Kingdom, or the Structure, Classification and Uses of Plants*, 2nd edn [London: Bradbury & Evans, 1847], xxvi, xxviii) had declared that the internal structure of plants was more significant when theorizing about classification than surfaces were. Another authoritative botanical work, George Bentham's *Handbook of the British Flora*, 5th edn, J. D. Hooker (rev.) (London: Reeve, 1887), xii, considered the forms of organs as chief focus of the morphological enquiry, but not the shape of the plant itself.

40. Phrenologists still espoused vitalism during the 1860s when Spencer was abandoning it. There was a general shift in his opinions: Spencer's enthusiasm for phrenology during the 1840s had been replaced by a commitment to stamping out this doctrine. His philosophical mentor, Hamilton, and the British Association for the Advancement of Science of which Whewell was a prominent member, had condemned the discipline. For Whewell's views on phrenology see Steven Shapin, "Phrenological Knowledge and Early Nineteenth-Century Edinburgh", *Annals of Science* 32(3) (May 1975), 219–43, 230, and Jack Morrell and Arnold Thackray, *Gentlemen of Science: Early Years of the British Association for the Advancement of Science* (Oxford: Clarendon Press, 1981), 278–9.

41. J. S. Henslow is now mostly known because of his connection with Darwin, but, in the mid-nineteenth century he was, with John Lindley and George Bentham, one of the serious voices in botany. Henslow draws a distinction between descriptive botany (which was the detailing and classification of external and internal structures considered as pieces of machinery) and physiological botany (the phenomena resulting from the presence of the living principle that operated in conjunction with the physical forces of attraction and affinity) (J. S. Henslow, *Descriptive and Physiological Botany* [London: Longman, 1839], 2). This referred to a respectable vitalist distinction that was untainted by the radical politics of phrenology.

42. Spencer, *The Principles of Biology*, vol. II, 157 n.

43. *Ibid.*, vol. I, 286.

44. *Ibid.*, vol. I, 289. The human species was, in Spencer's view, an exception to this, and would achieve "completion".

45. *Ibid.*, vol. II, 53. The use of Darwin's language in this passage should not obscure the characteristic mechanistic language of Spencer's botany. Spencer's attempt to explain the geometrical shapes of plants was not only evolutionary, but was partly based on a traditional botany, which can be examined under the heading "Organography and Glossology" in Henslow, *Descriptive and Physiological Botany*, 123–31.

46. See Rupke, *Richard Owen*, 202. Richards (*The Meaning of Evolution*, 140) is very clear on Owen's archetypes. Unfortunately, mid-nineteenth-century scientists were less clear. As Lyons has observed about the early Huxley, his "archetypes" were not dissimilar from Owen's "types" ("The Origins of T. H. Huxley's Saltationism", 478). Spencer's desire to cleanse "types" of idealistic connotations would have affected a number of scientific discourses in his period.

47. Spencer accepted utility in the sense of efficiency. The work of Milne-Edwards, Darwin and other contemporary scientists directed Spencer to the idea that nature tended towards economy, and that useless systems would disappear during the evolutionary process. Utility as efficiency was also the basis of Spencer's argument against design. That is, he thought that if there had been a Creator, He would not have planned to create organisms that possessed the inefficient arrangement of some existing species. Examples of this that particularly struck Spencer were animals that possessed more than a hundred sets of sexual organs.

48. Spencer, *The Principles of Biology*, vol. I, 354.

49. *Ibid*. This notion of the average benefit of Life allowed Spencer to cope with what he saw as an evolutionary moral problem: that the world had evolved to contain evolved lower types of organisms that would prey on higher types.

50. *Ibid.*, vol. I, 340, 354.

51. *Ibid.*, vol. I, 340.

52. *Ibid.*, vol. II, 343.

53. *Ibid*. Spencer was not proposing this as a theory about the development of osseous structures in general. On the contrary, the rickets example appeared to him to be a rare case in which there was a direct response to mechanical stress. Other examples, such as the evolution of the bones of the skull and of various dermal bones, he did not believe could be interpreted in this way. In those cases he thought that the movement towards equilibrium was indirect. In any case Spencer thought that both direct and indirect movements towards equilibrium would be aided by natural selection (*ibid.*, vol. II, 345).

54. If one considers Spencer solely as a biologist then it is obvious that his ideas placed him outside the change in evolutionary thinking initiated by Haeckel in the 1860s. On this and the cone of increasing diversity see Stephen Jay Gould, *Wonderful Life: The Burgess Shale and the Nature of History* (New York: Norton, 1989), 263–7. Bowler, *The Invention of Progress*, 182, seems mistaken to equate Spencer with Haeckel, and to stress the occasional progressive episode of a new vertebrate class that defined the main line of evolution towards mankind. For Spencer the key evolutionary factor for human beings was intelligence, without which they would be non-progressive in their development. For Spencer other animals were not in Haeckel's cone.

55. Spencer, *The Principles of Biology*, vol. II, 495.

56. Spencer, *The Principles of Sociology* (1877), vol. I, 628.

57. Spencer, *The Principles of Biology*, vol. II, 496.

58. Spencer's published views contain two positions on the subject. If one wishes to cite him as a liberal who had views that were similar to those of J. S. Mill then one can rely chiefly on his later political essays. This would make Spencer a relatively consistent individualist. This is the position of Mark Francis and John Morrow in *A History of English Political Thought in the Nineteenth Century* (London: Duckworth, 1994). However, if one is relying on Spencer's systematic philosophy as the palladium of his views on the future of the human species then individualism is unimportant. One could resolve this contradiction by saying that the loss of individualism was a positive outcome provided it had not been forced by legislation or by social coercion. This, however, is obviously unsatisfactory because Spencer was a sociologist and a polemicist who thought that coercion was a frequent aspect of social life.

59. Spencer, *The Principles of Biology*, vol. II, 497.

60. *Ibid.*

61. *Ibid.*

62. Spencer's statement in *The Principles of Biology* (vol. II, 449, 500 n.) that progress depends on laborious days and constant self-denial was simply patched in his text from his 1852 article on population. While keeping the same phrases he corrected them by saying that progress is *not* to be taken to mean a mentally laborious life (*ibid.*, vol. 11, 503). As can be seen from my analysis of *An Autobiography*, he had abandoned his earlier views on the virtue of industry.

63. *Ibid.*, vol. II, 501.

64. *Ibid.*, vol. II, 503, 506.

65. *Ibid.*, vol. II, 506.

66. *Ibid.*, vol. II, 506.

67. *Ibid.*, vol. II, 507.

68. *Ibid.*, vol. II, 508.

69. *Ibid.*, vol. I, 420–21.

70. Spencer's biochemistry was drawn from the work of Tyndall and Robert Bunsen.
71. *Ibid.*, vol. I, 41.
72. *Ibid.*

Chapter 14: Science and the classification of knowledge

1. I have used the term "non-functional" when describing Spencer's suspicion that natural objects displayed a supra-abundant or excessive amount of beauty. Adrian Desmond, when referring to a similar belief on the part of Huxley, calls it "anti-utilitarian" (*Huxley: The Devil's Disciple*, 236). This term, however, is to be avoided because the views of both Spencer and Huxley resemble Paley on this point, and Paley was a utilitarian.
2. Herschel, *A Preliminary Discourse*, 143.
3. M. J. Berkeley, *Introduction to Cryptogamic Botany* (London: H. Baillière, 1857), 6, 6 n. These plants included ferns and club-mosses.
4. *Ibid.*, 39. Berkeley and Hooker were being driven to rigorous theoretical enquiry by the continuing influence of Lindley's physiological analysis of plants. The intellectual standing of these botanists was deprecated by early historians of the subject such as Julius von Sachs and J. Reynolds Green. This has cast an unfortunate darkness over the study of a period when botany was a troubling paradigm for other sciences.
5. Louis Agassiz, *An Essay on Classification* (London: Trübner, 1859), 3.
6. Spencer, *The Principles of Biology*, vol. I, 292.
7. *Ibid.*, vol. I, 293.
8. The science community was already aware of the need for intelligent classification in natural history. Lindley (*The Vegetable Kingdom*, xxv) had warned against the vulgar who committed mistakes by focusing on unimportant features such as size and colour so that all trees became a class and all yellow-flowered composites marigolds. George Rolleston (*Scientific Papers and Addresses*, 2 vols [Oxford: Clarendon Press, 1884], vol. I, xxxii) thought that Darwin might have had more converts if his chapter on classification had been at the beginning of *On the Origin of Species*, instead of last but one.
9. In the mid-Victorian period standard textbooks such as Arthur Henfrey's *An Elementary Course of Botany: Structural, Physiological and Systematic* (London: John Van Voorst, 1857) had sections detailing both the artificial and natural schemes of classification. One should not overemphasize the battle between natural and artificial factions of nineteenth-century botanists. If one examines the ideas of Robert Brown, a practical botanist who avoided theory, it is clear that he claims that the artificial Linnaean system contributed to a more natural arrangement of plants by directing attention to their essential pasts (Robert Brown, "On the Natural Order of Plants called Proteacea", read 17 January, 1809, in *The Miscellaneous Botanical Works of Robert Brown*, 2 vols [London: Ray Society by Robert Hardwicke, 1867], vol. II, 5). Lindley and other mid-century botanists followed Brown in classifying in a natural system that absorbed the best of Linnaeus. Agassiz (*An Essay on Classification*, 3) broke ranks and confounded the artificial and natural systems.
10. Spencer, *The Principles of Biology*, vol. I, 297. See Arthur Henfrey, "Elementary Introduction to The Subject of Vegetable Physiology", *Journal of the Royal Agricultural Society of England* 18ii (1858), 5.
11. *Ibid.*, vol. I, 301–5.
12. *Ibid.*, vol. I, 303.
13. *Ibid.*, vol. I, 358–9.
14. It is an ongoing discussion as to whether a species represents something more than an empty category definable only in the language of formal relations rather than being the property of an organism. See V. H. Heywood, "The 'Species Aggregate' in Theory and Practice", in *Symposium on Biosystematics*, Regnum Vegetabile 27, V. H. Heywood & A. Löve (eds), 26–37 (Vienna: International Association for Plant Taxonomy, 1963), 26–7; Hermann Merxmüller, "Summary Lecture", in *Flowering Plants, Evolution and Classification of Higher Categories*, K. Kubitzki (ed.), 397–465 (Vienna: Springer, 1977), 397–8; and John Dupré, "On the Impossibility of a Monistic Account of Species", in *Species: New Interdisciplinary Essays*, R. A. Wilson (ed.), 3–22 (Cambridge, MA: MIT Press, 1999), 6–7. Spencer resembles those whom Dupré refers to as species pluralists.
15. Spencer, *The Principles of Biology*, vol. I, 59.
16. *Ibid.*, vol. II, 43.
17. Spencer, *The Principles of Biology*, vol. I, 222.

18. *Ibid.*, vol. I, 304–5.

19. *Ibid.*, vol. I, 305.

20. *Ibid.*

21. *Ibid.*, vol. I, 308, emphasis added.

22. G. H. Lewes, "Charles Darwin" [1868], in *Versatile Victorian: Selected Writings of George Henry Lewes*, Rosemary Ashton (ed.) (Bristol: Bristol Classical Press, 1992), 306–8. As a measure of Spencer's influence on Lewes one can compare the latter's statement that a "type" was nothing more than a subjective concept, with his earlier view that while a type had no objective reality – "it is the generalized expression of that which really exists" (Lewes, *The Life and Works of Goethe*, vol. II, 154).

23. Grant Allen, *Charles Darwin* (London : Longmans, 1886), 77. Allen did not wish to undermine Darwin's credit for discovering the theory of natural selection. In fact, he was very eloquent on Darwin's "genuine apostleship" on this doctrine (*ibid.*, 75, 85–6). However, if natural selection theory is detached from modification of species and applied to change in other areas, such as the human psyche, then it begins to look Spencerian. Allen's point here was that, seven years before the publication of Darwin's *On the Origin of Species*, Spencer had suggested that basing evolution on species change was too narrow (*ibid.*, 88–9).

24. Oken's vitalism reduced the essence of the organic genesis to three dynamic physical processes; "*Forming, chemicalizing, and electrifying or oxydizing processes*" (Lorenz Oken, *Elements of Physiophilosophy*, Alfred Fulk [trans.] [London: Ray Society, 1847], 184). In the early 1840s Spencer had been reductionist on the nature of the organic. It is tempting to speculate that Spencer's scientific ideas in *The Principles of Biology* are a throwback to the views he expressed in "Remarks upon the Theory of Reciprocal Dependence in the Animal and Vegetable Creations", *London, Edinburgh and Dublin Philosophical Magazine and Journal of Science* 24 (February 1844), 90–94, where he traces the "proximate cause of progressive development" to how plants and animals react with their environment. However, this speculation would be flawed for two reasons. First, Spencer's 1844 suggestion is not independent speculation. He is merely reacting favourably to (Jean Baptiste?) Dumas's two articles ("On the Chemical Status of Organised Beings", *London, Edinburgh, and Dublin Philosophical Magazine and Journal of Science*, third series, 19 [November 1841], 337–469; [December 1841], 456–69), where it was suggested that the animals should be seen as a giant apparatus for combustion, and neither he nor Dumas shows an interest in specific features of organic development. Secondly, Spencer's 1844 article is an exceptional piece; it appeared at a time when his writing output was almost entirely devoted to engineering, radical politics and phrenology. When he took up serious writing about science in the early 1850s he did not refer to the 1844 article, nor did his work resemble it in approach.

25. Spencer, *The Principles of Biology*, vol. I, 308, emphasis added.

26. Spencer, *The Principles of Biology*, vol. I, 309.

27. Herbert Spencer, *The Classification of the Sciences* (1864), reprinted in *Essays* (1874), vol. III, 10.

28. *Ibid.*

29. *Ibid.*, 25.

30. *Ibid.*, 26.

31. *Ibid.*

32. Spencer, *The Principles of Biology*, vol. II, 5 n.

33. *Ibid.*, vol. II, 213.

34. It has often been assumed that Spencer took his philosophy of science from Comte. This assumption apparently rests on the fact that some of his early supporters and friends, such as J. S. Mill, Eliot and Lewes, were, or had been, Comtean. The fact that Spencer strenuously denied this link as early as 1864 has been ignored. If one takes into account Spencer's acknowledged debt to Whewell – who was a trenchant critic of Comte – then Spencer's position can be more clearly understood.

35. Whewell, *History of the Inductive Sciences*, vol. III, 277.

36. Cuvier and the Jussieus are too well known to need an introduction, but Brown and Lindley have fallen into an obscurity that would have surprised Whewell. As he noted, Brown was much admired by Alexander von Humboldt and Lindley's *The Vegetable Kingdom* was the most able systematic botany of the period (*ibid.*, vol. III, 284–5).

37. *Ibid.*, vol. III, 279.

38. *Ibid.*, vol. III, 284.

39. The title of one of Lindley's books was *The Physiology of Plants* (although his *The Vegetable Kingdom*

was more frequently cited). The word "physiology" in the title was presumably an echo of Cuvier's famous book.

40. Cuvier's work appeared in an English translation as *The Animal Kingdom*.

41. Whewell, *History of the Inductive Sciences*, vol. III, 311.

42. Whewell, *The Philosophy of the Inductive Sciences*, vol. I, 575–6.

43. The bias against Adanson in Whewell's account extended on Cuvier's ideas, although Whewell also drew on Kurt Sprengel's account of Adanson. Whewell thought that Adanson's bold and ingenious attempt (1763) to create an artificial classification belonged to an earlier stage in the history of science (Whewell, *History of the Inductive Sciences*, vol. III, 378). Whewell distinguished between "natural" and "artificial" classification systems by saying that, in the former, botanists attempted to categorize some characteristics of plants as more important than others even before they clearly apprehended the rules of classification (Whewell, *The Philosophy of the Inductive Sciences*, vol. I, 488). In contrast to this, proponents of artificial systems did not attempt to designate a set of particular characteristics as more important than others and doubted that one could exactly specify a natural genus (*ibid.*, vol. I, 490). Whewell dealt with Adanson as a follower of Linnaeus (*ibid.*, vol. I, 500–501).

44. Whewell, *History of the Inductive Sciences*, vol. III, 278.

45. *Ibid.*, vol. III, 279.

46. In his generation Spencer was unusual in championing the merits of Adanson, although in a much later period, when botanists have computers to work with, this Frenchman has found many admirers.

47. Spencer, *The Principles of Biology*, vol. I, 292.

48. Whewell, *History of the Inductive Sciences*, vol. III, 366–7. This passage can also be found in the 1847 edition of Whewell's *History of the Inductive Sciences*, vol. III, 485–6, which was the one that had so marked an influence on Spencer's scientific views during the early 1850s. I have usually cited the 1857 edition because it appears to be more common, not because these are significant textual variations between it and the earlier one Spencer had consulted.

49. Whewell, *History of the Inductive Sciences*, vol. II, 365.

50. *Ibid.*, vol. III, 365–6. Henslow had been Darwin's teacher, but is important here as Whewell's close Cambridge friend and colleague.

51. *Ibid.*, vol. III, 285. The problem of whether the "architecture" of plants has diagnostic value in taxonomy continues to be a perplexing one. See P. B. Tomlinson, "Vegetative Morphology – Some Enigmas in Relation to Plant Systematics", in *Current Concepts in Plant Taxonomy for the Systematics Association*, V. H. Heywood & D. M. Moore (eds) (London: Academic Press, 1984), 55.

52. Whewell, *History of the Inductive Sciences*, vol. III, 360. This admiration was not unbounded. Like most of his contemporaries Whewell was critical of Goethe's efforts in the area of optics.

53. Whewell's admiration of the combination of poetry with science was confined to Goethe. He deprecated the value of Coleridge and thought that his use of Platonism was offensive ranting. See Mrs Stair Douglas, *The Life and Selections from the Correspondence of William Whewell, DD*, 2nd edn (London: Kegan Paul, 1882), 28–9, 94.

54. Whewell, *History of the Inductive Sciences*, vol. III, 361–2. In *The Principles of Biology*, Spencer denigrated Goethe's theory of plant development (as caused by the dominance of food – or waste-carrying channels in leaves) by saying it united "medieval physiology with Platonic philosophy" (Spencer, *The Principles of Biology*, vol. II, 164 n.). Before he had written *The Principles of Biology* Spencer was able to refer to Goethe and other Germans without denigrating them. See Spencer, "Progress: Its Law and Cause", 2.

55. Spencer objected to Comte's belief that sciences developed from simple to complex and suggested that this was why the Frenchman's description of the growth of knowledge was faulty.

56. See Spencer on beauty in Chapter 5.

57. Whewell, *Philosophy of the Inductive Sciences*, vol. I, 494.

58. *Ibid.*, vol. I, 12–13.

59. *Ibid.*, vol. I, 494–5.

60. *Ibid.*, vol. I, 495–6.

61. *Ibid.*, vol. II, 164 n. This criticism of Goethe is a harsh version of the one offered by Spencer's friend Lewes (*The Life and Works of Goethe*, vol. II, 143).

62. Spencer, *The Principles of Biology*, vol. II, 32.

63. *Ibid.*, vol. II, 33.

64. *Ibid.*, vol. II, 92.
65. *Ibid.*
66. *Ibid.*
67. *Ibid.*, vol. II, 208.
68. *Ibid.*, vol. II, 209.
69. *Ibid.*, vol. II, 213.
70. *Ibid.*, vol. II, 113.
71. *Ibid.*, vol. II, 114.
72. *Ibid.*, vol. II, 387.
73. *Ibid.*, vol. II, 117–21.
74. *Ibid.*, vol. II, 125.
75. *Ibid.*, vol. II, 169–70. See Henry Maudsley, *The Physiology and Pathology of the Human Mind* (London: Macmillan, 1867), 54. Maudsley refers to Gregarinida (a class of protozoans, parasitic chiefly in insects, molluscs and crustacea) as being the *lowest* animals of which he was aware, suggesting that he felt they were on the bottom rung of an evolutionary ladder. This is in contrast to Spencer, to whom their small size was merely something that prevented observation of internal organs. Like Spencer, Maudsley drew his information about simple organisms from T. H. Huxley, *Lectures on the Elements of Comparative Anatomy: On the Classification of Animals and on the Vertebrate Skull* (London: John Churchill & Sons, 1864), 6–9. Huxley had used Cuvier's system to categorize organisms, rather than an evolutionary tree or cone. The simple animals were at the top of the list, not the bottom.
76. Spencer, *The Principles of Biology*, vol. II, 128.
77. *Ibid.*, vol. II, 114.
78. *Ibid.*, vol. II, 145 n. Spencer had previously published this account of his "discovery" in January 1859 in his article "The Laws of Organic Form", *Medico-Chirurgical Review* 23 (January–April 1859), 189–202, esp. 191. Given that 1859 was also the year in which Darwin published *On the Origin of Species*, it would seem that Spencer, in *The Principles of Biology*, was claiming both that he had discovered scientific evolution, and that this discovery concerned the natural modification of organic forms. There was no hint that, during the 1850s, Spencer thought that his "discovery" emphasized either sexual selection or competition, even though he frequently recycled Darwin's views on these in *The Principles of Biology*. Spencer himself confused matters further by saying that he had generated a theory of natural selection, but this was not the same as Darwin's (see *The Principles of Biology*, vol. II, 500 n.).
79. Spencer's question resembles much later comments by plant taxonomists. See, for example, Tomlinson, "Vegetative Morphology", 55, and R. W. Snaydon, "Intraspecific Variation and its Taxonomic Implications", in *Current Concepts*, 203–31, esp. 212.
80. D'arcy Wentworth Thompson, *On Growth and Form*, J. T. Bonner (ed.) (Cambridge: Cambridge University Press, 1992). This work was first published in 1917, and has been often reprinted.
81. Spencer, *The Principles of Biology*, vol. II, 387.
82. *Ibid.*, vol. II, 400–402.
83. Since Spencer also argued that human evolution must be the same as other evolution (*ibid.*, vol. II, 494) this would suggest that while he thought that the cause of evolution was always the same, the process itself differed.
84. *Ibid.*, vol. II, 410. Spencer himself thought that his views on the relationship between individuation and genesis were more scientific and less moral than the ones he had offered in 1852 (*ibid.*, vol. II, 417, 472). As usual, though, he also obsessively claimed that his ideas were changeless and that they pre-dated those of his sources. An interesting example of his greed for priority can be seen when he remarks that when he had reviewed W. B. Carpenter in 1852 he had not then noticed that Carpenter had commented on the correlation between fertility and the size of eggs, but that in any case he had already possessed this idea for several years before this. Given how influential Carpenter was on Spencer during the early 1850s, and how undeveloped the latter's technical information on biology was before that time, Spencer's claim to priority here seems spurious.
85. The need to avoid stagnation caused Spencer to argue that, in some sense, human evolution must be the same as any other evolution (*ibid.*, vol. II, 494).

Chapter 15. Spencer's politics and the foundations of liberalism

1. Mill, *Three Essays on Religion*, 62. This citation is to "Nature" (the first of the essays), which Mill wrote between 1850 and 1858.

2. *Ibid.*
3. For example, Ellen Frankel Paul ("Herbert Spencer: Second Thoughts, A Response to Michael Taylor", in *Herbert Spencer: Critical Assessments*, vol. 2, J. Offer [ed.], 556–62 [London: Routledge, 2000], 558) wanted to leave Spencer outside the liberal camp because the "methodological holism" he embraced was incompatible with the political individualism she defines as liberalism. Not all modern scholars equate individualism with liberalism. The most prominent exception is Joseph Raz, whose *The Morality of Freedom* (Oxford: Clarendon Press, 1988; cf. 18) is a liberal critique of individualism.
4. Margaret Moore, *Foundations of Liberalism* (Oxford: Clarendon Press, 1993), 125.
5. *Ibid.*, 136.
6. Richard J. Arneson, "Perfectionism and Politics". *Ethics* 111(1) (October 2000), 37–63, esp. 42–4.
7. See for example, Colin Bird, *The Myth of Liberal Individualism* (Cambridge: Cambridge University Press, 1999), 60–61.
8. James Meadowcroft, *Conceptualizing the State: Innovation and Dispute in British Political Thought, 1880–1914* (Oxford: Oxford University Press, 1995), 213–14.
9. Such a deduction is made by Lightman ("Ideology, Evolution", 295), who, after the fashion of C. B. Macpherson (the author of *Possessive Individualism* and other books criticizing democratic consumer politics), sees Spencerianism as a doctrine of unrestrained competition and atomistic individualism. Since there is little evidence, recent scholars often post a warning that evidence for this doctrine is hard to find. See, for example, Michael Ruse, *Mystery of Mysteries: Is Evolution a Social Construction?* (Cambridge, MA: Harvard University Press, 1999), 78.
10. Spencer, "The Proper Sphere of Government", 7, 51.
11. Herbert Spencer, "Parliamentary Reform: The Dangers and the Safeguards", in *Essays* (1868), vol. II, 355–6. Spencer believed that contemporary death duties fall comparatively more heavily on the poor than the rich, and on real property rather than on personal property.
12. *Ibid.*, 370.
13. Herbert Spencer, "Specialized Administration" (1871), in *Essays* (1874), vol. III, 144. The charge that Spencer was an anarchist was pressed by Élie Halévy, *Thomas Hodgskin*, A. J. Taylor (trans.) (London: Ernest Benn, 1956), 25, 167, 171. Like many of Halévy's insights into nineteenth-century England, it was both interesting and misleading.
14. Spencer, "Specialized Administration", 145.
15. *Ibid.*
16. *Ibid.* In the preface to this essay, which he wrote in 1874 (*ibid.*, vii), Spencer added to this by saying that the body politic or national life will be regulated by non-state structures as well as by the state. The point of this addition was to demonstrate that he was not depending on appeals to the individual's reason even outside areas normally administered by government.
17. Spencer, "The Social Organism", 390–91.
18. *Ibid.*, 425.
19. Spencer, *Social Statics*, 12. It could be observed that the word "nation" does not have to be employed in a way that conflicts with individualism. For instance, a modern individualist might very well refer to Adam Smith's *An Inquiry into the Nature and Causes of the Wealth of Nations* in a sympathetic fashion. However, Spencer's focus on national *life* makes his philosophy resonate with corporate values rather than individualistic ones.
20. *Ibid.*, 14.
21. *Ibid.*, 18.
22. Mill, *A System of Logic*, vol. II, 515.
23. Spencer, *Social Statics*, 67.
24. *Ibid.*
25. *Ibid.*, 70.
26. *Ibid.*, 85.
27. Advocacy of land nationalization also caused Spencer embarrassment in the last two decades of the nineteenth century, when he was repeatedly forced to defend himself against libertarians who were concerned that he might be a closet socialist.
28. *Ibid.*, 115.
29. *Ibid.*, 117.
30. Herbert's uncle, Thomas Spencer, had published a pamphlet in which he suggested that the queen and

her thousand greatest subjects rid themselves of excess wealth in land, forests and revenues. Following this there would be no further inheritance of property by the wealthy in the name of "justice" (*The Parson's Dream and the Queen's Speech* [London: John Green, 1843], 3–9). As with Herbert, Thomas Spencer saw justice chiefly as a matter of equity, not fairness.

31. Spencer, *Social Statics*, 118. This was a highly emotive attack on his own early views. That is, Spencer had supported Lockean property theory in the early 1840s. See Spencer, "The Proper Sphere of Government", 16.

32. Spencer, *Social Statics*, 123.

33. *Ibid.*, 123.

34. *Ibid.*

35. *Ibid.*, 124.

36. Spencer, "Specialized Administration", 166. Victorians knew Humboldt's book under the title *The Sphere and Duties of Government*.

37. Spencer, *An Autobiography*, vol. I, 253, 378, and Duncan, *The Life and Letters*, 539. Spencer claimed to have read an 1887 translation of Kant's *Philosophy of Law* (Spencer, *The Principles of Ethics*, vol. II, 437–8). Rather oddly, Spencer suggested that Kant's views were closely allied with his own except on the subject of the *a priori*, which Spencer interpreted as social requirements that had to be conformed to in order for beneficial ends to be achieved.

38. Maccall and Spencer used the same publisher, John Chapman, but the former also published with the radical agitator G. J. Holyoake.

39. William Maccall, *The Newest Materialism* (London: Farrah, 1873), 96. This comment was written in 1862 in response to Spencer's *First Principles*.

40. William Maccall, *The Outlines of Individualism: A Lecture* (London: Holyoake, 1853), 3.

41. *Ibid.*, 7.

42. William Maccall, *The Elements of Individualism: A Series of Lectures*, 2 vols (London: John Chapman, 1847), vol. I, 191. Maccall's hostility to the nation was inconstant; he favoured it when it went to war and then he called on individuals to be heroic. He despised Quakers, Cobdens, Brights and Mialls as whimpering cowards who ignored Man's martial propensities (Maccall, *The Outlines of Individualism*, 18).

43. Maccall, *The Elements of Individualism*, vol. I, 191.

44. *Ibid.*, 193.

45. Maccall, *The Newest Materialism*, 18.

46. On Mill's adoption of Humboldt's man see Wilhelm von Humboldt, *The Limits of State Action*, J. W. Burrow (ed.) (Cambridge: Cambridge University Press, 1969), 16, n. 2.

47. J. S. Mill, *On Liberty*, 2nd edn (London: John W. Parker, 1859), 103. There is a technical debate about what Mill's notion of individuality entailed. It is beyond my scope here to engage in this discussion, but obviously I would agree with R. F. Ladenson ("Mill's Conception of Individuality", in *John Stuart Mill: Critical Assessments*, J. C. Wood [ed.], 521–33 [London: Routledge, 1991], vol. I, 528) that for Mill the culmination of individuality is the cultivation of reason, although I do not see why this commits him to oppose scholars who write about Mill's belief in the importance of human capabilities. For a balanced view of the social aspects of Mill's individuality see J. M. Robson, "Civilization and Culture", in *The Cambridge Companion to Mill*, Skorupski (ed.), 338–71, esp. 340–42.

48. Humboldt, *The Limits of State Action*, 16.

49. *Ibid.*, 18.

50. *Ibid.*, 17.

51. *Ibid.*, 23.

52. *Ibid.*, 27.

53. One significant difference is that Humboldt, unlike Mill, thought that state patronage was necessary if there was to be scientific progress. He also believed that one could pay too high a price for science if progress in this area was at the expense of freedom.

54. Gerald Gaus (*The Modern Liberal Theory of Man* [London: Croom Helm, 1983]), who largely ignores the romantic qualities of Mill's individual, sees the core of liberal individuality as "the positive development of one's nature, 'to be oneself'" (*ibid.*, 17). He also suggests that Mill only partly grasped the social implications of individuality and that those were only understood later by T. H. Green (*ibid.*, 270). It is, of course, unscholarly for a philosopher to explain a doctrine, such as liberalism, as if it had a core, or as if it would mature at some later period. It would be preferable

if philosophers contented themselves with the content and implications of statements and left other matters to historians.

55. Similarly, as a matter of expediency, Spencer thought that children would only learn effectively if they were not coerced. This educational liberalism was rooted in pedagogical experience, not in a theory of humanity.

56. Patrick Joyce, *Democratic Subjects: The Self and the Social in Nineteenth-Century England* (Cambridge: Cambridge University Press, 1994), 86.

57. See Ellen Frankel Paul, "Herbert Spencer: The Historicist as a Failed Prophet" and "Herbert Spencer: Second Thoughts", in *Herbert Spencer, Critical Assessments*, vol. 2, 528–47 and 556–62. In her frustration with Spencer's failure to advocate a collectivist programme that matched his social organism, Paul seems to imply that self-respecting liberalism should possess a rights doctrine and ignore empirical evidence.

58. Philip Pettit, *Republicanism*, refers to Spencer as a source of his anti-domination doctrine.

Chapter 16: The 1840s: Spencer's early radicalism

1. When writing for phrenologists rather than for his allies in the Complete Suffrage Union, Spencer once ventured the thought that primitive people had their "instincts" of veneration played on by construction of marvellous edifices such as cathedrals and temples. This suggested to him that the advocates of a state church were appealing to a primitive remnant of an older society (Herbert Spencer, "A Theory Concerning the Organ of Wonder", *Zoist: A Journal of Cerebral Physiology and Mesmerism, and Their Application to Human Welfare* 2 [April 1844–January 1845], 317–18). This seems to be the only time during the 1840s that the future psychologist appealed to the term "instinct" when referring to politics.

2. Herbert Spencer, "The Proper Sphere of Government", Letter I, in *Political Writings*, 4. (Originally published in the *Nonconformist* [15 June 1842]. The remaining eleven letters were published at intervals over the subsequent six months.)

3. *Ibid.*, 6. (Later, when he wrote "Specialized Administration", Spencer rejected the notion that politics was a matter of accurately reflecting interests.)

4. Spencer, "The Proper Sphere of Government", Letter X, in *Political Writings*, 50.

5. *Nonconformist* (16 November 1842), vol. II, 774, reviewed the fourth (and cheap) edition of Dymond's *Essays on the Principles of Morality and on the Private and Political Rights and Obligations to Mankind* (2 vols [London: Hamilton Adams, 1829) as a severe blow aimed at "that system of temporary expediency pervaded by the spirit of semi-infidelity, which is now so fashionable in literary and social circles". Spencer (*An Autobiography*, vol. I, 305–6) wrote that his father's admiration of Dymond's *Essays on the Principles of Morality* had apparently led to a sarcastic invitation for Herbert to write a better book. This refers to the period 1844 to 1846 and to the origins of *Social Statics*. However, Herbert was well aware of Dymond's work before this because part of "The Proper Sphere of Government" was written to defend Paley against Dymond's claim that "the declared will of God is the only possible standard of morals". Spencer recalled this last phrase even though he claimed no knowledge of its context (*An Autobiography*, vol. I, 306). Since "The Proper Sphere of Government" was reliant on the scriptures to support "expediencies" such as private property, Spencer could not plausibly claim that he was less orthodox than Dymond, nor that he had absorbed very much of Paley's political philosophy. The reality was that Herbert's ethical and religious convictions during the early 1840s had been offended by Dymond's excessively rational rejection of natural theology. However, Dymond's moderate views on political reform were congenial to readers of the *Nonconformist*, who were searching for a path between the extremes of Chartism and Toryism.

6. Dymond, *Essays on the Principles of Morality*, vol. II, 51.

7. *Ibid.*, 51–2. Dymond's "orthodox" treatment of submission was noticed with pleasure by Robert Southey when his work first appeared (*Quarterly Review* 44 [January/February 1831], 84).

8. The others were Joseph Sturge and Sharman Crawford. See F. C. Mather (ed.), *Chartism and Society: An Anthology of Documents* (London: Bell & Hyman, 1980), 221.

9. *Nonconformist* 2(85) (23 November 1842), 777.

10. *Ibid.* When Miall rejected the idea that government should pay attention to human nature he was probably rejecting the secular variety of political radicalism that wanted governments to acquire greater psychological knowledge so that legislation would not be at variance with the nature of men. George Combe, whose most popular work had appeared in a stereotyped edition by 1841,

had insisted that legislation is intended to regulate and direct the human faculties in their efforts at gratification. See George Combe, *The Constitution of Man*, 19th edn (Edinburgh: MacLachlan & Stewart, 1871), 299. Combe was more serious competition for a radical Christian than Bentham had been.

11. Spencer, "The Proper Sphere of Government", Letter III, in *Political Writings*, 14, 16. Spencer also thought that a national system of education would cause degeneration in the national character. See "The Proper Sphere of Government", Letter IX, in *Political Writings*, 49–50.

12. [Herbert Spencer], "Railway Administration", *Pilot* (28 September 1844). In addition, Spencer believed that in a few years – without government interference – railway proprietors would see their profits increased by freely giving to the public what was proposed would be provided artificially by government regulation. Ten years later he wrote another article on railways that suggested that he had been too optimistic, and that a history of the dishonesties in that industry since 1845 revealed the presence of deep-seated vice. In the later essay ("Railway Morals and Railway Policy" [1854], in *Essays* [1858], 59) he abandoned his watch-cry about over-legislation and called for an amendment to the laws of contract in order to lessen inefficiency and corruption in the unregulated market.

13. Spencer had been reading Samuel Laing's account of his tour of Sweden.

14. [Herbert Spencer], "Honesty is the Best Policy", *Pilot* (16 November 1844).

15. *Ibid*. The fact that this early use of organic metaphor resembles the kind of example of organic function being organized around nourishment means that it is close to the sort of biological metaphor Spencer later criticized when it appeared in Goethe's biological writings.

16. Spencer, "The Proper Sphere of Government", Letter VII, in *Political Writings*, 34.

17. *Ibid*.

18. Spencer mentioned Malthus at the beginning of his article "A Theory of Population", but this was only in the course of rebuking Archbishop Whately for forcing readers to guess that he was referring to Thomas Doubleday's work of demography, and, by implication, to the earlier work of Malthus. Spencer's reference to the history of demography was a display of pedantry. Unlike Darwin and Wallace, he was too fashionable to read Malthus's dated work in the middle of the nineteenth century.

19. Spencer, "A Theory of Population", 486, 489.

20. Spencer, "The Proper Sphere of Government", Letter IV, in *Political Writings*, 20.

21. *Ibid*.

22. David Weinstein, a commentator on Spencer's political philosophy, suggests that in the final analysis Spencer did not champion indefeasible moral rights. He holds that those had disappeared by the time Spencer published *The Principles of Ethics* (1879–93), and only exist in *Social Statics* (1851) (Weinstein, *Equal Freedom and Utility*, 214). While it is clear that Spencer increasingly distanced himself from rights language, I cannot agree that his early use of rights had ever supported an individualist political doctrine of a kind an American such as Weinstein would feel comfortable with.

23. Spencer, "The Proper Sphere of Government", Letter I, 5–6.

24. *Ibid*, Letter XII, 55.

25. The non-ethnographic material in Spencer's *Political Institutions* (Part V of *The Principles of Sociology*) chiefly relies on English nineteenth-century histories of Switzerland, the Republic of Venice, the Roman Empire, ancient Athens and so on. His treatment of this material was often historiographically unsatisfactory. It mostly consisted of lengthy borrowings punctuated by occasional comments that neither the intentionality of the historical actors nor the Whig suggestion that constitutional progress was accidental were acceptable. These comments did not lead Spencer to criticize his sources or to provide alternative accounts of specific constitutional change.

26. [Herbert Spencer], "The Great Social Law", *Pilot* (7 December 1844).

27. *Ibid*.

28. Spencer, "The Proper Sphere of Government", Letter I, 7.

29. The insistence that the state should carry out only its original function was, in Spencer's eyes, a consequence of the definition he had adopted. However, this argument about the original and sole function of the state suggests that at this period Spencer felt no discomfort in constructing an argument about the essence of something, in this case, the essential function. Later, during the 1860s, when he developed his elaborate philosophy of biology, he attempted to avoid referring to essences when describing the functions of organisms.

30. *Ibid.*, 6.
31. In "The Proper Sphere of Government", Letter V, in *Political Writings*, 24–5, Spencer had advised England to be the first nation to abandon war without waiting for others to do the same, and to rely on a system of peaceful arbitration. In *An Autobiography*, vol. I, 209–10, he described these early views on war as "utterly untenable" and recognized "the truth that, if the essential function of a government be that of maintaining the conditions under which individuals may carry on the business of life in security, this function includes, not only protection against internal enemies, but protection against external enemies".
32. Spencer, "The Proper Sphere of Government", Letter V, 23.
33. *Ibid.*, 24.
34. I do not think that there was a Victorian consensus on this subject. For example, in successive philosophical works J. S. Mill argued against, and in favour of, seeing protection as a fundamental duty of the state. See Francis & Morrow, *A History of English Political Thought*, 148, 154. It may be significant that Mill's scepticism about the need for protection was in *Principles of Political Economy*, 2 vols (London: John W. Parker, 1848), which was written in the same optimistic decade as Spencer's youthful "The Proper Sphere of Government".
35. Spencer, "The Proper Sphere of Government", Letter VI, in *Political Writings*, 29.
36. *Ibid.* Spencer's ideas here were quite conventional. British governors also tended to assume that British subjects outside the declared boundaries of a colony were not deserving of protection. See the description of Governor Gipps's attitude to the rights of British subjects in Australia during the 1840s in Mark Francis, *Governors and Settlers: Images of Authority in the British Colonies, 1820–1860* (Basingstoke: Macmillan, 1992), 170–72.
37. Spencer, "The Proper Sphere of Government", Letter VI, 29.
38. *Ibid.*, 31.
39. Spencer also attempted to refute an argument (one popularized again by Joseph Schumpeter after World War II) that military conflict invigorated the economy. The nineteenth-century variant depended, as Spencer himself did, on a metaphor linking society with an organism, to the effect that both enjoyed enhanced vitality through increased activity levels. Since Spencer shared the basis of this argument, he had no reply beyond the retort that the benefit resulting from military activity would only be temporary.
40. [Edward Miall], "Sense of Responsibility", *Nonconformist* 2(88) (14 December 1842), 825.
41. *Ibid.*
42. Spencer, "The Proper Sphere of Government", Letter VIII, in *Political Writings*, 43. This letter appeared on 26 October 1842: six weeks before Miall's "Sense of Responsibility".
43. Spencer, "The Proper Sphere of Government", Letter IX, 46. Spencer also reversed this argument in order to observe that many advocates of increased state intervention were offering homeopathic remedies. That is, restrictive customs duties were similar to diets; encouragement to home manufacturing was parallel to a stimulus to exercise; a national church was like a health tonic; a poor law was similar to ointment; and, finally, an income tax was a metaphor for an extensive bleeding (Spencer, "The Proper Sphere of Government", Letter XI, in *Political Writings*, 54).
44. This is not to say that all of Spencer's writings had the effect of reducing choice. For example, his educational doctrine, which consistently opposed the coercion of children, had the effect of increasing choice.
45. Spencer, "The Proper Sphere of Government", Letter II, in *Political Writings*, 6.
46. *Ibid.*, Letter IV, 16.
47. The land nationalization scheme that Spencer developed several years later while writing *Social Statics* came after he had stopped invoking God in his argument. That is, without the guidance of Christianity Spencer could advocate the public ownership of land as a matter of justice.
48. *Ibid.*, Letter II, 9. Spencer believed that, in the majority of cases, distress was self-induced, and was a divinely ordained punishment, suspension of which would remove the incentive individuals had for self-improvement (*ibid.*, Letter IV, 18).
49. *Ibid.*, Letter II, 10.
50. *Ibid.*, Letter III, 15. As a matter of fact, private benevolence operated in an erratic, rather than constant, fashion in the England of Spencer's youth. In some counties it was practised much more frequently than in others, and there has been a suggestion that stewards were more likely to be charitable than landowners. See David Roberts, *Paternalism in Early Victorian England* (London: Croom Helm, 1979), ch. V.

51. Spencer, "The Proper Sphere of Government", Letter IV, 17.

52. [Herbert Spencer], "The Great Social Law", *Pilot* (7 December 1844). Herbert's early views that people had no natural right to maintenance were clearly derivative from those of his uncle Thomas, one of the most successful pamphleteers of the period. On Thomas Spencer see Eileen Groth Lyon, *Politicians in the Pulpit: Christian Radicalism in Britain from the Fall of the Bastille to the Disintegration of Chartism* (Aldershot: Ashgate, 1999), 157–65.

53. Spencer, "The Proper Sphere of Government", Letter III, 16.

54. See M. W. Flinn, "Introduction", in *The Medical and Legal Aspects of Sanitary Reform*, Alexander P. Stewart and Edward Jenkins (Leicester: Leicester University Press, 1969), 10–11.

55. Spencer's early political theory and his comments on subjects such as poor relief were not parts of a doctrine that was systematically opposed to governmental growth. He would have seen no contradiction between his advocacy of reduced state involvement in areas such as education and increased activity in the provision of justice. Further, there is no evidence that contemporary opponents to governmental growth wished to have their views put in the kind of theoretical language Spencer always produced when writing about any subject. Instead, their hostility to governmental growth was phrased as opposition to centralization in a way that appealed to prejudice reinforced by traditional local habits and customs. Since Spencer consistently opposed idealization of the past he would have had no point of contact here. See William C. Lubenow, *The Politics of Government Growth: Early Victorian Attitudes Toward State Intervention, 1833–1848* (Newton Abbot: David and Charles, 1971), 44.

56. Spencer, "The Proper Sphere of Government", Letter X, 51.

57. *Ibid.*

58. *Ibid.*, 51–2. Later, in *Political Institutions*, 689, Spencer was hostile to contemporary despotism for encouraging a moral state of passive acceptance and expectancy. He believed that it had had evolutionary usefulness, but this was in ancient societies (*ibid.*, 361). However, he continued to believe that the administration of civil justice free of cost was the goal of political progress (*ibid.*, 748).

59. Herbert Spencer, Letter, *Nonconformist*, 2(80) (19 October 1842), 701.

60. Spencer, "The Proper Sphere of Government", Letter XI, 52.

61. *Ibid.*, 53.

62. In the late-eighteenth century and the early-nineteenth century debates on the need to reform suffrage were often viewed as a separate matter from civil liberty. On this subject see J. A. W. Gunn, *Beyond Liberty and Property: The Process of Self-Recognition in Eighteenth-Century Political Thought* (Montreal: McGill-Queen's University Press, 1983) and "Influence, Parties and the Constitution: Changing Attitudes, 1783–1832", *Historical Journal* 2 (1974); and Francis & Morrow, *A History of English Political Thought*, ch. 1. Paley and Dymond, the philosophical writers on whom Spencer was most reliant in "The Proper Sphere of Government", advocated more popular representation without adopting a rights theory that had implications for voting. There were some advocates for a rights-based and universal democratic voting system. For example, the *Pilot*, the short-lived Birmingham newspaper founded by Sturge, had as its editorial principles each man's perfect right to choose and the belief that no one was entitled to take any political action without consent. These ideas supported the *Pilot's* (28 September 1844) complaint that five million voteless citizens deserved the elective franchise as a right. Although Spencer was acquainted with Sturge and ultra-democrats there is no sign that he was ever affected by their views of suffrage.

63. When Dymond argued that civil liberty does not depend on the form of government, and believed that "all communities enjoy [civil liberty] who are properly governed", he supported this by referring to Paley's *The Principles of Moral and Political Philosophy: Essays on the Principle of Morality and on the Private and Political Rights and Obligations to Mankind*, 2 vols (London: Hamilton Adams, 1829), vol. II, 29. From this perspective civil liberty was a relative value, and differed from personal liberty. While this seems surprising from a later perspective, it is not as novel as Dymond's insistence that the possession of civil or political liberty was a right of a community, not of an individual. It was this belief that caused him to advise the inhabitants of Birmingham (an unrepresented city) to avoid protesting that a lack of representation meant that a man was not politically free: a particular privilege, such as voting, was not the test of liberty (*ibid.*, vol. II, 34–5).

64. [Edward Miall], *Nonconformist* (4 May 1842), 289.

65. [Edward Miall], "The Proper Extent of Organic Reform", *Nonconformist* (11 May 1842), 313.

66. *Nonconformist* (26 October 1842), 715.

67. Paul Crook, "Whiggery and America: Accommodating the Radical Threat". In *Essays in Honour of Malcolm I. Thomis: Radicalism and Revolution in Britain, 1775–1848*, Michael T. Davis (ed.), 191–206 (Basingstoke: Macmillan, 2000), 198–202.
68. Spencer, "The Proper Sphere of Government", Letter X, 50.
69. It was a common view of European legal authorities of the period that if subjects wished to acquire new citizenship they had first to seek permission to relinquish their existing citizenship from their current sovereign.
70. Spencer, "The Proper Sphere of Government", Letter IV, 19.
71. *Ibid.* Throughout his life Spencer continued to consider benefits in the distant future as off-setting immediate gratification. In "The Proper Sphere of Government" the future beneficiary was the national community, whereas after the publication of *Social Statics* (1851) it was the human species.
72. Dymond, *Essays on the Principles of Morality*, vol. I, 21; vol. II, 55–6. Since Dymond believed that the original impulse of human nature tended to the well-being of the species (*ibid.*, vol. I, 125), there was some measure of agreement between him and Paley. That is, both favoured the idea that happiness was ubiquitous and naturally possessed. Since this was the case, each could see the possession of happiness as simply due to the superabundance of this quality in the universe.
73. As a youth, Herbert Spencer was more moderate and orthodox than some in Derby. His defence of Paley contrasted sharply with William Strutt's deistical objection to the "benevolence of the Deity" at the Derby Philosophical Society (see Elliott, "Erasmus Darwin, Herbert Spencer", 16). Spencer's defence of Paley was a deliberate adoption of a modern religious stance against a more archaic faith in which morality was simply a given.
74. Spencer, "The Proper Sphere of Government", Letter IV, 19. Alex Tyrell regards Spencer as a rationalist (a sort of secularist anti-Christian) in the mid 1840s (*Joseph Sturge and the Moral Radical Party in Early Victorian Britain* [London: Christopher Helm, 1987], 146), but I think that this judgement accepts Spencer's own later reconstruction of his beliefs when he was attempting to construe his intellectual activities as consistent. The textual evidence of the 1840s, together with employment by stoutly Christian figures such as Miall and Sturge, suggest that he was still well within the boundaries of Christianity in this period.
75. Spencer, "The Proper Sphere of Government", Letter IV, 19–20.
76. *Ibid.*, Letter IX, 48.
77. *Ibid.*
78. This conclusion was not an early display of Lamarckianism on the part of Spencer. In the early 1840s he had yet to read much natural science literature.
79. *Ibid.*, 49.
80. See J. B. Schneewind, *Sidgwick's Ethics and Victorian Moral Philosophy* (Oxford: Clarendon Press, 1977).
81. Spencer, *An Autobiography*, vol. I, 498. This specific exercise of affection referred to 1856, but he borrowed children as late as the 1890s. The word "philoprogenitive" was one popularized by phrenologists in the early-nineteenth century.
82. Spencer, *Social Statics*, 476.
83. *Ibid.*, 199.
84. *Ibid.*, 65.Like Bentham, Spencer traced the origin of positive law to Sir Robert Filmer's doctrine that men are not naturally free. When writing *Social Statics* Spencer was still Christian enough to regard such a denial of rights as a "libel on the Deity". (*Ibid.*, 105-6.)
85. *Ibid.*, 415.
86. *Ibid.*, 62.
87. *Ibid.*, 416.
88. *Ibid.*, 238.
89. *Ibid.*, 205.
90. *Ibid.*, 208. In addition to this theoretical objection Spencer had a practical caveat to popular political power. He was concerned that "public opinion" would not uniformly support progressive causes. As evidence, he ironically noted that it had erected a half-dozen monuments to Wellington but none to icons of civilization such as Shakespeare, Newton or Bacon (*ibid.*, 107).
91. *Ibid.*, 209.
92. Spencer, "The Proper Sphere of Government", Letter IX, 48.
93. *Ibid.*
94. *Ibid.*

Chapter 17: Sociology as an ethical discipline

1. This statement refers to *The Data of Ethics* (London: Williams & Norgate, 1890) (or Part I of Volume I of *The Principles of Ethics*), which was originally published in 1879 shortly after the first volume of *The Principles of Sociology*.
2. David Miller (*Social Justice* [Oxford: Clarendon Press, 1976], 180–81) is typical of those philosophers when he constructs discussion of Spencer's views on absolute versus relative values as if *Social Statics* and *The Principles of Ethics* were the author's significant work on evolutionary philosophy. This is similar to assessing Bertrand Russell's logic by examining his *History of Western Philosophy* and one of his early anti-war pamphlets. Sociologists have developed a more sophisticated analysis of Spencer's theories of social evolution than philosophers. One of the most useful of these is the schematic account given by Robert G. Perrin ("Herbert Spencer's Four Theories of Social Evolution", in *Herbert Spencer: Critical Assessments*, vol. II, J. Offer [ed.], 508–27). Perrin's account is designed to aid current students of sociology, not to unpack Victorian distinctions in social justice.
3. Peter J. Bowler, *Evolution: The History of an Idea* (Berkeley, CA: University of California Press, 1984), 99.
4. Alain de Botton, *Status Anxiety* (New York: Pantheon Books, 2004), 69.
5. Spencer, *Political Institutions*, 680. *Political Institutions* was a separately published work, but it was also the concluding portion of Volume II of Spencer's *The Principles of Sociology*. The chapter cited in this note first appeared in the *Contemporary Review* in September 1881. See also *ibid.*, 94–5. Spencer believed that once production and distribution were negotiated in accordance to socialist and communist ideals the society concerned would cease to be industrial and become militant. As a psychologist Spencer was also conscious of the improbability of schemes based on universal brotherhood. He quoted, with relish, evidence from a South Australian commission showing that socialists found themselves unable to tolerate the widespread vices prevalent in rural communes (*The Principles of Sociology* [1896], vol. III, 568–70).
6. Spencer, *Political Institutions*, 681.
7. *Ibid.*, 690. This comment was not a criticism of the ancient historiography of Seutonius and Plutarch, but a dig at the historical heroes of Carlyle and J. S. Mill.
8. Herbert Spencer, "Bain on the Emotions and the Will", in *Essays*, second series (1863), 131. Much later, in *The Principles of Ethics*, vol. I, 394–6, Spencer rebuked those who used the misleading word "savage" to describe ethnic groups who were less ferocious than modern Europeans.
9. This was a simple antonym of "social man". Spencer did not favour any of the contemporary periodizations of humankind such as Tylor's primitive, barbaric and civilized stages. Conversely, Tylor proclaimed the independence of his views on cultural evolution from those of Spencer. See Robert Carneiro's "Introduction", in *The Evolution of Society: Selections from Herbert Spencer's Principles of Sociology*, Robert L. Carneiro (ed.), ix–lvii (Chicago, IL: University of Chicago Press, 1967).
10. Spencer, *The Principles of Psychology* (1872), vol. II, 521–35.
11. *Ibid.*, vol. II, 534.
12. *Ibid.*, vol. II, 602. Spencer believed that the feelings that impel and restrain us were still the same as those that operate on the savage. This idea made him jettison the notion that these had been a simple progression of morals, which was so dear to other Victorians.
13. *Ibid.*, vol. II, 537.
14. *Ibid.*, vol. II, 602.
15. *Ibid.* Also see the reference to Dymond in Herbert Spencer, *The Data of Ethics* (London: Williams & Norgate, 1890), 50.
16. This shift in Spencer's social science was overlooked by Stocking, who concentrated on the psychological evaluations of primitives in *The Principles of Sociology*, vol. I (*Race, Culture and Evolution*, 127; *Victorian Anthropology*, New York, 225). However, the change was noticed by Christopher Herbert (*Culture and Anomie: Ethnographic Imagination in the Nineteenth Century* [Chicago, IL: University of Chicago Press, 1991], 66–7). He credits Spencer's *The Principles of Sociology*, vol. I, with a paradigm shift from the traditional view that savages were figures of uncontrolled passions and promiscuity to the reverse, seeing them as rigidly controlled by custom. Spencer's imagination here is too captivated by sexuality, a subject on which he had no fixed evolutionary opinions. There was a change in Spencer's ethnography, but this was bound up with his condemnation of militarism and his belief in the interrupted evolution of individual moral characters.

17. Spencer's friend Lubbock, together with other prehistorians such as Daniel Wilson, focused on technological advantage, not emotional superiority, when contrasting ethnic groups.

18. Spencer, *Political Institutions*, 239 n.

19. *Ibid.*, 240 n. Spencer speculated that the missionaries who had brought "the religion of love" would be even funnier if they symbolically burnt a house there by way of demonstrating Christian practice. He made similar ironic comments about missionaries in India, where the teaching of rectitude was followed by the seizure of land and use of machine guns (*The Principles of Sociology* [1896], vol. III, 574).

20. Spencer, *An Autobiography*, vol. II, 353.

21. See, for example, Robert A. Stafford, *Scientist of Empire: Sir Roderick Murchison, Scientific Explanation and Victorian Imperialism* (Cambridge: Cambridge University Press, 1989).

22. Spencer observed that these benefits were not always appreciated by subject peoples. He wryly commented that extreme taxation in British India had caused some of its inhabitants to leave their homes and settle in the territory of the Nizam and in Gwalior.

23. Spencer, *Political Institutions*, 725.

24. *Ibid.*, 230.

25. While admitting there had been social benefits from war and conquest in the past, Spencer withheld admiration from ancient monarchs (*ibid.*, 231). When he objected to the epithet "great" being applied to past rulers whose conquests were gigantic crimes Spencer was referring to Victorian admiration of the great Alexander, Akbar and Charlemagne. He was also siding with Macaulay against J. S. Mill's devotion to great men as the causes of change. Karl Popper expressed hostility to ancient Greek philosophy in a way that parallels Spencer's ideas. However, there is no sign that Popper had read any social theorists except for Hegel and Marx.

26. Spencer, *The Study of Sociology*, 277.

27. Spencer, *Political Institutions*, 232.

28. *Ibid.*

29. *Ibid.*, 234–5, 719–21. Also see Herbert Spencer, *The Principles of Sociology* (1877), vol. I, 585, where, in addition to the hill tribes, he expresses admiration for the Pueblos, who seemed to be one of the few virtuous "pre-social" societies that had been peacefully enlarged.

30. Spencer, *Political Institutions*, 234.

31. Spencer's scepticism about the value of a social Darwinist explanation for empire was not unusual. On the contrary, Paul Crook convincingly argues that scientifically minded Victorians seldom accepted such rationales (*Darwinism, War and History: The Debate over the Biology of War from the "Origin of Species" to the First World War* [Cambridge: Cambridge University Press, 1994], 31; and "Historical Monkey Business: The Myth of a Darwinized British Imperial Discourse", *History* 84[276] [October 1999], 635–57).

32. Spencer, *Political Institutions*, 242.

33. *Ibid.*, 241.

34. *Ibid.*

35. See Spencer, *The Principles of Sociology*, vol. I, 611–13. He also cites his "The Social Organism" (1860) as a reply to Plato and Hobbes.

36. Spencer, *Political Institutions*, 242.

37. *Ibid.*

38. Spencer, *The Principles of Psychology* (1872), vol. II, 576, 662.

39. Spencer, *Political Institutions*, 234. Also see Spencer, *The Principles of Ethics*, vol. I, 405–9.

40. Spencer, *Political Institutions*, 234–5, 705, 719, 721–2. Later Spencer also endowed the hill tribes with the virtue of forgiveness (*The Principles of Sociology* [1896], vol. III, 576). The interesting omission from the list of "pre-social" virtues was chastity. Spencer could find no satisfactory correlation between chastity and social form or ethnicity. Chasteness might accompany ferocity and deceit, or it might not. This left Spencer to question the mores of his contemporaries. "Here are incongruities which appear quite irreconcilable with the ideas current among civilized peoples" (*The Principles of Ethics*, vol. I., 456).

41. Spencer, *Political Institutions*, 236.

42. *Ibid.*, 238. This sociological perception represented an advance from the harsher racial views of Spencer's *The Principles of Psychology* ([1872], vol. II, 576), where he restricted the possession of sociability to modern peoples.

43. Spencer, *Political Institutions*, 722.

44. *Ibid.*, 239–40 n.
45. Spencer's sociology is not divorced from history in the way Max Weber's is. That is, Spencer's militant and industrial types of society are not antinomies designed to destroy the plausibility of historical causality.
46. John Lubbock, *The Origin of Civilization and the Primitive Condition of Man: Mental and Social Condition of Savages* (London: Longmans, 1870), 303. Spencer (*Political Institutions*, 322) quotes Lubbock's idea in the sense that it was meant. Lubbock's intention was to overturn the European romantic vision that, in its infancy, the human species possessed freedom. Spencer shared this view, but not Lubbock's simplistic reliance on technology as an index for the level of civilization.
47. Spencer, *Political Institutions*, 710.
48. Spencer, *The Data of Ethics*, 40–41.
49. *Ibid.*, 110.
50. *Ibid.*, 111.
51. *Ibid.*, 193.
52. *Ibid.*, 215.
53. *Ibid.*, 84–5. This benefit was not greater economic activity, but increased social capital (*ibid.*, 90).
54. *Ibid.*, 57–8.
55. *Ibid.*, 99.
56. *Ibid.*, 100 n.
57. *Ibid.*, 14.
58. *Ibid.*, 19.
59. F. Max Muller, *Lectures on The Origin and Growth of Religion as Illustrated by the Religions of India: The Hibbert Lectures, Delivered in the Chapter House, Westminster Abbey in April, May, and June, 1878* (London: Longmans, 1882), 79.
60. Spencer, *The Data of Ethics*, 39.
61. *Ibid.*, 40–41.
62. *Ibid.*, 137.
63. *Ibid.*, 58.
64. *Ibid.*, 66.
65. *Ibid.*, 67.
66. *Ibid.*, 70.
67. *Ibid.*, 107.
68. *Ibid.*, vii.
69. *Ibid.*, 258. He defined these principles as those that pretended to be external, outside time or independent of the universe.
70. *Ibid.*, 133.
71. *Ibid.*, 137.
72. *Ibid.*, 137–8.
73. *Ibid.*, 183.
74. *Ibid.*, 204–5, 267.
75. *Ibid.*, 268.
76. *Ibid.*, 277.
77. Spencer believed that it was futile for ideal people – those who had the ability to control their behaviour and adjust it to the future state – to act as if they did not live under existing social conditions (*ibid.*, 279). This was a problem that troubled the late-nineteenth-century playwright George Bernard Shaw, but it left Spencer undisturbed.

Chapter 18: Sociology as political theory

1. Spencer, *Political Institutions*, 242. In the history of militarism Spencer is given credit for formulating many of the questions that were subsequently debated, even if few of the protagonists explicitly referred to him (V. R. Berghahn, *Militarism: The History of an International Debate, 1861–1979* [Cambridge: Cambridge University Press, 1981], 13). Fabrizio Battistelli ("War and Militarism in the Thought of Herbert Spencer", *International Journal of Comparative Sociology* 34 [September/December 1993], 192–209) has drawn attention to the way in which Spencer avoided the eighteenth-century habit of relying on classical sources when analysing violence. He suggests that it was Spencer's reliance on more recent ethnographics that transformed his views on war into modern ones.

2. Spencer, *Political Institutions*, 636. In *The Principles of Psychology* ([1872] vol. II, 577) Spencer had speculated that force had had an opposing tendency – called mutual aid or cooperation – that was based on sympathy, but that this had been repressed during the military period.
3. Shermer, *In Darwin's Shadow*, 160–74.
4. *Ibid.*, 159.
5. Spencer, *Political Institutions*, 256.
6. Spencer, *The Principles of Psychology* (1872), vol. II, 570.
7. *Ibid.*, vol. II, 571.
8. Spencer, *Political Institutions*, 644.
9. See Stephen Jay Gould, *The Lying Stones of Marrakech: Penultimate Reflections in Natural History* (London: Vintage, 2001), 257, and Steven Pinker, *The Blank Slate: The Modern Denial of Human Nature* (Harmondsworth: Penguin, 2002), 16.
10. Spencer, *The Principles of Sociology* (1877), vol. I, 540.
11. *Ibid.*
12. *Ibid.*, vol. I, 541–2.
13. Spencer's scenario for the origins of political structures is reminiscent of ancient Greece with its endemic wars between city-states that were organized on the basis of clans. This is not necessarily universal: there is no reason to believe that Sumerian social change began with violence or was clan based. It is likely that, in his attempt to answer Hobbes and other "classical" political theorists, Spencer unwittingly adopted a behaviourist account of ancient violence while consciously rejecting the social contract that was supposed to control it.
14. *Ibid.*, vol. I, 608.
15. Spencer's early theory of social organism ("The Social Organism") was an extended analogy between nerve physiology and political organization on the distribution of resources. Spencer did not discuss war, struggle or violence in the essay. Given that it was published the year after Darwin's *On the Origin of Species*, it would appear that Spencer's early attempts to apply evolutionary theory to human society were not even slightly affected by Darwin's theory of natural selection. In the last decade his life, after being assailed by critics for borrowing his evolutionary ethics from Darwin, Spencer's defence was that Darwinism referred to the natural selection theory that had appeared in *On the Origin of Species* in 1859, while his evolutionary doctrine pre-dated this. (Spencer, "General Preface" (June 1893), in *The Principles of Ethics*, vol. I, viii). He also claims here to be primarily interested in the application of evolution to human character, intelligence and society. This is a priority claim, but there is also an implication that Spencer's evolutionary theories are not restricted to natural selection.
16. Spencer, *Political Institutions*, 312.
17. *Ibid.*, 311.
18. *Ibid.*, 312.
19. *Ibid.* 316. Spencer drew his ideas of social divisions from his analogy with embryonic development although he illustrated them with empirical examples such as the political system of Dahomey. He believed that, following the development of specialized administrative functions, deliberative centres (parliaments) would emerge. He made this speculation on the basis of a parallel between a human cerebrum and the legislative assembly of a nation. (See *The Principles of Sociology* [1877], vol. I, 550–53.) In his 1860 essay "The Social Organism", Spencer had been very critical of comparisons between the organization of a society and the organization of the human body. This has been one of his early complaints against Plato and Hobbes. Then he had claimed that the correct procedure was to compare social organizations with the laws of organic growth in general.
20. Spencer, *Political Institutions*, 700, 241–2.
21. *Ibid.*, 327.
22. *Ibid.*, 321, emphasis added.
23. See Francis & Morrow, *A History of English Political Thought*, 215, 230 n.36.
24. Spencer, *Political Institutions*, 327.
25. On Mill's views on representative government see Francis & Morrow, *A History of English Political Thought*, 238. If Spencer had been better informed about parliamentary politics during the 1850s he might have grasped the significance of the idea of legitimate opposition which had just been coined by George Cornewall Lewis.
26. Spencer's early essay, "Representative Government – What is it Good For?" (1857) (in *Essays*, second series, 185–227) does not discuss political struggle, the value of opposition or what a modern theorist might label "contestation".

27. Spencer, *Political Institutions*, 326. Greta Jones has been taken to task by M. W. Taylor for observing that liberals, such as Spencer, had appropriated organicism (*Men Versus the State: Herbert Spencer and Late Victorian Individualism* [Oxford: Clarendon Press, 1992], 136–7). Alarmed by this judicious remark, Taylor purportedly demonstrated the opposite: that "to Individualists having made use of a formerly conservative instrument for their own purposes, Spencer's use of the social organism analogy was itself professionally conservative in its implications" (*ibid.*, 137). Taylor's response here not only reifies imaginary coherent ideologies of conservatism and individualism, but does considerable damage to a real ideology, liberalism. In a similar vein, James Meadowcroft (*Conceptualizing the State*, 61–2) had suggested that Spencer was conservative because Burke and Hugh Cecil used a growth image while J. S. Mill deprecated this practice. This sort of discourse would also make Marx and Engels into conservatives.

28. Spencer, *The Data of Ethics*, 51–2.

29. *Ibid.*

30. *Ibid.*, 123.

31. *Ibid.*, 114.

32. Spencer, *Political Institutions*, 328.

33. *Ibid.*, 361–2.

34. Spencer, *The Principles of Sociology* (1877), vol. I, 601–2.

35. Spencer, *Political Institutions*, 329.

36. *Ibid.*, 320.

37. *Ibid.*, 327.

38. Spencer seldom employed Darwinism or "survival of the fittest" language when writing about conflict and the process of political change during the feudal and early modern periods. Considering that graphic renditions of both medieval and dynastic violence were the stock-in-trade of nineteenth-century writers from Walter Scott to J. A. Froude, this omission seems deliberate.

39. *Ibid.*, 652.

40. *Ibid.*, 431.

41. *Ibid.*, 429.

42. *Ibid.*, 391.

43. *Ibid.*, 423. An account of the constitutional ideas of early-nineteenth-century Whigs can be found in Francis & Morrow, *A History of English Political Thought*, 9–26. During the 1850s Spencer had been sympathetic to Whiggish constitutionalism, and had drawn on authors such as Mackintosh. See, for example, *Social Statics*, 263.

44. Spencer, *Political Institutions*, 425.

45. *Ibid.*, 423.

46. *Ibid.*, 659.

47. *Ibid.*, 693.

48. *Ibid.*, 421.

49. *Ibid.*, 311.

50. *Ibid.*

51. Spencer's criticism of democratic practices was more severe when these were not restricted to national governance. In his popular sociological work, *The Study of Sociology*, 273–5, he used the example of the shareholders' meeting of a railway company to show that the "forms of free government" are useless even when the shareholders are well educated.

52. Spencer, *Political Institutions*, 312.

53. *Ibid.*, 313–15. Also see Spencer, *The Principles of Sociology*, vol. I, 549–51.

54. Spencer did have a place of ecclesiastical hierarchies in government. His "double compound societies" usually possessed such structures (*The Principles of Sociology*, vol. I, 573).

55. Spencer, *Political Institutions*, 316–17.

56. *Ibid.*, 317.

57. *Ibid.*, 661.

58. *Ibid.*, 696.

59. *Ibid.*, 321.

60. *Ibid.*, 323.

61. Spencer's contemporary, E. B. Tylor (*Primitive Culture: Research into the Development of Mythology, Philosophy, Religion, Art and Custom*, 2 vols [London: John Murray, 1871], vol. I, 42), accepted both progressive and degenerative analyses of social change. He regarded them as indicators of the

level of national life. Unlike Spencer he was not concerned with how advance or regression affected individual character or even individualism itself.

62. Spencer, *Political Institutions*, 661.
63. *Ibid.*
64. *Ibid.*, 662.
65. *Ibid.*
66. *Ibid.*, 665.
67. This was not dissimilar to the views of J. S. Mill's *Considerations on Representative Government*, 2nd edn (London: John W. Parker, 1861), where an emotional mass of citizens led by a politician occasionally break into the ordered and mechanistic work of professional administrators.
68. *Ibid.*, 622, emphasis added.
69. This sort of libertarian theory is often sourced in the Lockean idea that – in the state of nature – individuals own themselves as property. Spencer disagreed with this because he believed that less well-developed peoples were devoid of private ideas and sentiments and, by implication, had no private property. He would not give credence to ethnographic reports that bushmen and other indigenous groups had respect for private property (*ibid.*, 628). Spencer's views here are highly antithetical to libertarianism because, without a credible account of ownership in a state of nature, this would be difficult to sustain. This feature of Spencer's political theory has been overlooked because political writers have relied exclusively on *The Man "versus" the State*, where Spencer does suggest that there was a partial recognition of private property among indigenous peoples. See Spencer, *Political Writings*, 154.
70. *Ibid.*, 622.
71. *Ibid.*
72. *Ibid.*, 623. Spencer italicizes the word "consensus" repeatedly here and in surrounding passages.
73. *Ibid.*, 623.
74. *Ibid.*, 624.
75. Unlike most of the topical chapters in *Political Institutions*, Chapter IX, "Representative Bodies", was not previously published in a popular periodical. Spencer declined to take part in this popular debate because he had no fixed views on who should win. He did not believe that any system of representation – even a universal one – would have any effect except in representing the average nature of citizens, which was unlikely to be progressive. (See Spencer, *The Study of Sociology*, 289.) In this he may have been meditating on the fate of the young liberal intellectuals whose electoral defeats have been chronicled by Christopher Harvie, *The Lights of Liberalism: University Liberals and the Challenge of Democracy 1860–86* (London: Allen Lane, 1976).
76. *Ibid.*, 624. Also see *ibid.*, 626.
77. Spencer, *The Data of Ethics*, 108–9.
78. *Ibid.*, 137.
79. *Ibid.*, 146.
80. *Ibid.*, 146–7.
81. *Ibid.*
82. *Ibid.*
83. In *The Man "versus" the State* Spencer also suggested that Bismarck's leanings towards socialism indicated the true nature of that ideology. Given that he also claimed to be as alarmed as H. M. Hyndman was about social injustice (*Political Writings*, 162), there is no reason for supposing that Spencer had abandoned his progressive liberalism in reaction to socialism in the mid-1880s. Nor should one suppose that *The Man "versus" the State* is much more than a popularizing of the political discourse he had developed in his sociological writings during the previous decade.
84. Spencer, *Political Institutions*, 245.
85. *Ibid.*, 445–7.
86. *Ibid.*, 252.
87. *Ibid.*, 681.
88. *Ibid.*, 682.
89. Spencer, *The Study of Sociology*, 4.
90. Spencer, *Political Institutions*, 698.
91. Spencer, *The Data of Ethics*, 133.

Chapter 19: Progress *versus* democracy

1. Spencer, *An Autobiography*, vol. II, 365.
2. Spencer ("The Social Organism", 402) cited François Guizot as support for his views on representative government without any sense that the credibility of this source might be weakened by its association with the extremely high property qualifications of the July monarchy. Spencer had opened his essay by claiming to reiterate the ideas of the Whig orator, James Mackintosh (*ibid.*, 384).
3. Spencer, "Parliamentary Reform", 377. It was to this essay that Spencer appealed in order to demonstrate that the values incorporated in *The Study of Sociology* and *Political Institutions* had deep roots in his political thought (*An Autobiography*, vol. II, 366).
4. Spencer, "Parliamentary Reform", 378.
5. *Ibid.*
6. Spencer, *The Principles of Sociology* (1877), vol. I, 588–9.
7. *Ibid.*, vol. I, 605.
8. Spencer, "Progress: its Law and Cause", 58. Unlike the early-nineteenth-century Whigs from whom Spencer drew his metaphor for constitutional growth, he did not regard progress as accidental; it was beyond human control as a kind of "beneficent necessity".
9. Despotic tendencies in trade unions had concerned Spencer since 1860. He doubted if the franchise should be given to unionized workers who had so easily surrendered their liberties to their leaders (Spencer, "Parliamentary Reform", 361, 378). He was constant in his hostility to trade unions; he always saw them as despotic. See his "Introduction", in *A Plea for Liberty*, Thomas Mackay (ed.), 1–26 (New York: D. Appleton, 1891), 22–4.
10. Spencer, "Progress: Its Law and Cause", 12.
11. Spencer, *The Principles of Sociology* (1877), vol. I, 474.
12. *Ibid.*
13. *Ibid.*, vol. I, 525.
14. Spencer, "Parliamentary Reform", 363.
15. *Ibid.*, 354.
16. Herbert Spencer, "Representative Government" (1857), in *Essays* (1868), vol. II, 170–72. The corruption in Tammany Hall was as well known to mid-century Englishmen as it was to Americans. On English attitudes to American democracy see the work of D. P. Crook.
17. *Ibid.*, 166–7.
18. *Ibid.*, 168.
19. *Ibid.*
20. Herbert Spencer, *Ceremonial Institutions* (London: Williams & Norgate, 1879), 225. This work was Part IV of *The Principles of Sociology*; it also appeared at the beginning of volume II of that work.
21. Spencer, "Representative Government", 205.
22. *Ibid.* In *The Principles of Sociology* (vol. I, 55) Spencer also offered a more conventional rationale for restricting the scope of legislative action. By analogy between the cerebrum and a national assembly, the latter should not attend to routine actions, but leave them to specific agencies. Instead the assembly should occupy itself with the general requirements of balancing permanent interests.
23. Spencer, "Representative Government", 265. There is an irony here in that Spencer himself had been unable to perceive the need to protect society from external aggression when he published "The Proper Sphere of Government" fourteen years earlier.
24. *Ibid.*, 173.
25. *Ibid.*
26. *Ibid.*, 176–8.
27. *Ibid.*, 175.
28. *Ibid.*, 186.
29. Spencer, "Parliamentary Reform", 373.
30. *Ibid.*, 379.
31. *Ibid.*, 373.
32. *Ibid.*, 375.
33. Spencer, *The Principles of Sociology*, vol. I, 479.
34. *Ibid.*
35. *Ibid.*, 480.

36. Spencer, "Representative Government", 193.
37. *Ibid.*, 194.
38. *Ibid.*
39. Spencer (*ibid.*, 198) cited Leibniz and the geologist Charles Lyell for confirmation of his opinion that for change to have a permanent effect it must be slow.
40. Spencer, *The Principles of Sociology*, vol. I, 611. Spencer credited Plato, Hobbes and Comte as sources of the myth that societies could be artificially established. He noted that he had been making the same comments about Plato and Hobbes since 1860 (*ibid.*, 613). This is a reference to Spencer's "The Social Organism", 390–91.
41. Spencer, "Representative Government", 197.
42. *Ibid.*
43. *Ibid.*
44. *Ibid.*, 198.
45. *Ibid.*, 201.
46. *Ibid.*, 203.
47. T. H. Huxley, "Administrative Nihilism" (1871), in *Method and Results*, 251–89 (London: Macmillan, 1904), 269.
48. In 1870, the same year he published "Administrative Nihilism" as an address, Huxley was appointed to the Commission on Scientific Instruction together with the Duke of Devonshire, the President of the Iron and Steel Institute. According to one of his recent biographers, Huxley's political aim at this time was to increase the science and technology available to Britain's navel dockyards (Desmond, *Huxley: Evolution's High Priest*, 6).
49. Spencer, "Specialised Administration", 166.
50. *Ibid.*, 167.
51. *Ibid.* This insistence on the need for enhanced justice was one of the ideas that he had carried over from his earliest expression of political theory, "The Proper Sphere of Government".
52. Herbert Spencer, "Introduction", in *Essays* (1874), vol. III, vii.
53. Spencer, "Specialized Administration", 167.
54. *Ibid.*, 145
55. *Ibid.*, 144–5.
56. Spencer, *The Man "versus" the State* (1884), in *Political Writings*, 82.
57. Thomas Mackay, *A Plan for Liberty, An Argument Against Socialism and Socialistic Legislation, Consisting of An Introduction by Herbert Spencer and Essays by Various Writers* (New York: D. Appleton, 1891), ix. Mackay's adoption of the phrase "good-for-nothing" is from the end of the third essay of Spencer's *The Man "versus" the State*.
58. *Ibid.*
59. Spencer's intellectual position was increasingly isolated by the 1880s. He no longer felt admired by the scientific elite, and he speculated that this lack of empathy had something to do with his critical effort to apply the evolutionary analysis to contemporary society (see Spencer, *An Autobiography*, vol. II, 374). Politically his situation was increasingly exposed. *The Man "versus" the State* was ridiculed by liberal purists, and admired by the Roman Catholic newspaper the *Tablet*, which usually regarded Spencer as a dangerous influence. He began to fear that he would exasperate the Liberal Party so much that they would disregard everything he said (Duncan, *The Life and Letters*, 239, 242).
60. Spencer, *The Man "versus" the State*, 69.
61. *Ibid.*, 76.
62. *Ibid.*, 77–8.
63. *Ibid.*
64. *Ibid.*, 138.
65. *Ibid.*, 82.
66. *Ibid.*, 84.
67. *Ibid.*, 88–9.
68. *Ibid.*, 97.
69. *Ibid.*, 103–5.
70. *Ibid.*, 95–6.

Conclusion

1. See Michel Foucault, *The Archaeology of Knowledge*, A. M. Sheridan Smith (trans.) (London: Routledge, 1972).
2. Duncan, *The Life and Letters*, 533–76.
3. The original of the portrait on the cover of this book now resides in the National Portrait Gallery (NPG 1358). Although Spencer commissioned the picture for his American publisher, Youmans, he kept it to decorate his drawing room. Burgess, who was best known for his scenes of Spanish life, was chosen with the intention that the resulting work would have more novelty than if the sitter had been portrayed by a more conventional portrait painter.
4. Mrs Humphrey Ward, *Robert Elsmere*, new edn (London: Smith, Elder, 1888), 62–3.
5. Olive Schreiner, *The Story of an African Farm* (London: T. Fisher Unwin, 1925), 206; also see *ibid.*, 127, 208. Schreiner published this novel under the pseudonym "Ralf Iron" in 1883. The original of Waldo's stranger was a civil servant named Bertram, who, in 1871, gave the young novelist a copy of Spencer's *First Principles* when visiting her isolated mission house in Basutoland.
6. Spencer, *Political Institutions*, 321. I discuss Spencer's views on this subject in Chapter 18.

Bibliography

Primary sources

Manuscript material

The letters, journals and other manuscript material cited in this book can be found in archives at the institutions listed below. Many other archives were inspected, but yielded nothing of significance.

Alexander Turnbull Library, Wellington, New Zealand
Athenaeum Club, London
Birmingham University Library
Bodleian Library, Oxford
British Library (BL)
British Museum (BM)
Imperial College, London
Knox College, Galesburg, Illinois
London School of Economics and Political Science (LSE)
Longmans Archives, Harlow, Essex (incorporates the archives of the firm of J. W. Parker)
Manchester Central Library
Manchester Co-operative Union (National Co-operative Archive)
National Library of Scotland
National Library of Wales
Royal Institution of Great Britain
Royal Society of London
University College, London (UCL)

The letters of W. R. Greg were lent to me by Andrew Greg of Trinity College, Cambridge.

Newspapers and periodicals

The Athenaeum
Blackwood's Magazine
British and Foreign Medico-Chirurgical Review
The British Quarterly Review
The Contemporary Review
The Economist
The Ethical Review
The Fortnightly Review

Fraser's Magazine
The Leader
The London, Edinburgh and Dublin Philosophical Magazine and Journal of Science
Macmillan's Magazine
The Nineteenth Century
The Nonconformist
The North British Review
The Pilot (Birmingham)
The Quarterly Review
The Saturday Review
Vanity Fair
The Westminster Review
The Zoist: A Journal of Cerebral Physiology and Mesmerism, and their Application to Human Welfare

Books and pamphlets

Agassiz, Louis. *An Essay on Classification* (London: Trübner, 1859).

Allen, Grant. *Charles Darwin* (London: Longmans, 1886).

Amberley, Lord and Lady. *The Amberley Papers: The Letters and Diaries of Lord and Lady Amberley*, 2 vols, Bertrand Russell & Patricia Russell (eds) (London: Leonard and Virginia Woolf at the Hogarth Press, 1937).

Anon. [J. H. Leigh Hunt]. *Christianism: Or Belief and Unbelief Reconciled: Being Exercises and Mediations*. (Not for Sale; Only Seventy-Five Copies Printed, London, 1832).

Anon. [John Chapman]. *Human Nature. A Philosophical Exposition of the Divine Institution of Reward and Punishment, Which Obtains In the Physical, Intellectual, and Moral Constitution of Man; With An Introductory Essay. To Which Is Added A Series of Ethical Observations, Written During The Perusal Of The Rev. James Martineau's Recent Work, Entitled "Endeavours After a Christian Life"* (London: John Chapman, 1844).

Anon. *Essays and Reviews* [1860], 3rd edn (London: Longman, Green, Longman & Roberts, 1861).

Atkinson, H. G. & Harriet Martineau. *Letters on the Laws of Man's Nature and Development* (London: John Chapman, 1851).

Bailey, Samuel. *Letters on the Philosophy of the Human Mind*, 1st series (London: Longman, 1855).

Bailey, Samuel. *Letters on the Philosophy of the Human Mind*, 2nd series (London: Longman, 1858).

Bailey, Samuel. *Letters on the Philosophy of the Human Mind*, 3rd series (London: Longman, 1863).

Bain, Alexander. *John Stuart Mill: A Criticism, with Personal Recollections* (London: Longmans, Green, 1882).

Bain, Alexander. *Mental and Moral Science: A Compendium of Psychology and Ethics* (London: Longmans, 1868–72).

Bain, Alexander. *On the Applications of Science to Human Health and Well-Being* (London: J. J. Griffin, 1848).

Bain, Alexander. *On the Study of Character, Including An Estimate of Phrenology* (London: Parker, 1861).

Bain, Alexander. *The Senses and the Intellect* (London: J. W. Parker, 1855).

Barnett, Henrietta Octavia W. *Canon Barnett, His Life, Work and Friends, By His Wife*, 2 vols (London: John Murray, 1918).

Barratt, Alfred. *Physical Ethics or The Science of Action* (London: Williams & Norgate, 1869).

Barratt, Alfred. *Physical Metempiric* (London: Williams & Norgate, 1883).

Baynes, Thomas Spencer. *An Essay on the New Analytic of Logical Forms, Being that which Gained the Prize Proposed by Sir William Hamilton, in the Year 1846, for the Best Exposition of the New Doctrine Propounded in his Lectures: With an Historical Appendix* (Edinburgh: Sutherland & Knox, 1850).

Baynes, Thomas Spencer. *The Port-Royal Logic*, 2nd edn (Edinburgh: Sutherland & Knox, 1851).

Baynes, Thomas Spencer. "Sir William Hamilton". In *Edinburgh Essays by Members of the University, 1856*, 241–300 (Edinburgh: Adam & Charles Black, 1857).

Bentham, George. *Handbook of the British Flora*, 5th edn, J. D. Hooker (rev.) (London: Reeve, 1887).

Berkeley, M. J. *Introduction to Cryptogamic Botany* (London: H. Baillière, 1857).

Bray, Charles. *Phases of Opinion and Experience during a Long Life: An Autobiography* (London: Longmans, 1884).

Bray, Charles. *The Philosophy of Necessity: or, The Law of Consequences: As Applicable to Mental, Moral and Social Science*, 2 vols (London: Longman, Orme, Brown, Green & Longmans, 1841).

Bray, Charles. *The Philosophy of Necessity: or, Natural Law as Applicable to Moral, Mental and Social Science*, 2 vols (London: Longman, Green, Longman & Roberts, 1863).

Bray, Charles. *The Science of Man: A Bird's Eye View of the Wide and Fertile Field of Anthropology* (London: Longmans, 1868).

Brown, John. *Horae Subsecivae, Locke and Sydenham with Other Occasional Papers* (Edinburgh: Thomas Constable, 1858).

Brown, Robert. *The Miscellaneous Botanical Works of Robert Brown*, 2 vols (London: Ray Society by Robert Hardwicke, 1867).

Bryce, Mary R. L. *Memoir of John Veitch, L.L.D.* (Edinburgh: William Blackwood, 1896).

Burgon, J. W. *Lives of Twelve Good Men* 2 vols (London: John Murray, 1888).

Burke, Edmund. *A Philosophical Enquiry into the Origin of our Ideas of the Sublime and Beautiful* (London: Routledge & Kegan Paul, 1958).

Carlyle, Thomas. *Translations from the German: Wilhelm Meister's Apprenticeship and Travels*, 2 vols (London: Chapman & Hall, 1868).

Carlyle, Thomas. *The Correspondence of Thomas Carlyle and Ralph Waldo Emerson, 1834–1872*, C. E. Norton (ed.) (London: Chatto & Windus, 1883).

Carpenter, W. B. *Animal Physiology* (London: William S. Orr, 1843).

Carpenter, W. B. *A Manual of Physiology* (London: John Churchill, 1846).

Carpenter, W. B. *Memorial of William Benjamin Carpenter*, private circulation, printed by H. K. Lewis (Bodleian ref. 1891 e.339).

Carpenter, W. B. *Popular Cyclopaedia of Natural Science: Animal Physiology* (London: William S. Orr, 1843).

Carpenter, W. B. *Principles of General and Comparative Physiology, Intended as an Introduction to the Study of Human Physiology and as a Guide to the Philosophical Pursuit of Natural History*, 3rd edn (London: John Churchill, 1851).

Coleridge, Samuel Taylor. *The Friend*, 3rd edn, 3 vols (London: William Pickering, 1837).

Coleridge, Samuel Taylor. *Shorter Works and Fragments*, 2 vols, H. J. Jackson & J. R. de J. Jackson (eds) (Princeton, NJ: Princeton University Press, 1995).

Combe, George. *The Constitution of Man*, 19th edn (Edinburgh: MacLachlan & Stewart, 1871).

Coulton, G. G. *Fourscore Years: An Autobiography* (Cambridge: Cambridge University Press, 1944).

Courtney, Leonard. Funeral Address, 14 December 1903. *Ethical Review* (December 1906), *Herbert Spencer Memorial Number*.

Cuvier, Georges. *The Animal Kingdom, Arranged after its Organization, Forming a Natural History of Animals*, W. B. Carpenter & J. O. Westwood (eds), new edn (London: William S. Orr, 1849).

Darwin, Charles. *The Autobiography of Charles Darwin*, N. Barlow (ed.) (London: Collins, 1958).

Darwin, Charles. *The Descent of Man*, 2 vols (London: John Murray, 1871).

Darwin, Charles. *On the Origin of Species by Means of Natural Selection or the Preservation of Favoured Races in the Struggle for Life* (London: John Murray, 1859).

Darwin, Charles. *The Red Notebook of Charles Darwin, Bulletin of the British Museum (Natural History) Historical Series* 7 (24 April 1980).

Davis, Andrew Jackson. *The Principles of Nature, Her Divine Revelations, And a Voice to Mankind*, 2 vols (London: John Chapman, 1847).

Day, Thomas. *The History of Sandford and Merton: A Work Intended for the Use of Children*, 9th edn (London: John Stockdale, 1812).

Douglas, Mrs Stair. *The Life and Selections from the Correspondence of William Whewell, DD*, 2nd edn (London: Kegan Paul, 1882).

Dymond, Jonathan. *Essays on the Principles of Morality and on the Private and Political Rights and Obligations to Mankind*, 2 vols (London: Hamilton Adams, 1829).

Eliot, George. *Adam Bede* (Edinburgh: William Blackwood, [1859] 1901).

[Eliot, George.] *An Analytical Catalogue of Mr. Chapman's Publications* (London: John Chapman, 1852).

Eliot, George. *The George Eliot Letters*, G. S. Haight (ed.), 9 vols (New Haven, CT: Yale University Press, 1954–78).

Eliot, George. *George Eliot's Life, As Related In Her Letters and Journals*, arranged and edited by her husband, J. W. Cross, 3 vols (Edinburgh: William Blackwood & Son, 1885).

Ellis, Havelock. *The Nineteenth Century: A Dialogue in Utopia* (London: Grant Richards, 1900).

Emerson, Ralph Waldo. *English Traits* (Boston, MA: Houghton, Mifflin, 1884).

Emerson, R. W. *Essays, First Series* (1841) (Boston, MA: Houghton, Mifflin, 1884).

Ferrier, J. F. *Scottish Philosophy: The Old and the New* (Edinburgh: Sutherland & Knox, 1856).

Fraser, A. C. *Biographia Philosophica: A Retrospect* (Edinburgh: William Blackwood, 1904).

Fraser, A. C. *Rational Philosophy in History and in System: An Introduction to a Logical and Metaphysical Course* (Edinburgh: Thomas Constable, 1858).

Fraser, A. C. *Thomas Reid* (Edinburgh: Oliphant, Anderson & Ferrier, 1898).

Froude, J. A. *Thomas Carlyle, A History of the First Forty Years of His Life, 1795–1835*, 2 vols (London: Longmans, Green & Co., 1882).

Grant Duff, Mountstuart E. *Notes from a Diary, 1851–1872*, 2 vols (London: John Murray, 1897).

Greg, W. R. *The Creed of Christendom; Its Foundations and Superstructure* (London: John Chapman, 1851).

Greg, W. R. *Enigmas of Life* (with a Prefatory Memoir Edited by his Wife) (London: Kegan, Paul, Trench, Trübner, 1891).

Hamilton, Sir William, Bart. *Discussions on Philosophy and Literature, Education and University Reform Chiefly from the Edinburgh Review: Corrected, Vindicated and Enlarged, in Notes and Appendices* (London: Longman, Brown, Green & Longmans, 1852).

Hamilton, Sir William, Bart. *Lectures on Metaphysics and Logic*, 4 vols, H. L. Mansel & John Veitch (eds) (Edinburgh: William Blackwood, 1870).

Harrison, Frederic. *Autobiographic Memoirs*, 2 vols (London: Macmillan, 1911).

Henfrey, Arthur. *An Elementary Course of Botany: Structural, Physiological and Systematic* (London: John Van Voorst, 1857).

Henfrey, Arthur. "Elementary Introduction to The Subject of Vegetable Physiology", *Journal of the Royal Agricultural Society of England* 18ii (1858).

Hennell, Sara S. *Christianity and Infidelity: An Exposition of the Arguments on Both Sides. Arranged According to a Plan Proposed by George Baillie, Esq.* (London: Arthur, Hall, Virtue, 1857).

Hennell, Sara S. *Early Christian Anticipation of an Approaching End of the World, and its Bearing Upon the Character of Christianity as a Divine Revelation* (London: George Manwaring, 1860).

Hennell, Sara S. *Present Religion: As a Faith Owning Fellowship with Thought*, 3 vols (London: Trübner, 1887).

Hennell, Sara S. *Thoughts in Aid of Faith, Gathered Chiefly from Recent Works in Theology and Philosophy* (London: George Manwaring, 1860).

Henslow, J. S. *Descriptive and Physiological Botany* (London: Longman, 1839).

Herschel, J. F. W. *A Preliminary Discourse on the Study of Natural Philosophy* (London: Longman, 1831).

Hudson, W. H. *Herbert Spencer* (London: Archibald Constable, 1903).

Hudson, W. H. *Men, Books, and Birds* (London: Jonathan Cape, 1928).

Humboldt, Alexander von. *Kosmos: A General Survey of the Physical Phenomena of the Universe*, Augusta Pritchard (trans.), 2 vols (London: Hippolyte Ballière, 1845–8).

Humboldt, Wilhelm von. *The Limits of State Action*, J. W. Burrow (ed.) (Cambridge: Cambridge University Press, 1969).

Hunt, Leigh. *The Autobiography of Leigh Hunt*, J. E. Morpurgo (ed.) (London: Cresset Press, 1949).

Hunt, Leigh. *The Correspondence of Leigh Hunt, edited by his eldest son* [Thornton Hunt], 2 vols (London: Smith, Elder, 1862).

Hunt, Leigh. *The Religion of the Heart: A Manual of Faith And Duty* (London: John Chapman, 1853).

Huxley, T.H. *Collected Essays*, 9 vols (London: Macmillan, 1894–1909).

Huxley, T. H. *Lay Sermons, Addresses and Reviews* (London: Macmillan, 1870).

Huxley, T. H. *Lectures on the Elements of Comparative Anatomy: On the Classification of Animals and on the Vertebrate Skull* (London: John Churchill & Sons, 1864).

Huxley, T. H. *Method and Results* [1871] (London: Macmillan, 1904).

Huxley, T. H. *On The Education Value of the Natural History Sciences* (London: John Van Voorst, 1854).

Jones, Thomas Rymer. *The Natural History of Animals, Being the Substance of Three Courses of Lectures Delivered Before The Royal Institution of Great Britain*, 2 vols (London: John Van Voorst, 1845–52).

Kemp, T. Lindley. *Agricultural Physiology: Animal and Vegetable* (Edinburgh: Blackwood, 1850).

Kirby, William. *On the Power, Wisdom and Goodness of God, As Manifested in the Creation of Animals, and In their History, Habits and Instincts*, 2 vols (London: Henry G. Bohn, 1854).

Le Gallienne, R. *The Romantic '90s* (London: Robin Clark, 1993).

Lewes, G. H. *A Biographical History of Philosophy*, 4 vols (London: C. Knight, 1845–53).

Lewes, G. H. *Comte's Philosophy of the Sciences* (London: Henry G. Bohn, 1853).

Lewes, G. H. *The History of Philosophy*, 2 vols (London: Longmans, Green, 1871).

Lewes, G. H. *The Life and Works of Goethe: With Sketches of His Age and Contemporaries*, 2 vols (London: David Nutt, 1855).

Lewes, G. H. *The Physiology of Common Life*, 2 vols (Edinburgh: William Blackwood, 1859).

Lewes, G. H. *Versatile Victorian: Selected Writings of George Henry Lewes*, Rosemary Ashton (ed.) (Bristol: Bristol Classical Press, 1992).

Lindley, John. *The Vegetable Kingdom, or the Structure, Classification and Uses of Plants*, 2nd edn (London: Bradbury & Evans, 1847).

Lubbock, John. *Pre-Historic Times, As Illustrated by Ancient Remains, and the Manners and Customs of Modern Savages*, 5th edn (London: Williams & Norgate, 1890).

Lubbock, John. *The Origin of Civilization and the Primitive Condition of Man: Mental and Social Condition of Savages* (London: Longmans, 1870).

Lyell, Charles. *A Manual of Elementary Geology*, 5th edn (London: John Murray, 1855).

Lynn Linton, Eliza. *My Literary Life* (London: Hodder & Stoughton, 1899).

Maccall, William. *The Elements of Individualism: A Series of Lectures*, 2 vols (London: John Chapman, 1847).

Maccall, William. *The Newest Materialism* (London: Farrah, 1873).

Maccall, William. *The Outlines of Individualism: A Lecture* (London: Holyoake, 1853).

Mackay, R. W. *The Progress of the Intellect as exemplified in the religious development of the Greeks and Hebrews*, 2 vols (London: John Chapman, 1850).

Mackay, R. W. *A Sketch of the Rise and Progress of Christianity* (London: John Chapman, 1854).

Mackay, T. *A Plea for Liberty, An Argument Against Socialism and Socialistic Legislation, Consisting of An Introduction by Herbert Spencer and Essays by Various Writers* (New York: D. Appleton, 1891).

Mansel, H. L. *The Limits of Religious Thought Examined in Eight Lectures, Preached Before the University of Oxford, In the Year MDCCCLVIII, on the Foundation of the Late Rev. John Bampton, MA.* (Oxford: J. H. & Jas. Parker, 1858).

Mansel, H. L. *Metaphysics, or the Philosophy of Consciousness Phenomenal and Real* (Edinburgh: Adam & Charles Black, 1860).

Mansel, H. L. *The Philosophy of The Conditioned, Comprising Some Remarks on Sir William Hamilton's Philosophy and on Mr J. S. Mill's Examination of that Philosophy* (London: Alexander Strahan, 1866).

Mansel, H. L. *Psychology, The Test of Moral and Metaphysical Philosophy; An Inaugural Lecture Delivered in Magdalen College*, October 1855 (Oxford: William Graham, 1855).

Mantell, Gideon Algernon. *The Medals of Creation; Or First Lessons In Geology And In the Study of Organic Remains*, 2 vols (London: Henry G. Bohn, 1844).

Mantell, Gideon Algernon. *Petrifactions and their Teachings; Or, A Hand-Book to the Gallery of Organic Remains* (London: Henry G. Bohn, 1851).

Mantell, Gideon Algernon. *The Wonders of Geology*, 2 vols (London: Rolfe & Fletcher, 1838).

Martineau, James. *Endeavours After the Christian Life: A Volume of Discourses* (London: J. Green, 1843).

Martineau, James. *Essays, Philosophical and Theological* (London: Trübner, 1866).

Martineau, James. *The Rationale of Religious Enquiry: On the Questions Stated of Reason, the Bible, and the Church; in Six Lectures*, 3rd edn (London: John Chapman, 1845).

Masson, David. *Memories of London in the Forties* (Edinburgh: William Blackwood & Sons, 1908).

Masson, David. *Recent British Philosophy: A Review, with Criticisms; Including Some Comments on Mr. Mill's Answer to Sir William Hamilton* (London: Macmillan, 1865).

Maudsley, Henry. *The Physiology and Pathology of the Human Mind* (London: Macmillan, 1867).

McCosh, James. *The Intuitions of the Mind Inductively Investigated*, 2nd edn (London: Macmillian, 1865).

McCosh, James. *The Method of Divine Government, Physical and Moral*, 2nd edn (Edinburgh: Sutherland & Knox, 1850).

McCosh, James. *The Scottish Philosophy* (London: Macmillan, 1875).

Meinertzhagen, Georgina. *From Ploughshire to Parliament: A Short Memoir of the Potters of Tadcaster* (London: John Murray, 1908).

Mill, James. *Analysis of The Phenomena of The Human Mind*, a new edition with notes by Alexander Bain, Andrew Findlater and George Grote, 2 vols (London: Longmans, Green, Reader & Dyer, 1869).

Mill, James. "An Essay on Government" [1820]. In *Utilitarian Logic and Politics*, Jack Lively & John Rees (eds), 34–56 (Oxford: Clarendon Press, 1978).

Mill, John Stuart. *Autobiography*, 3rd edn (London: Longmans, Green, Reader & Dyer, 1874).

Mill, John Stuart. *The Collected Works of John Stuart Mill*, 33 vols, J. M. Robson (ed.) (Toronto: University of Toronto Press, 1963–91).

Mill, John Stuart. *Considerations on Representative Government*, 2nd edn (London: John W. Parker, 1861).

Mill, John Stuart. *Essays on Ethics, Religion and Society*, J. M. Robson (ed.) (Toronto: University of Toronto Press, 1969). [Volume X of the *Collected Works of John Stuart Mill*.]

Mill, John Stuart. *An Examination of Sir William Hamilton's Philosophy and of the Principal Philosophical Questions Discussed in his Writings*, 2nd edn (London: Longmans, Green, 1865).

Mill, John Stuart. *An Examination of Sir William Hamilton's Philosophy and of the Principal Philosophical Questions Discussed in His Writings*, 3rd edn (London: Longmans, Green, Reader & Dyer, 1867).

Mill, John Stuart. *Nature, The Utility of Religion and Theism being Three Essays on Religion*, 2nd edn (London: Longmans, 1874).

Mill, John Stuart. *On Liberty*, 2nd edn (London: John W. Parker, 1859).

Mill, John Stuart. *Principles of Political Economy*, 2 vols (London: John W. Parker, 1848).

Mill, John Stuart. *A System of Logic*, 4th edn, 2 vols (London: John W. Parker, 1856).

Milne-Edwards, Henri. *A Manual of Zoology*, 2nd edn, R. Knox (trans.) (London: Henry Renshaw, 1863).

Mozley, Thomas. *Reminiscences Chiefly of Oriel College and the Oxford Movement*, 2 vols (London: Longmans Green, 1882).

Mozley, Thomas. *Reminiscences, Chiefly of Towns, Villages, and Schools*, 2nd edn, 2 vols (London: Longmans Green, 1885).

Müller, F. Max. *Lectures on The Origin and Growth of Religion as Illustrated by the Religions of India: The Hibbert Lectures, Delivered in the Chapter House, Westminster Abbey in April, May, and June, 1878* (London: Longmans, 1882).

Newman, F. W. *A History of the Hebrew Monarchy*, 2nd edn (London: John Chapman, 1852).

Newman, F. W. *Lectures on Political Economy* (London: John Chapman, 1851).

Newman, F. W. *Phases of Faith; Or, Passages From the History of My Creed.* (London: John Chapman, 1850).

Newman, F. W. *The Soul, Her Sorrows and Her Aspirations: An Essay towards the natural history of the Soul as the true basis of theology*, 2nd edn (London: John Chapman, 1849).

Newman, F. W. *Theism, Doctrinal and Practical; or Didactic Religious Utterances* (London: John Chapman, 1858).

Oken, Lorenz. *Elements of Physiophilosophy*, Alfred Fulk (trans.) (London: Ray Society, 1847).

Paley, William. *Natural Theology, Or Evidence of the Existence and Attributes of the Deity* (London: C. J. G. & R. Rivington, 1830).

Paley, William. *The Principles of Moral and Political Philosophy: Essays on the Principle of Morality and on the Private and Political Rights and Obligations to Mankind*, 2 vols (London: Hamilton Adams, 1829).

Patmore, Coventry. *Memoirs and Correspondence*, 2 vols, Basil Champneys (ed.) (London: George Bell, 1900).

Pollock, Frederick. *Personal Remembrances of Sir Frederick Pollock, Second Baronet, Sometime Queen's Remembrancer*, 2 vols (London: Macmillan, 1887).

Raverat, Gwen. *Period Piece: A Cambridge Childhood* (London: Faber, 1955).

Reid, Thomas. *Essays on the Intellectual Powers of Man: A Critical Edition*, R. Derek Brookes & K. Haakonssen (eds) (University Park, PA: Pennsylvania State University Press, 2002).

Reid, Thomas. *The Works of Thomas Reid D.D. now Fully Collected with Selections from his Unpublished Letters*, 3 vols, Sir William Hamilton Bart. (ed.) (Edinburgh: Maclachlan, Stewart, 1846).

Reid, T. W. *The Life, Letters and Friendships of Richard Monckton Milnes, First Lord Houghton*, 2 vols (London: Cassell, 1890).

Ribot, T. H. (Paul). *English Psychology* (London: Henry S. King, 1873).

Ribot, T. H. (Paul). *Heredity, A Psychological Study of Its Phenomena, Laws, Causes and Consequences* (London: Henry S. King, 1875).

Rolleston, George. *Scientific Papers and Addresses*, 2 vols (Oxford: Clarendon Press, 1884).

Schiller, Friedrich. *The Philosophical and Aesthetic Letters and Essays of Schiller* (London: John Chapman, 1845).

Schreiner, Olive. *The Story of an African Farm* (London: T. Fisher Unwin, 1925).

Shelley, Percy Bysshe, *The Poems of Percy Bysshe Shelley*, T. Hutchinson (ed.) (Oxford: Oxford University Press, 1919).

Skelton, John. *The Table-Talk of Shirley: Reminiscences of and letters from Froude, Thackeray, Disraeli, Browning, Rossetti, Kingsley, Baynes, Huxley, Tyndall and others* (Edinburgh: William Blackwood & Sons, 1895).

Smith, George Barnett. *The Life and Speeches of the Right Hon. John Bright, MP* (London: Hodder & Stoughton, 1881).

Smith, Goldwin. *Reminiscences of Goldwin Smith*, Arnold Haultain (ed.) (New York: Macmillan, 1910).

Spencer, Herbert. *Aphorisms from the Writings of Herbert Spencer*, selected and arranged by Julia Raymond Gingell, 2nd edn (London: Chapman & Hall, 1894).

Spencer, Herbert. *An Autobiography*, 2 vols (London: Williams & Norgate, 1904).

Spencer, Herbert. "Bain on the Emotions and the Will", *British and Foreign Medico–Chirurgical Review* (January 1860). Reprinted in *Essays* (second series, 1863), 120–42; *Essays* (1868), vol. I, 300–24.

Spencer, Herbert. *Ceremonial Institutions* (London: Williams & Norgate, 1879) [Part IV of *The Principles of Sociology*].

Spencer, Herbert. *The Classification of the Sciences* (London: Williams & Norgate, 1864). Reprinted in *Essays* (1874), vol. III, 1–56.

Spencer, Herbert. *The Classification of the Sciences* (London: Williams & Norgate, 1871).

Spencer, Herbert. "A Criticism on Professor Owen's Theory of the Vertebrate Skeleton", *British and Foreign Medico–Chirurgical Review* (October 1858). Reprinted as Appendix B, *The Principles of Biology* (1898), vol. II, 517–35.

Spencer, Herbert. *The Data of Ethics* (London: Williams & Norgate, 1890) [Part I of *The Principles of Ethics*].

Spencer, Herbert. *Ecclesiastical Institutions* (London: Williams & Norgate, 1885) [Part VI of *The Principles of Sociology*].

Spencer, Herbert. *Education: Intellectual, Moral and Physical* (London: G. Manwaring, 1861).

Spencer, Herbert. "An Element in Method", in *The Principles of Psychology* (1855), 339–52. Reprinted as an essay in *Various Fragments* (1897), 4–14.

Spencer, Herbert. *Essays: Scientific, Political and Speculative* (London: Longman & Roberts, 1858).

Spencer, Herbert. *Essays: Scientific, Political, and Speculative*, second series (London: Williams & Norgate, 1863).

Spencer, Herbert. *Essays: Scientific, Political and Speculative*, 3 vols (London: Williams & Norgate, 1868–74).

Spencer, Herbert. *Facts and Comments* (London: Williams & Norgate, 1902).

Spencer, Herbert. *First Principles* (London: Williams & Norgate, 1862).

Spencer, Herbert. *First Principles*, 3rd edn (London: Williams & Norgate, 1875).

Spencer, Herbert. *First Principles*, 6th edn (London: Williams & Norgate, 1900).

Spencer, Herbert. "The Genesis of Science", *British Quarterly Review* 20 (July 1854), 108–62. Reprinted in *Essays* (1858), 158–227; *Essays* (1868), vol. I, 116–93.

Spencer, Herbert. "Gracefulness", *Leader* III, 144 (25 December 1852). Reprinted in *Essays* (1858), 406–11; *Essays* (1868), vol. II, 312–18.

Spencer, Herbert. *Herbert Spencer on the Americans, and The Americans on Herbert Spencer, Being a Full Report of His Interview, and of the Proceedings at the Farewell Banquet of Nov. 9, 1882* (New York: D. Appleton, 1882).

Spencer, Herbert. "Illogical Geology", *Universal Review* (July 1859). Reprinted in *Essays*, second series (1863), 57–104; *Essays* (1868), vol. I, 325–76.

Spencer, Herbert. "Introduction". In *A Plea for Liberty*, Thomas Mackay (ed.), 1–26 (New York: D. Appelton, 1891).

Spencer, Herbert. "The Laws of Organic Form", *British and Foreign Medico-Chirurgical Review* 23 (January–April 1859), 189–202.

Spencer, Herbert. *The Man "versus" the State*, D. Macrae (ed.) (Harmondsworth: Penguin, 1969).

Spencer, Herbert. "Manners and Fashion", *Westminster Review* 61 (new series vol. 5) (April 1854), 357–92. Reprinted in *Essays* (1858), 109–57; *Essays* (1868), vol. I, 61–115.

Spencer, Herbert. "Mills *versus* Hamilton – The Test of Truth", *Fortnightly Review* 1 (15 July 1865), 531–50. Reprinted in *Essays* (1868), vol. II, 383–413.

Spencer, Herbert. "The Nebular Hypothesis", *Westminster Review* 70 (new series vol. 14) (July 1858), 185–225. Reprinted in *Essays*, second series (1863), 1–56; *Essays* (1868), vol. I, 239–99.

Spencer, Herbert. "The Origin and Function of Music", *Fraser's Magazine* 56 (October 1857), 396–408. Reprinted in *Essays* (1858), 359–84; *Essays* (1868), vol. I, 210–38.

417

Spencer, Herbert. "Parliamentary Reform: The Dangers, and the Safeguards", *Westminster Review* 73 (new series vol. 17) (April 1860), 486–507. Reprinted in *Essays* (1868), vol. II, 353–82.

Spencer, Herbert. "Personal Beauty", *Leader* V, 212, 216 (15 April and 13 May 1854). Reprinted in *Essays* (1858), 417–29; *Essays* (1868), vol. II, 149–62.

Spencer, Herbert. *Political Institutions* (London: Williams & Norgate, 1882) (Part V of *The Principles of Sociology*).

Spencer, Herbert. *Spencer: Political Writings*, J. Offer (ed.) (Cambridge: Cambridge University Press, 1994).

Spencer, Herbert. *The Principles of Biology*, 2 vols (London: Williams & Norgate, 1864 and 1867).

Spencer, Herbert. *The Principles of Biology*, 2 vols (New York: D. Appleton & Co., 1898).

Spencer, Herbert. *The Principles of Ethics*, 2 vols (London: Williams & Norgate, 1892 and 1893).

Spencer, Herbert. *The Principles of Psychology* (London: Longman, 1855).

Spencer, Herbert. *The Principles of Psychology*, 2nd edn, 2 vols (London: Williams & Norgate, 1870 and 1872).

Spencer, Herbert. *The Principles of Sociology*, 2nd edn, vol. 1 (London: Williams & Norgate 1877).

Spencer, Herbert. *The Principles of Sociology*, 2 vols (London: Williams & Norgate, 1893).

Spencer, Herbert. *The Principles of Sociology*, 3 vols (London: Williams & Norgate, 1896).

Spencer, Herbert. "Progress: its Law and Cause", *Westminster Review* 67 (new series vol. 11) (April 1857), 445–85. Reprinted in *Essays* (1858), 1–54; *Essays* (1868), vol. I, 1–60.

Spencer, Herbert. "The Proper Sphere of Government", *Nonconformist* 2 (15 June–14 December 1842). Reprinted in *Political Writings* (1994), 1–57.

Spencer, Herbert. "Railway Morals and Railway Policy", *Edinburgh Review* 100 (October 1854), 420–61. Reprinted in *Essays* (1858), 55–108; *Essays* (1868), vol. II, 251–311.

Spencer, Herbert. "Representative Government – What Is it Good For?", *Westminster Review* 68 (new series vol. 12) (October 1857), 454–86. Reprinted in *Essays* (1858), 185–227; *Essays* (1868), vol. II, 163–209.

Spencer, Herbert. "The Social Organism", *Westminster Review* 73 (new series vol. 17) (January 1860), 90–121. Reprinted in *Essays*, second series (1863), 143–84; *Essays* (1868), vol. I, 384–428.

Spencer, Herbert. *Social Statics: Or the Conditions Essential to Human Happiness Specified, and the First of them Developed* (London: John Chapman, 1851).

Spencer, Herbert. *Social Statics, Abridged and Revised Together with Man vs the State* (London: Williams & Norgate, 1892).

Spencer, Herbert. "Specialized Administration", *Fortnightly Review* 16 (new series vol. 10) (December 1871), 627–54. Reprinted in *Essays* (1874), vol. III, 125–70.

Spencer, Herbert. *The Study of Sociology*, 6th edn (London: Henry S. King, 1877).

Spencer, Herbert. "A Theory Concerning the Organ of Wonder", *Zoist: A Journal of Cerebral Physiology and Mesmerism* II (April 1842–January 1845), 316–25.

Spencer, Herbert. "A Theory of Population, Deduced from the General Law of Animal Fertility", *Westminster Review* 57(112) (new series vol. 1) (January and April 1852), 468–501.

Spencer, Herbert. "The Universal Postulate", *Westminster Review* 60 (new series vol. 4) (October 1853), 513–50. Much of this essay was re-published under the heading "General Analysis", the introductory section of *The Principles of Psychology* (1855), 3–68.

Spencer, Herbert. "Use and Beauty", *Leader* III(93) (3 January 1852). Reprinted in *Essays* (1858), 385–9; *Essays* (1868), vol. I, 429–33.

Spencer, Herbert. "The Use of Anthropomorphism", *Leader* IV(189) (5 November 1853). Reprinted in *Essays* (1858), 430–35; *Essays* (1868), vol. I, 440–46.

Spencer, Herbert. *Various Fragments* (London: Williams & Norgate, 1897).

Spencer, Thomas. *The Parson's Dream and the Queen's Speech* (London: John Green, 1843).

Taine, H. *Notes on England*, W. F. Rae (trans.) (London: Strahan, 1873).

Taine, H. *On Intelligence* (London: L. Reeve, 1871).

Taylor, Anne. *Hymns for Infant Minds by the Authors of "Original Poems": Rhymes for the Nursery*, 13th edn (London: Holdsworth, 1821).

Taylor, Isaac. *Natural History of Enthusiasm*, 6th edn (London: Holdsworth & Ball, 1832).

Tylor, E. B. *Primitive Culture: Research into the Development of Mythology, Philosophy, Religion, Art and Custom*, 2 vols (London: John Murray, 1871).

Tyndall, John. *Address Delivered Before the British Association Assembled at Belfast* (London: Longmans, Green, 1874).

Tyndall, John. *Essays on the Use and Limit of the Imagination in Science* (London: Longmans, Green, 1870).

Tyndall, John. *Fragments of Science*, 5th edn (London: Longmans, Green, 1876).

"Two". *Home Life with Herbert Spencer* (Bristol: J. W. Arrowsmith, 1906).

Veitch, John. *Speculative Philosophy, An Introductory Lecture Delivered at the Opening of the Class of Logic and Rhetoric* (at Glasgow), 1 November 1864 (Edinburgh: William Blackwood, 1864).

Wallace, Alfred Russel. *Contributions to the Theory of Natural Selection* (London: Macmillan, 1871).

Wallace, Alfred Russel. *Man's Place in the Universe: A Study of the Results of Scientific Research in Relation to the Unity of Plurality of Worlds* (London: Chapman & Hall, 1903).

Wallace, Alfred Russel. *My Life: A Record of Events and Opinions*, 2 vols (London: Chapman & Hall, 1905).

Ward, Mrs Humphrey. *Robert Elsmere*, new edn (London: Smith, Elder, 1888).

Watts, Isaac. *Divine and Moral Songs, Attempted in Easy Language for The Use of Children* (Derby: Thomas Richardson, 1829).

Watts, Isaac. *Divine and Moral Songs* (Devizes: J. Harrison, c. 1830).

Webb, Beatrice. *My Apprenticeship* (London: Longmans, 1926).

Webb, Beatrice. *The Diary of Beatrice Webb*, N. MacKenzie & J. MacKenzie (eds) (London: Virago, in association with the London School of Economics, 1982).

Whewell, William. *History of the Inductive Sciences from the Earliest to the Present Time*, 3 vols (London: J. W. Parker, 1857).

Whewell, William. *The Philosophy of the Inductive Sciences*, 2 vols (London: J. W. Parker, 1847).

Whewell, William. *William Whewell, DD, Master of Trinity College, Cambridge: An Account of His Writings with Selections from his Literary and Scientific Correspondence*, 2 vols, I. Todhunter (ed.) (London: Macmillan, 1876).

White, J.B. *The Life of the Rev. Joseph Blanco White written by himself with portions of his correspondence*, 3 vols, J. H. Thom (ed.) (London: John Chapman, 1845).

White, William Hale. *The Autobiography of Mark Rutherford, Dissenting Minister* (London: Trübner, 1881).

White, William Hale. *Mark Rutherford's Deliverance: Being the Second Part of his Autobiography* (London: Trübner, 1885)

White, W. Hale. *The Early Life of Mark Rutherford, by Himself* (Oxford: Oxford University Press, 1913).

Secondary sources

Ames, W. *Prince Albert and Victorian Taste* (London: Chapman & Hall, 1967).

Anderson, N. F. *Women against Women in Victorian England: A Life of Eliza Lynn Linton* (Bloomington, IN: Indiana University Press, 1987).

Andreski, S. "Introductory Essay". In *Herbert Spencer: Structure, Function and Evolution*, S. Andreski (ed.), 7–32 (London: Michael Joseph, 1971).

Arneson, R. J. "Perfectionism and Politics". *Ethics* 111(1) (October 2000), 37–63.

Ashton, R. *George Eliot: A Life* (Harmondsworth: Penguin, 1997).

Ashton, Rosemary. *G. H. Lewes: A Life* (Oxford: Clarendon Press, 1991).

Battistelli, Fabrizio. "War and Militarism in the Thought of Herbert Spencer", *International Journal of Comparative Sociology* 34 (September/December 1993), 192–209.

Berghahn, V. R. *Militarism: The History of an International Debate, 1861–1979* (Cambridge: Cambridge University Press, 1981).

Berridge, V. "Victorian Opium Eating: Responses to Opiate Use in Nineteenth-Century England". *Victorian Studies* 21(4) (Summer 1978), 439–61.

Birch, D. (ed.). *Ruskin and The Dawn of the Modern* (Oxford: Clarendon Press, 1999).

Bird, C. *The Myth of Liberal Individualism* (Cambridge: Cambridge University Press, 1999).

Blunden, E. *Leigh Hunt* (London: Cobden-Sanderson, 1930).

Bolt, C. "The Ideas of British Suffragism". In *Votes for Women*, J. Purvis & S. Stanley Holton (eds), 34–56 (London: Routledge, 2000).

Bourdier, F. "Geoffrey Saint-Hilaire versus Cuvier: The Campaign for Paleontological Evolution (1825–1838)". In *Toward a History of Geology: Proceedings of the New Hampshire Inter-Disciplinary Conference on the History of Geology*, C. J. Schneer (ed.), 36–61 (Cambridge, MA: MIT Press, 1969).

Bowler, P. J. *Evolution: The History of an Idea* (Berkeley, CA: University of California Press, 1984).

Bowler, P. J. *The Invention of Progress: The Victorians and the Past* (Oxford: Blackwell, 1989).

Brewer, J. & E. Hellmuth (eds). *Rethinking Leviathan: The Eighteenth-Century State in Britain and Germany* (Oxford: Oxford University Press, 1999).

Brown, A. W. *The Metaphysical Society: Victorian Minds in Crisis 1869–1880* (New York: Columbia University Press, 1947).

Buchdahl, G. "Deductivist Versus Inductivist Approaches". In *William Whewell: A Composite Portrait*, M. Fisch & S. Schaffer (eds) (Oxford: Clarendon Press, 1991), 331–44.

Burkhardt, R. *The Spirit of System: Lamarck and Evolutionary Biology* (Cambridge, MA: Harvard University Press, 1995).

Burrow, J. W. *Evolution and Society: A Study in Victorial Social Theory* (Cambridge: Cambridge University Press, 1968).

Burrow, J.W. *Whigs and Liberals: Continuity and Change in English Political Thought* (Oxford: Clarendon Press, 1988).

Caine, B. *Destined to be Wives: The Sisters of Beatrice Webb* (Oxford: Clarendon Press, 1986).

Carneiro, R. L. (ed.). *The Evolution of Society: Selections from Herbert Spencer's Principles of Sociology* (Chicago, IL: University of Chicago Press, 1967).

Carpenter, J. E. *James Martineau, Theologian and Teacher: A Study of His Life and Thought* (London: Philip Green, 1905).

Carr, Robert. "The Religious Thought of John Stuart Mill: A Study In Reluctant Thought Scepticism". *Journal of The History of Ideas* 23(4) (October–December 1962), 475–95.

Cartwright, J. *Evolution and Human Behaviour: Darwinian Perspectives on Human Nature* (Cambridge, MA: MIT Press, 2000).

Casteras, S. P. "Pre-Raphaelite Challenges to Victorian Canons of Beauty". *Huntington Library Quarterly* 55 (February 1992), 13–35.

Coleman, W. *Biology in the Nineteenth Century: Problems of Form, Function and Transformation* (Cambridge: Cambridge University Press, 1977).

Collier, J. "Reminiscence of Herbert Spencer". In *Herbert Spencer: An Estimate and Review*, J. Royce (New York: Fox Duffield, 1904).

Collini, S. "Introduction". In *On Liberty and the Subjection of Women*, J. S. Mill (Cambridge: Cambridge University Press, 1989).

Collini, S. *Public Moralists, Political Thought and Intellectual Life in Britain, 1850–1930* (Oxford: Clarendon Press, 1991).

Collini, S., D. Winch & J. Burrow. *That Noble Science of Politics: A Study in Nineteenth-Century Intellectual History* (Cambridge: Cambridge University Press, 1983).

Colp, Ralph. *To Be an Invalid: The Illness of Charles Darwin* (Chicago, IL: University of Chicago Press, 1977).

Corsi, P. *Science and Religion: Baden Powell and the Anglican Debate, 1800–1860* (Cambridge: Cambridge University Press, 1988).

Corsi, P. "The Importance of French Transformist Ideas for the Second Volume of Lyell's Principles of Geology". *British Journal for the History of Science* 11(39) (1978), 221–44.

Corsi, Pietro. *The Age of Lamarck: Evolutionary Theories in France, 1790–1830* (Berkeley, CA: University of California Press, 1988).

Cosslett, A. T. "Science and Value: The Writings of John Tyndall". In *John Tyndall: Essays on a Natural Philosopher*, W. H. Brock, N. D. McMilland & R. C. Mollan (eds), 181–91 (Dublin: Royal Dublin Society, 1981).

Crook, D. P. *American Democracy in English Politics, 1815–1850* (Oxford: Clarendon Press, 1965).

Crook, D. P. *Darwinism, War and History: The Debate over the Biology of War from the "Origin of Species" to the First World War* (Cambridge: Cambridge University Press, 1994).

Crook, D. P. "Historical Monkey Business: The Myth of a Darwinized British Imperial Discourse". *History* 84(276) (October 1999), 635–57.

Crook, D. P. "Whiggery and America: Accommodating the Radical Threat". In *Essays in Honour of Malcolm I. Thomis: Radicalism and Revolution in Britain, 1775–1848*, M. T. Davis (ed.), 191–206 (Basingstoke: Macmillan, 2000).

Dale, P. A. "George Lewes' Scientific Aesthetic: Restructuring the Ideology of the Symbol". In *One Culture: Essays in Science and Literature*, G. Levine (ed.), 92–116 (Madison, WI: University of Wisconsin Press, 1987).

Davie, G. E. *The Democratic Intellect: Scotland and Her Universities in the Nineteenth Century* (Edinburgh: Edinburgh University Press, 1961).

de Botton, A. *Status Anxiety* (New York: Pantheon Books, 2004).

Desmond, A. *The Politics of Evolution: Morphology, Medicine, and Reform in Radical London* (Chicago, IL: University of Chicago Press, 1989).

Desmond, A. *Huxley: The Devil's Disciple* (London: Michael Joseph, 1994).

Desmond, A. *Huxley: Evolution's High Priest* (London: Michael Joseph, 1997).

Desmond, A. & J. Moore. *Darwin* (London: Michael Joseph, 1991).

di Gregerio, M. A. "The Dinosaur Connection: A Re-interpretation of T. H. Huxley's Evolutionary View". *Journal of the History of Biology* 15 (1978), 397–418.

Drummond, J. & C. B. Upton. *The Life and Letters of James Martineau*, 2 vols (London: James Nisbet, 1902).

Duncan, D. *The Life and Letters of Herbert Spencer* (London: Methuen, 1908).

Dunn, W. H. *James Anthony Froude: A Biography*, 2 vols (Oxford: Clarendon Press, 1961–63).

Dupré, J. "On the Impossibility of a Monistic Account of Species". In *Species, New Interdisciplinary Essays*, R. A. Wilson (ed.), 3–22 (Cambridge, MA: MIT Press, 1999).

Elkins, S. & E. McKitrick. *The Hofstadter Aegis: A Memorial* (New York: Knopf, 1974).

Elliott, P. "Erasmus Darwin, Herbert Spencer, and the Origins of the Evolutionary Worldview in British Provincial Scientific Culture, 1770–1850". *Isis* 94 (2003), 1–29.

Endersby, J. "Escaping Darwin's Shadow". *Journal of the History of Biology* 36(2) (2003), 385–403.

Eve, A. S. & C. H. Creasy. *Life and Work of John Tyndall* (London: Macmillan, 1945).

Everitt, A. *The Pattern of Rural Dissent: The Nineteenth Century* (Leicester: Leicester University Press, 1972).

Flinn, M. W. "Introduction". In *The Medical and Legal Aspects of Sanitary Reform*, A. P. Stewart & J. E. Jenkins (Leicester: Leicester University Press, 1969).

Flint, K. *The Woman Reader, 1837–1914* (Oxford: Clarendon Press, 1993).

Foucault, M. *The Archaeology of Knowledge*, A. M. Sheridan Smith (trans.) (London: Routledge, 1972).

Francis, M. *Governors and Settlers: Images of Authority in the British Colonies, 1820–1860* (Basingstoke: Macmillan, 1992).

Francis, M. "Herbert Spencer and the Mid-Victorian Scientists". *Metascience* 4 (1986), 2–21.

Francis, M. "Herbert Spencer and the Myth of Laissez-Faire". *Journal of the History of Ideas* 39(3) (1978), 317–28.

Francis, M. "Naturalism and William Paley", *History of European Ideas* 10(2) (1989), 203–20.

Francis, M. & J. Morrow. *A History of English Political Thought in the Nineteenth Century* (London: Duckworth, 1994).

Freeman, D. "The Evolutionary Theories of Charles Darwin and Herbert Spencer". In *Herbert Spencer: Critical Assessments*, vol. II, J. Offer (ed.), 5–69 (London: Routledge, 2000).

Gaus, Gerald. *The Modern Liberal Theory of Man* (London: Croom Helm, 1983).

Gilbert, S. M. & S. Gubar. *The Madwoman in the Attic: The Woman Writer and the Nineteenth-Century Literary Imagination* (New Haven, CT: Yale University Press, 1984).

Goldman, L. *Science, Reform and Politics in Victorian Britain: The Social Science Association 1857–1886* (Cambridge: Cambridge University Press, 2002).

Gooch, G. P. *Life of Lord Courtney* (London: Macmillan, 1920).

Gould, S. J. *The Lying Stones of Marrakech: Penultimate Reflections in Natural History* (London: Vintage, 2001).

Gould, S. J. *Wonderful Life: The Burgess Shale and the Nature of History* (New York: Norton, 1989).

Gray, T. S. "Herbert Spencer on Women". *International Journal of Women's Studies* 7 (1984), 217–31.

Gray, T. S. *The Political Philosophy of Herbert Spencer: Individualism and Organicism* (Aldershot: Avebury, 1996).

Green, J. C. *Evolution and its Impact on Western Thought* (New York: Mentor Books, 1961).

Gregory, R. L. *Mind in Science: A History of Explanations in Psychology and Physics* (London: Weidenfeld & Nicolson, 1981).

Gruber, H. E. *Darwin on Man: A Psychological Study of Scientific Creativity: Together with Darwin's Early and Unpublished Notebooks* (London: Wildwood House Books, 1974).

Gunn, J. A. W. *Beyond Liberty and Property: The Process of Self-Recognition in Eighteenth-Century Political Thought* (Montreal: McGill-Queen's University Press, 1983).

Gunn, J. A. W. "Eighteenth-Century Britain". In *Rethinking Leviathan: The Eighteenth-Century State in Britain and Germany*, J. Brewer & E. Hellmuth (eds), 99–125 (Oxford: Oxford University Press, 1999).

Gunn, J. A. W. "Influence, Parties and the Constitution: Changing Attitudes, 1783–1832". *Historical Journal* 2 (1974), 301–28.

Haight, G. S. *George Eliot and John Chapman, with Chapman's Diaries* (New Haven, CT: Yale University Press, 1940).

Haight, G. S. *George Eliot: A Biography* (Oxford: Oxford University Press, 1978).

Haines, B. "The Inter-Relations Between Social, Biological and Medical Thought, 1780–1850: Saint-Simon and Comte". *British Journal for the History of Science* 11i(37) (1978), 19–35.

Halévy, É. *Thomas Hodgskin*, A. J. Taylor (trans.) (London: Ernest Benn, 1956).

Harvie, C. *The Lights of Liberalism: University Liberals and the Challenge of Democracy 1860–86* (London: Allen Lane, 1976).

Heath, P. "Introduction". In *On The Syllogism and Other Logical Writings*, A. de Morgan, vii–xxx (London: Routledge & Kegan Paul, 1966).

Helmstadter, R. J. "W. R. Greg: A Manchester Creed". In *Victorian Faith in Crisis: Essays on Continuity and Change in Nineteenth-Century Religious Belief*, R. J. Helmstadter & B. Lightman (eds), 187–222 (Basingstoke: Macmillan, 1990).

Helmstadter, R. J. & B. Lightman (eds). *Victorian Faith in Crisis: Essays on Continuity and Change in Nineteenth-Century Religious Belief* (Basingstoke: Macmillan, 1990).

Herbert, C. *Culture and Anomie: Ethnographic Imagination in the Nineteenth Century* (Chicago, IL: University of Chicago Press, 1991).

Heywood, V. H. "The 'Species Aggregate' in Theory and Practice". In *Symposium on Biosystematics*, Regnum Vegetabile 27, V. H. Heywood & A. Löve (eds), 26–37 (Vienna: International Association for Plant Taxonomy, 1963).

Hilton, B. *The Age of Atonement: The Influence of Evangelicalism on Social and Economic Thought, 1785–1865* (Oxford: Clarendon Press, 1988).

Hodge, M. J. S. "Darwin and the Laws of the Animate Part of the Terrestrial System (1835–1837): On the Lyellian Origins of His Zoononical Explanatory Program". In *Studies in History of Biology*, vol. 6, W. Coleman & C. Limoges (eds) (Baltimore, MD: Johns Hopkins University Press, 1983).

Hofstadter, R. *Social Darwinism in American Thought*, 2nd rev. edn (Boston, MA: Beacon Press, 1955).

Holt, E. G. (ed.) *The Art of All Nations, 1850–73: The Emerging Role of Exhibitions and Critics* (Princeton, NJ: Princeton University Press, 1982).

Holt, R. V. *The Unitarian Contribution to Social Progress in England* (London: Lindsey Press, 1952).

Hont, I. *Jealousy of Trade: International Competition and the Nation-State in Historical Perspective* (Cambridge, MA: Harvard University Press, 2005).

Hughes, K. *George Eliot: The Last Victorian* (London: Fourth Estate, 1998).

Hull, D. L. *Darwin and his Critics: The Reception of Darwin's Theory of Evolution by the Scientific Community* (Cambridge, MA: Harvard University Press, 1973).

Huxley, L. *Life and Letters of T. H. Huxley*, 2nd edn, 3 vols (London: Macmillan, 1903).

Irvine, W. *Apes, Angels and Victorians: A Joint Biography of Darwin and Huxley* (London: Weidenfeld & Nicolson, 1955).

Jacobs, J. E. *The Voice of Harriet Taylor Mill* (Bloomington, IN: Indiana University Press, 2002).

Jensen, J. V. "The X Club: Fraternity of Victoria Scientists". *British Journal for The History of Science* 5(17), part 1 (June 1970), 63–72.

Jones, G. & R. Peel (eds). *Herbert Spencer, The Intellectual Legacy*. Proceedings of a conference organized by the Galton Institute, London 2003 (London: The Galton Institute, 2004).

Jones, H. F. *Samuel Butler*, 2 vols (London: Macmillan, 1919).

Joyce, P. *Democratic Subjects: The Self and the Social in Nineteenth-Century England* (Cambridge: Cambridge University Press, 1994).

Kaplan, F. *Thomas Carlyle: A Biography* (Ithaca, NY: Cornell University Press, 1983).

Karl, F. *George Eliot: A Biography* (London: HarperCollins, 1995).

Keegan, P. (ed.). *The New Penguin Book of English Verse* (London: Allen Lane, 2000).

Kennedy, J. G. *Herbert Spencer* (Boston, MA: Twayne Publishers, 1978).

Kent, C. *Brains and Numbers: Elitism, Comtism and Democracy in Mid-Victorian England* (Toronto: University of Toronto Press, 1978).

Kent, S. K. *Sex and Suffrage in Britain, 1860–1914* (Princeton, NJ: Princeton University Press, 1987).

Kitcher, P. "Mill, Mathematics, and the Naturalist Tradition". In *The Cambridge Companion to Mill*, J. Skorupski (ed.), 56–111 (Cambridge: Cambridge University Press, 1998).

Kuper, A. *The Invention of Primitive Society: Transformations of an Illusion* (London: Routledge, 1988).

Ladenson, R. F. "Mill's Conception of Individuality". In *John Stuart Mill: Critical Assessments*, vol. I, J. C. Wood (ed.), 521–33 (London: Routledge, 1991).

Layard, G. S. *Mrs Lynn Linton* (London: Methuen, 1901).

Leakey, R. E. *The Illustrated Origin of Species*, abridged and introduced by R. E. Leakey (London: Faber, 1979).

Ledger, S. "The New Woman and the Crisis of Victorianism". In *Cultural Politics at the Fin-De-Siècle*, S. Ledger & S. McCracken (eds), 22–44 (Cambridge: Cambridge University Press, 1992).

Ledger, S. & S. McCracken (eds). *Cultural Politics at the Fin-De-Siècle* (Cambridge: Cambridge University Press, 1995).

Lepenies, W. *Between Literature and Science: The Rise of Sociology* (Cambridge: Cambridge University Press, 1992).

Letwin, S. R. *The Pursuit of Certainty* (Cambridge: Cambridge University Press, 1968).

Levere, T. H. "Coleridge and the Sciences". In *Romanticism and the Sciences*, A. Cunningham & N Jardine (eds) (Cambridge: Cambridge University Press, 1990).

Levine, G. (ed.). *One Culture: Essays in Science and Literature* (Madison, WI: University of Wisconsin Press, 1987).

Levine, P. *Victorian Feminism, 1850–1900* (London: Hutchinson, 1987).

Lightman, B. "Henry Longueville Mansel and the Origins of Agnosticism". *History of European Ideas* 5(1) (1984), 45–64.

Lightman, B. "Ideology, Evolution and Late Victorian Agnostic Popularizers". In *History, Humanity and Evolution*, J. R. Moore (ed.), 285–309 (Cambridge: Cambridge University Press, 1989).

Lightman, B. *The Origins of Agnosticism, Victorian Unbelief and the Limits of Knowledge* (Baltimore, MD: Johns Hopkins University Press, 1987).

Lubenow, W. C. *The Politics of Government Growth: Early Victorian Attitudes Toward State Intervention, 1833–1848* (Newton Abbot: David & Charles, 1971).

Lyon, E. G. *Politicians in the Pulpit: Christian Radicalism in Britain from the Fall of the Bastille to the Disintegration of Chartism* (Aldershot: Ashgate, 1999).

Lyons, S. "The Origins of T. H. Huxley's Saltationism: History in Darwin's Shadow". *Journal of the History of Biology* 28 (1995), 463–94.

Macfarlane, Alan. *The Origins of English Individualism: The Family, Property and Social Transition* (Oxford: Blackwell, 1978).

MacLeod, R. M. "The X-Club, A Social Network of Science in Late-Victorian England". *Notes and Records of the Royal Society of London* 24(2) (April 1970), 305–22.

Macpherson, H. *Herbert Spencer: The Man and His Work*, 2nd edn (London: Chapman & Hall, 1901).

Manton, J. *Mary Carpenter and the Children of the Streets* (London: Heinemann Educational, 1976).

Marchant, J. *Alfred Russel Wallace: Letters and Reminiscences*, 2 vols (London: Cassell, 1916).

Martin, R. B. *Tennyson: The Unquiet Heart* (London: Faber, 1983).

Mason, M. *The Making of Victorian Sexual Attitudes* (Oxford: Oxford University Press, 1994).

Mather, F. C. (ed.). *Chartism and Society: An Anthology of Documents* (London: Bell & Hyman, 1980).

McMenemy, W. H. "The Water Doctors of Malvern, with Special Reference to the Years 1842 to 1872". *Proceedings of the Royal Society of Medicine*, Section on the History of Medicine, 46 (1952), 1–7.

Meadowcroft, J. *Conceptualizing the State: Innovation and Dispute in British Political Thought, 1880–1914* (Oxford: Oxford University Press, 1995).

Merxmüller, H. "Summary Lecture". In *Flowering Plants, Evolution and Classification of Higher Categories*, K. Kubitzki (ed.), 397–465 (Vienna: Springer, 1977).

Miller, D. *Social Justice* (Oxford: Clarendon Press, 1976).

Mineka, F. E. *The Dissidence of Dissent: The Monthly Repository 1806–1839* (Chapel Hill, NC: University of North Carolina Press, 1944).

Moore, J. R. *The Post-Darwinian Controversies: A Study of the Protestant Struggle to Come to Arms with Darwin in Great Britain and America, 1870–1900* (Cambridge: Cambridge University Press, 1979).

Moore, J. R. (ed.). *History, Humanity and Evolution* (Cambridge: Cambridge University Press, 1989).

Moore, J. R. & C. Chant. "The Metaphysics of Evolution". In *Science and Metaphysics in Victorian Britain*, 3–38 (Milton Keynes: Open University Press, 1981).

Moore, M. *Foundations of Liberalism* (Oxford: Clarendon Press, 1993).

Morrell, J. & A. Thackray. *Gentlemen of Science: Early Years of the British Association for the Advancement of Science* (Oxford: Clarendon Press, 1981).

Morrow, J. *Thomas Carlyle* (London: Hambledon Continuum, 2006).

Muggeridge, K. & R. Adam. *Beatrice Webb: A Life, 1853–1943* (London: Secker & Warburg, 1967).

Muggeridge, M. *Chronicles of Wasted Time: The Green Stick* (London: Collins, 1972).

Nagel, E. *John Stuart Mill's Philosophy of Scientific Method* (New York: Hafner, 1950).

Nicholson, H. *Small Talk* (London: Constable, 1937).

Paul, E. F. "Herbert Spencer: The Historicist as a Failed Prophet". In *Herbert Spencer: Critical Assessments*, vol. II, J. Offer (ed.), 528–47 (London: Routledge, 2000).

Paul, E. F. "Herbert Spencer: Second Thoughts, A Response to Michael Taylor". In *Herbert Spencer: Critical Assessments*, vol. II, J. Offer (ed.), 556–62 (London: Routledge, 2000).

Paxton, N. *George Eliot and Herbert Spencer: Feminism, Evolutionism and the Reconstruction of Gender* (Princeton, NJ: Princeton University Press, 1991).

Peel, J. D. Y. *Herbert Spencer: The Evolution of a Sociologist* (London: Heinemann Educational, 1971).

Perrin, R. G. "Herbert Spencer's Four Theories of Social Evolution" *Herbert Spencer, Critical Assessments*, vol. II, J. Offer (ed.), 508–27 (London: Routledge, 2000).

Pettit, P. *Republicanism: A Theory of Freedom and Government* (Oxford: Oxford University Press, 1997).

Phillips, M. "Macaulay, Scott and the Literary Challenge to Historiography". *Journal of the History of Ideas* 50 (1989), 117–33.

Pinker, S. *The Blank Slate: The Modern Denial of Human Nature* (Harmondsworth: Penguin, 2002).

Poovey, M. *Uneven Developments: The Ideological Work of Gender in Mid-Victorian England* (London: Virago, 1989).

Price, R. "Hydropathy in England 1840–1870". *Medical History* 25 (1981), 269–80.

Purvis, J. & S. Stanley Holton (eds). *Votes for Women* (London: Routledge, 2000).

Raz, J. *The Morality of Freedom* (Oxford: Clarendon Press, 1988).

Reed, J. R. "Learning to Punish: Victorian Children's Literature". In *Culture and Education in Victorian England*, P. Scott & P. Fletcher (eds) (Lewisburg, PA: Bucknell University Press, 1990).

Richards, R. J. *The Meaning of Evolution: The Morphological Constructions and Ideological Reconstruction of Darwin's Theory* (Chicago, IL: University of Chicago Press, 1992).

Ricks, C. "George Eliot: 'She Was Still Young'". In his Essays in Appreciation, 206–34 (Oxford: Oxford University Press, 1996).

Ricks, C. *Essays in Appreciation* (Oxford: Clarendon Press, 1996).

Robbins, W. *The Newman Brothers: An Essay In Comparative Intellectual Biography* (London: Heinemann, 1966).

Roberts, D. *Paternalism in Early Victorian England* (London: Croom Helm, 1979).

Robertson, J. *The Case for the Enlightenment: Scotland and Naples, 1680–1760* (Cambridge: Cambridge University Press, 2005).

Robson, J. M. "Civilization and Culture". In *The Cambridge Companion to Mill*, J. S. Skorupski (ed.), 338–71 (Cambridge: Cambridge University Press, 1998).

Rose, H. "Colonizing the Social Sciences". In *Alas, Poor Darwin: Arguments Against Evolutionary Psychology*, H. Rose & S. Rose (eds), 127–54 (New York: Harmony Books, 2000).

Rupke, N. *The Great Chain of History: William Buckland and the English School of Geology, 1814–1849* (Oxford: Clarendon Press, 1983).

Rupke, N. *Richard Owen: Victorian Naturalist* (New Haven, CT: Yale University Press, 1994).

Ruse, M. *Mystery of Mysteries: Is Evolution a Social Construction?* (Cambridge, MA: Harvard University Press, 1999).

Rylance, R. *Victorian Psychology and British Culture 1850–1880* (Oxford: Oxford University Press, 2000).

Sawyer, P. L. "Ruskin and Tyndall: The Poetry of Matter and the Poetry of Spirit". In *Victorian Science and Victorian Values*, J. Paradis & T. Postlewait (eds), 217–46 (New York: New York Academy of Science, 1981).

Scarre, G. *Logic and Reality in the Philosophy of John Stuart Mill* (Dordrecht: Kluwer, 1989).

Schneewind, J. B. *Sidgwick's Ethics and Victorian Moral Philosophy* (Oxford: Clarendon Press, 1977).

Secord, J. A. *Victorian Sensation: The Extraordinary Publication, Reception and Secret Authorship of "Vestiges of the Natural History of Creation"* (Chicago, IL: University of Chicago Press, 2000).

Shapin, S. "Phrenological Knowledge and Early Nineteenth-Century Edinburgh". *Annals of Science* 32(3) (May 1975), 219–43.

Shermer, M. *In Darwin's Shadow: The Life and Science of Alfred Russel Wallace – A Biographical Study on the Psychology of History* (Oxford: Oxford University Press, 2002).

Showalter, E. *Sexual Anarchy: Gender and Culture and the Fin-de-Siècle* (London: Virago, 1992).

Sieveking, I. G. *Memoirs and Letters of Francis W. Newman* (London: Kegan, Paul, Trench, Trübner, 1909).

Sitwell, E. *The English Eccentrics* (London: Faber, 1933).

Skinner, Q. *The Foundation of Modern Political Thought*, 2 vols (Cambridge: Cambridge University Press, 1978).

Skorupski, J. (ed.). *The Cambridge Companion to Mill* (Cambridge: Cambridge University Press, 1998).

Skorupski, J. *John Stuart Mill* (London: Routledge, 1989).

Smith, C. U. M. "Evolution and the Problem of Mind, Part I. Herbert Spencer". *Journal of the History of Biology* 15(1) (Spring 1982), 55–88.

Smith, R. "The Human Significance of Biology: Carpenter, Darwin, and the *Vera Causa*". In *Nature and the Victorian Imagination*, U. C. Knoepflmacher & G. B. Tennyson (eds), 220–28 (Berkeley, CA: University of California Press, 1977).

Snaydon, R. W. "Intraspecific Variation and its Taxonomic Implications". In *Current Concepts in Plant Taxonomy for the Systematics Association*, V. H. Heywood & D. M. Moore (eds), 203–31 (London: Academic Press, 1984).

Stafford, R. A. *Scientist of Empire: Sir Roderick Murchison, Scientific Explanation and Victorian Imperialism* (Cambridge: Cambridge University Press, 1989).

Steedman, C. *Childhood, Culture and Class in Britain: Margaret McMillan, 1860–1931* (London: Virago, 1990).

Stocking, Jr, G. W. *Race, Culture and Evolution: Essays in the History of Anthropology*, 2nd edn (Chicago, IL: University of Chicago Press, 1982).

Stocking, Jr, G. W. *Victorian Anthropology* (New York: Free Press, 1987).

Sturges, R. P. "The Membership of the Derby Philosophical Society, 1787–1802", *Midland History* 4(3&4) (1978), 212–29.

Sturrock, J. *The Language of Autobiography: Studies in the First Person Singular* (Cambridge: Cambridge University Press, 1993).

Taylor, M. W. *Men Versus the State: Herbert Spencer and Late Victorian Individualism* (Oxford: Clarendon Press, 1992).

Taylor, M. *The Decline of British Radicalism, 1847–1860* (Oxford: Clarendon Press, 1995).

Thomas, W. *The Philosophic Radicals: Nine Studies in Theory and Practice, 1817–1841* (Oxford: Clarendon Press, 1979).

Thompson, D'arcy W. *On Growth and Form*, J. T. Bonner (ed.) (Cambridge: Cambridge University Press, 1992).

Thomson, J. A. *Herbert Spencer* (London: Dent, 1906).

Tjoa, Hock Guan, *George Henry Lewes: A Victorian Mind* (Cambridge, MA: Harvard University Press, 1977).

Tomlinson, P. B. "Vegetative Morphology – Some Enigmas in Relation to Plant Systematics". In *Current Concepts in Plant Taxonomy for the Systematics Association*, V. H. Heywood & D. M. Moore (eds) (London: Academic Press, 1984).

Turner, E. S. *Taking the Cure* (London: Michael Joseph, 1967).

Turner, F. M. "John Tyndall and Victorian Scientific Naturalism". In *John Tyndall: Essays on a Natural Philosopher*, W. H. Brock, N. D. McMillan & R. C. Mollan (eds), 169–80 (Dublin: Royal Dublin Society, 1981).

Turner, F. M. *Contesting Cultural Authority: Essays in Victorian Intellectual Life* (Cambridge: Cambridge University Press, 1993).

Turner, F. M. *Between Science and Religion: The Reaction to Scientific Naturalism in Late Victorian England* (New Haven, CT: Yale University Press, 1974).

Tyrell, A. *Joseph Sturge and the Moral Radical Party in Early Victorian Britain* (London: Christopher Helm, 1987).

Upton, C. B. *The Life and Letters of James Martineau*, 2 vols (London: James Nisbet, 1902).

Walkowitz, J. R. *City of Dreadful Delight: Narratives of Sexual Danger in Late-Victorian London* (London: Virago, 1994).

Wallas, G. *Men and Ideas* (London: Allen & Unwin, 1940).

Webb, R. K. "The Faith of Nineteenth-Century Unitarians: A Curious Incident". In *Victorian Faith in Crisis: Essays on Continuity and Change in Nineteenth-Century Religious Belief*, R. J. Helmstadter & B. Lightman (eds), 126–49 (Basingstoke: Macmillan, 1990).

Weeks, D. & J. James. *Eccentrics: A Study of Sanity and Strangeness* (London: Phoenix, 1996).

Weinstein, D. *Equal Freedom and Utility: Herbert Spencer's Liberal Utilitarianism* (Cambridge: Cambridge University Press, 1998).

Wheeler, M. *Ruskin's God* (Cambridge: Cambridge University Press, 1999).

White, P. *Thomas Huxley: Making the "Man of Science"* (Cambridge: Cambridge University Press, 2003).

Willey, Basil. *Nineteenth Century Studies* (London: Chatto & Windus, 1949).

Wilson, R. J. *Darwinism and the American Intellectual: An Anthology* (Chicago, IL: Dorsey Press, 1989).

Wiltshire, D. *The Social and Political Thought of Herbert Spencer* (Oxford: Oxford University Press, 1978).

Yeo, R. *Refining Science: William Whewell, Natural Knowledge and Public Debate in Early Victorian Britain* (Cambridge: Cambridge University Press, 1993).

Yeo, R. "William Whewell, Natural Theology and the Philosophy of Science in Mid-Nineteenth-Century Britain". *Annals of Science* 36 (1979), 493–516.

Young, R. M. *Mind, Brain and Adaptation in the Nineteenth Century: Cerebral Localization and its Biological Context from Gall to Ferrier* (Oxford: Clarendon Press, 1970).

Index

spiritualism (spiritualist(s)) and 120, 122, 127, 136–8, 142, 151, 154, 197, 361, 365
  extreme forms of 165
  new 137, 151
  Spencer's 197, 331–2
  see also the New Reformation, the Unknown
Stephen, Leslie 116, 359
Stephens, J. F. 312
Stewart, Dugald 125, 160, 170, 171, 173–4, 182, 377
Stocking (Jr), George W. vii, 386, 402
Strauss, David Fredrich 116, 133, 165, 370, 371, 376
Stubbs, William 265, 302
Sturge, Joseph 397, 400, 401
Sturrock, John 20, 101
Sumner, W. G. 104, 189, 381
survival of the fittest
  differences between Darwin and Spencer to 192, 209
  Spencer's use of Darwin's 381–2,
  term coined by Spencer and used by Charles Darwin 3
Symonds, John Addington 196

Taine, Hyppolite 41, 381
Taylor, Anne 22, 23, 24
Taylor, Harriet 69, 354
Taylor, Isaac 363
Taylor, M. W. 395, 406
Tennyson, Alfred 29, 206, 289, 320, 348, 358, 359, 374
Thompson, D'arcy Wentworth 242, 394
Thoreau, Henry D. 104
Tjoa, Hock Guan 353, 385
Tocqueville, Alexis de 2, 297, 311
Troughton, Walter 94, 358, 360
Turner, Frank M. vii, 372
Turner, J. M. W. 89
Tylor, Edward Burnett 79–80, 306, 383, 386, 402, 406–7
Tyndall, Hector 131
Tyndall, John
  friend and supporter of Spencer 95, 130, 131, 146, 148, 152, 352, 363, 372, 391
  friendship with Hirst 127, 129, 130, 131, 364, 365
  and Huxley 133, 366–7, 371
  and materialism 129, 150, 151, 372
  religious views 42, 129–31, 149–52, 181, 206, 365, 372–3
  see also materialism (materialists)
Tyndall scholarship 364, 372

uncles 9, 40, 347
  lasting impact on Spencer 28, 29, 30, 34, 35, 38, 39, 71
  see also John Spencer, Richard Spencer, Thomas Spencer
Unknown, the 157, 163, 342, 383

Hirst, Hunt, Huxley and 119–20, 128, 148, 149–50
  sources drawn on by Spencer's notion of 118–19, 136, 138, 143, 165, 180–81, 331–2
  Spencer's belief in 5, 41, 45, 108, 123, 131, 156, 192, 193, 206, 294
  worship of 111, 131, 138
utilitarianism *see* Jeremy Bentham

Veitch, John 167, 375
  common-sense philosopher 157, 161
  and Hamilton 158–9, 375, 377–8
Venice 86, 256, 398
Vernet, Emile Jean Horace 81
vitalism 385, 386, 388, 389, 392
  early influence on Spencer 206, 212, 231, 234
  later discarded by Spencer 206, 212, 225, 235, 242
  Spencer and 203, 221
vitalist(s) 219, 233, 242, 389
  and Spencer 221, 225

Wallace, Alfred Russel 194, 219, 243, 294, 361, 373–4, 398
  admiration of Spencer 146, 154, 181, 373
Wallas, Graham 3
Ward, Mary Augusta 331
Watts, Isaac 22, 23, 24
Webb (*née* Potter), Beatrice 43–4, 94, 99, 155, 347, 354
  friendship with Spencer 25, 36, 41, 43, 59, 106–7, 348, 350
  see also Beatrice Potter
Webb, Sydney 43
Weber, Max 2, 78, 332, 341, 404
Weinstein, David 354–5, 398
Wesleyan 21–2, 30, 344, 345, 346
Whately, Richard 165, 398
Whewell, William 125, 162, 217, 342, 364, 376, 393
  ideas guided Spencer 10, 196, 198–9, 200, 207, 233, 234, 235, 363, 384, 392
  ideas opposed by Spencer 11, 178, 211, 228, 234, 235, 236, 237, 238, 379
  Mill's criticism of 10, 167, 170, 182, 196, 198, 207, 364, 384
  scientific philosophy of 10–11, 167, 233–8, 387, 389
White, Joseph Blanco 137, 367, 368
White, Paul 372
White, William Hale 25, 167, 366, 368, 370
Willey, Basil 373
Wilson, Daniel 403
Wilson, James 366
Wiltshire, David 343, 347, 354–5
Women *see* beauty, rights

Yeo, Richard 384
Youmans, Edward Livingston 60, 62, 97, 103, 352, 410